Molecular Genetics

A SERIES OF BOOKS IN BIOLOGY

EDITORS : Donald Kennedy and Roderic B. Park

FRONTISPIECE. (*overleaf*). Direct visualization of RNA transcription of genes. An electron micrograph of a portion of an extrachromosomal nucleolus isolated from an immature egg cell (oocyte) of a newt. Bottle-brush-like matrix elements (M) separated by matrix-free segments of the core axis (A) can be seen. The core axis is a protein-covered DNA double helix, and the thin fibrils of the matrix elements lying perpendicular to the core axis are protein-covered molecules of nascent ribosomal RNA being transcribed by RNA polymerase from the DNA template. The DNA core of each matrix consists of a gene carrying the nucleotide sequence of the ribosomal RNA, whose transcription evidently proceeds in the direction of lengthening perpendicular RNA fibrils. About a hundred RNA molecules are being transcribed simultaneously from a single gene. [From O. L. Miller and B. R. Beatty, *Science* **164**, 956 (1969). Copyright 1969 by American Association for the Advancement of Science.]

Molecular Genetics

AN INTRODUCTORY NARRATIVE

Gunther S. Stent

University of California, Berkeley

■▄■ W. H. FREEMAN AND COMPANY
San Francisco

Printed in the United States of America

Library of Congress Catalog Card Number: 75-115468
International Standard Book Number: 0-7167-0684-9

3 4 5 6 7 8 9 10

To the memory of my friends
David Powell Hackett and Jean Jacques Weigle

Preface

In the winter of 1954, Edward H. Adelberg and I began teaching an under-graduate course at the University of California that was supposed to bring the latter-day gospel of molecular genetics to the Berkeley students. It was an extraordinarily gratifying pedagogical undertaking to face an audience of innocents, who had not yet heard of the DNA double helix, and preach to them that a new era was dawning for the understanding of heredity. So enthusiastic were we in those days that we managed to give thirty lectures on what comparatively little was then known about mutation and genetic recombination in bacteria and their viruses. How times have changed! Molecular genetics has since grown from the esoteric specialty of a small, tightly knit vanguard into an elephantine academic discipline whose basic doctrines today form part of the primary school science curriculum. Through-out the period of the well-nigh explosive development of its subject, I have continued to teach this course, and had I not undertaken an annual pruning of the material, the number of lectures necessary to present it would by now have grown at least tenfold. This text presents the present scope and content of that course.

The evolutionary origin and essentially pedagogic purpose of this book are reflected in the narrative presentation of the material in the historical sequence in which it actually came to be known (and in the occasional burdening of the reader with long-abandoned theories). Not only did the text simply grow in this way, but also I happen to believe that an understanding of the essentials of molecular genetics can best be taught in an organic (rather than logical) manner. But in case my presentation should give the erroneous impression that it is an attempt at historiography, I must warn the reader that my "streamlined" account of past developments is intended to be neither a scholarly history nor a hagiography of molecular genetics.

In the first place, I have deliberately chosen to mention only a part of the diverse experimental materials that actually figured in the development of my subject. Without doing too much violence to what I believe to have been the actual sequence of events, I have for the most part tried to present the basic findings of molecular genetics as they came to be revealed through work done with the hemoglobins of humans and rabbits, with the tryptophan synthetase, the betagalactosidase, and the DNA-, RNA-, and protein-synthesizing machinery of the intestinal bacterium *Escherichia coli*, and with three or four viruses that grow on *E. coli*. Thus there are vast lacunae in my account: for instance, the two eukaryotic microbes that were the subjects of much intensive molecular genetic study, the bread mold *Neurospora crassa* and the yeast *Saccharomyces*, are mentioned only in passing; the bacterial enzymes active in the synthesis of arginine, histidine, and leucine and in the hydrolysis of organic phosphates, whose study contributed in a very important way to the understanding of the mechanics and regulation of protein synthesis, are given short shrift; and animal viruses, the study of whose reproduction paid enormous benefits in both practical and theoretical realms, have been passed over in near-total silence.

In the second place my story is historiographically defective as far as its mentioning of persons' names is concerned. The many investigators who happen to have worked with experimental material that I did not choose for presentation are completely missing from this account. Even the names of most of those who labored to bring forth the body of knowledge on which I *do* report have been left unmentioned, for I feared that providing a complete dramatis personae would prove tiresome for my readers. But neither did I want to opt for the other alternative of equitable scientific historiography—if you don't cite everybody, cite nobody—since I thought that every student ought to know the identity of at least *some* of the protagonists of the theater of molecular genetics. And so I have made an undoubtedly invidious selection of names, among which my own friends are probably over-represented.

I have attempted to present my material in such a manner that it is within the intellectual horizon of a reader who has completed two years of undergraduate science training. The only formal preparation that I have taken for granted is a year's study of general college chemistry, so that I am presuming at least a superficial familiarity with such concepts as atoms and molecules, weak and strong chemical bonds, chemical equilibrium, and oxidation-reduction and solution chemistry. I am not presuming prior college study in biology, and particularly not in genetics, although a command of these subjects, as well as of organic chemistry, should certainly help in the understanding of this text. Unfortunately, the undertaking to make the story of molecular genetics accessible to such a broad audience entailed the unavoidable drawback that some of the material presented here must be familiar to readers with prior college training in the life sciences. Although in composing my narrative I did have in mind a devoted reader who sticks with it from beginning to end, most veterans of a modern general biology course might prefer to begin their study of this text with Chapter 3, and readers in possession of the basic facts and terminology of biochemistry might even proceed directly to Chapter 5.

In order to open avenues to further and deeper study of molecular genetics, I have provided each chapter with a bibliography of pertinent literature. These bibliographies include three categories of references. The first category indicates the relevant chapters of three other books, which between them cover in greater detail much of the ground of this text. One is the collection of autobiographical and retrospective essays *Phage and the Origins of Molecular Biology* (J. Cairns, G. S. Stent, and J. D. Watson, eds, Cold Spring Harbor Laboratory of Quantitative Biology, New York, 1966), abbreviated as PATOOMB in the bibliographies. These essays, which were written by members of the Phage Group of which Max Delbrück was the central figure, trace out the intellectual, experimental, and personal developments that led to some of the main happenings of this story. The other two books are William Hayes' definitive text *The Genetics of Bacteria and their Viruses* (John Wiley & Sons, New York, 2nd edition, 1968) and my own earlier text *Molecular Biology of Bacterial Viruses* (W. H. Freeman and Company, San Francisco, 1963). These books are abbreviated as HAYES and MOBIBAV, respectively, in the bibliographies. The second category includes references to some of the original research papers in which the results of key experiments were first reported. The third category includes specialized texts, reviews, and monographs, in which the reader can find both further information and more extensive bibliographies.

Finally, I ought to mention here an outstanding book that covers more or less the same ground as this text: J. D. Watson's *Molecular Biology of the Gene* (2nd ed., W. A. Benjamin, New York, 1970). I have the highest regard for this deservedly successful introductory presentation; in my opinion, it has no peer in the literature of molecular genetics. Indeed, the only reason why I persevered in completing my own treatment of the same material is that I thought that some readers might profit more from my narrative approach than from Watson's sovereign didactics. But it is according to the high standards set by *Molecular Biology of the Gene* that I wish my own effort to be judged.

This text was completed during a sabbatical leave from the University of California, Berkeley. I thank the John Simon Guggenheim Memorial Foundation for its grant of a fellowship and Stephen W. Kuffler and John Nicholls of the Department of Neurobiology, Harvard Medical School, for hospitable accommodation in their laboratories. I am grateful to Robert S. Edgar, A. Dale Kaiser, and Charles Yanofsky for their critical readings of the manuscript and to Mrs. Margery Hoogs for her efforts to rectify its prose style.

August 1970   GUNTHER S. STENT

Contents

11. Phage Growth   298

12. Recombination   330

13. Genetic Fine Structure   362

14. Lysogeny and Transduction   399

15. DNA Transactions   441

16. DNA Transcription   462

Plates I through IV follow page 94

Molecular Genetics

FIGURE 1-1. Gregor Mendel (1822–1884). [Courtesy of the Moravian Museum, Brno.]

1. Heredity

GREGOR MENDEL

One summer evening in 1965, an enormous crowd, probably one of the largest in its 600-year history, packed the Church of the Assumption in the Moravian town of Brno to celebrate a memorial Mass for Gregor Mendel (Figure 1-1), one-time abbot of the Augustinian monastery to which that church had formerly belonged. It was not so much piety for a departed prelate that had brought together that largely noncommunicant crowd, but rather the wish to pay homage to the memory of the founder of genetics. For in Mendel's own church was then gathered an ecumenical body of geneticists who had come to Brno from all parts of the world at the invitation of the Czechoslovak Academy of Sciences to commemorate the 100th anniversary of the presentation of Mendel's paper "Experiments on Plant Hybrids," which in 1865 was reported to the Brno Society of Natural Science. This rite, in view of its timing, locale, and auspices, served also as a Te Deum for the official resurrection of genetics and for the rehabilitation of those geneticists who had survived nearly two decades of official suppression of "Mendelism-Morganism" in the Soviet Union and the Peoples' Republics under the political influence of the then recently deposed Soviet Lord of Biology, Trofim Lysenko. But this memorial Mass could have been thought to have had yet another symbolic meaning: a commencement exercise for the students of heredity, whose work actually began not a mere century ago, but ten thousand years earlier in neolithic times, and whose quest for the understanding of how like begets like was now about to reach its goal.

For the capacity of living organisms to pass on their own qualities to their offspring is so obvious that it no doubt ranks as one of man's earliest scientific observations. Indeed, it was precisely the recognition of heredity and of the possibility of selective breeding that enabled the Stone Age denizens of the Near East to develop some of our domestic animals and crop plants from wild prototypes. This first success in biotechnology brought about the dawn of civilization—the transition from nomadic, food-gathering societies to sedentary, agricultural-urban societies in the Fertile Crescent in about 8000 B.C. The practical know-how gathered in millenia of breeding experience was passed on as magical or religious canon. For example, the Biblical stricture "Thou shalt not let thy cattle gender with a diverse kind; thou shalt not sow thy field with mingled seed" indicates that the ancient Hebrews were aware of the importance of maintaining pure lines of animals and plants. By classical times, breeding rules were being applied also to the human stock, as is exemplified by the infanticide of defective offspring practiced in the Greek city states. These ancient rules and prescriptions for selection and breeding of stock were not significantly improved until the nineteenth century.

The philosophers of classical Greece gave some thought to hereditary processes. In the fifth century B.C., Hippocrates developed, or at least taught in his medical school, the first known theory of heredity. This theory held that a child possesses the qualities of its parent because the semen concentrates within it small representative elements from all parts—healthy and diseased—of the parental body. The corresponding parts of the filial embryo were then thought to be built up from the parental elements supplied by the semen. In accord with this view, Hippocrates believed in the hereditary transmissal of acquired characters. For instance, he thought that the trait of long-headedness arose through the archaic social custom of artificially distorting the normal globular soft skull of the newborn infant. Aristotle, less than a century later, showed the inadequacy of the Hippocratic view. Aristotle argued that the filial embryo cannot have been reconstituted from representative elements collected from the parental bodies because:

1. Parents (and also plants) produce offspring endowed with parental traits (for instance, grey hair) that are manifest only in the post-reproductive stage of life.

2. The body is only the wrapping of the embryo, and hence the Hippocratic theory leads to the absurd inference that parental clothes and shoes also send their representatives to the semen.

3. Children of crippled and mutilated parents do not always show the defects of their progenitors.

Aristotle proposed, therefore, that rather than supplying the constituent elements of the embryo, the semen of the father provides the *plans* according to which the unformed blood of the mother is to be shaped into the offspring. Thus Aristotle recognized that biological inheritance is not the passage through the generations of body part samples, but instead is attributable to the transmission of *information* for the embryonic development of the individual. This deep insight into the essence of heredity provided by Aristotle was forgotten for the next twenty-three centuries. What was remembered of Aristotelian reproductive biology consisted mainly of the description of fantastic hybrid matings between wildly different animal species. For instance, it was believed that a cross between a camel and a leopard had spawned the giraffe, and that eels come ashore to mate with snakes. The Renaissance, which had initiated the reawakening of interest in the physical sciences, and the rejection of dogmatic superstition, produced few new insights into heredity. Indeed, it saw the rise of a notion even less sophisticated than the Hippocratic doctrine, namely the *preformation* theory, which envisaged the process of individual development as merely the unfolding of a preformed tiny midget, or *homunculus*, present in the father's semen, or in the mother's blood. Hence this view necessarily led to the belief that all later generations of the human race had already been preformed, one into the other, in—depending on the relative roles assigned to male and female in this infinite recessional system of Chinese boxes—Adam or Eve. Not until Mendel provided his radically new insights did the new era dawn in which the mechanisms governing the self-reproduction of man and his fellow creatures became ultimately revealed.

For his "Experiments on Plant Hybrids" Mendel cultivated the common garden pea, *Pisum sativum* (Figure 1-2), in the garden of his Brno monastery. Mendel chose the pea plant for breeding experiments because its flower is so constructed as to render it naturally self-fertilizing: the pollen from another flower cannot gain access to the stigma, and hence the ovules of the flower are fertilized only by its own pollen. Nevertheless, it is possible to cross-fertilize a pea plant experimentally by opening the immature flower, removing its anther, which would bear the pollen, and later touching the stigma with pollen from another plant. Thus Mendel realized the possibility of breeding plants of exactly controlled descent. He had at his disposal various strains of *P. sativum* that differed from each other in a few, well-defined characters, such as seed color, seed morphology, or stem size. Each of these strains bred true, in that upon self-fertilization every progeny plant manifested the parental character. In his paper, Mendel reported as "Experiment 1" the following cross between two such strains. The ovules of flowers of a strain producing ordinary round seeds were cross-fertilized with pollen from flowers of a strain producing unusual, wrinkled seeds, and ovules of plants that produced wrinkled-seeds were cross-fertilized with pollen from plants that produced round seeds. Several hundred *first-filial-generation*

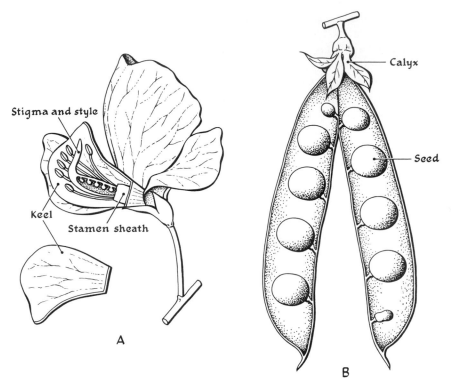

FIGURE 1-2.   A. The self-fertilizing flower of the garden pea. A portion of the keel, which encloses the reproductive organs, has been cut away to show the stamen, with its anthers bearing the male germ cells, or pollen, and the stigma, which receives the pollen from the anthers. The pollen grains then move along the style toward the ovary, where they fertilize the ovules, or female germ cells. B. The mature fruit (pod) containing the seeds that develop from the fertilized pea flower. [After J. B. Hill, H. W. Popp, and Alvin R. Grove, Jr., *Botany*, McGraw Hill, New York, 1967.]

hybrid seeds resulted from this cross, *all of which were round.* In the next year, Mendel planted 253 of these round hybrid seeds, allowed the resulting pea plants to self-fertilize and obtained 7,324 *second-filial-generation* seeds from them. He found that of these seeds 5,474 were round and 1,850 wrinkled, giving a ratio of round:wrinkled = 2.96:1 (Figure 1-3). Six more analogous *monohybrid* crosses between plants having other single character differences all gave the same general result:

1. Of the two alternative parental characters, only one appears in the first filial generation.

2. The character that vanishes in the first filial generation reappears among one-fourth of the members of the second filial generation.

From these observations Mendel made a brilliant deduction—one that must be placed among the most astute intellectual contributions to our understanding of nature. He deduced that the hereditary characters of the pea are carried and passed on to the progeny as discrete units. Each pea must possess a *homologous pair* of such units, of which it has received one from the pollen and one from the ovum whose union gave rise to the seed whence it sprung. Of the two homologous units that produce such alternative characters as round and wrinkled seed, one is *dominant*, and the other is *recessive*. Hence in the first filial generation of hybrids only the character of the dominant unit (round) is manifest, even though the cryptic recessive unit (wrinkled) is also present in every plant. Upon self-fertilization of flowers of the first filial generation, however, four kinds of seeds will arise in equal frequency (Figure 1-3), of which only one has drawn a pair of recessive units. This causes the character of the recessive unit (wrinkled) to be manifest in only one-third as many members of the second filial generation as the character of the dominant unit (round).

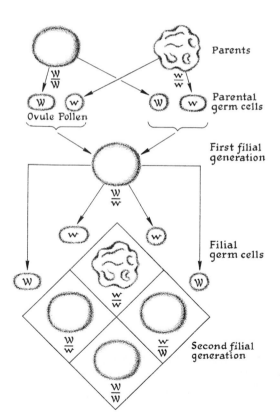

FIGURE 1-3.   Mendel's monohybrid cross. The ovules of flowers of a pure strain of the garden pea producing ordinary round seeds were cross-fertilized with pollen from flowers of a pure strain producing unusual, wrinkled seeds. The round-seed plant owes its character to the possession of a pair of *dominant* hereditary units labeled W/W and the wrinkled seed plant to a pair of *recessive* units labeled w/w. Plants of the first filial generation are all of the hybrid type W/w and hence show the dominant round-seed character. The flowers on plants grown from hybrid seeds of the first filial generation were allowed to self-fertilize. Since each such flower produces two types of ovules and two types of pollen, there will arise four types of seeds in the second filial generation: w/w, W/w, w/W, and W/W. Of these, only the w/w seeds show the recessive wrinkled character, whereas the other three types show the dominant round character.

Mendel's paper also reported the results of *dihybrid* crosses between a pair of pea strains that differed from each other in *two* alternative characters. The seeds of one strain were yellow and round, and the seeds of the other were green and wrinkled. The seeds of the first filial generation of this cross were all yellow-round, a result that was in agreement with the inference he had previously made on the basis of crosses between seeds differing in only one character—that yellow and round are dominant hereditary units *vis a vis* their recessive green and wrinkled alternatives. When the first filial generation of hybrid flowers grown from the seeds of this cross were allowed to self-fertilize, there appeared not only the two parental types but also two new *recombinant* types among the 556 peas of the second filial generation: 315 yellow-round, 101 yellow-wrinkled, 108 green-round and 32 green-wrinkled, or a ratio of $9:3:3:1$ of the four types (Figure 1-4). Mendel interpreted this result to mean that the two sets of hereditary units that determine seed color and seed morphology contributed to the first filial generation by the parental pollen and by the parental ovum are not necessarily transmitted together to the second filial generation. Instead, he inferred that there proceeds a random *segregation* of the parental units.

Mendel's revolutionary insights were, however, still too advanced for his times, and the publication of his results and conclusions in the *Journal of the Brno Society of Natural Science* remained unnoticed by the community of biologists for another thirty-five years. In particular, Charles Darwin, Mendel's most illustrious contemporary biologist, who had gained immediate fame for his theory of evolution, was never aware of Mendel's discovery of the hereditary units on which the natural selection propounded by Darwin actually operates. Indeed, Darwin's "pangenesis" concept of the mechanism of heredity, which envisaged that each part of the adult organism produces "gemmules," which are collected in the "seed" for transmission to the offspring, was more or less the same as that propounded by Hippocrates some twenty-three centuries earlier.

FIGURE 1-4.   Mendel's dihybrid cross. The ovules of flowers of a pure strain of the garden pea producing ordinary round and yellow seeds were cross-fertilized with pollen from flowers of a pure strain producing unusual wrinkled and green seeds. The plant producing round-yellow seed owes its character to the possession of two pairs of *dominant* hereditary units labeled W/W and Y/Y, and the plant producing wrinkled-green seed owes its character to two pairs of *recessive* units labeled w/w and y/y. Plants of the first filial generation are all of the hybrid type W/w, Y/y, and hence show the dominant round and yellow characters. The flowers on plants grown from hybrid seeds of the first filial generation were allowed to self-fertilize. Since each flower produces four types of ovules and four types of pollen, there will arise sixteen types of seeds in the second filial generation. Of these only one, the w/w, y/y type, shows the doubly recessive wrinkled and green characters, three show the recessive wrinkled and the dominant yellow character, three show the dominant round and the recessive green character, and nine show the doubly dominant round and yellow character.

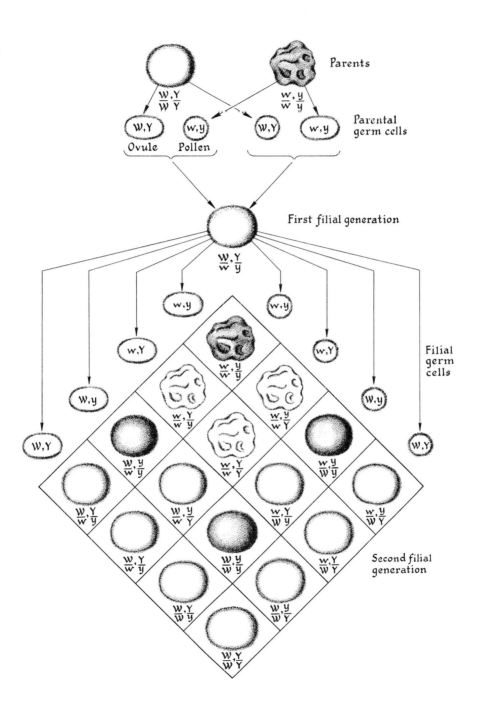

Parents

Parental germ cells

$\frac{w,Y}{w}\frac{}{Y}$ $\frac{w,y}{w}\frac{}{y}$

w,Y   w,y   w,Y   w,y

Ovule   Pollen

First filial generation

$\frac{w,Y}{w}\frac{Y}{y}$

w,y   w,y

Filial germ cells

w,Y   w,Y

w,y $\frac{w,y}{w}\frac{y}{y}$ w,y

$\frac{w,Y}{w}\frac{y}{y}$ $\frac{w,y}{w}\frac{y}{Y}$

w,Y $\frac{w,Y}{w}\frac{y}{y}$ $\frac{w,Y}{w}\frac{y}{Y}$ w,Y

$\frac{w,Y}{w}\frac{Y}{y}$ $\frac{w,y}{w}\frac{y}{Y}$ $\frac{w,Y}{w}\frac{y}{y}$ $\frac{w,y}{w}\frac{y}{Y}$

$\frac{w,Y}{w}\frac{y}{y}$ $\frac{w,y}{w}\frac{y}{y}$ $\frac{w,Y}{w}\frac{y}{Y}$

Second filial generation

$\frac{w,Y}{w}\frac{y}{y}$ $\frac{w,y}{w}\frac{y}{Y}$

$\frac{w,Y}{w}\frac{y}{Y}$

CHROMOSOMES

In the meantime, while Mendel's discovery lay dormant on the dusty shelves of some 120 libraries now known to have received the rather obscure journal in which he published his results, the problem of the mechanism of heredity was being attacked from another direction. At the time of the publication of Mendel's paper, the concept of the cell as the fundamental living unit had been established for about thirty years. But the structural elements of which cells are built up were only just then being recognized as a result of improvements in the design of microscopes and the application of stains by means of which various subcellular features could be given characteristic colors. The first insight brought by the study of cell structure and function, or *cytology*, was that the cell contains two distinct domains, the central *nucleus* and the peripheral *cytoplasm*. The boundary between nucleus and cytoplasm is formed by the *nuclear membrane*. Next, it was found that the nucleus is itself composed of two morphologically distinct parts: a granular area, the *chromatin*, which takes on an intense color upon treatment with certain stains, and the *nucleolus*, which does not take on that color. The cytoplasm for its part was seen to harbor a variety of distinct organelles, such as the *centrioles* and *vacuoles*. Thus, toward the end of the nineteenth century the general picture of the cell shown in Figure 1-5 had emerged.

There had been some controversy concerning the origin of cells. One mid-nineteenth century theory asserted, for instance, that new cells arise *de novo* from the assembly of subcellular components. But studies of plant and animal embryos were eventually to reveal that the cells of which the tissues of adult organisms are composed arise by a series of cell divisions that begin with the fertilized egg. This insight was generalized in the 1850's in Rudolf Virchow's dictum that every cell has sprung from another cell. With the recognition of the fertilized egg as the mother cell of the cell colony that makes up the multicellular organism, there arose also the speculation that despite their disparate sizes, sperm and egg make *equal* hereditary contributions to the individual that develops from their union. Cytological study of eggs and sperms showed that although the big egg possesses an enormous amount of cytoplasm, the tiny sperm is practically devoid of it. Yet the egg nucleus was found to have just about the same size as the sperm nucleus. From the postulated equality in hereditary contribution of egg and sperm and the gross inequality of their cytoplasms, it was then inferred that the nucleus rather than the cytoplasm must be the seat of cellular heredity. By the time of Mendel's death, in 1884, it had been discovered that the chromatin of the nucleus is composed of a countable number of thread-like particles, the *chromosomes* (Figure 1-6), and that the nuclei of egg and sperm contribute an equal number of chromosomes to the fertilized egg. Upon division of the fertilized egg and of all of its descendent cells, each member of this

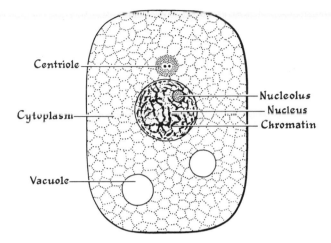

Centriole

Cytoplasm

Nucleolus
Nucleus
Chromatin

Vacuole

FIGURE 1-5.   Cartoon of the structure of a "typical" cell as
seen by visible light microscopy at the end of the nineteenth
century. [From "The Living Cell," by Jean Brachet, *Scientific
American*, September 1961. Copyright © 1961 by Scientific
American, Inc. All rights reserved.]

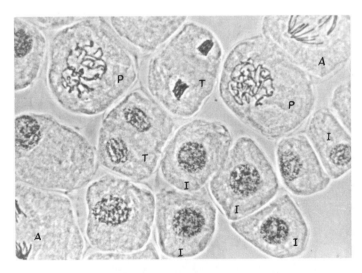

FIGURE 1-6.   Nuclei of dividing cells in the root of the bean plant *Vicia faba*. The
thread-like particles seen in the nuclei of the cells labeled P and A are chromosomes.
These cells are in the stages of *prophase* and *anaphase* of mitosis, respectively.
Chromosomes are not visible as individual threads in the nuclei of cells labeled I and T,
which are in the stages of *interphase* and *telophase* of mitosis, respectively (see Figure
1-7). [Courtesy of Walter Plaut.]

double set of chromosomes was seen to split longitudinally and be partitioned over the two daughter cells by an elaborate process that was given the name *mitosis* (Figure 1-7). Through mitosis each of the body cells of the developing embryo obtains the double set of chromosomes present in the fertilized egg. Microscopic examination of dividing cells in the ovaries and testes of adult animals soon revealed, however, that another process of chromosome partition, named *meiosis*, occurs in the formation of egg and sperm cells (Figure 1-8). In meiosis the number of chromosomes per body cell is halved, as a consequence of which the nuclei of the eggs or sperm produced by the animal once more contain the single set of chromosomes possessed by the egg and the sperm of its mother and father. Wilhelm Roux reasoned in the 1880's, in the ascendant Darwinian spirit of the time, that it is unlikely that these elaborate mitotic and meiotic processes of chromosome partition had evolved for no good reason, and proposed that they exist because the chromosomes constitute the hereditary material. Furthermore, without knowledge of Mendel's proof of their existence, he postulated a *linear arrangement of hereditary units* along the chromosomal threads. Roux's ideas were quickly

FIGURE 1-7.   Mechanism of chromosome distribution in ordinary cell division, or *mitosis*.
  Interphase (1). The nucleus is separated from the cytoplasm by the nuclear membrane. The chromosomes—two of paternal (black) and two of maternal (white) provenance— are in an extended form and invisible under an ordinary microscope. They are being replicated during interphase in preparation for the coming cell division. In the cytoplasm near the nuclear membrane are two "parent" centrioles, each paired with a smaller "daughter" centriole.
  Prophase (2, 3, and 4). The overt signs of the division process become manifest. The chromosomes condense into compact, visible structures. The two *sister chromosomes*, having arisen from the replication of a parent chromosome during the preceding interphase, appear as a sister chromosome pair joined at a morphologically distinct part called the *centromere*. The nuclear membrane disappears. The two centriole pairs move apart to the sites that are to become the poles of the daughter cells. During this movement, the centrioles radiate an *aster* of fibers in all directions. Fibers that make a bridge between the two centriole pairs form a structure called the *spindle*. The centromeres of the paired sister chromosomes become attached to spindle fibers.
  Metaphase (5). The chromosome pairs move to the equator of the spindle, and the joint centromere splits.
  Anaphase (6). The spindle fibers pull one member of each sister chromosome pair to opposite poles of the spindle. At this stage the process of cell-cleavage begins, which in animal cells derives from formation of a furrow from the cell periphery inwards, and in plant cells derives from the growth of a new cell wall from the inside outwards.
  Telophase (7 and 8). The sister chromosomes have reached opposite poles in their migration. The events of prophase now occur in reverse. Two nuclear membranes are formed, each enclosing one chromosome set for a daughter nucleus. The spindle fibers disappear, and the chromosomes distend and become invisible. In the cytoplasm, each centriole has been replicated, so that eight rather than four of them are now present. The cell-cleavage process goes to completion, which generates two daughter cells in their interphase. [From " How Cells Divide," by D. Mazia, *Scientific American*, September 1961). Copyright © 1961 by Scientific American, Inc. All rights reserved.]

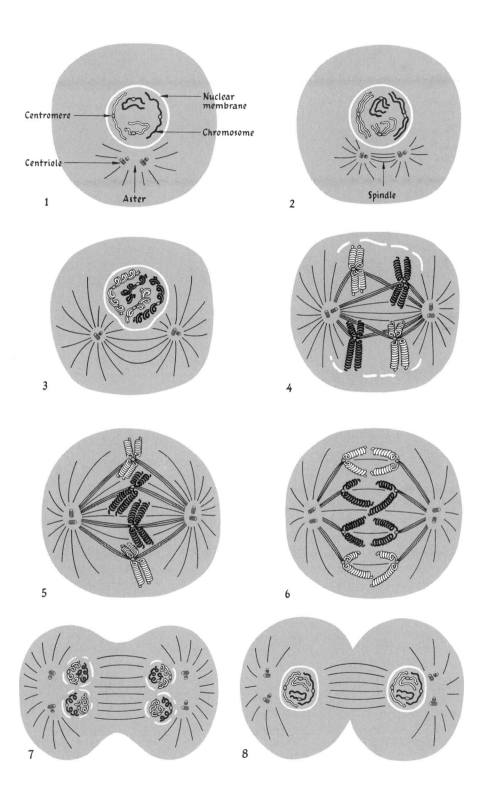

1

Centromere

Nuclear
membrane

Chromosome

Centriole

Aster

2

Spindle

3

4

5

6

7

8

adopted by August Weismann and elaborated into a complete theory of heredity and development. Weismann proposed that in multicellular organisms that reproduce sexually, the number of hereditary units is halved in the formation of egg and sperm, or pollen—that is, of the *germ cells*. The original number of hereditary units is then restored upon fusion of the nuclei of male and female germ cells in the fertilization process, to give rise to a new individual, whose hereditary material is half of paternal and half of maternal provenance. Unfortunately, Weismann envisaged that each of the many chromosomes present in the cell nucleus carries *all* of the hereditary units necessary for producing the entire individual. In *P. sativum*, whose nuclei contain fourteen chromosomes, this theory was clearly incompatible with Mendel's (then still unknown) inference that the pea plant is endowed with two, rather than fourteen, copies of each of its hereditary units. Weismann's theory became widely known and discussed in the closing years of the nineteenth century. Indeed, it stirred up a violent controversy that

FIGURE 1-8.   (See following two pages.) Mechanism of chromosome distribution in special cell division producing germ cells, or *meiosis*.

Prophase I. As in the prophase of mitosis, the two sister chromosomes that arose by replication of the parent chromosome in the preceding interphase condense into visible structures and form a chromosome pair joined at the centromere. The nuclear membrane disappears. The spindle is organized by the centrioles. In contrast to the prophase of mitosis, however, here the two *homologous sister chromosome pairs* (one pair being of paternal and the other of maternal provenance) come into point-to-point alignment, or *synapsis*.

Metaphase I. The two synapsed chromosome pairs move to the equator of the spindle, where their centromeres attach to the same spindle fiber. In contrast to the metaphase of mitosis, here the centromeres do not split.

Anaphase I. The spindle fibers pull each of the two synapsed homologous chromosome pairs to opposite poles of the spindle. That is to say, here the two sister chromosomes move together to the same pole. Cell cleavage then begins.

Telophase I. The homologous chromosome pairs have reached opposite poles in the migration. At this point, two cell nuclei have been generated, each of which has only a single rather than a double chromosome set, albeit each member of that single set being represented by a sister chromosome pair. The events of prophase are reversed and cell cleavage is completed.

Interkinesis. The two daughter cells now briefly remain in a state similar to the interphase of mitosis except that there is no chromosome replication (each chromosome being already represented by a sister chromosome pair anyway).

Prophase II, Metaphase II, Anaphase II, Telophase II, and *Interphase* (see p. 14). After interkinesis, the two daughter cells undergo a second cycle of cell division that is essentially an ordinary *mitosis*. That is to say, the sister chromosome pairs that had been formed in the interphase preceding the first division of meiosis condense and attach to the spindle in Prophase II: their joint centromere splits, and each sister is pulled to opposite poles of the spindle in Anaphase II; and four nuclei arise in Telophase II that contain a single chromosome set, each member of which is represented by only one chromosome. The four interphase cells that are present upon completion of Telophase II finally undergo maturation into either male or female germ cells.

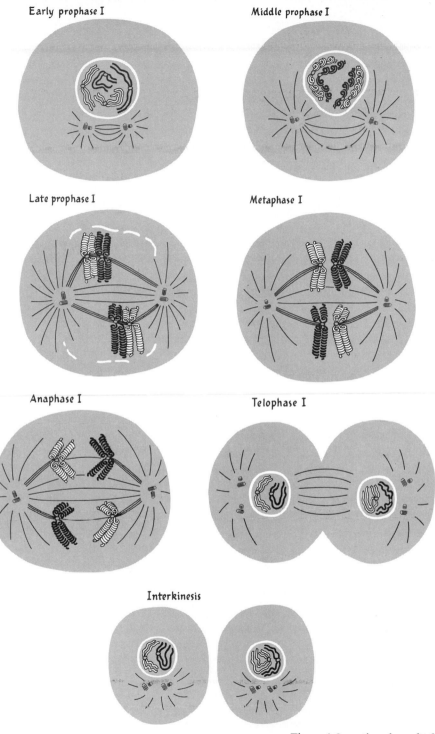

Early prophase I

Middle prophase I

Late prophase I

Metaphase I

Anaphase I

Telophase I

Interkinesis

Figure 1-8 continued overleaf

Prophase II

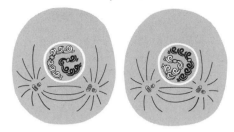

Metaphase II

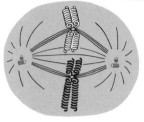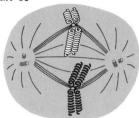

Anaphase II

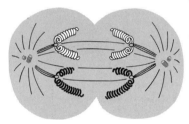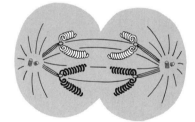

Telophase II

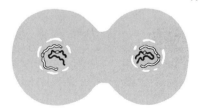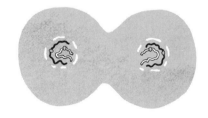

Interphase

Figure 1-8 continued

was to provide an incentive for carrying out the kind of quantitative breeding experiments that Mendel had already done 35 years earlier.

MENDELISM REDISCOVERED

One of the main protagonists in the debate raging about Weismann's theory was Hugo de Vries. He rejected some aspects of the theory but added to it the important notions that each of the postulated hereditary units controls a single character and that these units may be recombined in various ways in the offspring. To test this notion, he made first- and second-filial-generation crosses between strains of flowering plants that differed from each other in one or two characters. In this way, he rediscovered the 3:1 distribution of recessive characters among the second filial generation. He published these results in March 1900, and stated in passing that the same results and theoretical conclusions had been reported by Mendel in 1865. Two months earlier, however, in January of 1900, Carl Correns had already published a paper in which he reported on analogous experiments with inter-strain hybrids of maize; he, too, referred to Mendel's work. In May of 1900, after he had seen de Vries' account, Correns published another paper in which he reported that he had confirmed the results of Mendel's experiments with peas.

The rediscovery in 1900 of Mendel's paper caused tremendous excitement, since his laws of inheritance could now be understood in terms of the behavior of chromosomes in mitosis and meiosis. Evidently, each chromosome carries only a part of all the hereditary units necessary for producing the entire individual, so that the whole chromosome set present in the germ cell includes just one copy of each of these units. A cell carrying that single chromosome set was said to be in the *haploid* state. The individual that develops from the fertilized egg is thus endowed with a *pair* of homologous hereditary units—thanks to the two chromosome sets contributed by male and female germ cells. A cell carrying a double chromosome set is said to be in the *diploid* state. Upon reduction in meiosis of the two chromosome sets of its diploid body cells to a single set, the individual assigns but a single copy of each unit to the haploid germ cells with which it begets its own offspring.

An intensive study of heredity now began, and the invention of new terminology was one of its first fruits. First of all, the discipline itself was given the name *genetics*, and Mendel's hereditary unit became the *gene*. Two homologous genes representing alternative realizations of the same character, such as yellow and green seed color or round and wrinkled seed morphology, were called allelomorphs, a term later shortened to *alleles*. The individual that develops from the fertilized egg was called a *zygote*, a *homozygote* being one that carries a pair of identical alleles and a *hetero-zygote* being one that carries a pair of different alleles of a particular gene. The sum total of all of an individual's genes, and hence its entire chromosome set, came to be referred to as the *genome* of the individual.

In 1901 de Vries proposed that different alleles of the same gene arise by a sudden, discontinuous change of that gene—a process to which he gave the name *mutation*. He had based this idea on his finding that some plants exhibit variegated light-green–dark-green leaf patterns, a phenomenon that later studies showed could not be attributed to gene mutation at all. But the concept of gene mutation as a source of genetic diversity had nevertheless been born. In the first decade of this century genetic experiments designed to investigate the transmission of differences in a single character to the offspring of hybrid matings were extended to a wide diversity of organisms. This work showed that Mendelian analysis and the gene theory are both applicable not only to plants but also to animals, including man.

FIGURE 1-9.   T. H. Morgan (1866–1945). The handwriting in German script under the portrait is that of the geneticist Curt Stern, and states (in English translation): " Photo made by Bridges and Sturtevant (without Morgan's knowledge). The camera is hidden in Bridges' incubator. Taken in 1915 or 1916." [Courtesy of Curt Stern].

DROSOPHILA

A new phase in genetic research opened in 1910, when Thomas H. Morgan (Figure 1-9) turned to the study of the genetics of the fruit fly Drosophila (Figure 1-10). Morgan had realized that this organism would present many advantages for genetic experimentation over the plants and animals previously studied. It is so small and so simple to rear that many thousands of flies can be kept in the laboratory. Moreover, its generation time, from birth to reproductive maturity, is only about two weeks, thus allowing the experimenter to witness the appearance of many filial generations within a reasonably short period. Another advantage, not known to Morgan at the outset of his work, is that the Drosophila haploid genome is represented by only four chromosomes. Hence the diploid body cells contain $2 \times 4 = 8$ chromosomes each, compared to the 14 chromosomes of *P. sativum* or the 46 chromosomes of man.

One of the first fundamental questions that Morgan could settle decisively by means of Drosophila experimentation was the mechanism of sex determination in animals: What determines whether a fertilized egg is a male or female zygote? This problem had intrigued natural philosophers since ancient times. By 1900, detailed microscopic study of chromosome morphology had revealed a small but significant difference in the chromosome set of the

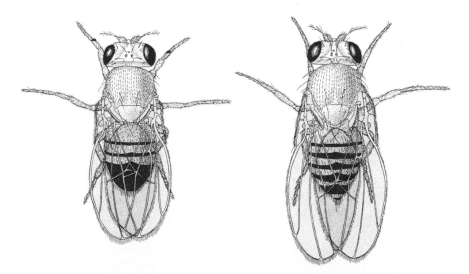

FIGURE 1-10.   The fruit fly Drosophila, a male at left and a female at right. [After A. H. Sturtevant and G. W. Beadle, *An Introduction to Genetics*, W. B. Saunders Co., Philadelphia, 1939.]

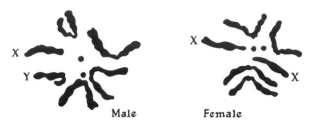

Male   Female

FIGURE 1-11.   The diploid chromosome complement of
Drosophila. The body cells of the female have four pairs
of homologous chromosomes, including a pair of X
chromosomes. The body cells of the male have the same
chromosome complement as those of the female, except
they carry an X-Y instead of an X-X chromosome pair.
[After T. Dobzhansky, *in* T. H. Morgan *The Scientific
Basis of Evolution*, W. W. Norton, New York, 1932.]

nucleus of male and female animals. For instance, the eight chromosomes
of the diploid body cells of Drosophila females can be clearly divided into
four pairs of homologous chromosomes on the basis of the similar appearance of each pair (Figure 1-11). In the diploid body cells of the male fly,
however, only three pairs of homologous chromosomes are manifest, the
remaining two chromosomes being quite dissimilar in their morphology.
These two unlike chromosomes found in the male were called the X and Y
chromosomes. Comparison of this unlike pair with the fourth like pair of
the female showed that she carries two X chromosomes, the Y chromosome
being absent from her body cells. These microscopic observations led to the
insight that the balance of X and Y *sex chromosomes* determines the sex
of the individual, XX being the female and XY being the male combination.
It thus became clear that sex is inherited like a simple Mendelian gene, in
that the female is an X/X homozygote and the male is an X/Y heterozygote.
Hence *all* the haploid eggs produced by meiosis in the female ovaries carry
an X chromosome, whereas of the haploid sperm produced by meiosis in
the male testes, half carry an X and half carry a Y chromosome (Figure 1-12).
Fertilization of an egg by X-carrying sperm will thus produce a female
zygote, and fertilization by a Y-carrying sperm will produce a male zygote.

   The large number of Drosophila individuals that can be reared in the
laboratory made possible the detection of a great diversity of hereditary
variants, or *mutants*. Thus by 1915 Morgan and his associates had found
85 different Drosophila mutant types, which differ from the normal, or
wild-type, fly in such characters as wing size, body color, eye color (Plate I),
bristle shape, and eye size. Each of these mutants, all of which were recognized
as single aberrant individuals among a brood of thousands of normal flies,
was inferred to owe its mutant character to a rare spontaneous mutation

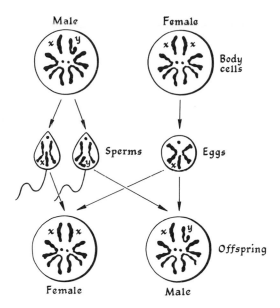

Male   Female

Body cells

Sperms   Eggs

Offspring

Female   Male

FIGURE 1-12.   The inheritance pattern of sex in Drosophila. Upon meiosis of the diploid X/Y body cells of the male, there arise two kinds of haploid germ-cells, or sperm: one carrying an X chromosome and the other carrying a Y chromosome, in addition to the three other chromosomes. Meiosis of the female X/X body cells gives rise to only one kind of haploid germ cell, or egg–namely, one carrying an X chromosome. The sex of the offspring is therefore determined by the kind of sperm that happens to have fertilized the egg. Since X- and Y-carrying sperm are produced in equal amounts, male and female offspring arise in equal frequency.

of the gene controlling that character. (In 1927 H. J. Muller, an early associate of Morgan's, was to show that X-irradiation of flies greatly raises the frequency of mutation of these genes above the spontaneous frequency.) The availability of these mutants made possible extensive crossing experiments designed to probe more deeply into the mechanisms of inheritance than had previously been possible. Crosses between doubly mutant flies carrying two mutant genes on two different chromosomes and normal flies carrying the corresponding wild-type alleles soon confirmed Mendel's results. The recessive characters disappeared in the first filial generation and reappeared in random recombination among the second filial generation. But when similar dihybrid crosses were carried out between flies whose two mutant genes were on the *same* chromosome, a new genetic perspective was opened. Since such *linked* genes form part of one single genetic structure, it might have been expected that they necessarily move together upon segregation of the diploid chromosome set in the meiosis responsible for production of the haploid germ cells. In agreement with this expectation it was found that the alleles of such linked genes do tend to reappear among the second filial generation in the same combination in which the parent flies introduced them into the cross. But Morgan discovered that despite their physical linkage some recombination does take place between genes on the same chromosome. That is, there arise flies among the second filial generation that carry on the *same* chromosome one gene whose allele was provided by one parent and another gene whose allele was provided by the other

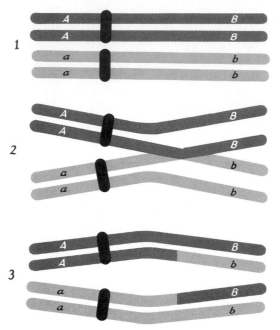

FIGURE 1-13. Schematic diagram of chromosome crossing over. (1) The two homologous pairs of sister chromosomes come into point-to-point alignment, or synapsis, in the prophase stage of the first of the two meiotic divisions that produce the germ cells. One sister chromosome pair carries alleles *A* and *B* of two different genes, whereas the other sister chromosome pair carries alleles *a* and *b* of those genes. (2) Two chromosomes break at corresponding places between the two genes present as different alleles and exchange fragments. (3) If this crossover were between chromosomes belonging to different sister pairs, two *recombinant* chromosomes will have been created, one of which carries the alleles *A* and *b* and the other the alleles *a* and *B*.

parent. Morgan interpreted this result in terms of the *crossing over* of homologous chromosomes, which F. A. Janssens had then recently observed during meiotic divisions of salamander sexual tissues (Figure 1-13). At the beginning of the first of the two meiotic divisions responsible for germ-cell production, each pair of homologous chromosomes of the diploid set comes into point-by-point alignment, or *synapsis*. In the course of this synaptic pairing, breaks occur at corresponding sites of the homologous chromosomes, and these breaks are in turn followed by an exchange and a cross-wise reunion of the fragments. This process evidently generates two *recombinant* chromosomes that incorporate part of the genes of one of the two homologous parental chromosomes and part of the genes of the other. Provided that the probability of making a break-and-reunion is constant per unit length of synapsed chromosomes, it follows that the smaller the distance separating any two genes on the same chromosome, the smaller the chance that a crossover will take place *between* them and hence result in a recombination of the parental alleles. This fundamental insight made possible a topographic survey of the location of the known mutant genes on the Drosophila chromosomes. By measuring the frequency of segregation of linked genes among the offspring of a great variety of mutant flies, Morgan and his associates were able to establish a *genetic map* of the four Drosophila chromosomes (Figure 1-14). This map indicated the chromosomal sites of the then known mutant genes.

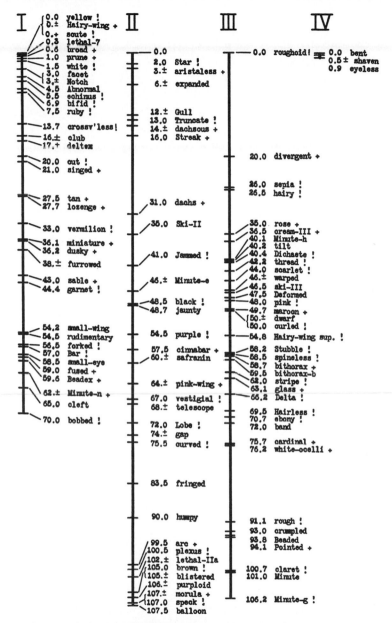

FIGURE 1-14.   The genetic map of the Drosophila chromosomes as charted by 1926. Each gene is named according to the abnormal character displayed by a known mutant allele; characters starting with a capital letter indicate a dominant mutant allele, and those starting with a lower case letter a recessive mutant allele. The "map distance," inferred from crossing-over frequencies, is given to the left of each character. Chromosome I is the X chromosome, and the poorly characterized chromosome IV is the tiny dot-like chromosome of which a pair is shown in Figure 1-11. No map of the Y chromosome was then available. [After T. H. Morgan, *The Theory of the Gene.* Yale Univ. Press, New Haven, 1926.]

CLASSICAL GENETICS

The work of Morgan and his associates, C. B. Bridges, A. H. Sturtevant, and H. J. Muller, brought about nearly universal acceptance of the Mendelian principles of heredity, though pockets of resistance to these views remained extant until very recent times, mainly among ideologically encumbered biologists and naive plant and animal breeders. This acceptance was to make possible great advances in the understanding of genetic processes, at the levels of individual cells, multicellular organisms, and populations. These advances, in turn, paved the way for theoretical insights, such as a quantitative analysis of the dynamics of organic evolution (a development which came to be known as "neo-Darwinism"), and brought forth tremendous practical benefits to agriculture and medicine. As far as agriculture was concerned, the principles of "Mendelism-Morganism" had at last made possible rational rather than rule-of-thumb breeding procedures. And through these procedures new varieties of traditional crop plants and domestic animals possessing economically important properties such as disease resistance, higher yield, capacity to grow in unfavorable climes and, more recently, toughness suitable for mechanical harvest, were produced. As far as medicine was concerned, the recognition of the genetic basis of many human disorders and diseases provided a rationale for taking measures for their prevention or relief.

During this entire development, however, the fundamental concept of genetics, the gene, remained largely devoid of any material content. Besides not having fathomed its physical nature, geneticists had been unable to explain how the gene manages to preside over specific cellular physiological processes from its nuclear throne or how it manages to achieve its own faithful replication in the cellular reproductive cycle. As late as 1950, in an essay written on the Golden Jubilee of the rediscovery of Mendel's work, H. J. Muller, by then one of the elder statesmen of genetics and a leading philosopher of the gene, described this state of affairs in the following terms: ". . . the real core of genetic theory still appears to lie in the deep unknown. That is, we have as yet no actual knowledge of the mechanism underlying that unique property which makes a gene a gene—its ability to cause the synthesis of another structure like itself, in which even the mutations of the original gene are copied. . . What must happen is that just that precise reaction is *selectively* caused to occur, out of a virtually infinite series of possible reactions, whereby materials taken from a common medium become synthetized into a pattern just like that of the structure which itself guides the reaction. We do not know of such things yet in chemistry."

The achievements that have been traced out here so far in the briefest form, and the body of knowledge they produced, are now generally referred to as *classical* genetics. It seems appropriate to reserve a special name for this period of genetic research, since there is one very important aspect that sets it apart from the *molecular* genetics that was to follow in its wake. The fundamental unit of classical genetics is an indivisible and abstract gene. The fundamental unit of molecular genetics, by contrast, is a concrete chemical molecule, the nucleotide, the gene being relegated to the role of a secondary unit aggregate comprising hundreds or thousands of nucleotides. For the classical geneticist, study of the detailed nature and physical identity of the gene, though undoubtedly of great intellectual interest, is not an essential part of his work. His theories on the mechanics of heredity and the experimental predictions to which these theories lead, are largely formal, and their success does not depend on knowledge of structures at the sub-microscopic, or molecular, level where the genes lie. Despite the lack of understanding of what Muller called its "real core," classical genetics *had* been fantastically successful. It had raised our understanding of the living world to previously unknown heights of sophistication.

THE BIRTH OF MOLECULAR GENETICS

By 1940 a new era was dawning for genetic research when a new crew, rather different in formation and motivation from the classical geneticists, began to take an interest in the nature of the gene. Many of these newcomers had little or no acquaintance with the body of genetic knowledge accumulated in the preceding decades, nor even with biology in general. Their training lay mainly in the physical sciences, and their interest was confined largely to solving one and only one problem: the physical basis of genetic information. There was nothing new, of course, in the phenomenon of physical scientists addressing themselves to the solution of biological problems— indeed many of the greatest contributions to nineteenth-century biology had been made by men trained in the physical sciences. Louis Pasteur, H. L. F. Helmholtz, and Mendel himself had been physical scientists. But there was a rather special philosophical twist that informed the men responsible for this particular development of the 1940's. For just when old-fashioned vitalism (the doctrine which held that, in the last analysis, the phenomena of life can be explained only by the existence of a mysterious "vital force," neither physical nor chemical in nature) was rapidly disappearing from intellectually enlightened circles, Niels Bohr (Figure 1-15) fashioned the idea that some biological phenomena might turn out *not* to be accountable wholly in terms of conventional physical concepts. In the wake of his for-mulation of the quantum theory of atomic structure Bohr developed the more general notion that the impossibility of describing the quantum of action from the purview of classical physics, and hence what he called its

FIGURE 1-15.   Niels Bohr
(1885–1962). [From
*Biographical Memoirs of
Fellows of the Royal Society*,
9, 37 (1963).]

"irrationality," is but a heuristic paradigm of how the encounter of what
appears to be a deep paradox eventually leads to a higher level of understanding. He presented this view in his address "Light and Life" before the
International Congress of Light Therapy in 1932. "At first," Bohr said,
"this situation [i.e., the introduction of an irrational element] might appear
very deplorable; but, as has often happened in the history of science, when
new discoveries have revealed an essential limitation of ideas the universal
applicability of which had never been disputed, we have been rewarded by
getting a wider view and a greater power of correlating phenomena which
before might even have appeared contradictory." In particular, Bohr thought
it would be well to keep this possibility in mind in the study of life: "The
recognition of the essential importance of fundamentally atomistic features
in the functions of living organisms is by no means sufficient for a compre-
hensive explanation of biological phenomena. The question at issue, therefore, is whether some fundamental traits are still missing in the analysis of
natural phenomena, before we can reach an understanding of life on the
basis of physical experience."

That *genetics* was, in fact, a domain of biological inquiry in which physical
and chemical explanations might turn out to be "insufficient" in Bohr's
sense was spelled out in 1935 by Bohr's pupil, Max Delbrück, in a paper

entitled "On the Nature of Gene Mutation and Gene Structure." Delbrück
pointed out that "whereas in physics all measurements must in principle
be traced back to measurements of place and time, there is hardly a case
in which the fundamental concept of genetics, the character difference, can
be expressed meaningfully in terms of absolute units." Thus, Delbrück
thought, one could take the view "that genetics is autonomous and must
not be mixed up with physicochemical conceptions." As Delbrück readily
admitted, "the refined [genetic] analysis of [the fruit fly] *Drosophila* has led
to [estimates of] gene sizes which are comparable to those of the largest
known molecules endowed with a specific structure. This result has led many
investigators to consider that the genes are nothing else than a particular
kind of molecule, except that their detailed structure is not yet known."
Nevertheless, Delbrück realized that one must keep in mind that there exists
here a significant departure from the chemical definition of the molecule:
"In chemistry we speak of a certain kind of molecule when we are faced
with a substance which reacts uniformly to chemical stimulation. In genetics,
however, we have by definition only a single representative of the relevant
'gene molecule,' in a chemically heterogeneous environment." In any case,
the main reason for thinking of the gene as a molecule in the first place was
its evident long-term *stability* in the face of outside influences. This stability,
Delbrück thought, could be accounted for only if each atom that makes
up the gene "molecule" were fixed in its mean position and electronic state,
so that only discontinuous, saltatory changes could occur in this arrangement, whenever an atom of the ensemble happened to acquire an energy
superior to the activation energy required to change its particular state.
These changes evidently would correspond to gene mutations.

In 1945, immediately after the conclusion of World War II, a little book
appeared which popularized these hitherto rather esoteric views and secured
for them a much wider audience. This was *What is Life?*, written by Erwin
Schrödinger (Figure 1-16). In his book, Schrödinger heralded the dawn of
a new epoch in biological research to his fellow physicists, whose knowledge
of biology was generally confined to a stale botanical and zoological lore.
Having one of the inventors of quantum mechanics address himself to the
question "What is life?" imposed upon them a confrontation with a fundamental problem worthy of their mettle. Since many of these physical scientists were suffering in the immediate post-war period from a general
professional malaise, they were eager to direct their efforts toward a new
frontier which, according to Schrödinger, was now ready for some exciting
developments.

Schrödinger begins his discussion with the comforting statement that "the
obvious inability of present-day physics and chemistry to account [for the
events that take place in a living organism] is no reason at all for doubting
that they can be accounted for by those sciences." Since, as Schrödinger

FIGURE 1-16.   Erwin Schrödinger (1887–1961). [From *Biographical Memoirs of Fellows of the Royal Society*, **7**, 221 (1961).]

points out next, organisms are large compared to atoms, there is no reason why they should not obey exact physical laws. And even the peculiar quality of living matter—namely, that it creates order out of disorder—does not put it beyond the pale of thermodynamics, whose Second Law asserts that in the Universe order decays into disorder. For life evidently feeds on the gigantic decay processes that occur in the Sun and is thus no more in violation of the Second Law than are rain clouds made up from water that sunlight has distilled from the oceans. To Schrödinger, the *real* problem in want of an explanation was heredity. For whereas the genes are evidently responsible for the order that an organism manifests, *their* dimensions are not so very large compared to those of atoms. How, then, Schrödinger wondered, do the genes resist the fluctuations to which they should be subject? How, for example, has the tiny gene of the Habsburg lip (Figure 1-17) managed to preserve for centuries its specific structure, and hence its information content, while being maintained at the body temperature of 310° above absolute zero? Building upon Delbrück's then ten-year-old proposal that genes are stable because their atoms stay put in energy wells, Schrödinger postulated that genes are able to preserve their structures because the chromosome that carries them is an *aperiodic crystal*. He suggested that this large aperiodic crystal is composed of a succession of a small number of isomeric elements,

FIGURE 1-17.   The Habsburg Lip. Inheritance of a royal mutant gene through the centuries. Upper left: Maximilian I (1459–1519). Upper right: Maximilian's grandson, Charles V (1500–1558). Lower left: Archduke Charles of Teschen (1771–1847). Lower right: Teschen's son, Archduke Albrecht (1817–1895). [by permission of The Picture Archives of the Austrian National Library.]

the exact nature of the succession constituting the *hereditary code*. Schrödinger illustrated the vast combinatorial possibilities of such a code by means of an example in which he used the two symbols of the Morse code as its isomeric elements. Schrödinger held that "we may safely assert that there is no alternative to [Delbrück's] molecular explanation of the hereditary substance. The physical aspect leaves no other possibility to account for its permanence. If the Delbrück picture should fail, we would have to give up further attempts." Furthermore, "from Delbrück's general picture of the hereditary substance it emerges that living matter, while not eluding the 'laws of physics' as established up to date, is likely to involve hitherto unknown 'other laws of physics,' which, however, once they have been revealed will form just as integral a part of this science as the former."

Inspired by the romantic notion of finding "other laws of physics" through the study of genetics, a number of physical scientists left the occupation for which they had been trained and addressed themselves to the problem of the nature of the gene. The entry of these new men into genetics and cognate fields in the 1940's produced a revolution in biology that, when the dust had cleared, left molecular biology as its legacy. As part of this revolution, molecular genetics was to develop out of classical genetics, and by the time of the Mendel Centennial in 1965 the nature of the gene was understood. Alas, the physicists were to be cheated out of their reward: no "other laws of physics" had turned up along the way. Instead, as the facts to be set forth in this book will show, the making and breaking of hydrogen bonds seem to be all there is to understanding the workings of the hereditary substance.

Bibliography

PATOOMB

Gunther S. Stent. Introduction: Waiting for the paradox.
Max Delbrück. A physicist looks at biology.
K. G. Zimmer. The target theory.

HAYES

Chapters 1 and 2.

ORIGINAL RESEARCH PAPERS

Correns, C. G. Mendel's Regel über das Verhalten der Nachkommenschaft der Rassenbastarde. *Ber. deutsch. botan. Gesellschaft* **18**, 158 (1900). Reprinted in English translation *in* The birth of genetics. *Genetics*, **35**, 33–41 (1950) and *in* C. Stern and Eva R. Sherwood, *The Origin of Genetics: A Mendel Source Book*. W. H. Freeman and Company, San Francisco. 1966.

Mendel, G. Versuche über Pflanzenhybriden. *Verh. naturforsch. Ver. Brünn*, **4**, 3 (1866). Reprinted in English translation *in* C. Stern and Eva R. Sherwood, *The Origin of Genetics: A Mendel Source Book.* W. H. Freeman and Company, San Francisco, 1966.

Morgan, T. H., A. H. Sturtevant, H. J. Muller, and C. B. Bridges. *The Mechanism of Mendelian Heredity.* Henry Holt, New York, 1915.

Muller, H. J. Artificial transmutation of the gene. *Science*, **46**, 84 (1927).

Weissman, A. *Essays on Heredity.* Translated by A. E. Shipley, S. Schönland, and others. Oxford Univ. Press, Vol. 1 and Vol. 2, 1891–1892.

SPECIALIZED TEXTS MONOGRAPHS AND REVIEWS

Bohr, N. Light and life. *Nature*, **131**, 421, 457 (1933). (An extensive discussion of the ideas presented here by Bohr and of their later misrepresentations and misuses to resurrect vitalism can be found in Chapter 8 of P. Frank, *Modern Science and its Philosophy*, Harvard Univ. Press, 1949.)

Dunn, L. C. (ed.). *Genetics in the 20th Century.* Macmillan, New York, 1951.

Fleming, D. Emigré physicists and the biological revolution. *Perspectives in American History*, **2**, 152–189 (1968).

Schrödinger, E. *What is Life?* Cambridge Univ. Press, New York, 1945.

Srb, A. M., R. D. Owen, and R. S. Edgar. *General Genetics* (2nd ed.). W. H. Freeman and Company, San Francisco, 1965.

Stern, C., and Eva R. Sherwood, *The Origins of Genetics: A Mendel Source Book.* W. H. Freeman and Company, San Francisco, 1966.

Sturtevant, A. H. *A History of Genetics*, Harper & Row, New York, 1965.

Wilson, E. B. *The Cell in Development and Heredity.* Macmillan, New York, 1896. 2nd ed., 1900; 3rd ed., 1925.

2. Cells

In order to fathom the "real core" of genetics, it proved necessary to consider the action of genes at the microscopic level of the single cell, rather than at the macroscopic level of the whole organism, which is made up of billions of highly differentiated cells. That is, before asking such questions as how the genes present in the parental Drosophila germ cells manage to give rise to an entire offspring fly, one had to inquire into the processes by means of which the genes preside over the formation of *cellular* structures and components in the successive cycles of cell growth and cell division. In other words, the fundamental biological problem of growth and reproduction had to be stated in terms of the gene-directed chemical synthesis of new cell material. Thus the account of this inquiry will begin with a brief consideration of the chemical nature of the cell.

CELL CHEMISTRY

In the 1830's, at the very time when the theory of the cell as the fundamental living unit was being worked out, the first food chemists recognized that three distinct classes of substances make up the bulk of living matter: the *lipids*, the *carbohydrates*, and the *proteins*. The lipids are compounds that are only sparingly soluble in water but are generally soluble in organic solvents, such as chloroform or ethanol. A typical lipid is composed of a molecule of glycerol, whose three hydroxyl groups are linked in ester bonds to fatty acids. The fatty acids are straight chains of 3 to 27 carbon atoms with a carboxyl group at the end of the chain. It is to the long fatty-acid chains that lipids owe their insolubility in water.

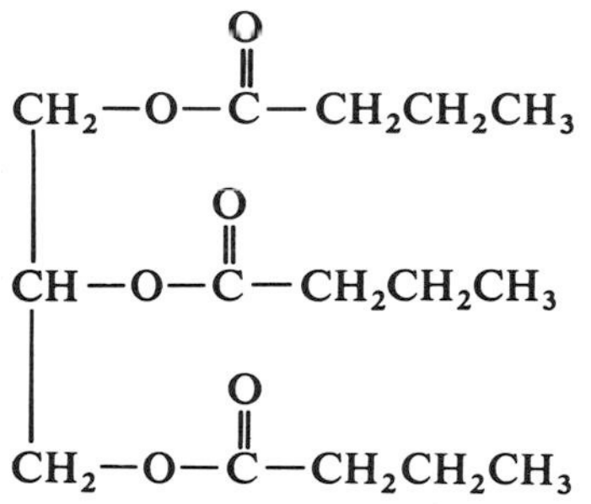

$$
\begin{array}{c}
\overset{\displaystyle O}{\underset{\displaystyle \|}{}} \\
CH_2-O-C-CH_2CH_2CH_3 \\
| \\
\overset{\displaystyle O}{\underset{\displaystyle \|}{}} \\
CH-O-C-CH_2CH_2CH_3 \\
| \\
\overset{\displaystyle O}{\underset{\displaystyle \|}{}} \\
CH_2-O-C-CH_2CH_2CH_3
\end{array}
$$

A lipid (butter fat)

In some lipids, the class called phospholipids, only two of the glycerol hydroxyls are esterified to fatty acids, whereas the third is esterified to phosphate. And in some phospholipids a nitrogenous base is linked to the glycerol hydroxyl through a phosphate diester bond:

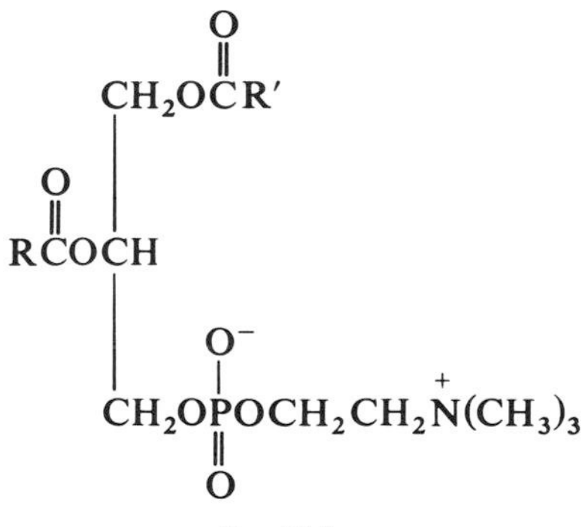

$$
\begin{array}{c}
\overset{\displaystyle O}{\underset{\displaystyle \|}{}} \\
CH_2OCR' \\
| \\
RCOCH \\
| \\
\overset{\displaystyle O^-}{\underset{\displaystyle |}{}} \\
CH_2OPOCH_2CH_2\overset{+}{N}(CH_3)_3 \\
\underset{\displaystyle O}{\overset{\displaystyle \|}{}}
\end{array}
$$

Lecithin

Because of the water-insolubility of the long-chain portions of their fatty-acid components, lipids readily aggregate in the aqueous environment of the cell to form much larger, two-dimensional macromolecular *surfaces*. These lipid surfaces constitute an important element of the membranes that define the physical boundary between nucleus and cytoplasm, and between cell and its surround.

The carbohydrates are compounds in which the elements carbon, hydrogen, and oxygen are present in the ratio $1:2:1$ and can thus be denoted by the general empirical formula $(CH_2O)_x$. The main carbohydrate cell constituents are *polysaccharides*, which consist of long chains of hundreds of linked sugar molecules. For instance, there are the two long chains of glucose

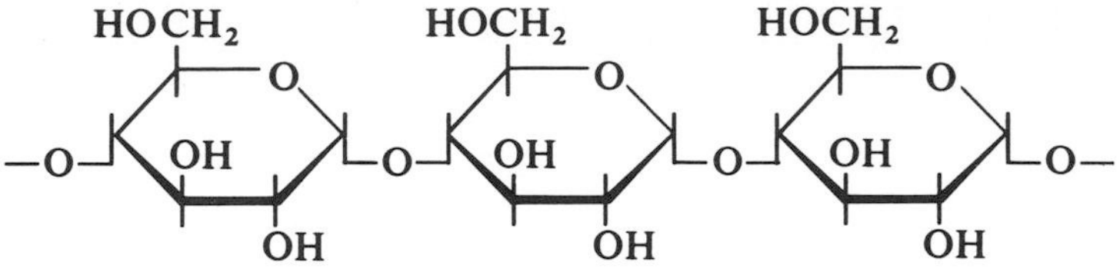

Starch

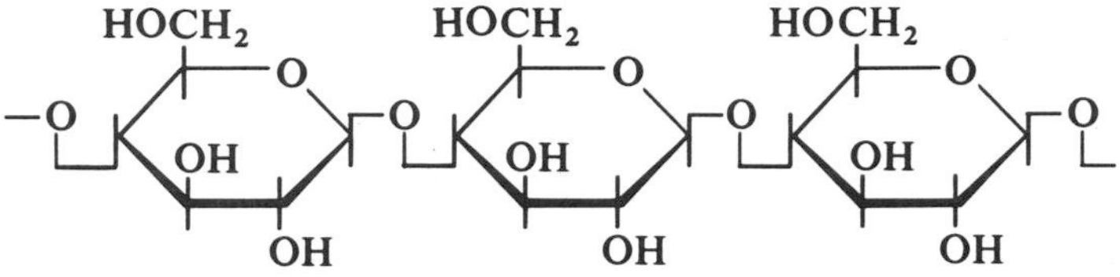

Cellulose

Although the structure of some other polysaccharides is more complex than that of either starch or cellulose, in that their chains are made up of two or more different sugars following each other in alternating sequence, starch and cellulose exemplify the two chief roles that polysaccharides play in the life of the cell—food reserve and structural member. Starch provides a store of glucose molecules on which the organism may later draw; cellulose gives rigidity to the walls of plant cells.

Despite their great biological importance, neither lipids nor polysaccharides will figure very often in the matters to be set forth in this text. The reason for this, as will be seen later, is that their relation to the action of genes is quite indirect, or of merely second order. Instead of the lipids and the polysaccharides, it is the proteins that are primarily connected with gene action and which will, therefore, loom large in what follows.

PROTEIN

The proteins are nitrogenous compounds which are generally water soluble in their native state but which become insoluble, or coagulate from their aqueous solution, upon being denatured by heat or exposure to strong acids. They make up from half to two-thirds the dry weight of most cells ("dry weight" refers to the weight of a cell after its internal water has been removed; water makes up about 80% of the total, or "wet weight" of most cells.) One of the earliest systematic investigators of proteins was G. J. Mulder, who first suggested the name "protein" (Greek *proteos*, "primary") in 1838. Six years later, Mulder wrote: "There is present in plants as well as animals a substance which . . . performs an important function in both. It is one of the very complex substances, which under various circumstances may alter their composition and serves . . . for the regulation of chemical metabolism. . . . It is without doubt the most important of the known components of living matter, and it would appear that, without it, life would not be possible. This substance has been named protein." In full accord with Mulder's appraisal of their fundamental importance, proteins will play a central role in the matters to be set forth in this text; indeed, it is actually toward an explanation of their formation that most of what is to be set forth here will be directed.

By the latter half of the nineteenth century it had been shown that whereas there exist many different kinds of proteins in nature, all proteins are large molecules which, upon hydrolysis, yield a class of simpler compounds, the *amino acids*. The common feature of all protein amino acids is shown in Figure 2-1: they contain a carbon atom, termed the α-carbon, to which both

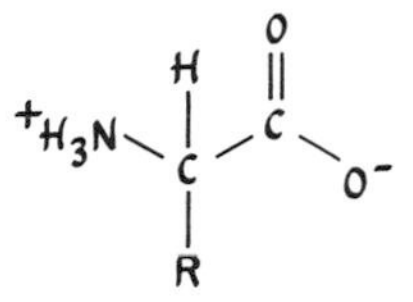

FIGURE 2-1. The general chemical formula of an amino acid. The letter R represents the side chain from whose particular structure different amino acids derive their individual character. The C atom to which the side chain is bonded is the α-carbon atom. The amino and carboxy groups are shown in their ionized forms to reflect the state of the molecule in water solution at physiological *p*H.

an amino and a carboxy group are attached. The remaining two valencies of the α-carbon atom are generally satisfied by a hydrogen atom and a *side chain*. It is to the particular structure of their side chains that different amino acids owe their individual character. The simplest amino acid, glycine, has no side chain at all; its α-carbon has but two hydrogen atoms. Glycine was also one of the first amino acids to be identified, its isolation from the protein gelatin dating back to 1820. By the end of the nineteenth century, all but four of the twenty amino acids (Figure 2-2) now known to make up proteins had been recognized. However, threonine, the last amino acid to be discovered, was identified only in 1935. Even though it was known that hydrolysis of some proteins yields amino acids that are not among the twenty shown in Figure 2-2, and despite the possibility that careful analysis of other proteins might in the future yield hitherto unknown amino acids, it became a basic tenet of molecular genetics in the middle 1950's that these and only these twenty *standard amino acids* are to be considered essential building

Monoaminomonocarboxylic

Glycine (Gly) | **Alanine (Ala)** | **Valine (Val)** | **Isoleucine (Ile)**

Leucine (Leu) | **Serine (Ser)** | **Threonine (Thr)**

Monoaminodicarboxylic

Aspartic acid (Asp) | **Glutamic acid (Glu)**

Diaminomonocarboxylic

Lysine (Lys) | **Arginine (Arg)**

Amides of monoaminodicarboxylic

Asparagine (Asn) | **Glutamine (Gln)**

Sulfur-containing

Cysteine (Cys) | **Methionine (Met)**

Aromatic

Phenylalanine (Phe) | **Tyrosine (Tyr)** | **Tryptophan (Trp)**

Heterocyclic

Histidine (His) | **Proline (Pro)**

blocks of *all* proteins. The presence of any nonstandard amino acid in some protein was declared as being not directly relevant to the central problem of protein formation. It is doubtful that the gene problem would have been solved by the time of the Mendel Centennial if this assumption—for its time, quite audacious and wholly unsupported—had not been made.

Inspection of the structural formulas shown in Figure 2-2 reveals that according to the nature of their side chains, the set of twenty standard amino acids (which are often designated by the three-letter abbreviations of their full names shown in Figure 2-2) can be classified into various sub-groups. Glycine, alanine, valine, leucine, isoleucine, serine, and threonine carry simple aliphatic side chains, and phenylalanine and tyrosine have simple aromatic side chains. Cysteine and methionine contain sulfur, and lysine, arginine, and histidine have a second amino group in aliphatic side chains. Aspartic acid and glutamic acid have a second carboxyl group in aliphatic side chains, and asparagine and glutamine are simple amides of aspartic and glutamic acids. Finally, tryptophan, proline, and histidine have heterocyclic side chains, proline being the only amino acid of the standard set whose α-amino group is a secondary rather than primary amine.

The way in which amino acids are linked together to make up the protein molecules from which they can be recovered upon hydrolysis was first outlined in 1902 by Emil Fischer. Fischer proposed that in proteins the α-amino group of one amino acid is joined to the α-carboxyl group of its neighbor amino acid by means of an amide linkage formed upon elimination of water. An amide bond that links two amino acids is called a *peptide bond*, and the amino acids so joined are said to constitute a *peptide*. Hence a protein molecule is a *polypeptide chain* composed of many *amino acid residues*, each residue joined to the next by a peptide bond (Figure 2-3).

As methods for the isolation, purification, and analysis of proteins improved during the first three decades of this century, two important facts emerged: each protein consists of a polypeptide chain of a definite length, the length for different proteins ranging from a few dozen to hundreds of amino-acid residues, and each protein contains different relative proportions of the twenty standard amino acids. From these facts the idea followed naturally that the individual character of a particular kind of protein depends on the number and kind of amino acids that make up its polypeptide chain.

NUCLEIC ACID

In addition to lipids, carbohydrates, and proteins, a fourth, and most important, class of substances present in living matter was discovered in 1868,

FIGURE 2-2.   The twenty standard amino acids of which natural proteins are composed. The diaminocarboxylic group includes also histidine.

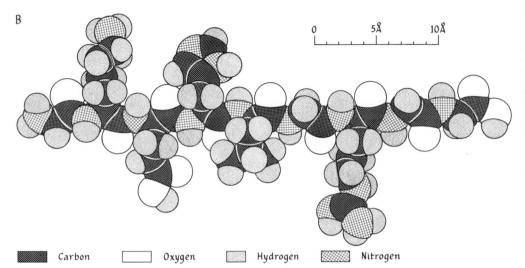

FIGURE 2-3.   Polypeptide chain. **A**. General chemical formula. **B**. A space-filling drawing of the polypeptide chain Gly-Lys-Glu-His-Leu-Gly-Arg-Gly-Gly. The relative sizes of the spheres correspond to the van der Waal's radii of the atoms they represent. [After E. J. DuPraw, *Cell and Molecular Biology*, Academic Press, New York, 1968.]

or within three years of Mendel's formulation of the gene concept. This class, the *nucleic acids*, turned out to constitute the genetic material, but nearly eighty years had to elapse before this was demonstrated to be so, and another ten years went by before that fact was generally appreciated. This long delay in the appreciation of the central role of nucleic acids in governing cellular processes is one of the ironies of the history of science, in that Friedrich Miescher (Figure 2-4) found them precisely with the purpose in mind of uncovering the chemical nature of the cell nucleus in the first place. And, as was set forth in Chapter 1, the notion that the cell nucleus is the seat of heredity had come to the fore at just about that time. In order to make a chemical study of the nucleus, Miescher undertook an analysis of cells, such as pus cells and salmon sperm, in which he knew the nucleus to represent a large fraction of the total cell mass. These analyses revealed that nuclei contain a hitherto unknown, phosphorus-rich, acid substance to which Miescher gave the name "nuclein." This novel substance was later rechristened "nucleic acid," the name by which it is now known.

FIGURE 2-4.   Friedrich Miescher (1844–1895). [From *Die histochemischen und physiologischen Arbeiten von F. Miescher*, F. C. W. Vogel, Leipzig, 1897.]

After the ubiquitous presence of nucleic acid in plant and animal kingdoms was demonstrated, the turn-of-the-century biochemist A. Kossel identified its constituents: the four nitrogenous bases *adenine, guanine, cytosine,* and *uracil* (the first two being *purines* and the second two being *pyrimidines*), a five-carbon sugar, and *phosphoric acid*. Further analytical work, done largely by P. A. Levene and by W. Jones in the 1920's, showed that there exist two fundamentally different kinds of nucleic acid, which are now called *ribonucleic acid*, or RNA, and *deoxyribonucleic acid*, or DNA (Figure 2-5).; RNA contains *ribose*, whereas DNA contains *deoxyribose* as its five-carbon sugar. Furthermore, DNA does not contain the pyrimidine uracil; instead of uracil it contains 5-methyluracil, or *thymine*. Nitrogenous base, sugar, and phosphoric acid are linked together in the manner shown in Figure 2-6 to form a *nucleotide*. By the 1930's it had been shown that nucleic acid molecules contain several such nucleotides linked through phosphate diester bonds between their sugars (Figure 2-7), but another ten years were to elapse before the enormously high chain length of nucleic acids became fully appreciated. As we now know, nucleic acid molecules consist of *polynucleotide* chains made up of thousands, and sometimes millions, of nucleotides in continuous chemical linkage.

The nucleotides of adenine, guanine, cytosine, uracil, and thymine are called *adenylic, guanylic, cytidylic, uridilic,* and *thymidylic acids*. The organic moiety of a nucleotide—the compound of nitrogenous base and sugar without added

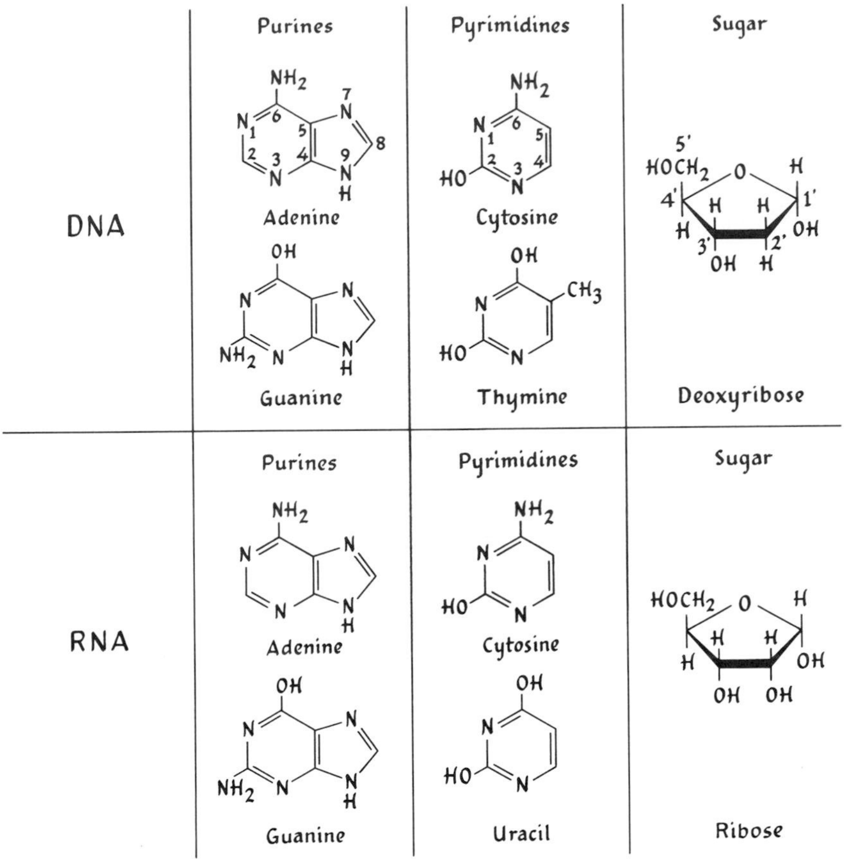

FIGURE 2-5.   The chemical building blocks of the nucleic acids DNA and RNA.

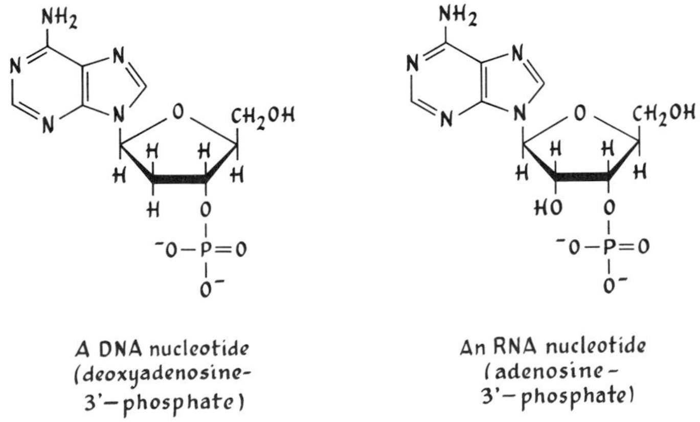

A DNA nucleotide
(deoxyadenosine-
3'−phosphate)

An RNA nucleotide
(adenosine−
3'−phosphate)

FIGURE 2-6.   The nucleotides of DNA and RNA.

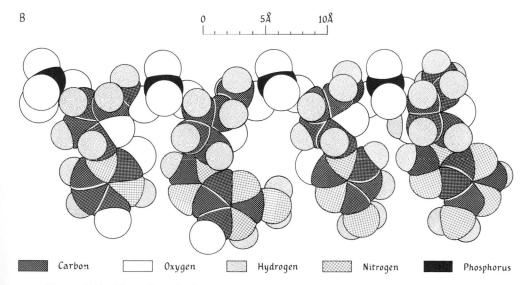

FIGURE 2-7.   The polynucleotide chains of DNA and RNA. **A.** Chemical formulas.
B. A space filling drawing of the RNA polynucleotide shown in Panel A. [After
E. J. Du Praw, *Cell and Molecular Biology*, Academic Press, New York, 1968.]

phosphate—is a *nucleoside*. The *ribo*nucleosides (those containing ribose) are called *adenosine, guanosine, cytidine,* and *uridine*; the corresponding *deoxyribo*nucleosides (those containing deoxyribose) are called *deoxyadenosine, deoxyguanosine, deoxycytidine,* and *thymidine*. Nucleotides are also called *nucleoside phosphates*; i.e., *adenosine monophosphate* (AMP), *guanosine monophosphate* (GMP), *cytidine monophosphate* (CMP), *uridine monophosphate* (UMP), and *thymidine monophosphate* (TMP).

When more refined techniques for the study of subcellular components were developed in the first decades of this century, it became possible to ascertain the intracellular localization of the two nucleic acid species, DNA and RNA. The DNA was identified as the substance responsible for the color produced by the cytological stain on the basis of which a part of the nucleus had been originally designated as "chromatin." Further study showed that DNA is, in fact, a major constituent of the chromosomes, in which it is combined with about three to four times its own weight of protein. In contrast to DNA, which is mainly found in the nucleus, the bulk of RNA is present in the cytoplasm. But a minor fraction of the cellular RNA exists within the nucleus, which contains the nucleolus, a region characterized by a particularly high concentration of RNA.

BACTERIA

The kind of "typical" cell that we have considered so far, and which is shown in the schematic diagram of Figure 1-5, is referred to as a *eukaryotic* cell. It is the basic unit not only of all higher, multicellular animals and plants but also of such lower, unicellular organisms as fungi, protozoa, and algae. Upon the invention in the 1930's of the electron microscope, with its limit of resolution one-hundredfold lower than that of the light microscope, and the subsequent development of refined techniques for staining and preparing specimens, a much more detailed view of the eukaryotic cell became available. Figure 2-8 shows an electron micrograph of the nucleus and surrounding cytoplasm of a cell of the bat. Here the nuclear membrane is clearly visible as a double-layered structure, as are the holes, or pores, in that membrane, through which nucleus and cytoplasm are in communication. The mitochondria in the cytoplasm, like the nucleus, can be seen to be enveloped by a membrane. Most important, the cytoplasmic structures, formerly termed "vacuoles" on the basis of light microscopy, are revealed to consist of a network of long, thin, membranous enclosures. This network, called the *endoplasmic reticulum*, is a complicated system of invaginations of the outer cell membrane. Thus the space inside a "vacuole." is actually in direct contact with the extracellular environment. The dark dots which can be seen to line the endoplasmic reticulum are the *ribosomes*—small

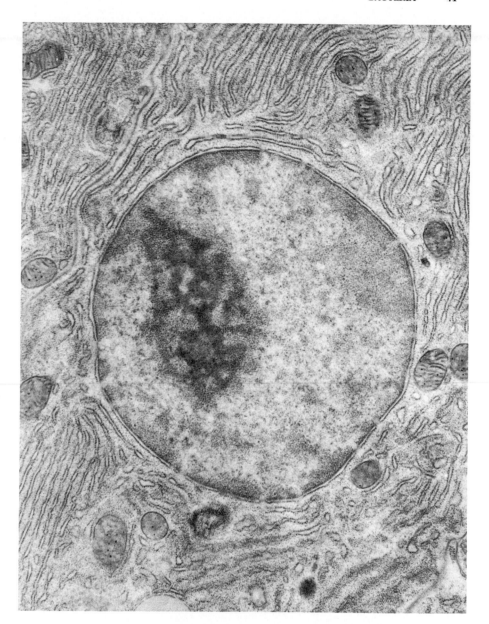

FIGURE 2-8.   Electron micrograph of a eukaryotic cell (from the pancreas of a bat).
The nucleus is the large round object in the center, which is surrounded by the
double-layered nuclear membrane. The smaller, round, striated structures in the
cytoplasm are mitochondria. The long, thin structures are the endoplasmic reticulum,
and the dark dots lining the reticulum are ribosomes. Enlargement × 18,000.
[Micrograph courtesy of Don Fawcett.]

particles composed of roughly equal proportions of RNA and protein, in which about two-thirds of the cytoplasmic RNA is localized.

Despite the apparent ubiquity of eukaryotic cells, their form of organization is not typical of the majority of living cells. That majority is made up of the humbler *prokaryotic* cells, which include the bacteria among their number. Since most of this text will be concerned with the insights revealed through experiments done with bacteria, some discussion of the nature of bacterial cells is now required. The reason for this preoccupation with bacteria at the expense of higher forms is that the new men who began to address themselves to the problem of heredity in the early 1940's differed from their predecessors not only in their training and general approach to biological problem-solving but also in their choice of experimental materials. Until the 1930's, virtually all genetic research had been done with plants and animals. Though multicellular, macroscopic organisms offer the convenience that the characters whose hereditary transmission is to be studied can generally be recognized at sight, plants and animals offer the disadvantage that the number of individuals that can be examined in any one breeding experiment is limited to a few hundred, or at most a few thousand. Furthermore, the life cycle of even the most rapidly reproducing plants and animals lasts weeks or months, so that completion of even a single experimental study of the distribution of parental characters among first and second generation progeny is comparatively lengthy. In order to overcome these limitations of higher forms as objects for intensive genetic study, the founders of molecular genetics turned to bacteria—microscopic organisms whose life cycle lasts for less than an hour, and of which billions of individuals can be raised overnight in a thimble-full of nutrient growth medium.

Bacteria were discovered among a new and previously unsuspected world of invisible living forms in the latter part of the seventeenth century, when Antony van Leeuwenhoek constructed the first microscope sufficiently powerful to render microorganisms visible to the human eye. Of these microorganisms Leeuwenhoek said, "there are more living in the scum on the teeth in a man's mouth than there are men in a whole kingdom." Although Leeuwenhoek's discovery excited great interest in its time, little further work was done on bacteria for another 200 years, until microbiology took its great upswing in the nineteenth century. The intellectual stimulus for this upswing was, in great part, the last stage of the venerable controversy concerning spontaneous generation. By that time it had already been shown that, contrary to former beliefs, mice and flies do not appear spontaneously in decaying matter, but it seemed still possible then that the microbes found in souring milk or rotten meat did in fact arise there *de novo*. At about the time that Mendel began to cross his peas, this controversy was finally settled by the experiments of Louis Pasteur and of John Tyndall, who showed that fermentation of a previously sterilized infusion is always the result of its subsequent infection by one or more ubiquitously present microorganisms.

Pasteur, Ferdinand Cohn, Robert Koch, and others who followed the trail blazed by the incisive experiments of these men, discovered during the latter part of the nineteenth century a vast ensemble of bacteria of the most diverse shapes, sizes, and functions. Before long, these discoveries were to alter the human condition, in that they led to the rationalization of the making of good wine, beer, and cheese, and to the conquest of infectious disease. Bacteria, it soon became recognized, play a vastly wider and far more fundamental role in the balance of the biosphere than merely being man's little friends and enemies. Indeed, macroscopic life as we know it would not be long for this world if it were not for bacterial intervention in the constant recycling of nitrogen and carbon between the atmosphere and organic matter.

THE PROKARYOTIC CELL

The fundamental chemical composition of bacteria is no different from that of higher cells: the bulk of a bacterium is also made up of lipids, carbohydrates, proteins, and the two nucleic acids RNA and DNA. But the manner in which these chemical components are organized to form a cell in the superkingdom of *prokaryota*, of which the bacteria form a part, differs in several important aspects from the cell structure of all higher forms, the *eukaryota*. First of all, the volume of a prokaryotic cell is from ten to ten thousand times smaller than that of a typical eukaryotic cell, and hence the total amount of these components per cell is correspondingly less. For instance, a typical bacterium contains on the order of 10^{-11} mg DNA, corresponding to 2×10^7 DNA nucleotides, or one thousandth the amount present in the nucleus of a mammalian cell. The eukaryotic cell, as we have seen, possesses a nucleus which is separated from the surrounding cytoplasm by a nuclear membrane. The nucleus contains the chromosomes, which are composed of DNA and protein, are the carriers of the genes, and are partitioned over the daughter cells by the elaborate processes of mitosis and meiosis. The cytoplasm, for its part, contains subcellular organelles. But the prokaryotic cell is constructed along simpler lines. Here nucleus and cytoplasm are not clearly delimited, there being no nuclear membrane. The DNA is not combined with protein to form any structures that resemble the eukaryotic chromosomes, and hence no mitotic or meiotic processes are manifest. Finally, no subcellular organelles are present that resemble mitochondria or centrioles. There can be little doubt that the simpler prokaryotes are the evolutionary antecedents of the more complex eukaryotes. Indeed, few later biological developments can have had greater impact on the later course of organic evolution than the ascension from prokaryotic to eukaryotic life in Precambrian times. For it made possible the eventual rise of multicellular organisms composed of highly differentiated cells that carry out specialized functions, and thus paved the way for the appearance of macroscopic creatures.

Bacteria are so small that they lie at the limit of resolution of ordinary light microscopes, their linear dimensions being only of the order of 1 μ. Thus it was long difficult to secure information about their internal structure through direct visual observation. Once the electron microscope was available, however, it had become possible to construct a detailed picture of the anatomy of the bacterial cell, as can be seen in the electron micrograph of Figure 2-9. It should be noted that the magnification of this electron micrograph is five times higher than that shown in Figure 2-8. Hence the size of the whole bacterial cell is not much greater than that of the mitochondrian organelles in the cytoplasm of the eukaryotic cell. Even though there is no

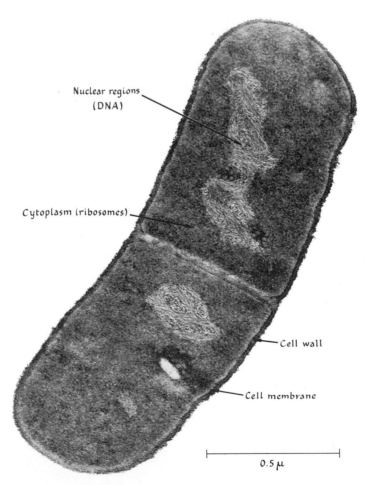

FIGURE 2-9.   Electron micrograph of a prokaryotic cell (*Bacillus subtilis*). [Micrograph courtesy of C. Robinow and J. Marak.]

genuine nucleus present in the prokaryotic cell, the DNA is evidently localized in a central body, which is embedded in an RNA-rich surround. As is true of the eukaryotic cell, most of the RNA in the prokaryotic cell is localized in *ribosomes*, which are visible in Figure 2-9 as a granular background over most of the cell. But no endoplasmic reticulum is visible here. Because of the formal analogy of the region to which the DNA is confined with the eukaryotic nucleus, this region is often referred to as the bacterial "nucleus"; and the remainder of the cell is usually called the bacterial "cytoplasm." This paradoxical extension of eukaryotic terminology to bacteria, whose very lack of these architectural features is supposed to set them apart from higher forms, has become so well-established in molecular genetics that it will not be possible to avoid such loose talk in this text.

The bacterium can be seen to be bounded by a *cell wall*—a rigid structure of considerable chemical complexity, containing polysaccharides, proteins, and lipids. The precise arrangement of these cell-wall components varies among different bacterial types, and endows the cell with a high degree of surface specificity. The cell wall is responsible for the characteristic shape of the bacterium—spherical, straight rod, or curved rod—and provides the strength necessary to prevent the bursting of the cell from internal osmotic pressure. Pressed tightly against the inside of the cell wall is the delicate *cell membrane*—the permeability barrier of the bacterium, which encloses the *protoplast*, or the remainder of the prokaryotic cell. As can be seen in the electron micrograph of Figure 2-10, the bacterial nucleus (i.e., its DNA) is connected with the cell membrane.

The bacteria can be divided into two subgroups, according to the manner in which the cell wall reacts with the stain devised by the Danish physician Cristian Gram: *gram-positive* bacteria retain the color of the stain in their cell envelope, whereas *gram-negative* bacteria do not. The reaction with the gram stain seems to be correlated with numerous other bacterial properties, so that it appears that gram-positive and gram-negative bacteria constitute separate "natural" subgroups. Gram-positive bacteria offer an experimental advantage over gram-negatives for the study of their intracellular components: the gram-positive cell wall can be digested completely by a simple chemical treatment. If this digestion is carried out in an ordinary medium, the protoplast bursts, because once deprived of its structural support by the cell wall, the delicate cell membrane cannot withstand the internal osmotic pressure. But if digestion of the cell wall is carried out in a medium of high external osmotic pressure, such as a concentrated solution of sucrose, then the bacterium without its cell wall survives as a spherical protoplast. This protoplast can still perform most of the functions of the intact bacterium, including growth and division. It is not possible, however, to remove the cell wall of gram-negative bacteria completely by such treatments; nevertheless, enough of the structural wall members can be digested away so that

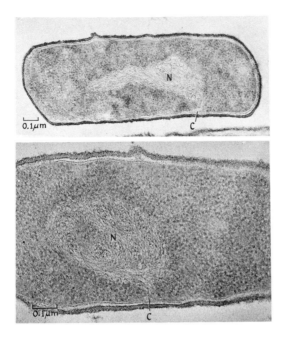

FIGURE 2-10.   Electron micrographs of *Bacillus subtilis*. The nuclear body (N) can be seen to be directly connected with the cell membrane (C). [From A. Ryter and F. Jacob, *Ann. Inst. Pasteur* **107**, 384 (1964).]

the cell either bursts in a medium of low osmotic pressure or rounds up into a sphere in a medium of high osmotic pressure. This partial removal of the cell wall is useful for some experiments to be considered later, since it increases the permeability of the cell, particularly for the penetration of macromolecular substances from the outside.

Some Bacterial Species

Of the vast number of bacterial species extant in nature, less than half a dozen will find mention in these pages. And even among this select group, one species, *Escherichia coli*, will demand our attention to the near exclusion of the others. This gram-negative bacterium, whose generic name derives from its discovery by Theodor Escherich and whose specific name derives from its presence in the human intestine, is by now undoubtedly the best-understood cell in all creation. That *E. coli* is nonpathogenic for man is probably one of the main reasons for its popularity as an experimental object among molecular biologists and biochemists. Only one other genus of gram-negative intestinal bacteria, closely similar to *Escherichia* in morphology and physiology, will find mention here: *Salmonella*. This genus has to be treated with some respect in the laboratory, since one of its species, *Salmonella typhi*, causes typhoid fever in man. In order to avoid probable decimation of the experimenters studying the genetics of *Salmonella*, avirulent strains of species such as *Salmonella typhimurium* have been generally employed, whose viru-lent strains are pathogenic for the mouse rather than for man. Despite its lurking threat of potential pathogenicity, *Salmonella* still enjoys some con-siderable popularity among molecular geneticists because one of the most important processes of genetic transfer between bacterial cells happened to be discovered with that genus, as will be shown later. Finally, the names of two gram-positive bacterial species will figure in later chapters: *Streptococcus pneumoniae* and *Bacillus subtilis*. The former ("pneumococcus," as it is also called) is a pathogenic bacterium that infects the respiratory tract of man and other mammals, where it causes pneumonia. The latter is a harmless bacterium that inhabits the soil, where it quietly carries out the decom-position of dead organic matter. As far as their cell morphology is concerned, *E. coli*, *Salmonella typhimurium*, and *B. subtilis* are all straight rods approxi-mately 2μ long and 1μ in diameter. *Streptococcus pneumoniae*, however, is a sphere, and usually exists in the form of chains of spheres held together by a mucoid substance on the cell wall.

BACTERIAL GROWTH

Rod-shaped bacteria like *E. coli* reproduce themselves by elongation into a longer rod without increasing their diameter. When such a bacterium has attained twice its original length, the elongated rod constricts in the middle and splits into two identical daughter cells. This event is the bacterial cell division. The two newly born daughter cells continue to elongate, and when each has attained twice its original length both divide in turn to yield four granddaughter cells (Figure 2-11). Since bacteria, like all other living cells, are composed mainly of the elements carbon, hydrogen, nitrogen, oxygen, sulfur, and phosphorus, they must assimilate these elements from their environment for this reproductive process by growth and cell division. Bacterial species, however, differ greatly with respect to the kind of chemical compounds that they can utilize as sources of raw material from which to fashion a pair of daughter cells. These nutritional variations generally reflect adaptations to the natural habitat: bacteria that subsist autonomously on rocky ledges or in desert wastelands have to be able to get along on the simplest of substrates, whereas bacteria that live in intimate contact with living tissues and feed parasitically on components of their hosts can afford to be more fastidious.

Though *E. coli* inhabits an environment as rich in organic foodstuffs as the human intestine, its nutritional requirements are relatively modest; Table 2-1 shows the composition of a simple chemically defined liquid medium that supports its growth. The only organic component of this defined, or "synthetic," medium is the sugar glucose. The mixture of sodium and potassium phosphates serves not only as the source of phosphorous,

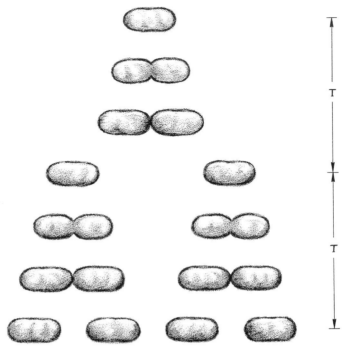

FIGURE 2-11.   Pattern of growth and reproduction of a bacterium. The generation time, T, is the time elapsed between the moment a bacterium has been formed and the moment at which it splits into two daughter cells.

sodium, and potassium, but also functions as a buffer that maintains the medium at a near-neutral pH, in the face of production of either hydroxyl or hydrogen ions by the chemical activities of the growing bacterial culture.

TABLE 2-1.   A Synthetic Growth Medium for *E. coli*

Constituent	Amount
NH_4Cl	1.0 g
$MgSO_4$	0.13 g
KH_2PO_4	3.0 g
Na_2HPO_4	6.0 g
Glucose	4.0 g
Water	1000 ml

Although this simple medium is perfectly adequate, cultures of *E. coli* are generally maintained in a nutrient broth extracted from meat or yeast. In such nutrient broth, *E. coli* bacteria grow more rapidly than in the synthetic medium, since the bacteria find ready-made in broth many of the substances, such as amino acids and purines and pyrimidines, that must be homemade in the synthetic medium. As will be demonstrated, however, there are many situations in which cultivation of the bacteria must be carried out in synthetic medium, since the presence of one or another component of the complex nutrient broth would interfere with certain experimental observations.

Bacteria are grown not only in liquid media, but also on *solid* nutrient surfaces. Such solid media are prepared by adding from 1 to 2% of agar to nutrient broth or to synthetic media similar to the one formulated in Table 2-1. The most common vessel used in the cultivation of bacteria on solid agar is the *petri plate*, shown in Figure 2-12. Each bacterial cell placed on such an agar surface grows and divides, and thus gives rise to a *colony* of descendant bacteria. After a few hours' growth, this colony contains so many cells that it is visible to the naked eye (Figure 2-12). The millions of

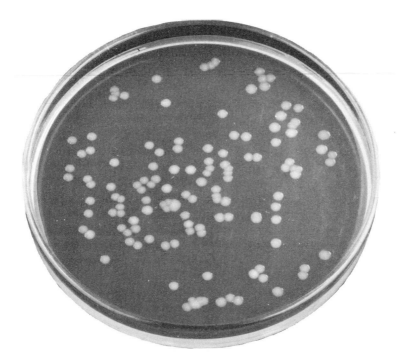

FIGURE 2-12.   An agar-filled petri plate on which individual bacteria have grown into macroscopic colonies. [From *Molecular Biology of Bacterial Viruses*, by G. S. Stent. W. H. Freeman and Company. Copyright © 1963.]

bacteria present in such a colony constitute a *clone*—a term that designates a population of individuals, or cells, all of whom have descended by a sequence of cell divisions from a *single* ancestor. (The cell ensemble that constitutes a multicellular higher plant or animal is thus a clone descended from the fertilized egg.)

Some bacterial species, such as *Bacillus subtilis* (but not *E. coli*), undergo a special, asymmetric type of cell division when their cultures have exhausted the available nutrients. As a prelude to this division, much of the solid matter of the bacterium, particularly its DNA, collects at one extremity of the cell, amounting to a volume that is as little as 10 per cent of that of the whole bacterium. These concentrated and highly dehydrated protoplasmic constituents are then walled off from the bulk of the cellular space to form a *spore* (Figure 2-13). The spore is very resistant to heating, drying, and other potential environmental stresses, and thus can survive under conditions that would kill the ordinary, or vegetative, bacterium. The spore can remain dormant for long periods—centuries or even millenia, until it encounters a growth environment more favorable than that which caused the vegetative bacterium to form the spore. This favorable environment provides a signal for the spore to *germinate*—that is, to take up water and to reconstitute the vegetative bacterium, which then resumes its growth.

The temperature at which bacteria manifest optimal growth usually reflects the conditions of their natural habitat. The optimal temperature for *E. coli*, and other bacteria that inhabit warm-blooded animals, is about 37°C. Cultures of *E. coli* are therefore generally grown at this temperature, which is maintained either as the air temperature of an incubator chamber or as the water temperature of a constant-temperature bath. When the temperature of incubation is lower than 37°C, the rate of bacterial growth is less than that

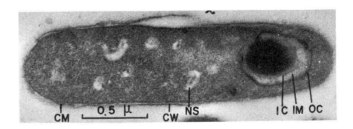

FIGURE 2-13. Electron micrograph of a bacterium in the process of forming a spore. The spore is enveloped by an outer spore coat (OC), an intermediate space (IM), and an inner coat (IC). The wall (CW), the membrane (CM), and the nuclear region (NS) of the vegetative bacterium are still intact. [From A. Takagi, T. Kawata, and S. Yamamoto, *Journal of Bacteriology*, **80**, 37, 1961.]

at the optimum temperature. Thus most experimental cultures of *E. coli* are grown at 37°C, though, as we shall see, the use of suboptimal or supraoptimal temperatures is also of some importance in bacterial genetics. *Unless explicitly stated otherwise, however, it can be assumed that the temperature of incubation was near 37°C in all the investigations to be described in the following chapters.*

CELL COUNTS

The most direct way of estimating the number of bacteria is to count the number of cells visible under the microscope. This is done with the aid of a special counting chamber, which brings accurately known, very small culture volumes (on the order of 10^{-6} ml) into the microscopic field. This method is known as the *total cell count*. Accurate total cell counts can be made only on cultures containing more than 10^7 bacteria/ml, since at lower cell densities too few bacteria appear in the microscope field. Recently, an electronic device, the Coulter counter, has become available that performs an automatic total cell count by detecting the passage of each bacterium in a culture volume through an electric field.

Another method consists in spreading a known volume of the culture fluid on the surface of a nutrient agar plate and then incubating the plate. Since each bacterium gives rise to a single bacterial colony, one need count only the number of colonies that appear on the agar after incubation and then reckon back to the original concentration of bacteria in the culture by dividing this count by the culture volume spread on the plate. This method is called the *colony count*. It also goes under the name of *viable count*, since only living cells (those capable of giving rise to an indefinite succession of descendants) will figure here, in contrast to the total cell count, which encompasses both viable and nonviable cells. The colony-count method can be used for counting low concentrations of cells, since even a single viable bacterium can be detected in this way; the colony count, however, is equally applicable to high cell concentrations, since dense bacterial suspensions can first be diluted as necessary so that the optimal number of about 100 colonies per plate can be spread on the agar.

Finally, the number of bacteria can also be estimated from the *turbidity* of the culture. Concentrated bacterial suspensions appear turbid because light entering the culture fluid is scattered by the cells, the fraction of the incident light scattered being more or less proportional to the total mass of bacteria present. The turbidity can be determined either by a *photometer*, which measures the amount of incident light transmitted, or by a *nephelometer*, which measures the amount of incident light scattered. Turbidity can be related to cell number by carrying out on the same growing culture of bacteria a parallel series of turbidimetric measurements, total cell counts, and colony counts. The normalization obtained in this way, however, is valid only for the precise physiological conditions of the test culture, since the mass per cell is not invariant. The turbidity of a rapidly growing culture of *E. coli* becomes just perceptible to the naked eye at a density of about 10^7 cells per ml; it is only above this cell concentration that turbidimetry finds application in counting bacteria.

DYNAMICS OF GROWTH

We may now consider how the size of the bacterial population increases while the simple life cycle depicted in Figure 2-11 obtains for its members. In that life cycle, the time period elapsing between the moment a cell is born by division of its parent cell and the moment at which it itself divides is called the *generation period*, or the *division period*. Now let the number of bacteria in a culture be N_0. One generation period later, all of these N_0 have divided in two, yielding $2N_0$ bacteria. The $2N_0$ bacteria continue to grow and divide, so that at the end of the second generation period $2 \times 2N_0$ bacteria are present. At the end of the third generation period the number has increased to $2 \times 2 \times 2N_0$, and so on. Thus it follows that the number N of bacteria present after g generation periods of growth is given by the equation

$$N = 2^g N_0 \qquad (2\text{-}1)$$

Now let the generation period be T minutes; then after t minutes have elapsed,

$$g = t/T \qquad (2\text{-}2)$$

By substituting equation (2-2) into equation (2-1) we obtain

$$N = 2^{t/T} N_0 \qquad (2\text{-}3)$$

Evidently the size N of the bacterial population depends upon the elapsed time t as an exponent to which the number 2 is raised. Hence the population is said to grow *exponentially*. Equation (2-3) indicates how experimental data on the dynamics of bacterial growth are to be analyzed, for if N is determined at various times t (by total cell count, by viable colony count, or by turbidimetry), then the generation period T can be calculated. For this purpose,

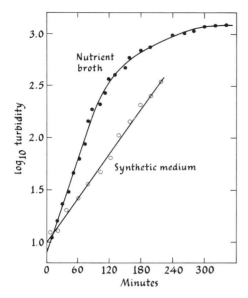

FIGURE 2-14.   The growth of two well-aerated cultures of *E. coli*, one in nutrient broth and the other in a synthetic medium containing glucose as its only source of carbon atoms. The ordinate shows the logarithm of the turbidity of the culture, in arbitrary units. A turbidity of 100 units corresponds to a bacterial density of about 10^8 cells/ml. [From *Molecular Biology of Bacterial Viruses*, by G. S. Stent. W. H. Freeman and Company. Copyright © 1963.]

it is convenient to express equation (2-3) in logarithmic rather than exponential terms:

$$\log_{10}N = (t/T) \log_{10} 2 + \log_{10} N_0$$

$$= \frac{0.301}{T}t + \log_{10} N_0 \qquad (2\text{-}4)$$

Thus, according to equation (2-4), a plot of $\log_{10}N$ against t yields a straight line, whose slope is $0.301/T$ and whose intercept at $t = 0$ is $\log_{10} N_0$. Figure 2-14 presents two such plots, which represent the growth of two *E. coli* cultures, one growing in the glucose-salts medium of Table 2-1 and the other growing in a medium containing nutrient broth. Measurement of the slopes of the straight-line segments of the two curves indicates that in the glucose-salts medium $T = 50$ minutes, and that in nutrient broth $T = 20$ minutes. Hence, in meat broth the bacteria grow more than twice as fast as in glucose-salts medium. At the faster rate of growth, an experimenter can easily witness 6 generation periods in two hours, a reproductive sequence for which 120 years would be necessary in the human family tree.

It is to their capacity for rapid multiplication that bacteria owe their numerical preponderance among living forms. Nevertheless, there are natural curbs to population explosions among bacterial cultures. Regardless of the size of the vessel, a bacterial culture cannot continue exponential growth at a generation time of 20 minutes for very long; if this were possible, a single *E. coli* bacterium would in 24 hours give rise to 2^{72}, or about 10^{22}, descendants, weighing some ten thousand metric tons; after another 24 hours of exponential growth, the weight of the descendants would amount to several times that of the earth. That our planet has not been converted into one microbial mass is due not only to the exhaustion by bacteria of the nutrients that support their growth, but to the deterioration of their environment through their excretion, at an ever-increasing rate, of products toxic to themselves. This environmental deterioration, both in nature and in the laboratory, soon causes the growth of bacteria to slow down from the maximum rate afforded by the most favorable conditions until their capacity to reproduce finally falls to zero. At this stage, when there is no further increase in the net number of cells, the culture is said to have attained the *stationary phase* (see Figure 2-14). Cultures of *E. coli* grown either in the simple synthetic medium or in nutrient broth enter the stationary phase when the bacterial concentration reaches the range of $2 \times 10^9 - 5 \times 10^9$ cells/ml.

RESTATEMENT OF THE FUNDAMENTAL PROBLEM

The fundamental problem of biological reproduction posed at the beginning of this chapter can now be restated in terms of the life of the bacterial cell. As has been shown, bacteria multiply at the expense of whatever medium they depend upon for growth, and they synthesize from the growth medium the lipids, polysaccharides, proteins, and nucleic acids that make up their cell structures. Since a bacterium produces the equivalent of itself in each generation period, it follows that all of its structures and components are duplicated and distributed with sufficient precision to yield two daughter cells identical to each other and to the parent cell. We may now ask how the bacterium manages to form its complex constituents from such simple chemicals as glucose, ammonia, inorganic salts, and water—and how the orderly processes of reproduction and division are preserved over countless generations.

The eventual answer to this question must take into account that the formation of this new cell material has to proceed at a hierarchy of levels. At the lowest level, the bacterium synthesizes from simple nutrient chemicals the building blocks of the lipids, polysaccharides, proteins, and nucleic acids. Since the variety of such building blocks is rather limited—the fatty acids, sugars, amino acids, and purines and pyrimidines comprise only about 60 different kinds of molecules—the ensemble of chemical reactions necessary to produce that variety need not be of an incomprehensible complexity. At the next higher level, the bacterium joins these building blocks into macro molecules through dehydration reactions (the reverse of hydrolysis). For instance, formation of the peptide bond in the joining of amino acids into a polypeptide chain involves the dehydration reaction

Finally, at the highest level, the macromolecular components are joined into even larger cell structures, such as the cell wall, the cell membrane, and the ribosomes. According to this view, the levels of synthesis can be summarized as follows:

glucose, ammonia, inorganic salts		amino acids, nucleotides, sugars, fatty acids	$-H_2O$	proteins, nucleic acids, polysaccharides (> 1500 atoms per molecule), lipids		large cell structures
(< 25 atoms per molecule)		(< 60 atoms per molecule)		(about 150 atoms per molecule)		

In the next chapter we will proceed to an examination of the lowest of these levels.

Bibliography

SPECIALIZED TEXTS, MONOGRAPHS AND REVIEWS

Loewy, A. and P. Siekevitz. *Cell Structure and Function*, Holt, Rinehart and Winston, New York, 1962.

Mahler, H. R., and E. H. Cordes. *Biological Chemistry*. Harper & Row, New York, 1966. Chapters 1, 2, and 9.

Sistrom, W. *Microbial Life*. Holt, Rinehart and Winston, New York, 1962.

Stanier, R. Y., M. Doudoroff, and E. A. Adelberg. *The Microbial World* (3rd ed.), Prentice-Hall, Englewood Cliffs, N. J., 1970.

Swanson, C. P. *The Cell* (2nd ed.). Prentice-Hall, Englewood Cliffs, N. J., 1964.

3. Metabolism

Consider a well-aerated culture of bacteria growing in a chemically defined medium like that formulated in Table 2-1. In such a culture the synthesis of all cell components proceeds from the glucose, ammonia, and inorganic salts supplied as nutrients. Suppose that after some amount of growth of the culture, the bacteria were collected from the culture fluid and analyzed for the total weight of carbon they contain. This weight would be a measure of the total quantity of organic cell material that has grown. Suppose, furthermore, that the weight of glucose carbon consumed by the bacteria is ascertained by subtracting from the weight of glucose carbon initially added that which remains in the growth medium. A comparison of these two measurements would show that, of the glucose carbon consumed, only about half has been converted into cell carbon and the other half has been oxidized to CO_2, according to the reaction

$$C_6H_{12}O_6 + 6O_2 \longrightarrow 6CO_2 + 6H_2O$$

This bipartite utilization of the glucose carbon illustrates a fundamental aspect of cell growth: for the synthesis of their components, cells must extract from their environment not only the atoms from which these components are to be assembled, but also the energy needed to drive the chemical reactions involved in the assembly process. Thus cells carry out two general classes of vital reaction sequences: *degradative* sequences that convert glucose to CO_2 and yield the approximately 9,000 calories of energy per gram of glucose carbon that is oxidized; and *synthetic* sequences that convert glucose into cell components and consume the energy made available by the degradative sequences. The joint occurrence of these two classes of reaction sequences constitutes the *metabolism* of the cell. We turn now to the degradative reaction sequences.

FERMENTATION

The discovery that grape juice, left to itself, may turn into an intoxicating and pleasant-tasting beverage dates back to neolithic times. As was set forth in Chapter 1, the first exploitation of hereditary principles in plant and animal breeding had provided the economic basis for the dawn of civilization, but the development of the vintner's art was needed to adapt the human psyche for life in the civilized world that was to spring from the agricultural-urban revolution. Wine is mentioned in ancient Egyptian writings, where the invention of viticulture is attributed to the god Osiris; the Bible gives credit to Noah for this great benefaction to mankind; and the Greeks thanked their god Dionysus for it. Throughout the ages the process of alcoholic fermentation, with its attendant bubbling and frothing, excited wonderment, but, just like heredity, fermentation came to be understood only in the nineteenth century. At the very beginning of that century J. L. Gay-Lussac had deduced that the overall fermentation reaction can be described as

$$C_6H_{12}O_6 \longrightarrow 2C_2H_6O + 2CO_2$$
$$\text{Glucose} \qquad\qquad \text{Ethanol}$$

In the 1830's Theodor Schwann, one of the founders of cell theory, demonstrated that the yeast (meaning froth) connected with fermentation are living cells. He proposed that yeast cells grow and multiply at the expense of the sugar present in the grape juice and leave behind the alcohol. About twenty years later, Pasteur, whose interest in industrial fermentation processes was one of the main reasons for his entrance into microbiology, provided the deep insight that fermentation represents an incomplete oxidation of glucose by "life without air." This is in contrast to "life with air," whose fundamental reaction, as A. L. Lavoisier had already realized late in the eighteenth century, results in the complete oxidation of organic matter to carbon dioxide and water. Pasteur considered fermentation to be so complex a reaction sequence that he thought it could be performed only by living cells. But in 1897, Edward Buchner showed that this is not true. Buchner ground a paste of brewer's yeast with abrasive sand and collected the juice released from the broken cells. Upon addition of glucose to the juice from the yeast cells, Buchner observed the cell-free production of ethanol and CO_2.

The discovery of cell-free fermentation had a profound effect on the future course of biological research: it had now become possible to study biochemical reactions in test tube solutions rather than in the living cell. With this development biochemistry began to separate as a distinct scientific discipline from physiology, which had until then been the exclusive province of the chemistry of life. In the next fifty years, biochemists sought to find the answers to two questions pertaining to the transformation of glucose into ethanol and CO_2: what are the chemical intermediates in this reaction sequence, and what is the nature of the agents present in the cell extract that allow these reactions to occur so rapidly and so specifically?

Because of their capacity to promote fermentation, the agents extracted from yeast were first called "ferments." Eventually, however, it became clear that yeast extract contains a great diversity of agents that are capable of promoting, or *catalyzing*, a wide spectrum of chemical reactions, of which those pertaining to fermentation are but a small sample. These catalytic agents were then given the name *enzymes* (meaning "in yeast"). It was soon found that enzymes can be extracted from, and are therefore present in, all living cells. One by one, the metabolic transformations catalyzed by the enzymes were identified and characterized by study of cell-free reaction mixtures. These studies showed that although the overall cell metabolism appears at first sight to consist of a reaction network of stupefying complexity, it is not, after all, beyond the pale of comprehension. That is, the transformation of glucose, ammonia, and inorganic salts into the 60 or so building blocks of cell components—the lowest level in the hierarchy of synthetic processes posited at the close of the last chapter—appears to comprise about one or two thousand different chemical reactions. But this wide panorama of metabolic reactions actually represents no more than half a dozen basically different types of chemical processes, which are used over and over again, in different sequences and on different molecules. Among these basic processes are the removal or addition of hydrogen (oxidation-reduction), carbon dioxide (decarboxylation-carboxylation), and water (dehydration-hydration). Most of these one or two thousand reaction steps owe their occurrence to the presence of a particular enzyme that specifically catalyzes one and only one basic chemical transformation of one and only one set of reactants.

GLYCOLYSIS

The fermentative reaction sequence leading from glucose to ethanol and CO_2, historically the first metabolic transformation to come under study, turned out to be central to all metabolism. Buchner's experiments had shown that glucose added to yeast disappears concomitantly with the production of ethanol and CO_2. It was soon found that addition of such chemicals as arsenate or mercuric salts to the reaction mixture causes ethanol and CO_2 production to stop—without, however, arresting the disappearance of glucose. The inhibitory effect of these chemicals derives from their capacity to combine with and abolish the catalytic action of one or another of the enzymes involved at some intermediate stage of the fermentative reaction sequence. Further investigation of this phenomenon led to the finding that chemical intermediates of the reaction sequence leading from glucose to ethanol accumulate in the inhibited extract. Their accumulation made it possible to isolate and identify these intermediates. Once they had been identified, the intermediates were synthesized in the laboratory and added to the extract, in place of glucose. The eventual production of ethanol and CO_2 in such extracts then confirmed the inference that the synthetic chemical added in place of glucose is, in fact, an intermediate of the fermentation sequence.

Arthur Harden and W. J. Young, two of the first biochemists to follow Buchner's lead, soon found that the presence of inorganic phosphate is necessary for fermentation, and this finding, in turn, led to the discovery that the first step in the reaction sequence is the phosphorylation of glucose to form glucose-6-phosphate

Glucose-6-phosphate

But as O. Meyerhof showed in the late 1920's this step in the reaction sequence, which is catalyzed by an enzyme that he called *hexokinase*, does not require inorganic phosphate at all. Instead, it consists in the transfer to glucose of the terminal, or γ, phosphate of adenosine triphosphate, or ATP, thus leaving behind adenosine diphosphate, or ADP, according to the reaction shown in Figure 3-1. Adenosine triphosphate had only just then been discovered as a naturally occurring compound in muscle tissues, and the central importance of that compound for the energetic aspects of general metabolism was not to emerge until the next decade. By 1940 F. Lipmann and H. Kalckar had shown that ATP is none other than the "vital force" that drives those metabolic reactions that require the input of energy.

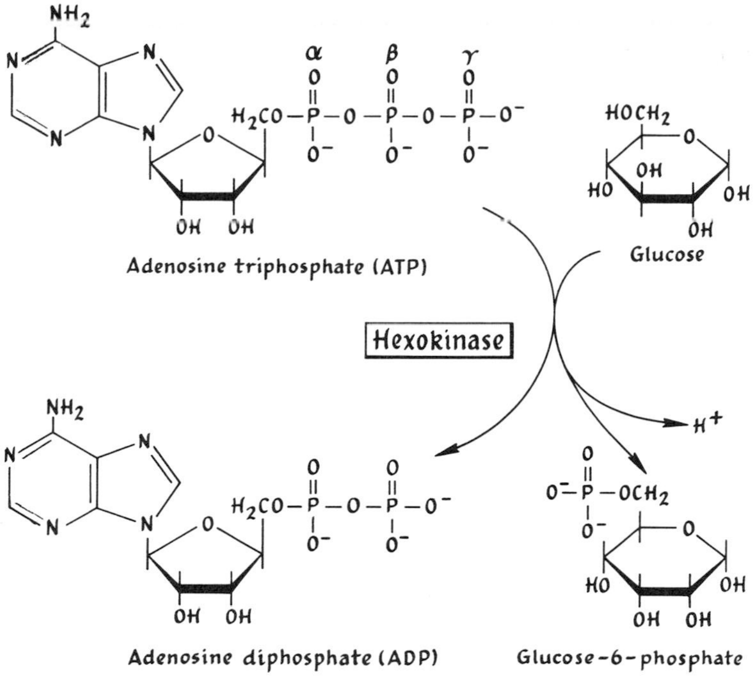

FIGURE 3-1.   Phosphorylation of glucose by adenosine triphosphate (ATP) catalyzed by hexokinase.

Before proceeding further in this account of the fermentation sequence, let us stop briefly to consider some properties of ATP. As can be seen in Figure 3-1, ATP is a ribonucleotide, and because its structure includes the purine base adenine it resembles the adenosine monophosphate, or AMP, building block of RNA. The presence of four closely spaced negative charges on ATP strongly favors the hydrolysis of the γ phosphate. That is to say, the reaction (where P represents inorganic phosphate)

$$ATP + H_2O \rightleftharpoons ADP + P + 2H^+$$

tends to go to the right. If we now consider the reaction

$$glucose + P + 2H^+ \rightleftharpoons glucose\text{-}6\text{-}phosphate + H_2O$$

we find that it tends to go to the left, and hence the direct phosphorylation of glucose by inorganic phosphate is not a reaction that can proceed in good yield. Nevertheless, the tendency of this reaction to go to the left is not as great as the tendency of the hydrolysis of ATP to go to the right.

Thus in a system in which these two reactions can be coupled, such as that offered by the enzyme hexokinase, the overall reaction

$$\text{glucose} + \text{ATP} \rightleftharpoons \text{glucose-6-phosphate} + \text{ADP}$$

will tend to go to the right. That is, a good yield of glucose-6-phosphate can be produced at the expense of the hydrolysis of ATP. This principle of driving a reaction that does not proceed spontaneously by coupling it with another reaction that *does* is central to the energetics of metabolism. Moreover, the hydrolysis of ATP is the chief (though by no means only) reaction to which the other reactions are coupled. Thus ATP is the common carrier of chemical energy in the molecular transformations that take place in living cells. The phosphorylation of ADP by inorganic phosphate to yield ATP is also the major route by which the phosphate component of the growth medium enters the metabolic chain. It is from its initial position as the γ-phosphate of ATP that freshly assimilated phosphorus starts out on its manifold intracellular metamorphoses.

By 1940, the chemical conversion of glucose into ethanol and CO_2 had been worked out, according to the scheme shown (in abbreviated form) in Figure 3-2. As can be seen there, the phosphorylated glucose is transformed into its fructose isomer. Next, fructose undergoes a second phosphorylation at the expense of the hydrolysis of a second molecule of ATP to yield ADP. The resulting fructose-1,6-phosphate is cleaved into two three-carbon fragments, yielding (after isomerization of one of the fragments) two molecules of glyceraldehyde phosphate. It is at this stage that inorganic phosphate finally enters the reaction sequence, which will not proceed beyond the formation of glyceraldehyde phosphate in the absence of inorganic phosphate. The glyceraldehyde skeleton is then oxidized to glyceric acid, and the chemical energy made available by this oxidation is utilized by the enzyme *glyceraldehyde-phosphate dehydrogenase*, which catalyzes this reaction and drives two others. One of these two is the phosphorylation of phosphoglyceric acid by inorganic phosphate to yield diphosphoglyceric acid.

The other of the two is the reduction of another carrier of metabolic energy, which we must now briefly consider. Harden and Young's pioneering studies had shown that yeast extract contains some heat-stable, relatively small organic molecules whose presence is required for fermentation. One kind of such essential molecule was later recognized as ATP, without which, as we have already seen, glucose will not be phosphorylated at the very outset of its fermentation. But another kind of essential molecule was identified in the early 1930's as *nicotinamide adenine dinucleotide*, or NAD^+, whose structure is shown in Figure 3-3. As can be seen, the structure of NAD^+ consists of nicotinamide (which has a positively charged nitrogen in its six-membered ring) linked to phosphorylated ribose—a structure formally

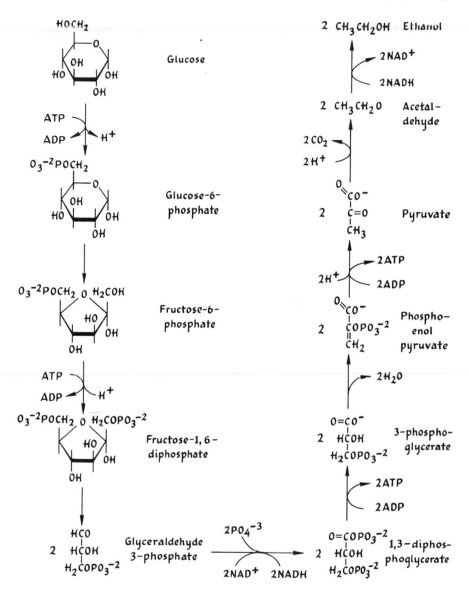

FIGURE 3-2.   Fermentation of glucose, or glycolysis.

analogous to a ribonucleotide (see Figure 2-6). This ribonucleotide is linked through a phosphate diester bond to adenosine monophosphate, or AMP, to form the dinucleotide from which NAD$^+$ derives its name. Upon the transfer to NAD$^+$ of two hydrogen atoms made available by the oxidation of a molecule such as glyceraldehyde to glyceric acid in the last two steps of the

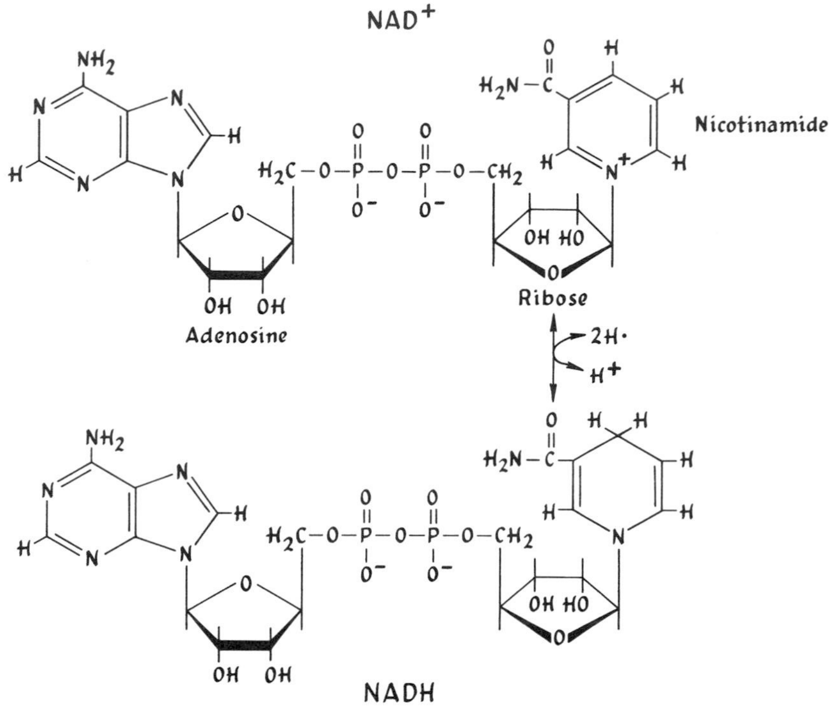

FIGURE 3-3.   Nicotinamide adenine dinucleotide in oxidized (NAD$^+$) and reduced (NADH) form.

fermentation process the nicotinamide part of NAD+ is reduced by saturation of one of the double bonds of the six-membered ring, as shown in Figure 3-3. This reduced form of NAD$^+$ is designated as NADH. The NADH so formed can later be reoxidized to NAD$^+$, with attendant reduction of some other molecule, as exemplified by the terminal stage of fermentation, to be considered next.

In the next step of the fermentation sequence, the inorganic phosphate previously incorporated into the organic ester bond of diphosphoglyceric acid is transferred to ADP, yielding ATP and phosphoglyceric acid as products. This reaction proceeds because the hydrolysis reaction

$$-\overset{\overset{\displaystyle O}{\|}}{C}-O-\overset{\overset{\displaystyle O}{\|}}{\underset{\underset{\displaystyle O^-}{|}}{P}}-O^- + H_2O \rightleftharpoons -\overset{\overset{\displaystyle O}{\|}}{C}-O^- + {}^-O-\overset{\overset{\displaystyle O}{\|}}{\underset{\underset{\displaystyle O^-}{|}}{P}}-O^- + 2H^+$$

has a greater tendency to go to the right than does the reaction

$$\text{ATP} + \text{H}_2\text{O} \rightleftharpoons \text{ADP} + \text{P} + 2\text{H}^+$$

Hence the tendency of the overall reaction

$$\text{ADP} + \underset{\underset{\text{O}^-}{|}}{-\overset{\overset{\text{O}}{\|}}{\text{C}}-\text{O}-\overset{\overset{\text{O}}{\|}}{\text{P}}-\text{O}^-} \rightleftharpoons \text{ATP} + -\overset{\overset{\text{O}}{\|}}{\text{C}}-\text{O}^-$$

is to go to the right. Since this reaction concerns both molecules of diphospho-glyceric acid derived from the cleavage of fructose-1,6-diphosphate, two mole-cules of ADP are phosphorylated at this step per input molecule of glucose. Hence the two molecules of ATP which were consumed in the preceding phosphorylation steps are restored to the metabolic economy. The resulting phosphoglyceric acid is rearranged and dehydrated to yield phosphoenol pyruvic acid, a transformation which neither requires nor yields chemical energy. But this rearrangement yields another organic phosphate ester of the type

$$-\overset{\overset{\text{O}}{\|}}{\text{C}}-\text{O}-\underset{\underset{\text{O}^-}{|}}{\overset{\overset{\text{O}}{\|}}{\text{P}}}-\text{O}^-$$

whose hydrolysis in the next step to yield pyruvic acid is coupled to the phosphorylation of ADP by inorganic phosphate to yield ATP. At this stage, there is a net gain of two molecules of ATP at the expense of oxidizing and cleaving one molecule of glucose to yield two molecules of pyruvic acid. The fermentation sequence is now completed by decarboxylation of pyruvic acid to yield acetaldehyde and CO_2. Finally, acetaldehyde is reduced to ethanol, a step which proceeds thanks to the reoxidation to NAD^+ of one molecule of NADH. Thus the overall reaction of fermentation is not that written by Gay-Lussac, but has to be represented as

$$\text{C}_6\text{H}_{12}\text{O}_6 + 2\text{ADP} + 2\text{P} \longrightarrow 2\text{C}_2\text{H}_6\text{O} + 2\text{CO}_2 + 2\text{ATP}$$

That the fermentation reaction sequence shown in Figure 3-2 has a much wider biological significance than the production of alcohol by yeast was first shown by Meyerhof in the early 1930's. He discovered that essentially the same process takes place in the muscle tissues of mammals, except for the very last step. Muscle cells draw chemical energy from their glucose reserve (which they hold in the form of glycogen) by breaking it down to pyruvic acid with attendant production of ATP. In muscle tissue, however, the pyruvic acid is ultimately converted to lactate ($CH_3CHOHCOO^-$) rather than to ethanol and CO_2. Not long after Meyerhof's observation was made known, it was shown that bacteria maintained in the absence of air also convert glucose to lactate through the same reaction sequence. Indeed, this general scheme—to which has been assigned the name *glycolysis* (glyco = sugar; glykis = sweet), or glycolytic pathway—has been found to be one of the principal ways in which glucose enters into cellular metabolism.

COMPLETE OXIDATION

The steps of the glycolytic pathway up to and including the formation of pyruvic acid proceed whether the cell is in contact with molecular oxygen or not. In most types of cells, however, including yeast and bacteria, the final conversion of pyruvic acid to ethanol or lactic acid proceeds in abundant yield only in the *absence* of oxygen. In the presence of oxygen, or for "life with air," pyruvic acid is subjected to further oxidative processes whose yield in chemical energy ultimately results in the generation of more ATP. The most important of these processes, the *citric acid*, or *tricarboxylic acid*, *cycle*, is shown in Figure 3-4 in a form even more highly abbreviated than the compressed version of glycolysis presented in Figure 3-2. This oxidation process of pyruvic acid is also called the "Krebs Cycle," after Hans Krebs, who first demonstrated its cyclical nature in 1937. Pyruvic acid first loses its carboxyl group as CO_2, and the remaining oxidized two-carbon fragment condenses with the four-carbon oxaloacetic acid to form the six-carbon citric acid. This oxidation is coupled with the reduction of one molecule of NAD^+ to NADH. Citric acid is then rearranged, oxidized, and decarboxylated to yield another molecule of CO_2 and the five-carbon α-ketoglutaric acid. This oxidation of citric acid is coupled with the reduction of one more molecule of NAD^+ to NADH. The resulting α-ketoglutaric acid is in turn oxidized and decarboxylated to yield a third molecule of CO_2 and the four-carbon oxaloacetic acid in a series of steps to one of which is coupled the phosphorylation by inorganic phosphate of one molecule of ADP to yield one molecule of ATP. This oxidation of α-ketoglutaric acid is also coupled with the reduction of three more molecules of NAD^+ to NADH. By the time this stage of the process is reached, the pyruvic acid molecule that entered the cycle has been completely oxidized by successive decarboxylation of its three carbon atoms, giving rise to three molecules of CO_2. The remaining molecule of oxaloacetic acid then condenses with another molecule of pyruvic acid furnished by glycolysis and thus re-enters the citric acid cycle. The complete oxidation of pyruvic acid has consumed three molecules of water and resulted in the reduction of five molecules of NAD^+ to NADH

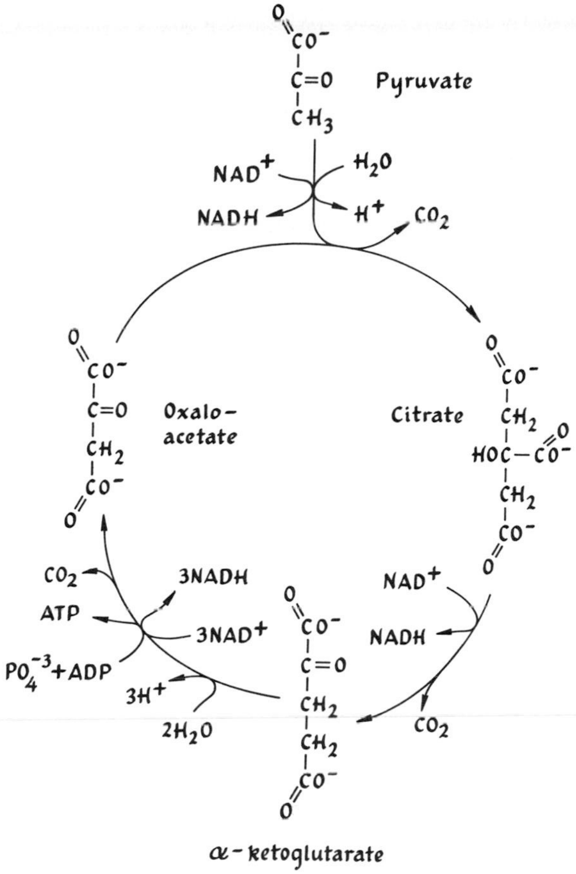

FIGURE 3-4.   The citric acid cycle in highly abbreviated form.

and in the phosphorylation of one molecule of ADP to yield ATP. Since each original molecule of glucose gave rise to two molecules of pyruvic acid in glycolysis, the net chemical reaction of glycolysis and citric acid cycle can thus be written

$$C_6H_{12}O_6 + 12NAD^+ + 4P + 4ADP + 6H_2O$$

$$\longrightarrow 6CO_2 + 12NADH + 4ATP + 12H^+$$

The oxidation of glucose now enters its third and final episode, *oxidative phosphorylation*, in which molecular oxygen, the oxidant of the overall reaction, finally becomes involved. Oxydative phosphorylation is carried out by a complicated enzymatic ensemble, collectively termed the *respiratory enzymes*. In this reaction sequence molecular oxygen reoxidizes the 12 NADH molecules produced by glycolysis and the citric acid cycle to yield NAD^+ and water. The respiratory enzymes that catalyze this reaction channel the chemical energy made available by the reoxidation of NADH (through a mechanism not yet understood in all of its details) into the phosphorylation of ADP by inorganic phosphate to yield a total of 34 molecules of ATP, according to the overall reaction

$$12NADH + 12H^+ + 6O_2 + 34P + 34ADP$$
$$\longrightarrow 12NAD^+ + 12H_2O + 34ATP$$

Since glycolysis and citric acid cycle together yield four molecules of ATP before oxidative phosphorylation commences, it follows that the complete oxidation of one molecule of glucose provides enough chemical energy to convert 38 molecules of ADP into ATP, for which conversion 38 molecules of inorganic phosphate are assimilated from the growth medium. Hence by summing the overall reactions previously written for glycolysis, citric acid cycle, and oxidative phosphorylation, we arrive at the following grand equation, which describes the extraction of chemical energy from glucose by "life with air":

$$C_6H_{12}O_6 + 6O_2 + 38ADP + 38P$$
$$\longrightarrow 6CO_2 + 6H_2O + 38ATP$$

METABOLISM OF SUGARS OTHER THAN GLUCOSE

Many bacterial species grow perfectly well in a synthetic medium in which the glucose of the formula described in Table 2-1 has been replaced by some other carbohydrate. For instance, *E. coli* is able to utilize for its growth the five-carbon sugars *arabinose* and *xylose*, the six-carbon sugar *galactose* (and the six-carbon hexaalcohol *mannitol*), and the six-carbon sugar disaccharides *maltose* and *lactose* (Figure 3-5). In order to make such surrogate carbon and energy sources palatable to an enzymatic ensemble designed for the processing of glucose, the bacteria first convert each of these carbohydrates into either glucose or one of the later intermediate metabolites of the glycolytic pathway. These preliminary conversions are themselves carried out by a special ensemble of enzymes. We shall now consider two examples of such preprocessing of carbohydrates by enzymes that figured in important ways in the development of molecular genetics.

As is shown in Figure 3-6, the first step in the utilization of galactose consists in its phosphorylation by ATP to yield galactose-1-phosphate and ADP. This step is catalyzed by the enzyme *galactokinase*. The second step

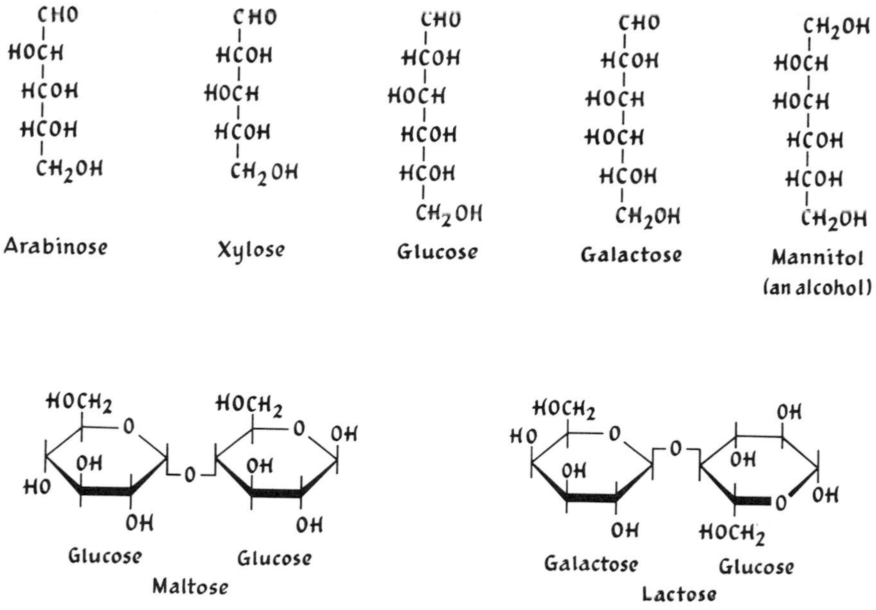

FIGURE 3-5.   Some sugars and an alcohol that can serve *E. coli* as source of carbon atoms and energy.

is an exchange reaction catalyzed by the enzyme *uridyl transferase*, in which the galactose-1-phosphate reacts with uridine diphosphate glucose (UDPG) and glucose-1-phosphate. The third and final step, catalyzed by the enzyme *UDP Gal epimerase*, consists in the intramolecular rearrangement of the galactose molecule to yield UDPG. The overall reaction represented by these three steps (in which the uridine diphosphate has entered merely catalytically) is therefore

$$ATP + galactose \rightleftarrows ADP + glucose\text{-}1\text{-}phosphate$$

The glucose-1-phosphate produced in this way enters the glycolytic pathway after its conversion to glucose-6-phosphate by an enzyme not specifically associated with galactose utilization.

In order to utilize the disaccharide lactose (glucosyl-β-galactoside), the bacteria first hydrolyze it into its two monosaccharide constituents, according to the reaction

$$lactose \longrightarrow galactose + glucose$$

This hydrolysis is catalyzed by the enzyme *lactase*. Since the lactase in *E. coli* is capable also of hydrolyzing β-galactosides other than lactose, it is described more precisely as a *β-galactosidase*. Once lactose has been hydrolyzed, its glucose moiety can enter glycolysis directly, whereas its galactose moiety must first be converted into glucose-1-phosphate via the reaction sequence shown in Figure 3-6 before being offered for degradation to the enzymes of the glycolytic pathway.

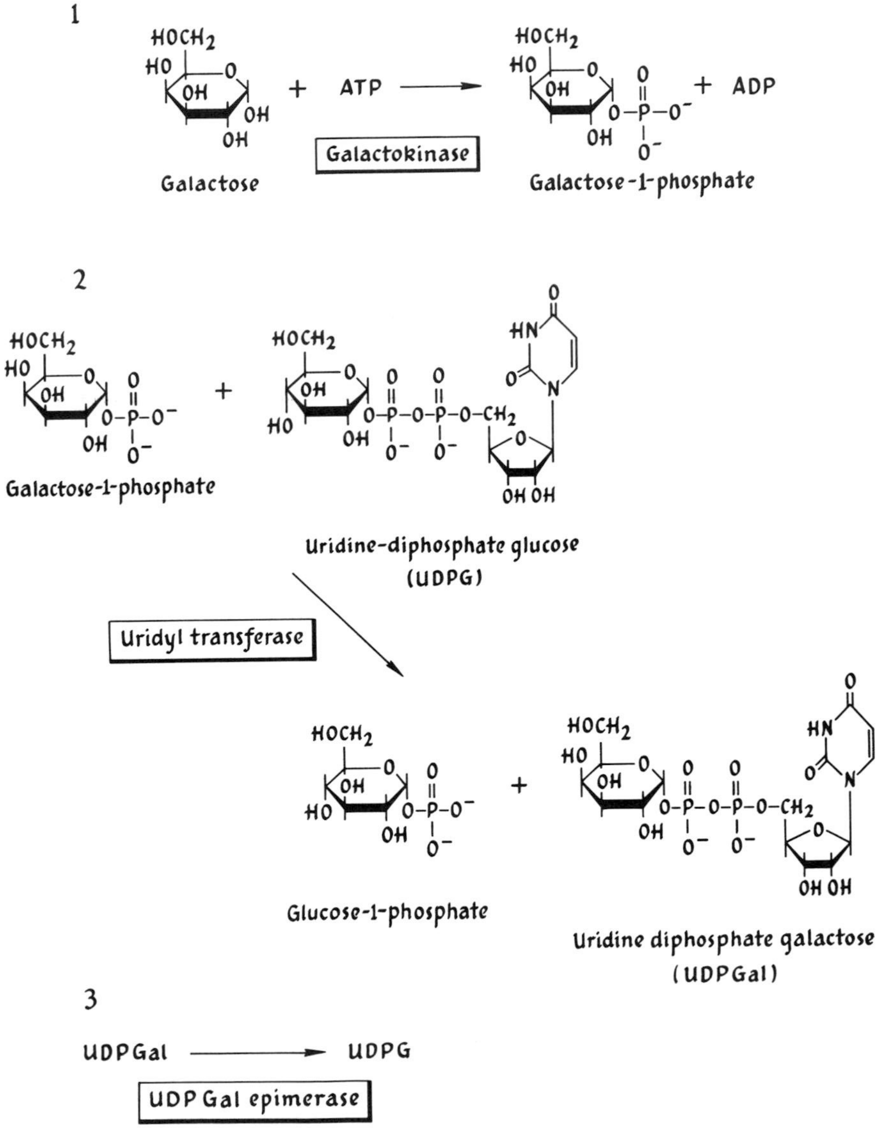

FIGURE 3-6.   The preprocessing of galactose for glycolysis.

BUILDING-BLOCK SYNTHESIS

After having considered the main *degradative* reaction sequences by means of which the cell, and in particular the *bacterial* cell on whose life most of these chapters will be focused, converts into ATP the energy realized by the oxidation of glucose to CO_2, we may now proceed to examine briefly the synthetic reaction sequences by means of which the building blocks of cells are constructed from the other half of the glucose that is consumed but not oxidized completely to CO_2. These synthetic activities now require expenditure of the chemical energy stored in the form of ATP by the degradative activities. The synthetic aspect of metabolism proved more difficult to unveil than the degradative aspect. The reason for this is that whereas degradative reactions such as those that constitute the glycolytic pathway and the citric acid cycle could be worked out by study of chemical reactions that take place in cell extracts, the complicated synthetic sequences cannot in general be resolved by such means. For instance, whereas yeast extract does break down glucose to ethanol and CO_2, it does not convert glucose into appreciable amounts of amino acids. Hence the biosynthesis of the building blocks of cells had to be studied mainly through experiments carried out with *intact, living* cells, and this endeavor could not really get very far before isotopic tracer elements, such as ^{32}P, ^{14}C, ^{3}H and ^{15}N, became generally available for biochemical experimentation in the 1940's. After that time, it was feasible to introduce an isotopically labeled compound into a cell or a cell extract and then follow the course taken by its marked atoms on their metabolic path toward conversion into cell building blocks. As early as the 1930's, however, the possibility of using the metabolic defects caused by gene mutations for an analysis of synthetic pathways had already been recognized. This procedure will be the subject of discussion in later chapters. By about 1955, the general sequence of cellular building-block biosynthesis had been more or less completely worked out, thanks to the combined application of radioactive tracers and genetic techniques. Alas, even a superficial bird's-eye view of the awesome panorama of that network of known metabolic reactions would carry us far beyond the scope of this text. Hence we must confine ourselves here to a cursory review of a few representative pathways of amino acid and nucleotide synthesis, the understanding of which is essential for appreciating the genetic experiments to be recounted in the following chapters. In any case, it can be stated as a general principle that the point of departure of most of these synthetic pathways is one or another of the intermediate compounds of the degradative reaction sequence of glycolysis and citric acid cycle.

This principle is exemplified by the pathways of amino acid biosynthesis. As is shown in Figure 3-7, the twenty protein amino acids can be grouped

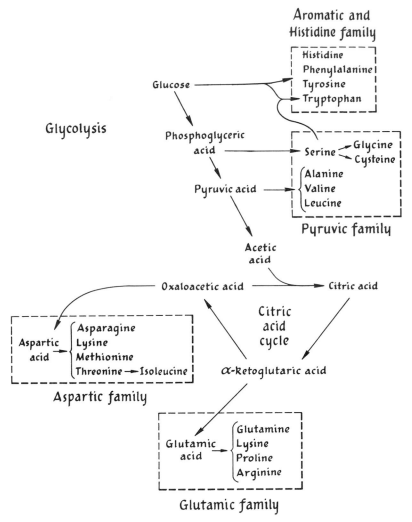

FIGURE 3-7.   Biosynthetic origin of the four amino acid families.

into four families, according to the metabolic provenance of their skeletons of carbon atoms. Synthesis of the *aromatic and histidine family* departs from glucose at the top, and that of the *pyruvic family* from phosphoglyceric acid in the middle and from pyruvic acid at the bottom of the glycolytic pathway. Synthesis of the *glutamic family* departs from α-ketoglutaric acid in the middle, and that of the *aspartic family* from oxaloacetic acid at the end of the citric acid cycle. We may now consider the synthetic pathways of representative members of these amino acid families.

SYNTHESIS OF THE AMINO ACIDS GLUTAMIC ACID, ALANINE,
ASPARTIC ACID, SERINE, AND GLUTAMINE

The formation of glutamic acid is central to amino acid biosynthesis because
it is in its synthesis that inorganic ammonia is first elevated to the status of
an organic amine. It was seen in Figure 3-4 that midway in the citric acid
cycle there arises the five-carbon dicarboxylic acid α-ketoglutaric acid. As
is shown in Figure 3-8, the enzyme *glutamic dehydrogenase* catalyzes the
replacement of the α-keto group of α-ketoglutaric acid by an inorganic
ammonium ion, thus giving rise to glutamic acid. This amination involves
a reduction of the α-keto group, which is accomplished by coupling the
reaction to the oxidation of one molecule of NADH to NAD^+.

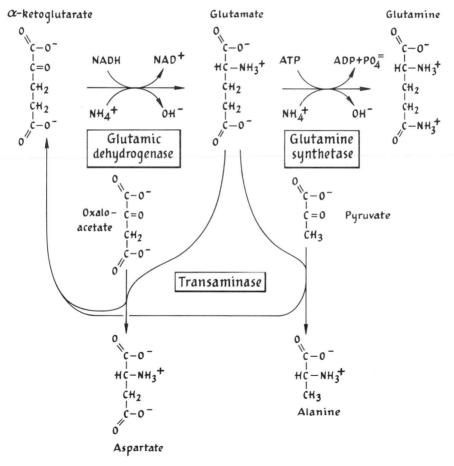

FIGURE 3-8.   Biosynthetic pathway of the amino acids glutamic acid, glutamine,
aspartic acid, and alanine.

It is to be noted that whenever an intermediate such as α-ketoglutaric acid is removed from the citric acid cycle, an equivalent number of oxalo-acetic acid molecules must be restored to the cycle if the cycle is to continue operating in the long run. This replenishment of oxaloacetic acid is accomplished by the addition of CO_2 to pyruvic acid, according to the reaction

$$
\begin{array}{ccc}
\underset{\text{CH}_3}{\overset{\text{O}}{\overset{\|}{\underset{|}{\text{C}-\text{O}^-}}}}\underset{}{\overset{}{\underset{}{\text{C}=\text{O}}}} + \text{ATP} + \text{CO}_2 & \rightleftarrows & \underset{\underset{\text{O}}{\overset{\|}{\text{C}-\text{O}^-}}}{\overset{\text{O}}{\overset{\|}{\text{C}-\text{O}^-}}}\quad + \text{ADP} + \text{P} + \text{H}^+
\end{array}
$$

Pyruvic acid   Oxaloacetic acid

The energetic requirements of this reaction are met by the concomitant hydrolysis of one molecule of ATP to yield ADP and inorganic phosphate.

We may now estimate the chemical energy consumed by the synthesis of one molecule of glutamic acid. First, the removal of one molecule of α-keto-glutaric acid from the citric acid cycle, to provide the carbon skeleton for glutamic acid, involves the loss of the three molecules of NADH and the one molecule of ATP that would have accrued had the oxidation of α-keto-glutaric acid to oxaloacetic acid been completed. Second, the reductive amination of the α-keto group of α-ketoglutaric acid by inorganic ammonium ion involves the loss of one more molecule of NADH. Third, the removal of one molecule of α-ketoglutaric acid from the cycle requires replenishment by one molecule of oxaloacetic acid. This replenishment involves the loss of the energy that would have been realized by the complete oxidation of the precursor pyruvic acid molecule in the citric acid cycle, or of five molecules of NADH and one molecule of ATP. Furthermore, one molecule of ATP is expended upon the carboxylation of pyruvic acid to form oxaloacetic acid. Thus the synthesis of one glutamic acid molecule results in a total deficit of nine molecules of NADH and three molecules of ATP. But since that deficit of nine molecules of NADH could have yielded $(9/12) \times 34 = 25.5$ molecules of ATP upon oxidative phosphorylation, the cost to the metabolic economy of the synthesis of one molecule of glutamic acid thus amounts to $25.5 + 3 = 28.5$ molecules of ATP.

The α-amino group of glutamic acid is available for the synthesis of other nitrogenous cell building blocks, either through direct structural transformations of glutamic acid itself or through *transamination reactions.* In trans-amination reactions the α-amino group of glutamic acid is transferred to another α-keto acid, without the expenditure of further chemical energy.

Two typical transamination reactions are the formation of alanine from pyruvic acid (end-product of glycolysis) and of aspartic acid from oxaloacetic acid (end-product of the citric acid cycle), both of which are shown in Figure 3-8. The enzymes that catalyze such reactions are called *transaminases*. After transfer of the amino group from glutamic acid, the resulting α-keto-glutaric acid can either accept another ammonium group by reductive amination to reconstitute itself as glutamic acid or return to the citric acid cycle whence it came to complete the oxidation of its progenitor glucose molecule.

The energetic cost of synthesizing alanine and aspartic acid may now be calculated. The utilization of pyruvic acid as the carbon skeleton of alanine results in loss of the total energy realized by the oxidation of pyruvic acid in the citric acid cycle, or five molecules of NADH and one molecule of ATP. To that must be added the one molecule of NADH consumed by the amination of the glutamic acid amino-donor molecule. Thus, per molecule of alanine formed, the metabolic economy incurs a loss of $34 \times (6/12) + 1 = 18$ molecules of ATP. The carbon skeleton of aspartic acid is derived from oxaloacetic acid at the end of the citric acid cycle, the cost to the metabolic economy being only the amount of oxaloacetic acid necessary to replenish the cycle; as has already been shown, this amounts to five molecules of NADH and two molecules of ATP. In addition, one molecule of NADH is consumed in the reductive amination of glutamic acid, so that, per molecule of aspartic acid, there is an expenditure of $(6/12) 34 + 2 = 19$ molecules of ATP.

Serine is synthesized via a slightly more complicated synthetic pathway. The starting point, midway in glycolysis, is 3-phosphoglyceric acid. As is shown in Figure 3-9, phosphoglyceric acid is oxidized to the α-keto acid phosphohydroxy pyruvate with the attendant reduction of one molecule of NAD^+ to yield NADH. A transamination exchanges the α-amino group of glutamic acid for the α-keto group of phosphohydroxy pyruvic acid and gives rise to phosphoserine and α-ketoglutaric acid. Hydrolysis of the phosphate ester of phosphoserine finally yields serine and inorganic phosphate. The energetic cost of synthesizing one molecule of serine arises from the loss of the one molecule of ATP that would have been available from the eventual transfer of the phosphate group of phosphoglyceric acid to ADP and from the loss of five molecules of NADH and one molecule of ATP realizable from the complete oxidation of the pyruvic acid derivate of phosphoglyceric acid in the citric acid cycle. The loss of one molecule of NADH incurred in the reductive amination of glutamic acid need not be taken into account here, since it is offset by the gain in one molecule of NADH produced upon oxidation of phosphoglyceric to phosphohydroxy pyruvic acid. The cost of synthesis of one molecule of serine can, therefore, be figured as $(5/12) \times 34 + 2 = 16.2$ molecules of ATP.

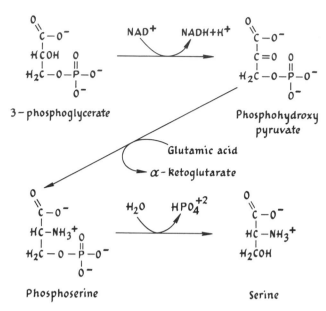

FIGURE 3-9.   Biosynthetic pathway of the amino acid serine.

A second pathway of assimilation of inorganic nitrogen into organic nitrogen also involves glutamic acid—namely, the amidation of the γ-carboxyl of glutamic acid by ammonium ions to yield glutamine (Figure 3-8). This reaction is catalyzed by the enzyme *glutamine synthetase*, and requires the expenditure (via hydrolysis) of one molecule of ATP to yield ADP and inorganic phosphate. The glutamine nitrogen atoms assimilated in this way may then proceed in a series of further metabolic reactions and have as one of their principal ultimate destinations two of the four purine ring nitrogens of adenine and guanine. They also furnish the nitrogen atom of the indole ring of the amino acid tryptophan, whose synthesis will be considered shortly. The cost per molecule of synthesizing glutamine evidently exceeds by just one molecule of ATP that of glutamic acid, and hence can be reckoned as 29.5 molecules of ATP.

SYNTHESIS OF THE NUCLEOTIDE URIDINE MONOPHOSPHATE

Aspartic acid, whose synthesis was outlined in Figure 3-8, is the starting point for the construction of the pyrimidine ring portion of the nucleotide uridine monophosphate (UMP), as is shown in Figure 3-10. In the first step, the α-amino group of aspartic acid condenses with CO_2 and with the γ-amide

FIGURE 3-10.   **A.** Biosynthetic pathway of the nucleotide uridine monophosphate (UMP). **B.** Biosynthetic pathway of the nucleotide component ribose-5-phosphate.

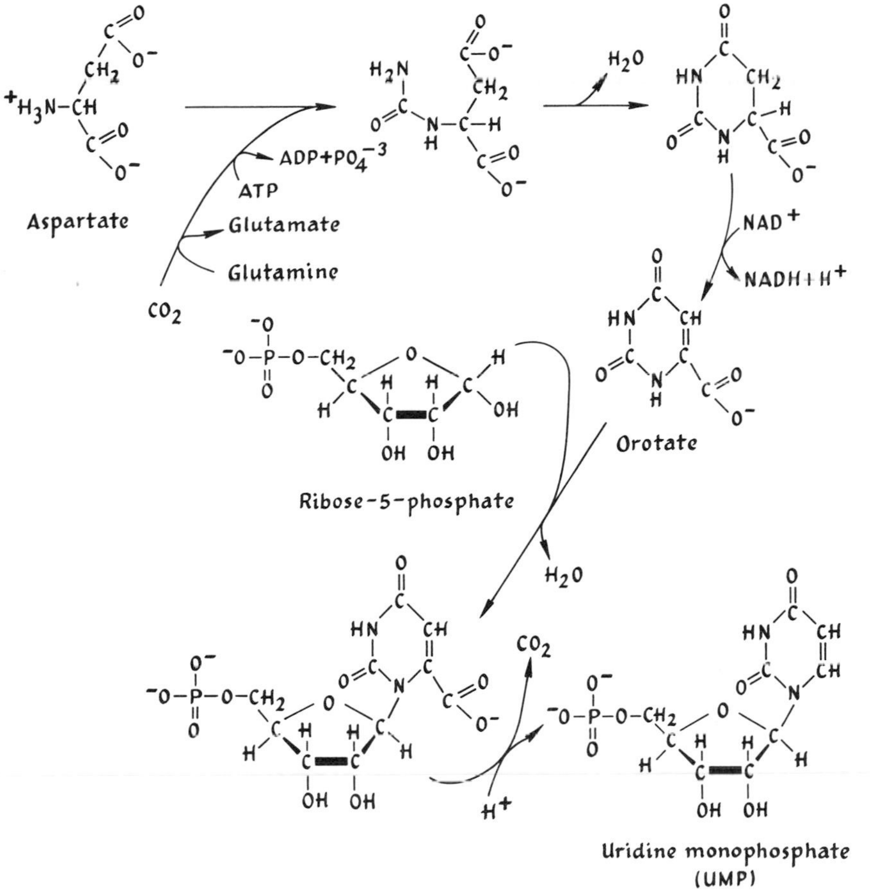
Aspartate
ATP
ADP+PO4-3
Glutamate
Glutamine
CO2
H2O
NAD+
NADH+H+
Orotate
Ribose-5-phosphate
H2O
CO2
H+
Uridine monophosphate
(UMP)

A

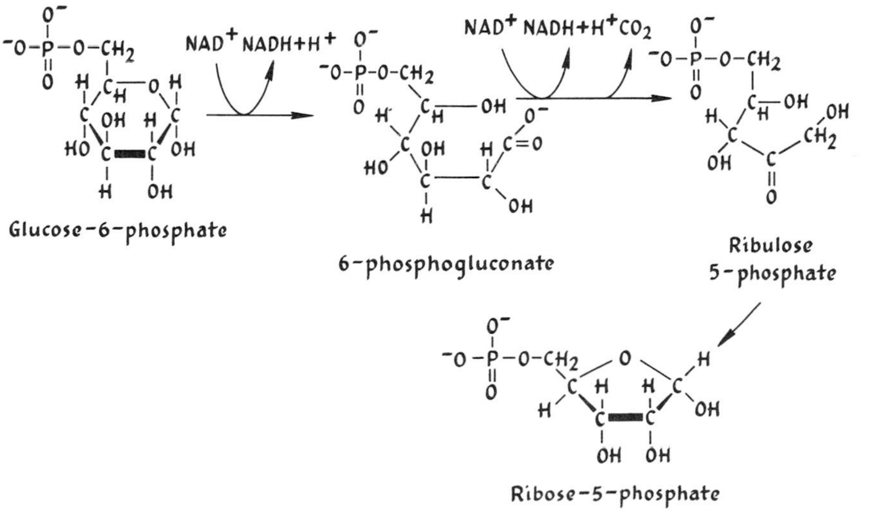
NAD+ NADH+H+
NAD+ NADH+H+ CO2
Glucose-6-phosphate
6-phosphogluconate
Ribulose
5-phosphate
Ribose-5-phosphate

B

group of glutamine to yield an open structure which contains all the atoms of the future pyrimidine ring. The ring then closes and is oxidized to yield orotic acid. This compound is then condensed with ribose-5-phosphate to yield the nucleotide oritidine-5-phosphate. Finally, decarboxylation of orotidine-5-phosphate converts the nucleotide into UMP. The formation of ribose-5-phosphate is shown in the lower part of Figure 3-10. As can be seen, its synthesis begins with glucose-6-phosphate, the very first intermediate of the glycolytic pathway.

As is evident in Figure 3-10, the five steps leading from aspartic acid to UMP represent none but the already familiar reactions of oxidation-reduction, hydration-dehydration, and carboxylation-decarboxylation. It is once more evident here that the large chemical transformation of glucose, ammonia and inorganic phosphate into a compound as complicated as the nucleotide UMP is achieved as the sum of many stepwise transformations of only a few different basic types. By following the same principles as those employed in the energy cost analysis of amino acid synthesis it can be estimated that synthesis of one molecule of UMP involves the cell metabolism in an investment of 53.5 molecules of ATP.

Synthesis of the Amino Acid Tryptophan

Finally we shall consider a synthetic pathway even more complicated than that of UMP, namely the biosynthesis of the amino acid tryptophan. The pathway is so complicated because of the twenty standard amino acids tryptophan carries the biggest and most elaborate side chain. As can be seen in Figure 3-11, tryptophan synthesis involves in its afferent pathways erythrose-4 phosphate, twice phosphoenol pyruvic acid, as well as the phosphorylated derivative of ribose-5-phosphate, phosphoribosyl pyrophosphate (PRPP) and the two amino acids glutamine and serine. Four of the six carbon atoms of the aromatic ring of the tryptophan side chain are derived from erythrose-4-phosphate (itself derived from glucose-6-phosphate), and the other two are derived from phosphoenol pyruvic acid, by means of a series of transformations which ultimately yield the intermediate chorismic acid. From chorismic acid, which figures as a precursor not only of tryptophan but also of the other two amino acids with aromatic side chains, tyrosine and phenylalanine, there are five more steps to tryptophan. In contrast to the preceding reaction sequence leading to chorismic acid, these last five steps pertain only to tryptophan synthesis. The first of these steps converts chorismic acid to anthranilic acid, a reaction which is catalyzed by the enzyme *anthranilate synthetase*. This reaction involves cleavage of the pyruvic acid side chain from the 6-carbon ring and transfer of an amino group to

FIGURE 3-11.   Biosynthetic pathway of the amino acid tryptophan. A. Formation of tryptophan precursors. B. The final steps in tryptophan synthesis.

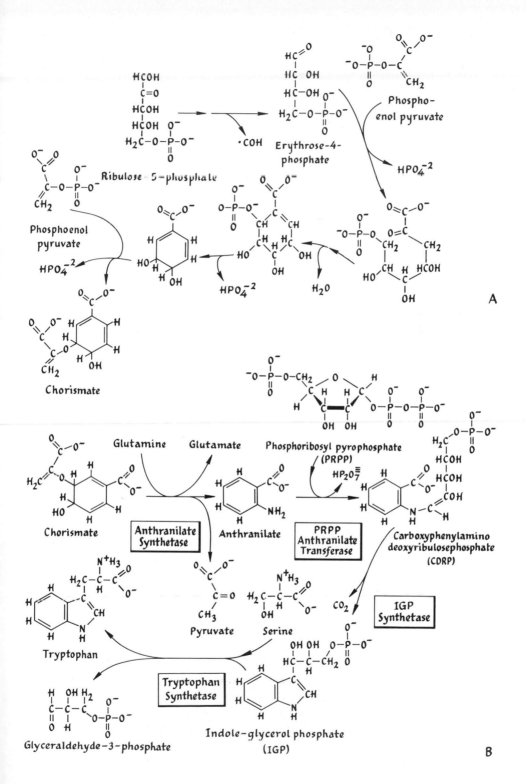

the ring from glutamine. For the second step, another enzyme, *phosphoribosyl anthranilate transferase*, catalyzes the condensation of anthranilic acid with PRPP to yield CDRP (carboxyphenylamino deoxyribulose phosphate). The third step consists of the transformation of this condensation product into indole glycerol phosphate (IGP), which at this stage possesses the hetero-cyclic indole ring of the tryptophan side chain. This step is catalyzed by the enzyme *IGP synthetase*. The fourth and last step consists of the replace-ment of the glycerol phosphate moiety of IGP by serine. This reaction is catalyzed by the enzyme *tryptophan synthetase*, and yields triose phosphate and tryptophan as its products. The total chemical energy that has to be expended on this elaborate synthetic sequence can be estimated to amount to 80 molecules of ATP, which makes tryptophan the most expensive of the twenty standard amino acids.

SUMMARY STATEMENT ON METABOLISM

The facts set forth in this chapter now allow statement of some generaliza-tions pertaining to the chemistry of cellular self-reproduction:

1. The 60 different basic cellular building blocks, in particular the amino acids and the nucleotides, are synthesized from glucose and ammonia in a limited set of chemical reactions numbering about one to two thousand. Each step in this reaction sequence leading from nutrient to cell building blocks proceeds through the agency of a specific enzyme.

2. The chemical energy required for these syntheses is secured by the stepwise oxidative degradation of glucose to CO_2 through glycolysis and citric acid cycle. This energy is first invested in the phosphorylation of ADP by inorganic phosphate to produce ATP, whose coupled hydrolysis to regenerate ADP is later used to drive synthetic reactions. The various chemical intermediates of the degradative reaction sequence serve also as starting materials for the synthetic pathways leading to the construction of building blocks. The partition between degradative and synthetic pathways is such that about half the glucose carbon atoms consumed by a growing bacterial culture wind up in the cellular building blocks and half wind up as CO_2.

3. When a carbon compound other than glucose serves as nutrient for cell growth the compound is first converted into glucose or into one of the intermediates of the glycolytic pathway by means of a few reaction steps. In this way, the standard reaction sequence leading to synthesis of the cell building blocks from glucose can work without necessity for modification if a substitute carbon and energy source is to be metabolized.

4. Each reaction step accomplishes only a small chemical change, such

as addition or removal of H_2O, hydrogen, CO_2, or phosphate. It is the aggregate reaction sequence of small steps, which accomplishes the large transformations of glucose into the cellular building blocks.

These insights into the chemical synthesis of cell components show that the role of enzymes is absolutely central in the drama of cellular growth. That is, it now appears that the character of a cell—the building blocks of which its structures are composed and the chemical processes it carries out— is determined by the ensemble of enzymes it contains. Hence the fundamental problem of cellular self-reproduction must be fathomed in terms of finding an explanation for the *function* and *formation* of enzymes. The next chapter provides the solution of the first of these two parts of our fundamental problem, in that it tells what enzymes are and how they manage to catalyze specific chemical reaction steps. As for the solution of the second of these two parts of the problem, it will require most of the remainder of this text to give an account of how the cell manages to form the enzymes that make it what it is.

Bibliography

PATOOMB

Herman M. Kalckar. High energy phosphate bonds: Optional or obligatory?

SPECIALIZED TEXTS, MONOGRAPHS, AND REVIEWS

Cohen, Georges N. *Biosynthesis of Small Molecules.* Harper & Row, New York, 1967.

Lehninger, A. L. *Bioenergetics.* Benjamin, New York, 1965.

Mahler, H. R., and E. H. Cordes. *Biological Chemistry.* Harper & Row, New York, 1966. Chapters 5, 10, 12, 13, 14, and 15.

4. Enzymes

In 1836, sixty years before Buchner's discovery of cell-free fermentation, J. J. Berzelius coined the term *catalysis* to describe the phenomenon by which some agents are capable—at low temperatures—of accelerating chemical reactions which ordinarily proceed at appreciable rates only at high temperatures. Among the chemical reactions which Berzelius considered to be of catalytic nature was the fermentation of glucose to ethanol and CO_2. With prophetic insight he wrote: "We have justifiable reasons to suppose that, in living plants and animals, thousands of catalytic processes take place between the tissues and the fluids and result in the formation of the great number of dissimilar chemical compounds, for whose formation out of the common raw material, plant juice or blood, no probable cause could be assigned. The cause will perhaps in the future be discovered in the catalytic power of organic tissues of which the organs of the living body consist." By the middle of the nineteenth century, M. Traube proposed that this catalytic power of organic tissues resides in its *proteins*, and by the end of the century the catalysts had been given the name *enzymes*. And, as the preceding chapter set forth, the biochemical studies that Buchner's discovery of cell-free fermentation made possible eventually showed that the grand ensemble of cellular metabolic pathways depends on the presence of one to two thousand different enzymes, each of which catalyzes a single reaction step. In this chapter we shall inquire into the nature of these enzymes; that is, we shall now seek answers to the questions of what enzymes are and how they work.

The study of enzymes required development of quantitative methods for measuring their catalytic activity. A substance acted upon by an enzyme is called its *substrate*, and the activity of an enzyme is expressed as the amount of substrate converted into the reaction product per unit of time. For example, let us consider the activity of β-galactosidase, the enzyme that catalyzes the hydrolysis of its lactose disaccharide substrate to yield glucose and galactose according to the reaction

The specificity of action of this enzyme is such that it catalyzes the hydrolysis of β-galactosides other than lactose. In particular, the colorless compound orthonitrophenylgalactoside (ONPG)

is a substrate of this enzyme, and yields upon its hydrolysis galactose and the intensely yellow orthonitrophenol. Thus the catalytic activity of β-galactosidase enzyme present in a cell extract can be measured by adding a known volume of the extract to a solution of ONPG of known concentration and then determining the intensity of yellow color generated in the reaction

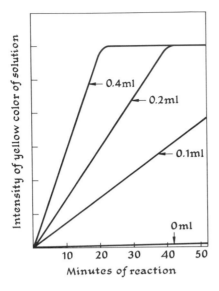

FIGURE 4-1.   Activity assay of β-galactosidase. Various volumes of an extract of *E. coli* are added to samples of a 0.003 *M* solution of ONPG maintained at 28°C. The intensity of yellow color developed (indicated in arbitrary units on the ordinate) is determined at the time indicated on the abscissa by measuring the amount of light of wavelength 420 mμ absorbed by the solution. Each curve has been labeled according to the volume of extract added.

mixture at various times. Since that intensity is proportional to the number of substrate molecules hydrolyzed, the enzyme activity can be expressed in terms of the rate of color production. Figure 4-1 shows the result of such an activity assay. It can be seen there that very little spontaneous hydrolysis of ONPG occurs in the neutral water solution; no appreciable color appears in the sample to which *none* of the cell extract was added. In contrast, a strong yellow color develops in the sample which received 0.4 ml of the extract; the color has reached its maximum intensity within 20 minutes, by which time hydrolysis of ONPG has gone to completion. Furthermore, it can be seen that the initial rate of color production in that sample is, respectively, two and four times that observed in the samples which received only 0.2 ml and 0.1 ml of the extract. That is to say, the rate of substrate hydrolysis is proportional to the concentration of β-galactosidase enzyme present in the reaction mixture. Hence measurement of the catalytic activity offers a means of assaying the amount of enzyme contained in a cell extract. The ratio of catalytic activity to the total dry weight of organic material present in the extract represents the *specific activity* of the enzyme preparation.

ENZYME STRUCTURE

With the development of methods of assaying specific enzyme activity at the beginning of this century, it became feasible to attempt to purify and isolate from the cell extract the particular enzyme responsible for catalysis of a given reaction, in order to ascertain the chemical nature of the enzyme. The general method followed in attempting to isolate enzymes consists in fractionating the extract by some chemical procedure and then assaying the fractions obtained for their specific enzyme activity. Any fraction in which the specific enzyme activity is found to be higher than in the original, unfractionated extract can be considered to have been *enriched* in its enzyme content. That fraction can then be subjected to a second fractionation procedure, in the hope that one of its second-generation fractions will manifest a yet higher specific activity, and hence represent a further enrichment. Thus fractionation can be repeated successively until, finally, a fraction with *maximum* specific activity is attained, which ideally ought to consist of the pure enzyme. Obtaining pure enzymes by this method was easier conceived than done, however, because it turned out that enzymes must be handled by methods far less drastic than those that were known to turn-of-the-century organic chemists. Thus new and very gentle fractionation methods had to be developed before enzyme purification could really get under way; these methods included precipitation of the enzyme from extracts by adding organic solvents or high concentrations of salts to the water solution, and reversible adsorption of the enzyme from the extract onto the ionic surfaces of solid materials, such as silica gels. In any case, the task of obtaining in pure form one of the many enzymes present in a total cell extract was a formidable one, since even a relatively abundant enzyme rarely constitutes more than 1%, and generally only 0.1%, of the total organic material of the cell. Few organic compounds had until then been obtained in pure form from biological sources of comparable impurity. Consequently it was not until 1926 that James Sumner first obtained an enzyme in pure form. This was the enzyme *urease*, which catalyzes the hydrolysis of urea to yield carbon dioxide and ammonia. Sumner had used a 32% solution of acetone in water to precipitate much of the organic material in a jack bean extract rich in urease activity, while leaving most of the urease activity in solution. After filtering off this precipitate and letting the filtrate stand overnight in the cold, he found that crystals of protein had formed in the filtrate. Study of these protein crystals revealed that they were pure urease enzyme endowed with a specific enzymic activity enormously greater than that of the original jack bean extract. It was thus proven that the enzyme urease is a protein and that a polypeptide chain is capable of acting as a catalyst in the facilitation of a chemical reaction. In the years that have elapsed since Sumner's discovery, hundreds of other enzymes have been crystallized, including the *E. coli* β-galactosidase. All of these enzymes turned out to be proteins, although many of them were found to contain some other organic molecules in addition to their polypeptide chains.

Thus Traube's mid-nineteenth century proposition that the catalytic power of living cells resides in their proteins came to be fully confirmed. At the time of Sumner's crystallization of urease, this proposition was badly in need of confirmation, because in the 1920's the view of R. Willstätter held sway that enzymes are *not* proteins. Willstätter had recently purified several yeast enzymes and obtained catalytically active solutions, which according to his chemical analyses appeared to be free of protein. And, since Willstätter and his contemporaries did not appreciate the enormous catalytic power which resides in a single enzyme molecule, they failed to realize that a solution of an enzyme may be too dilute to register its protein content by ordinary chemical tests while still being catalytically active. Sumner's proof was, therefore, an important milestone along the road to understanding the chemical basis of cell function; the quest for knowing what enzymes are and how they work now became a problem of analyzing protein structure.

PRIMARY STRUCTURE

The determination of the chemical structure of a protein must begin with a quantitative analysis of the amino acid composition of its polypeptide chains. For this purpose, the pure and, if available, crystalline protein is generally subjected to acid hydrolysis, in order to hydrolyze all of the peptide bonds in which its constituent amino acids are linked. The relative amounts of the twenty standard amino acids liberated by such hydrolysis can then be assayed by the method of ion-exchange resin chromatography developed in the early 1950's by W. H. Stein and S. Moore (Figure 4-2). The results of such an analysis of two *E. coli* enzymes, β-galactosidase and tryptophan synthetase, are shown in Table 4-1. (As will be shown shortly, tryptophan synthetase is composed of two different polypeptide chains, called A protein and B protein. The data presented here pertain only to the A protein.) As can be seen from the table, the two enzymes differ in their relative content of the twenty standard amino acids. For instance, tryptophan synthetase is nearly twice as rich in alanine but only half as rich in methionine as β-galactosidase, and contains no tryptophan at all. Since the finding that enzymes differ in their amino acid composition is valid in general, it could be concluded that the functional character of different enzymes—that is, the specificity of their catalytic power—is reflected in the amino acid composition of their polypeptide chains.

The next step in determining protein structure consists in measuring the length of the polypeptide chain, in order to ascertain how many amino acids are linked together to make up the polymer molecule. For this purpose, a

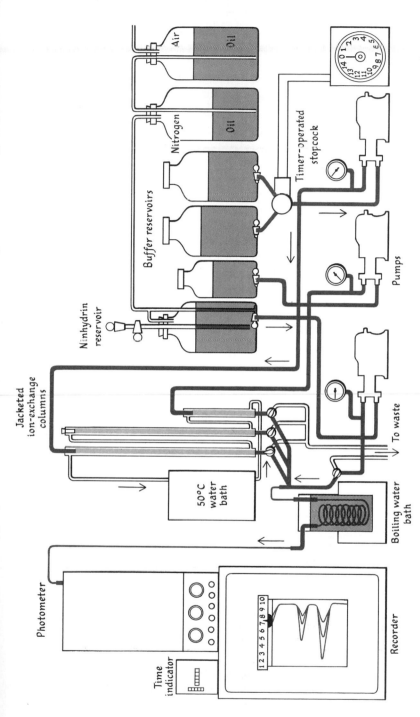

FIGURE 4-2. Quantitative amino acid analysis of proteins. A. The Moore-Stein automatic amino acid analyzer. The solution containing the hydrolyzed enzyme protein is placed at the top of one of the ion-exchange resin columns. The two pumps at lower right drive salt solutions through the column; the third pump delivers a color reagent (nihydrin) to the effluent emerging from the bottom of the column. The mixture passes through a boiling water bath, where the reaction between amino acids and color reagent produces a blue color. The intensity of the blue color is measured as the effluent passes through a photometer and is registered continuously on an automatic recorder (see part B on pp. 88 and 89).

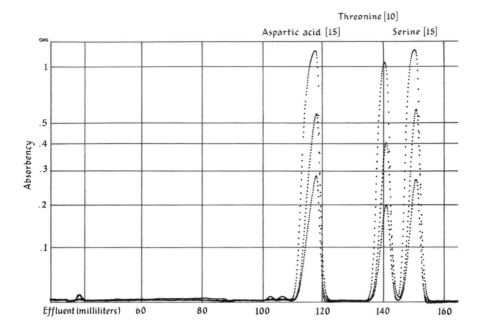

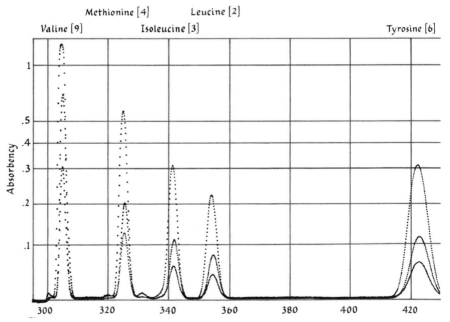

FIGURE 4-2 (cont.)   **B**. The result of an amino acid analysis of hydrolyzed ribonuclease
(RNAse). The names of the amino acids are written above their corresponding peaks
of color intensity, which emerge successively from the column. The number of residues
of each amino acid present is calculated from the intensity of color developed.
(Proline gives proportionately less color than other amino acids.) Histidine, lysine,
arginine, and ammonia are resolved by a special column and recorded separately.

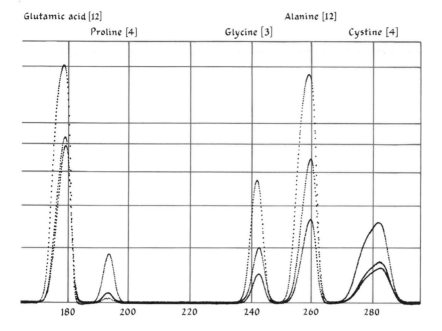

Glutamic acid [12]
Proline [4]
Alanine [12]
Glycine [3]
Cystine [4]

180   200   220   240   260   280

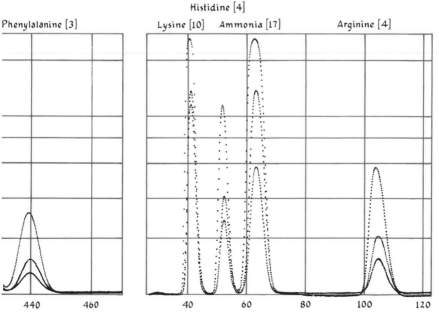

Histidine [4]
Phenylalanine [3]
Lysine [10]   Ammonia [17]
Arginine [4]

440   460   40   60   80   100   120

The ammonia derives from the hydrolysis of asparagine and glutamine. The abscissa shows the volume of solution pumped through the column; the ordinate shows the intensity of blue color. The three sets of curves represent measurements of the color intensity at different wavelengths and depth of solution. [Parts **A** and **B** from "The Chemical Structure of Proteins," by W. H. Stein and S. Moore. *Scientific American*, February 1961. Copyright © 1961 by Scientific American, Inc. All rights reserved.]

minimum chain length can be calculated from compositional data, such as those presented in Table 4-1, by taking into account that the polypeptide chain must contain *at least one* residue of the least abundant amino acid. Thus tryptophan synthetase contains 1.1 mole percent of the least abundant

TABLE 4-1.   Amino Acid Composition of Three Proteins

Amino acid	*E. coli* β-galactosidase		*E. coli* tryptophan synthetase (A protein)		Bovine insulin	
	Mole per-cent	Number per chain	Mole per-cent	Number per chain	Mole per-cent	Number per chain
Ala	8.0	93	15.0	40	6.0	3
Arg	6.3	74	4.1	11	1.9	1
Asn	10.5	123	8.2	22	6.0	3
Asp					—	0
Cys	1.6	19	1.9	5	11.8	6
Gln	12.1	142	10.9	29	6.0	3
Glu					7.8	4
Gly	7.2	85	7.1	19	7.8	4
His	3.1	36	1.5	4	3.7	2
Ile	4.1	48	7.1	19	1.9	1
Leu	9.4	110	10.1	27	11.7	6
Lys	2.5	29	4.9	13	2.0	1
Met	2.1	24	1.1	3	—	0
Phe	3.8	45	4.5	12	6.0	3
Pro	5.7	67	7.1	19	1.9	1
Ser	5.7	67	4.1	11	6.0	3
Thr	5.5	65	3.3	9	2.0	1
Trp	3.0	35	—	0	—	0
Tyr	3.1	36	2.7	7	7.7	4
Val	6.3	75	6.4	17	9.8	5
Total	100.0	1173	100.0	267	100.0	51

amino acid, methionine, and hence the polypeptide chain must contain a total of at least $100/1.1 = 91$ amino acid residues. If methionine is present more than once, then the actual chain length is a multiple of that minimum chain length of 91 amino acid residues. An approximation of the actual chain length can be obtained by physicochemical methods—for example, by measuring the sedimentation rate of the polypeptide chain in a centrifuge or by measuring the amount of light which its solution scatters. (The longer the chain, the faster its sedimentation rate and the more the light scattered per unit of mass.) The approximate chain length obtained in this way is an indication of the number of times the minimum chain length has to be multiplied to obtain an accurate value for the actual chain length. Such procedures showed that the enzymes β-galactosidase and tryptophan synthetase are composed of polypeptide chains containing, respectively, 1173 and 267 amino acids.

We may now inquire into the precise sequence in which these 1173 or 267 amino acids are joined in their polypeptide chains, or into what is called the *primary structure* of the protein molecule. From the traditional viewpoint of structural analysis of organic compounds as it was practiced until the 1940's, the mere thought of unraveling the chemical structure of a molecule as large as these polypeptide chains was enough to stagger the imagination. But in 1949, Frederick Sanger set out to determine the exact amino acid sequence of the polypeptide chain of bovine *insulin*, whose amino acid composition is shown in Table 4-1. Insulin is not an enzyme, but a *hormone* that regulates the function of some of the enzymes active in the glucose metabolism of vertebrate animals. Sanger chose insulin for his studies because it is a relatively short polypeptide, consisting of only 51 amino acids. Furthermore, large quantities of beef insulin of crystalline purity were available to him for analysis. Sanger soon found that the 51 amino acids of insulin are present in two different polypeptide chains, an A chain containing 21 amino acids and a B chain containing 30 amino acids. The A and B chains are held in covalent linkage by two disulfide (—S—S—) bonds formed between the sulfur atoms of pairs of cysteine residues in both chains.

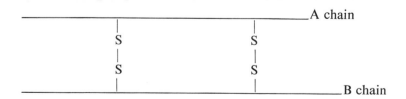

Sanger was able to split and separate the two chains, and thus reduce his task to finding the sequence of the 21 amino acids of the A chain and of the 30 amino acids of the B chain.

Sanger's first approach to sequence analysis was to employ the reagent fluorodinitrobenzene, which reacts with the free α-amino group of the amino acid at one end of the polypeptide chain. This end is called the "amino terminus" of the chain. The reaction with fluorodinitrobenzene produces the strongly colored dinitrophenyl, or DNP, derivative of the amino-terminal amino acid (Figure 4-3). At the other end of the chain, there is present an amino acid with a free α-carboxy group, and this end is called the "carboxy terminus." The α-amino and α-carboxy groups of all the other amino acids form part of peptide bonds, and hence are not "free." After its reaction with fluorodinitrobenzene, the polypeptide chain can be subjected to complete acid hydrolysis to yield its constituent amino acids. From among these amino acids, the strongly colored DNP derivative of the amino-terminal amino acid can be readily isolated and identified by chromatography. In this manner, it was possible to ascertain that glycine and phenylalanine are, respectively, the amino-terminal amino acids of the A and B chains. The fluorodinitrobenzene-treated polypeptide chain can also be subjected to *incomplete* acid hydrolysis, to yield a variety of peptide *fragments*. Among these fragments there will be some which bear the readily identifiable DNP-derivative of the amino-terminal amino acid. After their chromatographic separation, these DNP-labeled amino-terminal fragments can be subjected to complete acid hydrolysis in order to ascertain their amino acid composition. The amino acid sequence of the amino-terminus can be inferred from the hierarchic pattern of the amino acid composition of DNP-labeled fragments of various chain lengths. For instance, the following table shows the amino acid composition of four different DNP-labeled peptide fragments recovered by Sanger after partial hydrolysis of the fluorodinitrobenzene-treated B chain of insulin. These compositional data allowed Sanger to conclude that the amino-terminal amino acid sequence of the B chain is phenylalanine-valine-aspartic acid-glutamine.

Peptide fragment	Hydrolysis products of fragment	Inferred primary structure of fragment
B1	DNP-Phe	DNP-Phe
B2	DNP-Phe + Val	DNP-Phe-Val
B3	DNP-Phe + Val + Asp	DNP-Phe-Val-Asp
B4	DNP-Phe + Val + Asp + Gln	DNP-Phe-Val-Asp-Gln

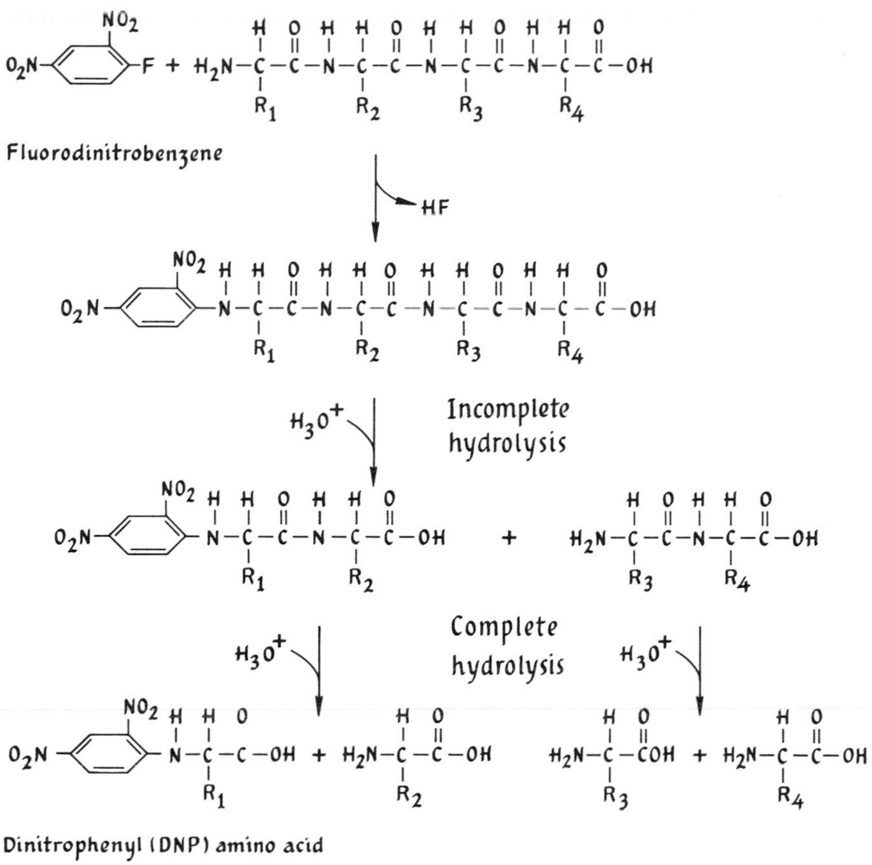

Dinitrophenyl (DNP) amino acid

FIGURE 4-3.   The fluorodinitrobenzene method of labeling with dinitrophenyl (DNP) the amino-terminal amino acid of a polypeptide chain.

Unfortunately, the fluorodinitrobenzene method can yield the sequence of only the first few amino acids following the amino terminus. From the sixth or seventh amino acid onward the analytical results become too ambiguous to permit reliable structural inferences. Hence, though this procedure can be used for the sequence analysis of short polypeptides four or five amino acids long, it is not directly applicable to the entire sequence of 21 and 30 amino acids of the A and B chains of insulin, not to speak of the hundreds of amino acids in such enzymes as β-galactosidase and tryptophan synthetase.

Sanger's next step, therefore, was to split both chains into a variety of polypeptide fragments, each fragment containing no more than five amino acids. He achieved this splitting by use of certain enzymes extracted from the digestive juices of the bovine stomach which catalyze specifically the hydrolysis of peptide bonds that link particular amino acids. Sanger then separated these fragments from each other by chromatography and determined *their* amino acid sequences by the fluorodinitrobenzene method. Since many of the fragments turned out to contain an amino acid sequence that overlapped in part with the amino acid sequence of some other fragment, it was possible to arrange these fragments in a unique linear order according to their position in the intact polypeptide chain. From this order, and from the amino acid sequence of the fragments, Sanger finally managed to reconstruct, by 1955, the exact amino acid sequence of the whole insulin molecule. By way of example, we may consider the peptide fragments from which Sanger reconstructed the sequence of the first eight amino acids of the insulin A chain:

Dipeptides	Ile-Val	Gln-Cys
	Val-Glu	
	Glu-Gln	
Larger peptides	Ile-Val-Glu	Cys-Cys-Ala
	Gly-Ile-Val-Glu	
		Gln-Cys-Cys-Ala
	Ile-Val-Glu-Gln	

Sequence of octapeptide	Gly-Ile-Val-Glu-Gln-Cys-Cys-Ala

The complete amino acid sequence of both chains of insulin—that is, their primary structure—and the way in which they are joined are shown in Figure 4-4 (see p. 95). It can be seen there that in addition to the two interchain disulfide bonds there exists also one *intra*chain disulfide bond that links the two other cysteine residues of the A chain.

Sanger thus demonstrated that insulin is made up of two polypeptide chains, each representing a definite sequence of the set of amino acid monomers. The techniques provided by Sanger's pioneering work could now be extended to determination of the primary structure of longer polypeptide chains, in particular to the analysis of enzyme proteins. Though many technical improvements have been made and automated equipment has come into use in the meantime, analyzing the complete amino acid sequence of an enzyme protein containing hundreds of amino acids still remains a formidable undertaking, requiring some three or four man-years of hard work. One of the longest polypeptide chains for which the amino acid sequence has now been worked out is the A protein of *E. coli* tryptophan synthetase, whose primary structure is shown in Figure 4-5.

Studies of the primary structure of enzymes have thus led to the general conclusion that a given enzyme represents a unique permutation of the twenty kinds of amino acids. For instance, every molecule of the A protein of *E. coli* tryptophan synthetase has the sequence shown in Figure 4-5, and

PLATE I. Some eye-color mutants of Drosophila.
[After E. M. Wallace, *in* A. H. Sturtevant and G. W. Beadle, *An Introduction to Genetics*, Saunders, Philadelphia, 1939.]

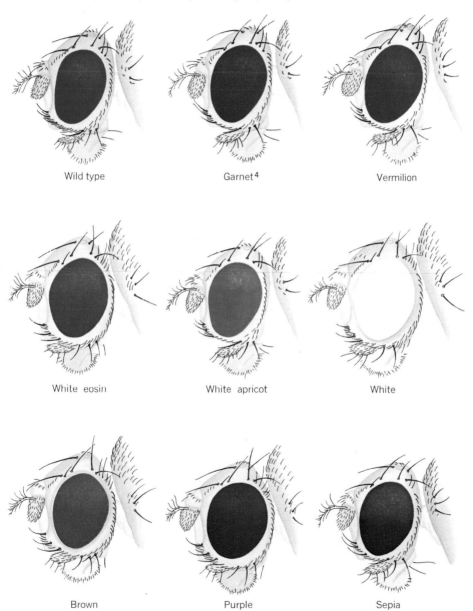

Wild type   Garnet[4]   Vermilion

White eosin   White apricot   White

Brown   Purple   Sepia

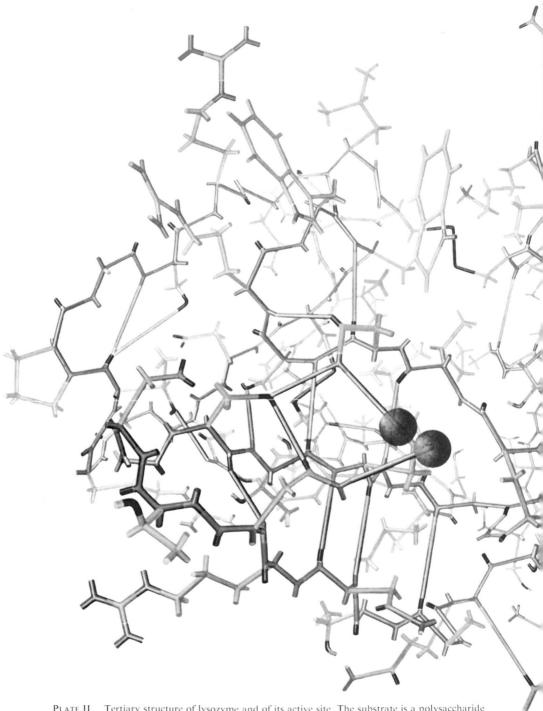

PLATE II.   Tertiary structure of lysozyme and of its active site. The substrate is a polysaccharide chain consisting of six residues of glucose, each carrying an acetylated amino group. The enzyme catalyzes the hydrolysis of the —C—O—C— bond linking the second and third glucose residues of the chain. The four red balls represent active-site amino acid oxygen atoms. They play an active role in the hydrolysis of the glucose-glucose bond, which lies between them when the substrate is in the cleft provided for it by the enzyme protein. A complex network of hydrogen bonds holds the substrate in the cleft.

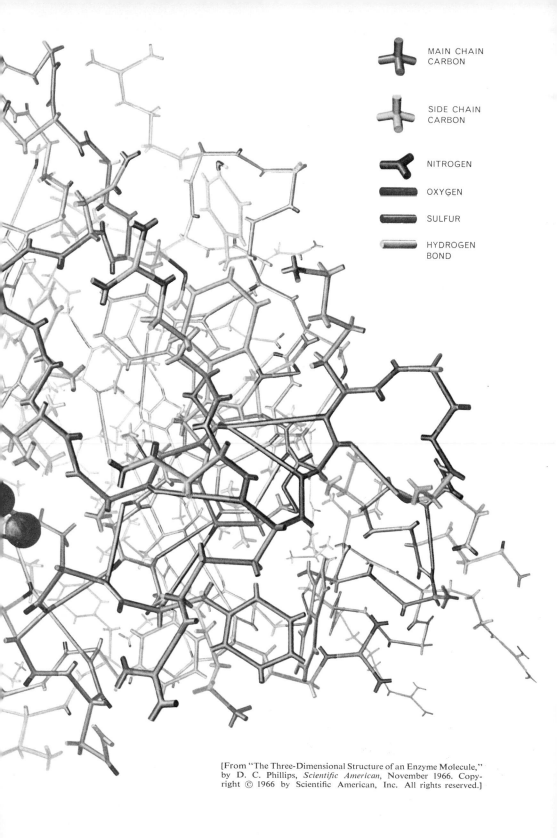

MAIN CHAIN
CARBON

SIDE CHAIN
CARBON

NITROGEN

OXYGEN

SULFUR

HYDROGEN
BOND

[From "The Three-Dimensional Structure of an Enzyme Molecule,"
by D. C. Phillips, *Scientific American*, November 1966. Copyright © 1966 by Scientific American, Inc. All rights reserved.]

PLATE III. Dark red wild-type Lac$^+$ and white mutant Lac$^-$ colonies of *E. coli* plated on EMB-lactose indicator agar. Some red-white *sectored* colonies also appear, in which the white bacteria cover a fourth to half the colony area. The sectored colonies are descended from unstable Lac$^-$/F-Lac$^+$ partial diploids that have lost their F-*lac* sex factor and reverted to the F$^-$Lac$^-$ type upon the first or second cell division (see Chap. 10). Colonies of wild-type Gal$^+$, mutant Gal$^-$, and unstable partial diploid Gal$^-$/λ *dg*Gal$^+$ transductant *E. coli* would present similar red, white, and sectored aspects (see Chap. 14). [Courtesy of John Roth.]

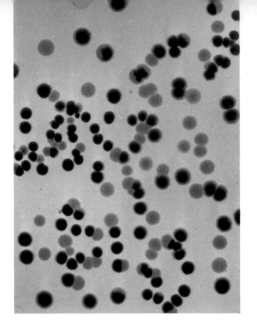

PLATE IV. Puffs on the giant chromosome IV in the salivary gland of *Chironomus*. (Above) The DNA has been stained brown and the protein green. (Below) The DNA has been stained blue and the RNA reddish-violet. The puffs are evidently rich in RNA and protein. [Courtesy of Ulrich Clever (top); Claus Pelling (bottom).]

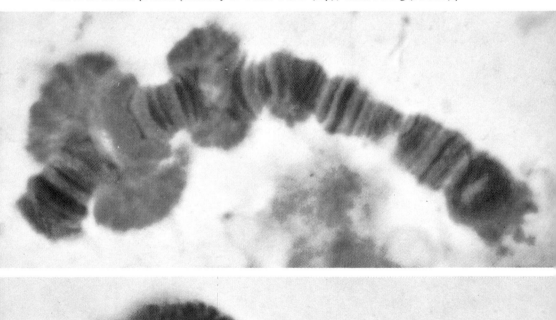

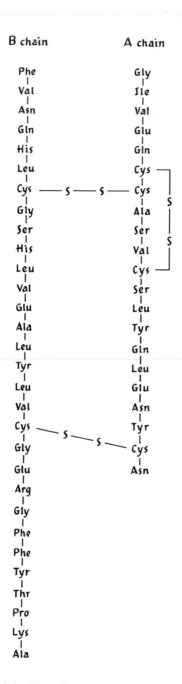

FIGURE 4-4. The primary structure of beef insulin.

```
1                          10                              20
Met-Gln-Arg-Tyr-Glu-Ser-Leu-Phe-Ala-Gln-Leu-Lys-Glu-Arg-Lys-Glu-Gly-Ala-Phe-Val-Pro-

                           30                              40
Phe-Val-Thr-Leu-Gly-Asp-Pro-Gly-Ile-Glu-Gln-Ser-Leu-Lys-Ile-Asp-Thr-Leu-Ile-Glu-Ala-

                           50                              60
Gly-Ala-Asp-Ala-Leu-Glu-Leu-Gly-Ile-Pro-Phe-Ser-Asp-Pro-Leu-Ala-Asp-Gly-Pro-Thr-Ile-

                           70                              80
Gln-Asn-Ala-Thr-Leu-Arg-Ala-Phe-Ala-Ala-Gly-Val-Thr-Pro-Ala-Gln-Cys-Phe-Glu-Met-Leu-

                           90                              100
Ala-Leu-Ile-Arg-Gln-Lys-His-Pro-Thr-Ile-Pro-Ile-Gly-Leu-Leu-Met-Tyr-Ala-Asn-Leu-Val-

                           110                             120
Phe-Asn-Lys-Gly-Ile-Asp-Glu-Phe-Tyr-Ala-Gln-Cys-Glu-Lys-Val-Gly-Val-Asp-Ser-Val-Leu-

                           130                             140
Val-Ala-Asp-Val-Pro-Val-Gln-Glu-Ser-Ala-Pro-Phe-Arg-Gln-Ala-Ala-Leu-Arg-His-Asn-Val-

                           150                             160
Ala-Pro-Ile-Phe-Ile-Cys-Pro-Pro-Asn-Ala-Asp-Asp-Asp-Leu-Leu-Arg-Gln-Ile-Ala-Ser-Tyr-

                           170                             180
Gly-Arg-Gly-Tyr-Thr-Tyr-Leu-Leu-Ser-Arg-Ala-Gly-Val-Thr-Gly-Ala-Glu-Asn-Arg-Ala-

                           190                             200
Ala-Leu-Pro-Leu-Asn-His-Leu-Val-Ala-Lys-Leu-Lys-Glu-Tyr-Asn-Ala-Ala-Pro-Pro-Leu-Gln-

210                        220                             230
Gly-Phe-Gly-Ile-Ser-Ala-Pro-Asp-Gln-Val-Lys-Ala-Ala-Ile-Asp-Ala-Gly-Ala-Ala-Gly-Ala-

                           240                             250
Ile-Ser-Gly-Ser-Ala-Ile-Val-Lys-Ile-Ile-Glu-Gln-His-Asn-Ile-Glu-Pro-Glu-Lys-Met-

                           260                             267
Leu-Ala-Ala-Leu-Lys-Val-Phe-Val-Gln-Pro-Met-Lys-Ala-Ala-Thr-Arg-Ser
```

FIGURE 4-5.   The primary structure of the tryptophan synthetase A protein of *E. coli*.

no molecule of any other *E. coli* protein species has that same sequence. There is no need to worry that the demand for a unique primary structure might place an inconvenient restriction on the total number of different enzymes which could exist in nature, a restriction which might lead to a tiresome sameness of living forms. For of a polypeptide chain of 200 amino acids, which can have any one of the 20 standard amino acids at each of the 200 positions in its chain, there can exist $20^{200} = 10^{260}$ different variants.

SECONDARY STRUCTURE

The determination of primary structure gives but a one-dimensional picture of the polypeptide chain, and provides no indication of the three-dimensional conformation that the protein molecule assumes in space. Yet it is the three-dimensional conformation that is actually responsible for the catalytic action with which the enzyme is endowed. Since there exists the possibility of free rotation in space of the atoms linked to the α-carbon atom of each amino acid in the polypeptide chain, the total number of theoretically possible three-dimensional conformations of that chain is very large. We shall now consider the efforts that were undertaken to ascertain which one of the myriad possible conformations that a polypeptide chain of given primary structure actually assumes. These efforts were the work of a second group of physical scientists whose movement into biological research in the 1930's and 1940's was more or less concurrent with that of the founders of molecular genetics (Chapter 1). In contrast to the physicist-geneticists who had drawn their inspiration from Niels Bohr and whose quest was motivated by the desire to understand the physical basis of the hereditary storage of biological information, the interest of this second group of men was focused on the three-dimensional structure—that is, on the *form*—of biological molecules. This group of structural analysts, among whose preoccupations genetics played at most a peripheral role, can be considered as having descended from W. H. Bragg and W. L. Bragg.

The Braggs, father and son, had invented X-ray crystallography in 1912. The principle of this method is that a beam of parallel X-rays incident on a crystalline (that is, regular and repetitive) array of atoms is diffracted in a pattern which is characteristic of the atomic weight and spatial arrangement of these atoms. Thus, by close examination of the pattern of X-rays scattered by a crystal as recorded on a photographic plate, a trained crystallographer can infer the spatial arrangement of the atoms which gave rise to that diffraction pattern. The reason why individual atoms can be "seen" by means of an X-ray camera, but not by means of an electron microscope (not to speak of an ordinary microscope), is that the wavelength of X-rays is about 1 Å, or the same order of magnitude as atomic diameters, whereas the wavelength of the electron beam of the electron microscope is about ten times too long for this purpose. The first X-ray diffraction analyses were carried out on simple inorganic salts, such as NaCl (Figure 4-6). Gradually,

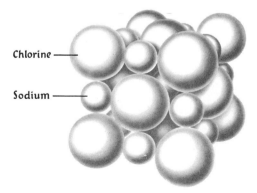

Chlorine ——

Sodium ——

FIGURE 4-6. The crystal lattice structure of NaCl as inferred from X-ray diffraction analysis.

FIGURE 4-7.   Linus Pauling
(b. 1901). [Courtesy of
L. Pauling.]

X-ray crystallographic methods were extended to organic molecules, first to the simple ones, and then to ever more complicated molecules, until finally crystals of such enormous macromolecules as nucleic acids and proteins were placed before X-ray cameras. As can be imagined, the X-ray diffraction patterns of such macromolecules are of fantastic complexity, and their analysis and transformation into the spatial arrangement of the thousands of atoms of which these macromolecules are made is a task so gigantic that even the determination of the primary structure of polypeptides seems simple by comparison.

Though X-ray analyses of the spatial conformation of polypeptides got underway in the 1930's, through the efforts of such pupils of the Braggs as W. T. Astbury and J. D. Bernal, the first great success in the crystallographic determination of protein structure was attained only in 1951. In that year, Linus Pauling (Figure 4-7) managed to work out a *secondary structure* of proteins. The term "secondary structure" refers to the spatial conformation of the "backbone" of the polypeptide chain:

$$
\begin{array}{ccccccc}
 & H & O & & H & O & \\
 & | & \| & | & | & \| & | \\
-N & -C & -C & -N & -C & -C- \\
 & | & & & | & &
\end{array}
$$

Pauling's success was due in part to a novel approach to structural determination, in which guesswork and model-building played a much greater role than in the more straightforward, analytical procedure of more conventional crystallographers. Pauling had decided some years earlier that it ought to be possible to deduce the structure of the polypeptide chain from a knowledge of the exact spatial conformation of the peptide bond. He therefore concentrated his X-ray crystallographic analyses on the determination of the lengths and angles of the bonds that link the backbone atoms of amino acids and small peptides (Figure 4-8). After obtaining these data, Pauling was able to construct the theoretical model of a regular polypeptide backbone, shown in Figure 4-9. This secondary structure, called the α-helix, derives its stability from the formation of hydrogen bonds between the hydrogen atom of the α-amino group of every amino acid residue and the oxygen atom of the α-carboxyl group of the fourth amino acid residue down the chain. The α-helix makes one full turn for every 5.4 Å and contains 3.6 amino acids per turn.

Subsequent structural studies of both natural and synthetic polypeptides by various physical methods, principally X-ray crystallography, showed that the α-helix certainly does exist in nature. It is the secondary structure assumed by the backbone of polyglycine, which is the simplest of all polypeptides and whose α-carbon atoms have no side chains. The α-helix configuration is assumed also by *parts* of the polypeptide chain of many natural proteins, though hardly any proteins are wholly helical. There are three reasons why most natural proteins also contain nonhelical regions as part of their secondary structure. First, the presence of any proline residue (Figure 2-2), whose α-amino group cannot participate in hydrogen bonding, necessarily causes

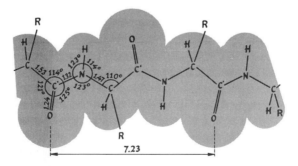

FIGURE 4-8. The bond angles (in degrees) and bond lengths (in angstroms) of the peptide bond linking amino acids with side chains (labeled R) in a polypeptide chain. [From "The Structure of Protein Molecules," by L. Pauling, R. B. Corey, and R. Hayward, *Scientific American*, July 1954. Copyright © 1954 by Scientific American, Inc. All rights reserved.]

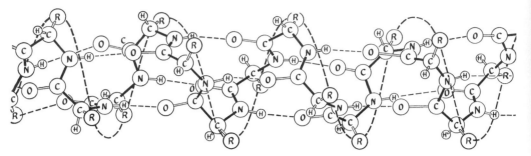

FIGURE 4-9.   Pauling's α-helix structure of the polypeptide chain backbone. The atoms

of each repeating

$$-\overset{|}{C}-\overset{\overset{H}{|}}{N}-\overset{\overset{O}{||}}{C}-\overset{|}{C}-$$

unit lie in a plane. The change in angle between

one unit and the next occurs at the α-carbon atom (to which the side chain R is
attached). The helix is held rigidly by the hydrogen bond (straight broken line) between
the α-amino hydrogen of one amino acid residue and the α-carboxyl oxygen of the
fourth amino acid residue down the chain. The wavy broken line traces the turns of the
helix. [From " Proteins," by P. Doty. *Scientific American*, September 1957. Copyright
© 1957 by Scientific American, Inc. All rights reserved.]

a local interruption in the helix. Second, formation of intramolecular disulfide
bonds between cysteine residues, such as those present in the A chain of
insulin, can distort the helix. Third, and possibly most important, the
character of the amino acid side chains attached to the backbone introduces
additional chemical and spatial factors that determine whether formation
of an α-helix is feasible at any point of the chain. Indeed, the very diverse
functional specificities manifested by the enzymes is a consequence of their
irregular secondary structure, for a collection of perfect α-helices would be
a rather monotonous molecular ensemble.

TERTIARY STRUCTURE

The complete three-dimensional conformation of *all* the atoms making up
the polypeptide chain is called the *tertiary structure* of the protein. This
tertiary structure is, of course, even more dependent on the nature of the
amino acid side chains, and hence on the primary structure, than is the
secondary structure of the backbone. The determination of tertiary structure
by X-ray crystallography turned out to be an extremely difficult undertaking.
In 1937, Max Perutz, then a student of J. D. Bernal, set out to ascertain the
structure of the protein *hemoglobin* by X-ray crystallography for his Ph.D.
thesis. Fortunately for him he was awarded his degree before working out
the structure, because he was to labor on the completion of this project for
the next 25 years. Hemoglobin, which will figure again in the discussion of

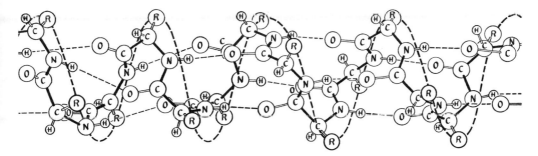

Chapters 17 and 18, is not an enzyme in the strict sense of the word. It is the principal protein present in red blood cells of vertebrate animals. It combines reversibly with molecular oxygen and thus carries oxygen from the lung to the tissues. The hemoglobin molecule consists of four polypeptide chains, each about 150 amino acids in length. These four chains are actually two pairs of identical primary structure, one pair being called the α-chains and the other the β-chains. Moreover, the hemoglobin molecule contains four atoms of iron. Each of these iron atoms lies in the center of a molecule of *heme* ($C_{33}N_4O_4H_{32}$), which is a flat molecule composed of four connected heterocyclic organic rings. Each of the four atoms of iron embedded in the heme can take up one molecule of oxygen.

Because of the limitations of the tools available at the time Perutz began the task he had cut out for himself, his initial progress in interpreting the X-ray diffraction pictures was very slow. But his progress became more rapid in the early 1950's, after he obtained X-ray photographs of hemoglobin molecules into which mercury atoms had been introduced artificially in ligature to the SH-groups of the side chain of two cysteine amino acids. The prominent X-ray reflections of the mercury atoms, he found, could serve as landmarks for the positioning of other atoms in that large protein molecule. Furthermore, Perutz had by that time been joined by John C. Kendrew, who had set out to analyze the tertiary structure of *myoglobin*—an oxygen-carrying protein present in the muscle cells of both vertebrate and invertebrate animals. Myoglobin is a simpler version of hemoglobin, in that the myoglobin molecule consists of but a single polypeptide chain of about the same length as the α- and β-chains of hemoglobin, has but a single atom of iron embedded in one heme molecule, and can take up one molecule of oxygen. Kendrew was one of the first to use the high-speed computing machines that were only just then becoming generally available in research laboratories to analyze X-ray diffraction data. By combining the heavy-atom landmark technique with the enormously increased capacity for data processing, Kendrew managed by 1957 to work out the structure of the myoglobin protein molecule shown in Figure 4-10. The limit of resolution of this analysis is 6 Å, which suffices only to indicate the shape of the polypeptide backbone and the positions of iron and heme, but does not yet show the position of the

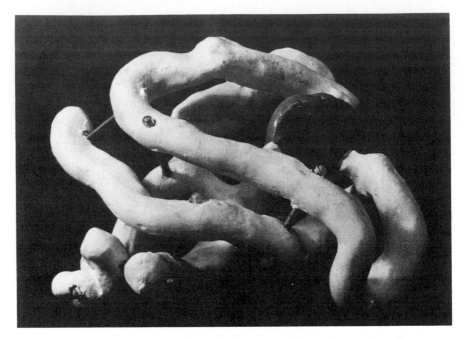

FIGURE 4-10.   The first determination of the tertiary structure of a protein molecule. Kendrew's 1957 model of the structure of myoglobin at a resolution of 6 Å. The heme group is the flat section at upper right. [This photograph, by George Rodger of Magnum Photos, Inc., appeared in "The Three-Dimensional Structure of a Protein Molecule," by J. C. Kendrew, *Scientific American*, December, 1961.]

atoms of the amino acid side chains. The reaction to seeing the three-dimensional conformation of a protein molecule for the first time was one of disappointment, since the strangely contorted *Gestalt* of the myoglobin polypeptide did not seem to convey any particular significance to the beholder. Within the next two years, however, Perutz was able to complete his structural analysis of hemoglobin to the same 6-Å limit of resolution. He then found that both α- and β-polypeptide chains of hemoglobin manifest essentially the same shape as the myoglobin polypeptide (Figure 4-11). In the intervening years, it has become clear that that contorted shape appears to be a fundamental pattern of nature—one that is often encountered in the analysis of tertiary protein structure. By 1960, after three more years of extensive work, Kendrew was able to refine his analysis of myoglobin structure down to the level of resolution of 2 Å. This undertaking required computational analysis of no less than 10,000 spots on X-ray photographs of myoglobin crystals. The result of this labor is shown in Figure 4-12. At last it was possible to assign a spatial position to each one of the several thousand atoms which make up a large protein molecule.

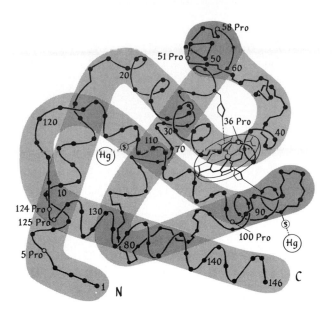

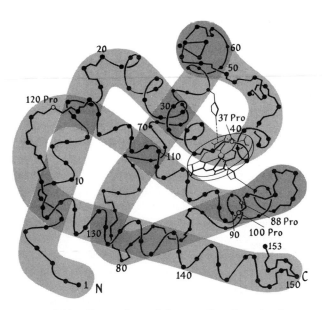

FIGURE 4-11.   Comparison of the overall tertiary structures of the β-polypeptide chain of hemoglobin (top) and of the single polypeptide chain of myoglobin (bottom). Every tenth amino acid residue is marked, as are the proline residues, which often coincide with turns of the chain. Circles marked Hg show where mercury atoms can be attached to —SH groups. [From "The Hemoglobin Molecule," by M. F. Perutz, *Scientific American*, November 1964. Copyright © 1964 by Scientific American, Inc. All rights reserved.]

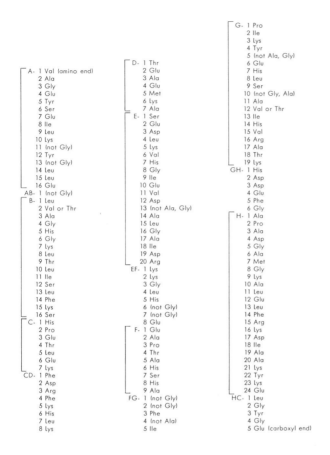

A- 1 Val (amino end)
2 Ala
3 Gly
4 Glu
5 Tyr
6 Ser
7 Glu
8 Ile
9 Leu
10 Lys
11 (not Gly)
12 Tyr
13 (not Gly)
14 Leu
15 Leu
16 Glu
AB- 1 (not Gly)
B- 1 Leu
2 Val or Thr
3 Ala
4 Gly
5 His
6 Gly
7 Lys
8 Leu
9 Thr
10 Leu
11 Ile
12 Ser
13 Leu
14 Phe
15 Lys
16 Ser
C- 1 His
2 Pro
3 Glu
4 Thr
5 Leu
6 Glu
7 Lys
CD- 1 Phe
2 Asp
3 Arg
4 Phe
5 Lys
6 His
7 Leu
8 Lys

D- 1 Thr
2 Glu
3 Ala
4 Glu
5 Met
6 Lys
7 Ala
E- 1 Ser
2 Glu
3 Asp
4 Leu
5 Lys
6 Val
7 His
8 Gly
9 Ile
10 Glu
11 Val
12 Asp
13 (not Ala, Gly)
14 Ala
15 Leu
16 Gly
17 Ala
18 Ile
19 Asp
20 Arg
EF- 1 Lys
2 Lys
3 Gly
4 Leu
5 His
6 (not Gly)
7 (not Gly)
8 Glu
F- 1 Glu
2 Ala
3 Pro
4 Thr
5 Ala
6 His
7 Ser
8 His
9 Ala
FG- 1 (not Gly)
2 (not Gly)
3 Phe
4 (not Ala)
5 Ile

G- 1 Pro
2 Ile
3 Lys
4 Tyr
5 (not Ala, Gly)
6 Glu
7 His
8 Leu
9 Ser
10 (not Gly, Ala)
11 Ala
12 Val or Thr
13 Ile
14 His
15 Val
16 Arg
17 Ala
18 Thr
19 Lys
GH- 1 His
2 Asp
3 Asp
4 Glu
5 Phe
6 Gly
H- 1 Ala
2 Pro
3 Ala
4 Asp
5 Gly
6 Ala
7 Met
8 Gly
9 Gly
10 Ala
11 Leu
12 Glu
13 Leu
14 Phe
15 Arg
16 Lys
17 Asp
18 Ile
19 Ala
20 Ala
21 Lys
22 Tyr
23 Lys
24 Glu
HC- 1 Leu
2 Gly
3 Tyr
4 Gly
5 Glu (carboxyl end)

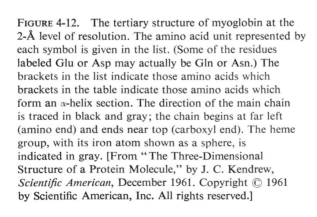

FIGURE 4-12. The tertiary structure of myoglobin at the 2-Å level of resolution. The amino acid unit represented by each symbol is given in the list. (Some of the residues labeled Glu or Asp may actually be Gln or Asn.) The brackets in the list indicate those amino acids which brackets in the table indicate those amino acids which form an α-helix section. The direction of the main chain is traced in black and gray; the chain begins at far left (amino end) and ends near top (carboxyl end). The heme group, with its iron atom shown as a sphere, is indicated in gray. [From "The Three-Dimensional Structure of a Protein Molecule," by J. C. Kendrew, *Scientific American*, December 1961. Copyright © 1961 by Scientific American, Inc. All rights reserved.]

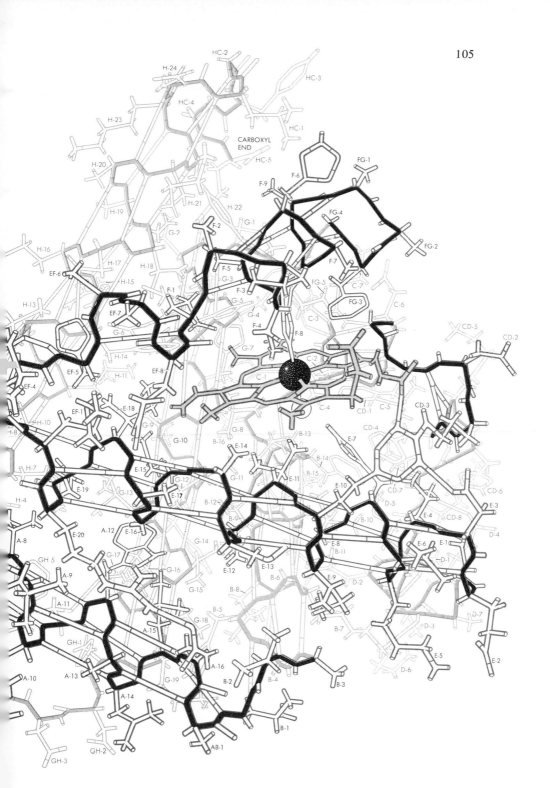

Upon contemplation of this more detailed representation of the myoglobin molecule, some generalizations about the tertiary structure of proteins do emerge. First, it can be seen that the long polypeptide chain is coiled and folded so as to produce a compact, nearly spherical macromolecule. Along the length of that chain there is a succession of helical segments (corresponding to Pauling's α-helix) interrupted by corners and irregular, nonhelical segments. As could be expected, the corners are made up by proline residues, and in the irregular regions are found amino acid residues whose side chains engage in a variety of chemical interactions—attractive, and repulsive— with both the backbone and with other neighboring side chains.

Second, the character of the amino acid side chains also has a decisive influence on the overall conformation of the polypeptide chain. The most important fact in connection with the effect of side chain character on tertiary structure is that the protein does its work in an aqueous intracellular environment. Thus the structure of myoglobin revealed that in assuming their three-dimensional conformation, polypeptide chains coil, twist, and bend in such a manner as to bring into their dry center most of those amino acids whose side chains are "hydrophobic" (bond poorly with water), such as isoleucine, valine, proline, and phenylalanine, and bring into their wet periphery most of those amino acids whose side groups are "hydrophilic" (bond well with water), such as glutamate, lysine, and threonine. The tertiary structure of a protein is therefore that one of many possible three-dimensional conformations which, because of the variety of weak chemical interactions between the atoms of the backbone and those of the diverse side chains, provides for maximal stability in water.

QUATERNARY STRUCTURE

The preceding accounts of the primary, secondary, and tertiary structures of polypeptide chains still leave for discussion one yet higher level of protein structure. Although some proteins, such as myoglobin, consist of a single polypeptide chain, the majority of protein species are aggregates of two or more polypeptide chains. Hemoglobin is a typical example, in that its molecule consists of two α- and two β-polypeptide chains. The spatial conformation assumed by such an aggregate of polypeptide chains is called the *quaternary* structure. Figure 4-13 shows a representation of the quaternary

FIGURE 4-13.   The quaternary structure of the hemoglobin molecule, as inferred from X-ray diffraction studies, seen from above (top) and side (bottom). The two α-polypeptide chains are shown as light gray blocks and the two β-polypeptide chains as dark gray blocks. The letter N in the top view identifies the amino ends and the letter C the carboxyl ends of the two α-chains. Each chain enfolds a heme group, shown as a cross-hatched disk. [From "The Hemoglobin Molecule," by M. F. Perutz, *Scientific American*, November 1964. Copyright © 1964 by Scientific American, Inc. All rights reserved.]

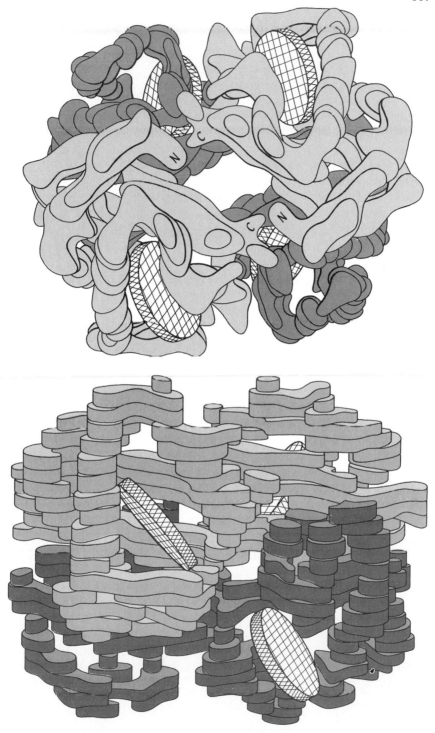

structure of the hemoglobin molecule, which Perutz was able to glean from his X-ray crystallographic analysis. The quaternary structure of *E. coli* β-galactosidase is somewhat simpler, in that its molecule is an aggregate of four *identical* polypeptide chains, each 1173 amino acids long. The *E. coli* tryptophan synthetase molecule is also an aggregate of four polypeptide chains, but its quaternary structure resembles that of hemoglobin, in that it is composed of two pairs of dissimilar chains: one of these chains is the A protein, whose primary structure of 267 amino acids was shown in Figure 4-5, whereas the other chain is the somewhat longer B protein, which contains about 450 amino acids.

At the present time, rather little is known of the nature of the intermolecular bonds responsible for the formation of specific polypeptide chain aggregates in quaternary protein structure, though it appears likely that the same kinds of interactions which figure in the determination of secondary and tertiary structure—hydrogen bonding, electrostatic attraction, and congregation of hydrophobic amino acid side chains—are also involved here. Much more complicated quaternary protein structures than those of the two-by-two aggregate of hemoglobin α- and β-chains or of tryptophan synthetase A- and B-proteins are now known. Indeed, it can be considered a general principle of cell organization that all large proteinaceous structures are quaternary aggregates of polypeptide chains, of which few chains contain more than 1000 amino acids each.

It has already been asserted that the chemical character of the amino acid side chains exerts a profound effect on the three-dimensional conformation of the protein molecule. This assertion will now be rephrased in terms of one of the most fundamental principles of molecular biology: *the primary structure of the polypeptide chain is solely responsible for secondary, tertiary, and quaternary protein structure.* As will be seen in later chapters, the discovery of this principle in the 1950's was to make possible the solution of the problem of how the cell manages to synthesize its enzymes. At first, the notion of the determination of protein conformation by primary structure was merely held as an article of faith, but presently it could be substantiated by direct experiments. In these experiments an enzyme was treated with chemical reagents in such a manner that the native quaternary and tertiary structure of its polypeptide chains was completely disrupted and all of its catalytic activity abolished, though the primary structure of the chains remained intact. The disruptive reagents were then removed, and it was found that the enzyme slowly regained its original tertiary and quaternary structure, as well as its catalytic activity. Thus it could be concluded that once a polypeptide chain of a given amino acid sequence has been synthesized in the cell, the exact spatial conformation necessary for catalytic activity develops spontaneously, without there being any need for intervention by further agencies (Figure 4-14).

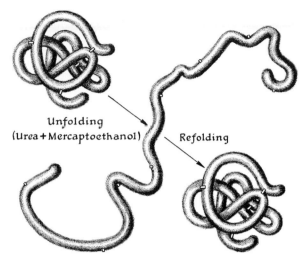

FIGURE 4-14.   Test of the theory that the primary structure of a protein determines its tertiary structure. A native, enzymatically active protein molecule containing several intramolecular disulfide bonds (indicated here by rectangles) is caused to unfold by dissolving it in an 8 M urea solution containing mercaptoethanol. (The mercaptoethanol reduces each —S—S— bond to a pair of —SH groups, and the urea breaks the intramolecular hydrogen bonds.) After these agents are removed, the protein molecule undergoes spontaneous reoxidation and refolding to regenerate the original tertiary structure and enzymatic activity. [After C. J. Epstein, R. F. Goldberger, and C. B. Anfinsen, *Cold Spring Harbor Symp. Quant. Biol.* **28**, 439 (1963).]

ENZYME ACTION

We may now consider how a polypeptide chain can act as the catalyst of a specific chemical reaction. Since the turn of this century it has been realized that an enzyme must temporarily form a chemical bond, or enter into a transient complex with the substrates whose reaction it catalyzes. The notion of an enzyme-substrate complex has greatly facilitated the analysis of enzyme action. In particular, it has provided an understanding of the *rate* at which an enzyme-catalyzed reaction proceeds by taking account of two quite separate reaction steps: the formation of the enzyme-substrate complex, and the actual chemical reaction which transforms the substrates held in the complex into reaction products. An enzymic reaction may therefore be written as

$$\underset{\text{enzyme}}{\overset{\text{substrate}}{+}} \xrightleftharpoons[\text{Affinity}]{} \underset{\text{complex}}{\overset{\text{substrate-}}{\text{enzyme}}} \xrightarrow[\text{number}]{\text{Turnover}} \underset{\text{enzyme}}{\overset{\text{product}}{+}}$$

AFFINITY AND TURNOVER NUMBER

The extent to which the first of these steps occurs depends on the fraction of enzyme molecules which bind substrate at any given substrate concentration, or on the *affinity* of the enzyme for its substrate. The rate at which the second of these steps occurs depends on the average time that elapses before a substrate molecule which is held in a complex is finally transformed into the reaction product. This average waiting time is more conveniently expressed as its *reciprocal* value, which indicates the number of complexed substrate molecules converted per enzyme molecule per unit of time, or the *turnover number* of the enzyme. Thus the higher the affinity of enzyme for its substrate and the higher its turnover number, the greater the reaction rate. The overall rate of any enzyme catalyzed reaction is, therefore, the product of the number of enzyme-substrate complexes present times the turnover number.

Substrate affinity and turnover number of an enzyme—for instance, of β-galactosidase—can be measured by means of the following kind of experiment. A series of reaction mixtures is incubated at 28°C, all containing the same concentration of purified β-galactosidase enzyme (say .01 μg/ml, or 10^{10} enzyme molecules/ml) and variable concentrations of ONPG substrate (as high as 10^{-3} M, or 6×10^{17} ONPG molecules/ml). The rate of enzyme catalysis is then determined in each reaction mixture by measuring the amount of ONPG hydrolyzed (or the intensity of yellow color generated) at various times. The result of such an experiment is presented in Figure 4-15, where the number of ONPG molecules hydrolyzed during the first ten minutes of reaction is plotted against the initial ONPG concentration in the reaction mixture. As can be seen, the rate of ONPG hydrolysis increases proportionately with the ONPG concentration until a limit of 2.4×10^{16} molecules per 10 min, or 4×10^{13} molecules per second, is reached, beyond which there occurs no increase in rate, even though the ONPG concentration is raised still further. This limiting rate of hydrolysis is attributable to the continuous saturation of all the 10^{10} molecules/ml of β-galactosidase in the reaction mixture with their substrate at the highest ONPG concentrations. Hence that limiting rate must reflect the turnover number of the β-galactosidase-lactose complex, which can be estimated from these data to be

$$\frac{4 \times 10^{13} \text{ ONPG molecules/sec/ml}}{10^{10} \text{ enzyme molecules/ml}}$$

$$= 4000 \text{ ONPG molecules/enzyme molecule/sec}$$

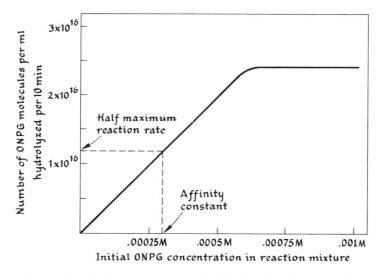

FIGURE 4-15. Kinetics of ONPG hydrolysis by β-galactosidase. A series of reaction mixtures is incubated at 28°C, all containing 0.01 μg/ml pure β-galactosidase enzyme. Each mixture receives the initial concentration on ONPG indicated on the abscissa. The intensity of yellow color developed is measured after 10 minutes of reaction time, and from this measurement is calculated the number of ONPG molecules per ml hydrolyzed, shown on the ordinate.

Some enzymes manifest an even greater turnover number than β-galactosidase, and can transform as many as 10^5 substrate molecules/second. Considering that they occur at the relatively low temperature of 28°C, these rates are very high indeed; in fact, they would be considered nothing short of explosive in the study of ordinary organic chemistry.

The affinity of β-galactosidase for its ONPG substrate can now be estimated from the submaximal rates of hydrolysis at lower ONPG concentrations. As can be seen in Figure 4-15, the rate of the reaction at first rises proportionally with the ONPG concentration. Evidently this means that at the lowest ONPG inputs only a small fraction of the enzyme molecules are complexed with their substrate at any moment and that this fraction increases linearly with the substrate concentration. The slope of this rising straight line is thus a measure of the affinity of β-galactosidase for its ONPG substrate. The affinity is generally expressed, however, in terms of an *affinity constant*, which is defined as that substrate concentration at which half the maximum reaction rate is attained (i.e., at which half the enzyme molecules are complexed with substrate). As can be seen from the data of Figure 4-15, the affinity constant of β-galactosidase for ONPG is 3×10^{-4} M.

The concepts of enzyme-substrate complex and turnover number also help to explain the high degree of *specificity* that enzymes manifest for the reactions they catalyze. As was set forth earlier, the one to two thousand different cell enzymes catalyze no more than about a half-dozen fundamentally different changes of chemical bonds, but each enzyme catalyzes only one of the myriad possible chemical transformations produced by any such fundamental bond change. Indeed, the central role of enzymes in cellular life depends as much on their capacity for highly selective choices of substrates and chemical transformations as on their capacity for intense catalytic activity. By way of demonstrating this selectivity, we may consider the data of Table 4-2, which gives both turnover number and affinity constant for the hydrolysis by β-galactosidase of a variety of galactosides. As can be seen, both turnover number and affinity constant vary over a wide range. It is to be noted that among the compounds tested here, lactose, the natural substrate, is by no means the "best" substrate of β-galactosidase. Furthermore, it can be seen that of the two compounds in this series one may have the higher affinity for the enzyme and the other the higher turnover number. The sulfur analog of ONPG, thio(o-nitrophenyl)-β-galactoside, shows this point most dramatically: it has a much higher enzyme affinity (or lower affinity constant) than lactose, but is not hydrolyzed at all. (The disparity in the data for the hydrolysis of ONPG of Figure 4-15 and Table 4-2 is attributable to the difference in temperature at which these measurements were made)

THE ACTIVE SITE

The specificity of substrate recognition, in both formation of the complex and catalytic chemical transformation, is due to the existence of a special domain on the surface of the enzyme—namely, its *active site*. Considering the size of a substrate molecule such as lactose, which is recognized at the active site, it may be estimated that the active site corresponds to an area of about 400 Å^2, or to only a few percent of the total surface area of the enzyme. Many enzymes, particularly those composed of only a single polypeptide chain, possess only a single active site. But the β-galactosidase protein has four such active sites—one furnished by each of the four identical polypeptide chains that aggregate to form its quaternary structure. The first detailed molecular view of an active site was gained only in 1964, upon the X-ray-diffraction analysis by David C. Phillips of the tertiary structure of the crystalline enzyme shown in Plate II. It can be seen there that the substrate can be embedded in a shallow cleft of the enzyme surface, surrounded by about 20 amino acids of the polypeptide chain. It is this group of amino acids which make up the active site and whose side chains form weak chemical bonds with the substrate. The affinity of the enzyme for its substrate reflects precisely the formation of these bonds. It is to be noted, however, that the amino acids that constitute the active site do not by any means occupy contiguous positions on the polypeptide chain. Instead, they are

TABLE 4-2.   Affinity Constant and Turnover Number for the Hydrolysis by *E. coli* β-Galactosidase of Various β-Galactosides at 20°C

β-galactoside substrate	Structure of residue linked to galactose	Affinity constant M	Turnover number*
o-nitrophenyl (ONPG)		1.61×10^{-4}	2600
p-nitrophenyl		5.13×10^{-5}	320
Methylsalicylate		2.5×10^{-3}	61
Phenyl		1.47×10^{-3}	151
Glucosyl (lactose)		1.9×10^{-3}	95
Ethyl	$-O-CH_2CH_3$	—	0.9
Thio(o-nitrophenyl)		1.20×10^{-4}†	0

After K. Wallenfels and P. Malhotra, *in* P. D. Boyer, H. Lardy, and K. Myrbäck (eds.), *The Enzymes* (2nd ed.), Academic Press, New York, 1960, p. 490.

*Molecules of substrate hydrolyzed per enzyme molecule per second, reckoning that 1 mg of enzyme protein contains 6.9×10^{14} β-galactosidase molecules.

†Concentration at which thio(o-nitrophenyl)-β-galactoside produces half-maximal inhibition of the hydrolysis of lactose substrate.

brought together from widely separated parts of the primary structure by the contortions of the polypeptide chain that produce the tertiary structure. That is to say, the active site is an attribute of the three-dimensional conformation of the enzyme. The specificity of substrate recognition afforded by the active site is a consequence of the nature and exact spatial position of the side chains of its amino acids, which form a receptacle whose shape accommodates precisely the "perfect" substrate and provides the necessary opportunities for weak chemical bonding between enzyme and substrate.

MECHANISM OF CATALYSIS

Granted that the active site can hold the substrate, thanks to the formation of weak chemical bonds, how does it manage to catalyze the chemical transformation that is, after all, the main object of this exercise? Before fashioning a reply to this question we must recall that there exist two aspects to any chemical reaction, namely its rate and the equilibrium concentration of its products and reactants. Thus at room temperature and neutral pH the spontaneous hydrolysis of a 1 M water solution of lactose proceeds so slowly that less than 50% of the lactose molecules are hydrolyzed per year. The reaction continues at this slow but steady pace until about 99.9% of the lactose has been hydrolyzed. No further net hydrolysis of lactose occurs after this point because by the time this stage of the reaction is reached, a sufficiently high concentration of the hydrolysis products, glucose and galactose, will have built up in the solution so that their combination to reform lactose will proceed just as fast as the hydrolysis of the remaining 0.001 M lactose. At this stage, the reaction is said to have reached equilibrium. A most fundamental feature of all catalysis, including that promoted by enzymes, is that the presence of a catalyst can affect only the *rate* at which the reaction reaches its equilibrium but never the state of that equilibrium. Thus addition of β-galactosidase to the lactose solution will cause hydrolysis to reach equilibrium in as many minutes as the spontaneous reaction requires years, but upon attaining this catalytic equilibrium the residual lactose level in the solution is still exactly the same as in the spontaneous equilibrium. The reason for this is that the enzyme accelerates the hydrolysis of lactose into glucose and galactose just as much as it accelerates the eventual reformation of lactose from high concentrations of the reaction products. Thus enzymes do not perform chemical miracles whose occurrence would be impossible in their absence; they merely speed up the rate of spontaneous chemical reactions which would not otherwise have reached equilibrium in our lifetime.

Why does the spontaneous hydrolysis of lactose proceed so slowly? One theory of chemical reaction rates, the *transition state* theory, provides the following explanation. Since lactose has many covalent chemical bonds that can both rotate and stretch, the molecule can assume a variety of conformational states, between which it rapidly alternates in solution. In some of these states, distortions occur in the chemical bonds—for example, in one form the galactose ring is bent out of shape, and a strain is placed on the —C—O—C— bond that links it to glucose. Although the occurrence of this particular distorted conformation is much less frequent than that of strain-free forms—say, only one of 10^4 lactose molecules may assume it at any one instant—it is just the strain placed on the —C—O—C— bond in that conformation which renders it especially susceptible to hydrolytic attack by water. Since the much more frequent strain-free conformations of lactose are virtually resistant to hydrolytic attack, it is only those rare lactose molecules which happen to have assumed the strained state that are eligible for hydrolysis. Once in the strained state, the lactose molecule may either return to one of the strain-free states, and thus survive as lactose, or it may be hydrolyzed and fall apart into glucose and galactose. Hence the low spontaneous rate of hydrolysis devolves from the rarity of the only reactive conformation, or transition state, in which the lactose molecule is eligible for hydrolysis. It is to be noted that when galactose and glucose combine in the reverse of the hydrolysis reaction, they too must assume the reactive conformations that allow the nascent lactose molecule either to pass into one of the stable conformations, and thus remain on the scene as lactose, or to be immediately hydrolyzed and regenerate its glucose and galactose parent molecules. Thus the transition-state theory asserts that the rate of a reaction is proportional to the frequency of the reactive transition states of the reactants.

The foregoing considerations now permit a ready explanation of enzymic catalysis, in that combination of the substrate with the active site can be envisaged to raise the frequency of occurrence of the rare transition states. From this point of view, the active site of the β-galactosidase enzyme would be thought to have a preferential affinity for lactose molecules in their transition state. Hence β-galactosidase would selectively remove from solution the rare lactose molecules in the hydrolytic transition state and hold them in their reactive form for a much longer time than they would exist in free solution. A turnover number of 4000 lactose molecules/enzyme molecule/second for β-galactosidase can then be interpreted to mean that once a lactose molecule in the transition state is selected by the active site of the enzyme, 1/4000 second passes on the average before it is hydrolyzed into galactose and glucose. Since the transition state figures also in the reverse reaction, which reconstitutes lactose from galactose and glucose, the selective stabilization of that transition state by the enzyme also accelerates the reverse reaction. Thus, as has already been stated, the presence of the enzyme does not affect the equilibrium state of the hydrolytic reaction.

CONTROL OF ENZYME ACTIVITY

Growth of any cell depends upon a flow of chemical intermediates through a complex network of interdependent reaction pathways; this flow generates a *balanced* supply of chemical energy and of the 60 cellular building blocks. That is to say, the one to two thousand enzymes that catalyze the individual reaction steps of this panorama of pathways are so closely coordinated in their relative rates of function that both undersupply and oversupply of chemical energy and building blocks, as well as of chemical intermediates of these blocks, are avoided. It is because of this harmonious enzymatic interplay that free building blocks and chemical intermediates constitute less than 10% of the total dry weight of the cell. How is this astonishing dynamic coordination of enzyme action achieved? One of the first clues concerning the mechanism by which cells ensure the balanced synthesis of their building blocks was obtained in 1953 by Aaron Novick and Leo Szilard through their study of the tryptophan metabolism of *E. coli*. According to the schema presented in Figure 3-11, the terminal stage in the biosynthesis of tryptophan comprises the following enzyme-catalyzed reaction steps:

glutamine
chorismate
$\longrightarrow$ anthranilate $\longrightarrow$
indole-glycerol
phosphate
(IGP)
serine
$\searrow$
$\longrightarrow$ tryptophan

Anthranilate
synthetase

IGP
synthetase

Tryptophan
synthetase

When Novick and Szilard examined the rate of synthesis of IGP in *E. coli* cultures maintained under various conditions, they discovered that addition of high concentrations of tryptophan to the growth medium causes an immediate halt to IGP synthesis (and hence of tryptophan synthesis) and that addition of low tryptophan concentrations permits some residual IGP (and tryptophan) synthesis, albeit at a reduced rate. To explain this finding they proposed that tryptophan must inhibit the catalytic function of an enzyme that catalyzes formation of IGP or one of its precursors. Or, more exactly, they envisaged that the rate of function of that enzyme bears an inverse relation to the intracellular tryptophan concentration; that is, the higher the concentration of tryptophan present, the slower the rate of IGP synthesis, and hence of tryptophan synthesis. Later studies showed that the putative tryptophan-sensitive enzyme is, in fact, *anthranilate synthetase*, which catalyzes the first step of the terminal stage of the pathway. These later studies involved experiments with cell extracts of *E. coli* containing anthranilate synthetase and its substrates, chorismate and glutamine, in which it was found that the formation of the anthranilate reaction product

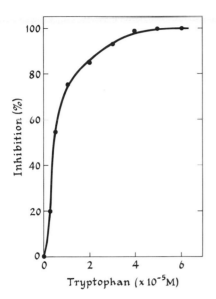

FIGURE 4-16.   Effect of tryptophan on the catalytic activity of *E. coli* anthranilate synthetase. The ordinate shows the percent reduction produced by the tryptophan concentration shown on the ordinate in the initial rate of synthesis of anthranilate in a solution containing 10^{-2} M glutamine and 10^{-5} M chorismate. The presence of 2×10^{-4} M indole, serine, tyrosine, phenylalanine, or histidine instead of tryptophan produced no appreciable inhibition of anthranilate synthetase activity. [After J. Ito and C. Yanofsky, *J. Bacteriol.* **97**, 734 (1969)].

stops as soon as tryptophan is added to the reaction mixture. The result of such an experiment is shown in Figure 4-16. The inhibition of anthranilate synthetase function is produced *only* by tryptophan, and not by any other cell building block.

This highly specific inhibition of the function of an enzyme in its pathway thus provides a close and highly flexible control over the synthesis of tryptophan. As long as tryptophan is continuously incorporated into polypeptides for the synthesis of new proteins in a rapidly growing *E. coli* cell the intracellular tryptophan concentration stays at a relatively low level. Under these conditions, anthranilate synthetase remains active and allows for a steady production of new tryptophan through the pathway. But if growth, and hence protein synthesis, slows down for some reason, or if the cell is brought into the presence of an abundant exogenous tryptophan supply, the intracellular tryptophan concentration rises to a high level. The function of anthranilate synthetase is inhibited, and no further synthesis of tryptophan (or of any of the tryptophan-specific pathway intermediates) occurs until the intracellular tryptophan concentration has once more fallen to the "balanced" level. This mechanism of automatic control is called *feedback inhibition*.

The following experiment, which was first carried out in the late 1950's, shows that control of enzyme activity by feedback inhibition is a general feature of bacterial metabolism. A culture of *E. coli* is grown in a synthetic medium like the one formulated in Table 2-1, except that it contains radioactively labeled ^{14}C-glucose as the only carbon source. Under these conditions all building blocks must contain ^{14}C-atoms at the same specific radioactivity as the glucose nutrient from which they have been synthesized. The experiment consists in adding to that simple ^{14}C-glucose growth medium some particular cell building block in ready-made, non radioactive form (e.g., the amino acid serine), allowing the bacteria to grow in that particularly supplemented medium, and then examining the specific radioactivity of the carbon atoms of the 60 or so building blocks that were synthesized. The result of such an experiment is that all the building blocks are fully ^{14}C-labeled, except that building block which was added in nonradioactive form to the medium and those building blocks for which the added building block is a biosynthetic intermediate. For instance, if serine is the added exogenous building block, then the serine residues in the *E. coli* protein are found to be almost entirely nonradioactive, as is the portion of the tryptophan residues,

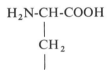

which, according to the pathway shown in Figure 3-11, is derived from serine. That is to say, the presence of serine in the growth medium feedback inhibits some step of the endogenous biosynthesis of serine from the ^{14}C-glucose nutrient, causing the exogenous amino acid to be used as the exclusive source of metabolic serine.

This automatic control makes it impossible for any pathway to remain disharmoniously hyperactive for long in the metabolic interplay of reactions. If a pathway happens to have generated an oversupply of its end-product, then feedback inhibition shuts down the pathway until utilization of the end-product has caused it to fall to the appropriate level. Hence we see that the ensemble of metabolic pathways is self-adjusting, or *homeostatic*, because of a regulatory device built into some of the enzymes that occupy strategic positions in the reaction network. This process allows the uniform maintenance of each building block at that intracellular concentration which is appropriate for its conversion into the macromolecular state and matches rate of building block synthesis with rate of building block utilization.

In vitro studies of the mechanism of feedback inhibition with purified enzymes have shown that there exist two basic ways in which an inhibitor can affect the activity of the enzyme it controls. One of these ways is for the inhibitor to impair the formation of the enzyme-substrate complex; in other words, the inhibitor reduces the affinity of the active site for its substrate. The other way is for the inhibitor to impair the transformation of the enzyme-bound substrate into product; in other words, the inhibitor reduces the turnover number. Either way, the inhibitor forms a complex with the enzyme at a specific site on the enzyme surface; this specific site has a high affinity for the inhibitor and is entirely distinct from the active site (The nonidentity of active and inhibitor-affined sites almost follows from first principles. Consider, for example, chorismic acid and glutamine, the substrates accommodated by the active site of the enzyme anthranilate synthetase; these bear no structural resemblance at all to the enzyme inhibitor tryptophan.) Combination of the inhibitor with its affined site causes a change in tertiary and/or quaternary structure of the enzyme, with the result that the active site becomes deformed and loses its substrate affinity or catalytic power, as is illustrated schematically in Figure 4-17. Enzymes which

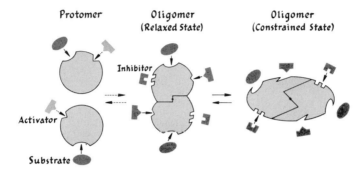

FIGURE 4-17.   Allosteric model of the regulation of enzyme function. The quaternary structure (oligomer) of a hypothetical enzyme is made up of two identical polypeptide chains (protomers). The oligomer can exist in a "relaxed" state (middle) or a "constrained" state (right). In the relaxed state it binds substrate and "activator" molecules; in the constrained state it binds inhibitor molecules. The binding of substrate and activator molecules stabilizes the "relaxed" state, whereas the binding of inhibitor molecules stabilizes the "constrained," and hence catalytically inactive, state. [From "The Control of Biochemical Reactions," by J. Changeux, *Scientific American*, April 1965. Copyright © 1965 by Scientific American, Inc. All rights reserved.]

have the property that their spatial conformation, and hence the catalytic power of their active sites, changes upon interaction of the protein with an inhibitor molecule at a second site were given the name *allosteric* enzymes (Greek *allo*, "other," and *stere*, "solid") by Jacques Monod, Francois Jacob, and J. P. Changeux. Just as for the active site, the specific spatial conformation of the inhibitor-affined allosteric site is also determined by the primary structure of the polypeptide chain. We now reach a further important generalization concerning the paramountcy of primary structure: not only the quality and diversity but also the homeostatic balance of cell metabolism is governed by the amino acid sequences of its protein molecules.

Bibliography

ORIGINAL RESEARCH PAPERS

Kendrew, J. C. Myoglobin and the structure of proteins. *Science*, **139**, 1259 (1963)

Monod, J., J. P. Changeux, and F. Jacob. Allosteric proteins and cellular control systems. *J. Mol. Biol.* **6**, 306 (1963).

Pauling, L., R. B. Corey, and H. R. Branson. The structure of proteins: Two hydrogen-bonded helical configurations of the polypeptide chain. *Proc. Natl. Acad. Sci. Wash.*, **37**, 205 (1951).

Perutz, M. F. X-ray analysis of hemoglobin. *Science*, **140**, 863 (1963).

Sanger, F. The structure of insulin. *In* D. E. Green (ed.), *Currents in Biochemical Research.* Interscience, New York, 1956.

Umbarger, H. E. Feedback control by endproduct inhibition. *Cold Spring Harbor Symp. Quant. Biol.*, **26**, 301 (1961).

Yanofsky, C. The tryptophan synthetase system. *Bacteriol. Reviews*, **24**, 221–245 (1961).

SPECIALIZED TEXTS, MONOGRAPHS, AND REVIEWS

Bernhard, Sydney A. *The Structure and Function of Enzymes.* Benjamin, New York, 1968.

Mahler, H. R., and E. H. Cordes. *Biological Chemistry.* Harper & Row, New York. 1966. Chapters 3, 6, and 7.

Perutz, M. F. *Proteins and Nucleic Acids: Structure and Function.* Elsevier, Amsterdam, 1962

5. Genes

The considerations of the preceding two chapters have shown how the ensemble of cellular proteins functions to make the cell what it is—an engine built of highly specific structural members and enzymes which carries out a complex network of catalytically facilitated metabolic reactions. The basic problem of cellular self-reproduction can now be restated in terms of this question: How, in the course of the cell generation period, does the entire apparatus of cellular proteins double, so that each of the two daughter cells generated by fission of the parent cell can be endowed with its own complete enzyme outfit? In view of the fundamental principle enunciated in the preceding chapter, which asserts that the primary structure of the polypeptide chain is solely responsible for secondary, tertiary and quaternary protein structure, this question can, in turn, be reduced to the following query: How is that particular sequence of the twenty protein amino acids assembled which makes up the primary structure of any given one of the one to two thousand different enzyme molecules? The amino acid building blocks themselves are, of course, synthesized by the metabolic pathways, examples of which we considered in Chapter 3. And the dehydration reaction by means of which these amino acids are joined by peptide bonds into polypeptide chains is, as we may readily imagine, catalyzed by one or more specific cellular enzymes. Nevertheless, when we try to imagine how at each step of the assembly process of a particular polypeptide chain one and only one kind of amino acid is to be selected from the twenty kinds available for insertion into the chain, we encounter a difficulty.

THE ENZYME-CANNOT-MAKE-ENZYME PARADOX

If we were to imagine that the ordered amino acid assembly is the work of yet another "ordering enzyme," then we are obliged to postulate that to each particular protein of given primary structure there corresponds a specific "ordering enzyme" which "knows" how to assemble that particular protein. But if that "ordering enzyme" also turns out to be a protein of specific amino acid sequence, it becomes apparent that, instead of providing an answer, we have merely generated a paradox. Obviously, the postulated "ordering enzyme" would require postulation of yet another "ordering enzyme" for its own formation, which, in turn, would require postulation of a third "ordering enzyme," and so on *ad infinitum*. Thus it becomes apparent that the *informational* element in the synthesis of proteins (i.e., the "knowledge" required for assembly of particular amino acid sequences) cannot itself reside in ordinary enzymes but has to be carried by other cellular elements. These informational elements must possess the capacity not only for directing the assembly of amino acids into predetermined polypeptide sequences but also for achieving their own self-replication. For if the two daughter cells generated by fission of the parent cell are to spawn their own daughter cells, each daughter must be endowed not only with the complete parental outfit of enzymes but also with the complete set of parental informational elements.

The enzyme-cannot-make-enzyme paradox thus leads to the insight that cells owe their character to the possession of self-reproducing informational elements that govern enzyme synthesis. But since the governance of cell character has been previously attributed to the hereditary units, or *genes*, we may equate these informational elements with genes. In other words, it follows from first principles that the question posed in Chapter 1—"How does the gene manage to preside over specific cellular physiological processes from its nuclear throne?"—can be answered as follows; the gene directs the assembly of amino acids into a polypeptide chain of given primary structure. Alas, this *a priori* argument came to be made only in the 1950's, long after the relationship between genes and the synthesis of enzymes had, in fact, already been deduced from quite different premises. But the historical fact that hindsight was involved in "predicting," half a century after the rediscovery of Mendel's papers, the existence of genes from considerations of protein structure and synthesis should not detract from the intrinsic theoretical interest of this argument. Its elaboration finally freed the concept of the gene from its previously unavoidable dependence on character differences. The Mendelian gene can be conceptualized only if two different allelic versions of it—for instance, round and wrinkled seeds—are available for crossbreeding experiments. But the gene as determinant of protein structure draws conceptual content from the existence of a polypeptide chain of given amino acid sequence.

THE ONE-GENE–ONE-ENZYME THEORY

The first connection between genes and enzymes was made within a few years of the rediscovery of Mendelism and the discovery of cell-free fermentation. In 1902, Archibald Garrod concluded from a study of family pedigrees that alkaptonuria, an arthritic condition of man accompanied by the excretion of wine-colored urine, is a hereditary disease. Garrod inferred also that this condition is attributable to an alteration in the normal nitrogen metabolism which causes excretion of a dark substance in place of the normal urine constituent urea. In 1908, Garrod proposed that alkaptonuric individuals are homozygous for a recessive gene and that possession of this gene engenders the failure of some enzymatically catalyzed metabolic reaction—a failure which in turn causes accumulation and excretion of the substrate normally destroyed by that reaction. Garrod coined the phrase "inborn errors of metabolism" to describe the hereditary failure of gene-controlled enzymatic reactions. But Garrod's notions, like those of Mendel, seem to have been so far ahead of their time that they had little influence in the market place of genetic ideas until their rediscovery thirty years later.

The continuity of study of the biochemical effects of genes began with a series of investigations on two recessive eye-color mutations of Drosophila, *vermilion*, and *cinnabar* (see Plate I). The eyes of flies homozygous for either one of these mutant genes are not as deeply red as those of the wild type fly, because *vermilion* and *cinnabar* mutant flies fail to form a brown pigment responsible for the normal eye color. In 1935, G. W. Beadle and B. Ephrussi implanted embryonic eye tissue from larva of *vermilion* and *cinnabar* mutants into larva of normal Drosophila flies and observed that upon metamorphosis of these larva into mature flies the implanted eye tissue developed into supernumerary eyes with normal eye color. It could be concluded, therefore, that the body tissues of the normal flies supply some substance which the *vermilion* and *cinnabar* mutant eye tissues are unable to synthesize but which they can convert into the brown eye pigment. Beadle and Ephrussi then implanted the same embryonic mutant tissues into the larva of *vermilion* and *cinnabar* mutant flies and observed that *vermilion* eye tissue implanted into *cinnabar* host larva developed the normal eye color, whereas *cinnabar* eye tissue implanted into *vermilion* host larva developed the mutant *cinnabar* eye color. Beadle and Ephrussi inferred from these observations that the synthesis of the brown eye pigment arises by the metabolic chain

precursor ⟶ substance I ⟶ substance II ⟶ brown pigment

The *vermilion* mutant would thus carry a block in the reaction that converts the precursor to substance I, whereas the *cinnabar* mutant would carry a block in the reaction that converts substance I to substance II. Thus in the wild-type host larva both mutant eye-tissue transplants are provided with substance II, which they can convert to the brown pigment. The *vermilion* mutant eye-tissue transplant in the *cinnabar* host larva is provided with substance I, which it can convert to substance II and to brown pigment. But the *cinnabar* mutant eye-tissue transplant in the *vermilion* host larva is not provided there with the substance II that it lacks, and hence fails to form the brown pigment. Within a few years biochemical studies showed that the "precursor" is the amino acid tryptophan and that substances I and II are formylkynurenin and hydroxykynurenin. The genetically controlled metabolic eye color sequence could thus be written as shown in Figure 5-1.

The stage was now set for formulating more clearly the physiological role of genes. The normal wild-type allele of the *vermilion* Drosophila gene could be envisaged as presiding over the formation of an enzyme which catalyzes

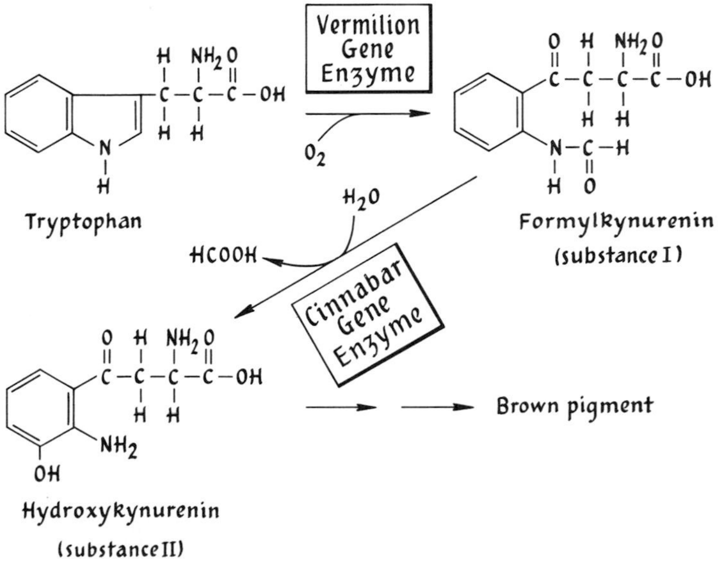

FIGURE 5-1. Some steps in the biochemical reaction sequence leading to the synthesis of the brown eye pigment of *Drosophila*.

the conversion of tryptophan to formylkynurenin. The mutant allele, by contrast, has lost the capacity to form that enzyme. Hence the tissues of a *homozygous* mutant fly carrying the *vermilion* mutant gene on both of its homologous chromosomes lack the enzyme, and the metabolism of such flies is blocked at the reaction step normally catalyzed by that enzyme. The tissues of a *heterozygous* fly, carrying one mutant and one wild-type allele of the *vermilion* gene, would contain the enzyme, however and hence *are* capable of forming formylkynurenin. Similarly, the mutant allele of the *cinnabar* gene has lost the capacity to form the enzyme that catalyzes the conversion of formylkynurenin to hydroxykynurenin, the enzyme that is normally formed under the direction of the wild-type allele. From this viewpoint, the *recessive* character of both the *vermilion* and *cinnabar* mutations is accounted for by the absence of an enzymatic function which the dominant wild-type allele can supply.

In 1940, Beadle and E. L. Tatum (Figure 5-2), who had previously collaborated in efforts to establish the chemical identity of substances I and II in the synthesis of Drosophila eye pigment, developed a new experimental approach for the study of the genetic control of metabolic reactions. They had become discouraged over the difficulties they had encountered with Drosophila as an object for biochemical studies and turned their attention

FIGURE 5-2.   (Left) George W. Beadle (b. 1903). (Right) Edward L. Tatum (b. 1909). [Courtesy News and Publication Service, Stanford University].

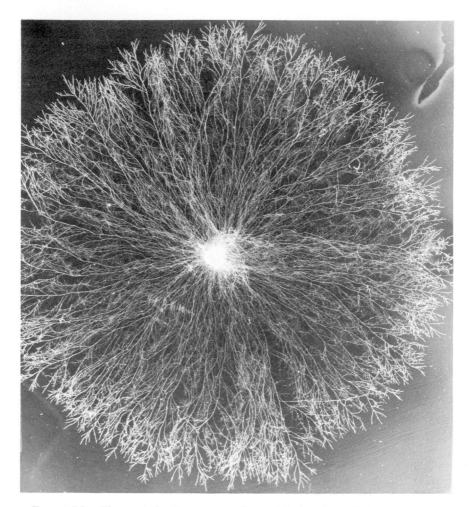

FIGURE 5-3.   The vegetative form, or *mycelium*, of the bread mold Neurospora
(magnified 3.7 times). The branched filaments that make up the mycelium contain
haploid nuclei, arisen by a series of mitotic divisions from a haploid gamete of one of
two opposite mating types. Sexual reproduction ensues upon the union of vegetative
cells from mycelia of opposite mating type, leading to fusion of two haploid nuclei.
The diploid nuclei produced by such fusion at once undergo meiosis to regenerate
haploid gametes. [Courtesy of Alfred Sussman.]

to a fungus, the bread mold *Neurospora crassa* (Figure 5-3). Their choice of
this fungus had been guided by the findings made in the 1930's that Neurospora can be handled as easily in genetic cross-breeding experiments as
Drosophila and that it can be grown in a simple synthetic medium, provided
that the vitamin biotin is furnished to it as the only complex organic supplement. According to Beadle: "With the new organism our approach could

be basically different. Through control of the constituents of the culture medium we could search for mutations in genes concerned with the synthesis of already known chemical substances of biological importance. We soon found ourselves with so many mutant strains unable to synthesize vitamins, amino acids and other essential components of protoplasm that we could not decide which ones to work on first." The procedure used in the isolation of these mutant strains was to inoculate with single Neurospora spores a large number of test tubes containing a *complete* growth medium, which term refers to the simple synthetic growth medium *fortified with yeast extract*. Samples of the fungal growth appearing on the complete medium were then transferred to test tubes containing *minimal* medium, which term refers to the simple synthetic medium fortified only with biotin. In this way Beadle and Tatum found that though the wild-type fungus can grow on both complete and minimal medium, about one in two hundred Neurospora spores gives rise to a fungus which can grow only on the complete medium. The inference was made, therefore, that the unusual fungi which fail to grow upon transfer to minimal medium are the descendants of mutant spores in whose hereditary material a gene mutation has blocked an essential metabolic pathway. Each such mutant is able to grow on the complete medium because it finds ready-made in the yeast extract the very substance whose capacity for synthesis from glucose and inorganic salts it has lost. The nutritional factors which the mutant might find in the complete medium are mainly the twenty standard amino acids, two or three vitamins, and the purine and pyrimidine bases. Hence, in order to identify the exact nature of the growth requirement of any particular mutant, samples of the fungal growth are inoculated into a series of minimal media to which various of these putative growth factors have been added. The particular growth requirement is then inferred to be that substance (or those substances) whose addition to the minimal medium promotes growth of the mutant.

Genetic crosses between the Neurospora wild type and the many mutants isolated and characterized in this manner revealed that most of them owe their growth factor requirement to the mutation of a single gene in the Neurospora genome. Furthermore, detailed biochemical study of the aberrant metabolism of the mutants showed that most of them carry a block at a single step in the reaction sequence leading up to the synthesis of the amino acid, vitamin, purine, or pyrimidine required for growth. On the basis of these facts, Beadle and Tatum now promulgated the *one-gene–one-enzyme theory*. This theory asserts that each gene has only one primary function, which usually consists in directing the formation of one and only one enzyme, and thus in controlling the single chemical reaction catalyzed by that one enzyme. Though, as has already been noted, the idea that genes control single metabolic functions had been adumbrated by Garrod some thirty years earlier, Beadle and Tatum's clear formulation and strong experimental evidence in favor of the one-gene–one-enzyme theory had a profound impact on subsequent thought about the "real core" of genetics. For the belief that in presiding over cellular metabolism each gene has but one function gave hope of ultimately finding out what that one function might actually be. It is important to note, however, that the one-gene–one-enzyme theory of the 1940's still belongs to that first phase of genetic research which was categorized as "classical genetics" in Chapter 1. For the "gene" to which that theory attributes the power to direct the formation of a single enzyme is still the indivisible, formal, abstract Mendelian hereditary unit, whose physical nature has no relevance to the interpretation of the experimental results. Furthermore, that theory had nothing to say about how the gene actually manages to direct the formation of the enzyme under its dominion. Above all, it did not include the idea that the gene directs the assembly of amino acids into a polypeptide chain of given primary structure. Elucidation of the physical nature of the gene and of its role as the informational element of the enzyme-cannot-make-enzyme paradox was to be the work of *molecular* genetics, the birth of which, it should be noted, was rendered valuable midwife service by the one-gene–one-enzyme theory.

BACTERIAL MUTANTS

SUGAR FERMENTATION

Beadle and Tatum founded a school of biochemical geneticists whose work with Neurospora and other fungi during the next decade was to make many decisive contributions to the understanding of the genetic control of cellular metabolism. But by the mid-1950's the main thrust of gene-enzyme studies had shifted to bacteria, and especially to *E. coli*. It must not be thought, however, that *E. coli* is a Johnny-come-lately on the biochemical-genetic scene. On the contrary, one of the earliest observations concerning the genetic control of metabolic processes pertained to the fermentation of lactose by *E. coli*. In 1907, R. Massini isolated a mutant strain of *E. coli* which he called *mutabile* and which, in contrast to ordinary *E. coli* strains, is unable to ferment lactose. The metabolic defect of this *lactose-negative* strain could be readily observed by plating the strain on a special sugar-utilization indicator medium called *EMB agar*. This agar contains a nutrient broth medium containing the dyes eosin yellow and methylene blue and the sugar whose fermentation is to be tested. On EMB agar, a bacterium able to ferment the sugar produces a dark red colony, whereas a nonfermenting bacterium produces a white colony (Plate III). (The red and white colors reflect alternative states of the indicator dyes, which in turn signal the respective chemical changes produced in the agar during bacterial growth with or without fermentation of the test sugar.) Thus when the lactose-negative *E. coli mutabile* is plated on EMB-lactose agar, its colonies are

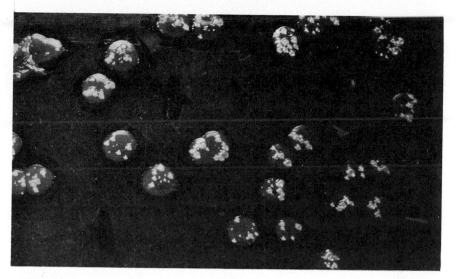

FIGURE 5-4.   Lactose-positive (Lac$^+$) papillae on lactose-negative (Lac$^-$) colonies of *E. coli* [From W. Braun, *Bacterial Genetics* (2nd ed.), W. B. Saunders, Philadelphia, 1965.]

white. Upon prolonged incubation of these plates, however, isolated dark-red spots, or papillae, appear on the colonies (Figure 5-4). Upon picking and replating the bacteria present in the red papillae on EMB-lactose agar, Massini found that these bacteria had regained the capacity to ferment lactose—that is, they had become lactose-positive. Thus during the growth of the colony initiated by a lactose-negative bacterium, *mutations* of the type lactose-negative → lactose-positive had taken place, and these gave rise to subclones of bacteria to whom the capacity for lactose fermentation, characteristic of ordinary *E. coli*, had been restored. But, as will be discussed in more detail in Chapter 6, it did not seem possible in Massini's day to prove conclusively the genetic nature of these changes. Nevertheless, despite the uncertainty concerning its very existence, I. M. Lewis had, by 1934, already measured the frequency of the lactose-negative → lactose-positive mutation in *E. coli mutabile* as being on the order of one per hundred thousand cell generations.

Once the true mutational basis of bacterial variation had been established in the 1940's, Joshua Lederberg found lactose-negative mutants of ordinary, or *wild type*, lactose-positive *E. coli* by inspecting about half a million *E. coli* colonies appearing on some thousand EMB-lactose agar plates. Almost all of the half-million colonies were red, of course, but there appeared about 300 white colonies of lactose-negative mutants.

At this point it is necessary to introduce some nomenclature of *bacterial* genetics, which in its present guise differs from that employed in the genetics of higher, or eukaryotic, organisms, including Neurospora. In bacterial genetics the overt character, or *phenotype* (from the Greek *phainein*, "to show") of an individual is indicated by a three-letter abbreviation in roman type, the first letter of which is capitalized. To that three-letter symbol is affixed a superscript that indicates the status of the bacterial property for which the abbreviation stands. Thus the lactose-positive *E. coli* wild type is designated as Lac$^+$, whereas a lactose-negative mutant is designated as Lac$^-$. The gene which is held to be in control of the phenotypic character under consideration is designated by a three-letter abbreviation in italic lower-case letters. Thus the gene that determines the property of lactose fermentation is designated as *lac*.

Lederberg then isolated the 300 Lac$^-$ *E. coli* mutant colonies and attempted to establish the biochemical basis of their inability to ferment lactose. Although the reason for the Lac$^-$ character of some of these mutants did not seem very clear at first, other mutants left no doubt as to the cause of their Lac$^-$ behavior: these mutants, it turned out, lack the enzyme β-galacto-sidase, which as was seen in Chapter 3, catalyzes the first step in lactose metabolism—namely, the hydrolysis of lactose into galactose and glucose. Hence, in these mutants the Lac$^+ \to$ Lac$^-$ mutation seems to pertain to *lac* gene which controls the synthesis of β-galactosidase. Similar mutant hunts on EMB-galactose agar led to the isolation of Gal$^-$ mutants of *E. coli* that are unable to utilize galactose as a carbon source. Subsequent enzymatic analysis of an ensemble of Gal$^-$ mutants revealed that such mutants fall into at least three classes. One Gal$^-$ mutant class lacks galactokinase, the first enzyme of the pathway of galactose metabolism shown in Figure 3-6, but does possess uridyl transferase and UDP Gal epimerase. The second and third Gal$^-$ mutant classes lack, respectively, uridyl transferase and UDP Gal epimerase, but do possess the other two enzymes. Thus it was concluded that the synthesis of each of these three galactose-fermentation enzymes is controlled by one of three separate *gal* genes and that mutation in any one of these *gal* genes renders the bacterium incapable of forming the correspond-ing enzyme, without affecting its capacity to form the other two enzymes. Included among this ensemble of Gal$^-$ mutants, however, were some that lack all three enzymes. Evidently the character of these mutants could *not* be explained merely in terms of the loss of capacity to form a given enzyme by mutation of one gene. Since the manner in which the mutation of one gene can affect the formation of more than one enzyme went unexplained for some years, discussion of this phenomenon is deferred until Chapter 18.

Use of the EMB indicator agar also led to the isolation of *E. coli* mutants unable to utilize such sugars and alcohols as arabinose (Ara$^-$), xylose (Xyl$^-$), manitol (Mtl$^-$), or maltose (Mal$^-$) as their source of carbon and energy. Enzymatic analysis of these sugar-fermentation mutants revealed that most of them lack the ability to form one or another of the enzymes of the Ara$^+$

Xyl+ Mtl+ Mal+ *E. coli* wild type that are involved in the preprocessing of the relevent sugar or alcohol before its breakdown via the glycolytic pathway.

AUXOTROPHY

A few years after his work with Beadle had shown that powerful insights could be gained through the study of Neurospora mutants that have clearly defined nutritional requirements, Tatum began a systematic search for analogous nutritional mutants in *E. coli*. In order to isolate rare bacterial mutants that have a nutritional requirement for their growth with which the parent *E. coli* wild type can dispense, Tatum followed essentially the same procedure that had yielded the Neurospora mutants. For this purpose large numbers of *E. coli* cells were plated on a nutrient broth agar—that is, on complete medium. The colonies that came up after overnight incubation on the complete medium were then picked, and part of each bacterial clone was tested to ascertain whether it would grow on a minimal medium—that is, agar containing a synthetic medium like the one formulated in Table 2-1. In this way Tatum found that, whereas the overwhelming majority of colonies consisted of bacterial clones capable of growing on both media, about one percent of the colonies that grew on the complete medium were made up of bacteria that are unable to grow on the minimal medium. That is, these rare colonies were made up of clones of nutritional *E. coli* mutants, which require for their growth some factor present in the complete medium but absent from the minimal medium. Such bacterial mutants came to be known as *auxotrophs*, from the Greek *aux-*, "increase" (not from the Latin *auxi-*, "help"), and *trophe*, "food." The normal *E. coli* wild type, which is able to grow on both media and hence can dispense with all exogenous nutrients other than those furnished to it in the minimal medium, is said to be a *prototroph*, from *protos*, "first," or "minimal."

The exact nature of the growth requirement of any particular auxotrophic mutant strain could be established by means of the same procedure previously used in the Neurospora work. In this procedure, samples of the auxotrophic bacterial clone are inoculated into a series of minimal media to which various putative growth factors, such as amino acids, vitamins, purines, and pyrimidines, have been added. The growth requirement is then inferred to be that substance whose addition to the minimal medium is necessary and sufficient to allow growth of the auxotroph. Though Tatum was not able to identify the exact nature of the growth requirement of every auxotrophic mutant he had isolated, he did succeed in establishing that many of his *E. coli* auxotrophs responded to the addition of just one *single* factor to the minimal medium. For instance, one mutant required for its growth only threonine, another only proline, another only tryptophan, another only thiamin, and another only adenine. The phenotypes of these auxotrophic mutants are designated as Thr⁻, Pro⁻, Trp⁻, Thi⁻, and Ade⁻, respectively, in contradistinction to the Thr⁺ Pro⁺ Trp⁺ Thi⁺ Ade⁺ *E. coli* prototrophic wild type. It is to be noted that the minus or plus superscripts affixed to these symbols mean inability or ability to *synthesize* the substance represented by the three-letter abbreviation, and not, as in the case of sugar-fermentation mutants, inability or ability to *utilize* the substance as a source of carbon and energy. (Unfortunately, this is not the only instance of nomenclatural ambiguity that the student of molecular genetics will encounter.)

These properties of *E. coli* auxotrophs offered further support for the one-gene–one-enzyme theory. The restoration of growth by the addition of a single amino acid, vitamin, or purine to an auxotroph suggested that the mutation which created the auxotroph out of the prototrophic wild type pertained to a bacterial gene that controls a single enzyme catalyzing a step in the biosynthesis of the growth factor.

The procedure Tatum had used for the isolation of his auxotrophs can be described only as hard labor, in that it involved the random hand-picking and retesting of thousands of bacterial colonies. It was good news for bacterial geneticists, therefore, when in 1948, both B. D. Davis and J. Lederberg published a method that allows direct selection of auxotrophs. The basis of this method is that the antiobiotic penicillin kills bacteria only as long as the bacterial cells are growing in its presence. Since penicillin interferes with the synthesis of the bacterial cell wall, bacteria growing in the presence of the antibiotic outgrow their integument and ultimately burst. But it obviously makes no difference to nongrowing bacteria, which are not forming any cell walls anyway, whether their cell wall synthesis is inhibited by penicillin or not. Thus, in order to select a small fraction of auxotrophic mutants from among a very large number of individuals of their prototrophic sib, a bacterial culture is inoculated into a *minimal* medium containing penicillin. Under this condition, all the prototrophs of the culture will grow and eventually be killed by the penicillin. But any auxotrophs, which may be present in the culture and which lack some growth factor not supplied to them in the minimal medium, will be spared. After the prototrophs have been killed, the penicillin is removed from the culture medium, and the few surviving cells are plated on complete medium agar. In principle (though, alas, not always in fact) the only colonies that appear on the complete medium agar after penicillin treatment should be clones of auxotrophs, whose specific growth requirements can then be identified, as has already been described.

The availability of the penicillin selection technique (and the subsequent development of the *replica plating method* to be described in Chapter 6) has meanwhile allowed the ready isolation of vast numbers of *E. coli* auxotrophs. Among these auxotrophs specific growth requirements were found for both purines and pyrimidines and for seventeen of the twenty standard amino acids; the only three amino acids for which no auxotrophic mutants are known are alanine, asparagine and glutamine.

By means of these selective methods it is possible also to build *polyauxotrophic* mutant strains that have more than one growth requirement. For instance, once one has isolated a histidine-requiring, or His⁻ mutant, one could subject this mutant to a second cycle of penicillin selection, this time adding histidine to the minimal medium during penicillin treatment. In this way, all His⁻ bacteria requiring *only* histidine would grow and, hence, die. Among the surviving bacteria, however, one could expect to find mutants that carry an additional auxotrophic mutation; for instance, a mutant might be found that also requires leucine, in which case it would be a His⁻ Leu⁻ *double* auxotroph. The process of building strains that carry a variety of genetic markers can thus be continued indefinitely. For instance, by plating a large number of cells of the His⁻ Leu⁻ double auxotroph on EMB-lactose agar and picking one of the rare white colonies, one would have secured a His⁻ Leu⁻ Lac⁻ triple mutant.

TEMPERATURE-SENSITIVE MUTANTS

The 11th Cold Spring Harbor Symposium on Quantitative Biology, held in the summer of 1946, was dedicated to "Heredity and Variation in Microorganisms." This symposium turned out to be a memorable event in the history of molecular genetics, since it was here that the discoveries of sexuality in both bacteria and viruses were first announced, topics that will be treated in later chapters of this text. But the single topic of greatest interest to the audience present at that meeting was not either of these quite unexpected discoveries but rather the apparent triumph of the one-gene–one-enzyme theory. Several speakers presented their analyses of auxotrophic mutants in fungi and bacteria, which showed that, in strong support of the theory, growth could be restored to most such mutants by adding a single metabolite to the minimal medium. After one of these presentations, Max Delbrück rose to point out that although these data were certainly compatible with the thesis that every gene controls the formation of a single enzyme that, in turn, catalyzes a single reaction step in the grand metabolic ensemble, they could not really be considered to offer any *proof* of the validity of that theory. For by the very method of isolation of the auxotrophs, a strong bias had been introduced in favor of finding exactly the kind of mutant everyone seemed so pleased to have found. That is, if there *were* genes that controlled not one but very many enzymes, then it seemed likely that at least one of those enzymes would be concerned with an *indispensable* function. Such an indispensable function would be one that cannot be bypassed by the presence of any of the relatively simple substances in the complete medium.

In other words, even if one-gene–many-enzymes were the rule, no mutations in such genes would have been discovered, since the corresponding mutant cells would not have presented themselves as colonies on the complete medium agar in the first place. Delbrück summed up his criticism by requesting the champions of the one-gene–one-enzyme theory to propose experiments by means of which the theory could be *disproved*, since "if such methods are not readily available, then the mass of compatible evidence carries no weight whatsoever in supporting the thesis."

Delbrück's request was easier made than met, since it seemed well-nigh impossible to determine whether mutations in genes that control indispensable functions make up a large or a small fraction of all mutations that actually do occur. Nevertheless, Beadle and Tatum's pupil Norman Horowitz did manage to think of a method by which this point might be settled—namely, by the use of *temperature-sensitive* mutants. Such mutants are only able to grow at temperatures lower than that at which the wild type is able to grow. Temperature-sensitive mutants were already known to exist in Neurospora when Horowitz and his collaborator U. Leupold began a systematic survey of temperature-sensitive mutants in *E. coli*. For this purpose, wild-type *E. coli* were plated on minimal medium agar and incubated for 48 hours at 40°C, during which time all bacteria able to grow at this temperature gave rise to visible colonies. The agar plate was then photographed and incubated for an additional 120 hours at 25°C. During this low-temperature incubation period a few additional colonies came up, the locations of which were established by comparing the final colony pattern on the plate with the photograph taken at the end of the high-temperature incubation period. These additional colonies were evidently clones of temperature-sensitive mutants of *E. coli*. These clones could be isolated and their behavior studied at the higher, or *restrictive*, temperature. In this way it was found that among 161 temperature-sensitive mutants, only 37 were products of mutations in genes that control indispensable functions, in that these mutants could *not* grow on complete media at the restrictive temperature. All of the remainder, and hence the majority of the mutant isolate, were able to grow on complete media at the restrictive temperature. That is, growth could be restored to the majority of the mutants by one or more of the simple substances present in nutrient broth, substances required only at the restrictive temperature, but not at the lower, or permissive, temperature. Further analysis showed that at the restrictive temperature these temperature-sensitive mutants resembled the previously known *E. coli* auxotrophs, in that—for most of them—addition to the minimal medium of one or another of the standard amino acids, vitamins, or purine and pyrimidine bases sufficed to promote their growth. Horowitz and Leupold concluded, therefore, that whereas mutations in genes controlling indispensable functions certainly *do* occur in *E. coli*, their frequency does not, after all, seem to be so great as to invalidate the earlier support provided for the one-gene–one-enzyme theory by the isolation of ordinary, single-growth-requirement auxotrophs.

In 1951, when Horowitz and Leupold published these findings, the actual molecular basis of the temperature sensitivity of their mutants was still quite unclear. They did not even trouble to state explicitly in their paper whether they thought that it is the mutant gene itself or the enzyme controlled by that gene whose function is temperature sensitive, even though this distinction ought to have been taken into account in the interpretation of their data as alleged support for the one-gene–one-enzyme theory. Probably because of that lack of understanding of their true nature, temperature-sensitive mutants did not play a very important role for the next ten years in the development of molecular genetics. When an explanation of their behavior *was* finally available, temperature-sensitive mutants were suddenly rediscovered, and as we shall see presently, they came to provide one of the most powerful tools for the global characterization of the genetic material of bacteria and bacterial viruses. By that time, however, the once hotly contested one-gene–one-enzyme theory had long since been replaced by much more specific and detailed notions.

PATHWAY ANALYSIS

We may now focus more closely on the nature of auxotrophic mutants. In terms of the one-gene–one-enzyme theory, we have already seen that the growth requirement of an auxotroph is attributable to the mutation of a gene that controls an enzyme involved in the biosynthesis of that growth factor. But since the biosynthesis of amino acids, vitamins, and purines and pyrimidines from the carbon skeletons provided by glycolysis and citric acid cycle is achieved by the tandem action of a series of enzymes, each of which would be controlled by its own gene, auxotrophy for a single growth factor could conceivably result from mutation in any one of several different genes. The early recognition of this principle by the very first students of auxotrophic mutants of Neurospora and *E. coli* proved to be of great importance, not only for the ultimate understanding of the structure and function of the genetic material, but also for the elucidation of the biosynthetic pathways themselves. For the purpose of this discussion, we shall confine ourselves to the example of Trp⁻ auxotrophs that require tryptophan for growth, though entirely analogous data collected over nearly two decades by a host of workers are now available for almost every other growth factor that has been identified.

In the mid-1950's Charles Yanofsky (Figure 5-5) began to isolate a collection of Trp⁻ auxotrophic mutants of the prototrophic Trp⁺ *E. coli* wild type. During the succeeding ten years this collection gradually grew to

FIGURE 5-5.   Charles Yanofsky
(b. 1925). [Courtesy C. Yanofsky.]

comprise a very large number of members. One of the first studies made with
the mutants of that collection was to test whether any of these Trp⁻ auxotrophs were able to grow on a minimal medium supplemented with substances
other than tryptophan. The result of this test is summarized in Table 5-1,
where it is seen that Trp⁻ auxotrophs actually fall into three distinct classes
with respect to their growth requirement: class TrpB⁻, which grows only on

TABLE 5-1.   Properties of Tryptophan Auxotrophs of *E. coli*

| Group | Growth on minimal medium plus | | | | Substance accumulated |
	No supplement	Anthranilic acid	Indole	Tryptophan	
Trp⁺	+	+	+	+	none
TrpE⁻	−	+	+	+	none
TrpD⁻	−	−	+	+	anthranilic acid
TrpC⁻	−	−	+	+	CDRP*
TrpA⁻	−	−	+	+	IGP†
TrpB⁻	−	−	−	+	indole

* CDRP = carboxyphenylamino deoxyribulose phosphate.
† IGP = indole glycerol phosphate.

tryptophan; class TrpA⁻, TrpC⁻, TrpD⁻, which grows *either* on tryptophan *or* indole; and class TrpE⁻, which grows on tryptophan *or* indole *or* anthranilic acid. (The Trp⁺ prototroph, of course, grows also in the absence of any supplement.) The terminal stage of the pathway of tryptophan biosynthesis (see Figure 3-11)

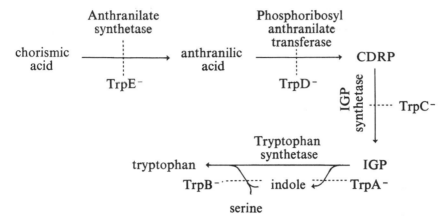

provides an explanation for the existence of these three classes. Class TrpB⁻ auxotrophs evidently carry mutations in a gene that controls the last enzyme of the pathway, tryptophan synthetase, since loss of that enzyme can be remedied only by supplying tryptophan itself. Class TrpE⁻ auxotrophs must carry mutations in a gene that controls the first enzyme of the pathway, anthranilate synthetase, since these auxotrophs evidently possess all the enzymes necessary for converting anthranilic acid into tryptophan. The existence of the third class, TrpA⁻, TrpC⁻, TrpD⁻, caused some confusion for a time. Evidently these auxotrophs carry mutations in genes controlling enzymes that catalyze reaction steps prior to the last step, since they are able to convert indole into tryptophan. But why should they be able to do so, when indole is not shown as an intermediate of the biosynthetic pathway set forth in Figure 3-11? This mystery was cleared up when Yanofsky discovered that each of the two kinds of polypeptide chains, the A protein and the B protein, which constitute the quaternary structure of *E. coli* tryptophan synthetase, is an enzyme in its own right. In the quaternary tryptophan synthetase aggregate of four polypeptide chains the two A-protein chains catalyze the first part of the overall reaction, yielding indole as a transitory intermediate and triose phosphate, whereas the two B-protein chains catalyze the condensation of the indole ring with serine to yield tryptophan (Figure 5-6). Thus the mutants of the TrpA⁻, TrpC⁻, TrpD⁻ class must possess a functional tryptophan synthetase B protein that enables them to convert exogenous indole into tryptophan. It should be noted that neither CDRP nor IGP can be tested as supplemental growth factors in the classification

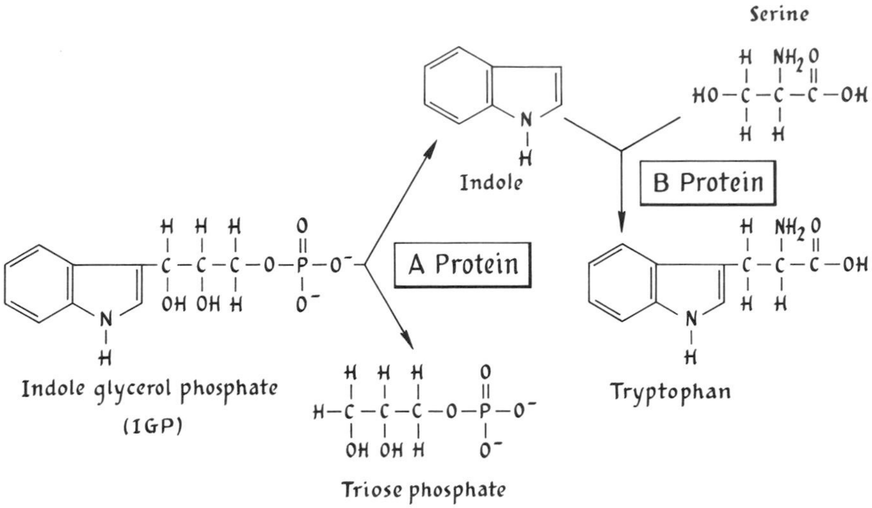

FIGURE 5-6.   The terminal reaction sequence in tryptophan biosynthesis catalyzed by the *E. coli* tryptophan synthetase enzyme complex consisting of A-protein and B-protein.

of Trp$^-$ auxotrophs, since these phosphorylated pathway-intermediates cannot enter the bacterial cell from the growth medium. Thus the general principle underlying this analysis of the growth response of the three classes of Trp$^-$ mutants is that different auxotrophic mutants are blocked at different steps of the tryptophan biosynthetic pathway, and that any auxotroph can convert into tryptophan only those intermediates whose synthesis would normally occur after the step blocked by the gene mutation.

A corollary of this principle is that since the auxotroph cannot convert the metabolic intermediate produced immediately before the blocked step into the next substance in the chain, then that metabolic intermediate should accumulate in the mutant cell. Biochemical analysis of cultures of the various Trp$^-$ auxotrophs confirms this expectation, as is shown in Table 5-1. Class TrpB$^-$ can be seen to accumulate indole. This finding can only mean that class TrpB$^-$ auxotrophs lack a catalytically active tryptophan synthetase B protein but possess an active A protein. The active A-protein component of the tryptophan synthetase enzyme complex of the TrpB$^-$ auxotrophs causes the conversion of IGP into triose phosphate and indole, which in the absence of an active B-protein enzyme, accumulates instead of being converted into tryptophan. Class TrpE$^-$ mutants, which have been previously inferred to lack an active anthranilate synthetase enzyme, do not accumulate any noticeable product. The reason for the failure to find, in this mutant class, an accumulation of chorismic acid—the intermediate immediately preceding the blocked reaction step—is that chorismic acid is a precursor also for the synthesis of cell building blocks other than tryptophan. Hence even in the absence of an active anthranilate synthetase enzyme, chorismic acid continues to be converted into its other metabolic derivatives. As can be seen in Table 5-1, the third class of auxotrophs, which can grow either on indole or on tryptophan, is resolved into three distinct subclasses on the basis of the metabolic accumulation products. Class $TrpA^-$ evidently lacks an active tryptophan synthetase A protein, and hence accumulates IGP; class $TrpC^-$ evidently lacks an active IGP synthetase, and hence accumulates CDRP; and class $TrpD^-$ evidently lacks an active phosphoribosyl anthranilate transferase, and hence accumulates anthranilic acid.

It must be noted that the Trp^- auxotrophs accumulate appreciable quantities of the intermediate preceding the blocked reaction step only while they are being starved for tryptophan: accumulation stops as soon as the mutants are supplied with the exogenous, ready-made tryptophan they require for growth. The explanation of this phenomenon is to be sought in the *feedback-inhibition* that tryptophan exerts on the metabolic pathway leading to its own synthesis. As was set forth in Chapter 4, the catalytic activity of anthranilate synthetase, the first enzyme of the terminal, tryptophan-specific reaction sequence of the pathway, is inhibited by tryptophan. Hence while growing in the presence of tryptophan the entire pathway is shut down, and every bacterium, even the Trp^+ prototroph, behaves as if it belonged to the $TrpE^-$ class, which lacks anthranilate synthetase activity.

The accumulation of sequential intermediates in the biosynthetic pathway by different classes of Trp^- auxotrophs can be demonstrated also by means of *cross-feeding* experiments. Figure 5-7 presents the result of an experiment in which samples of cultures of $TrpE^-$, $TrpB^-$, and $TrpD^-$ auxotrophs were streaked across an agar plate containing minimal agar and a very small amount of tryptophan growth factor. All three cultures grew until the added tryptophan had been exhausted, producing the three faintly visible colonial streaks on the agar surface. But at the point of contact between the $TrpD^-$ and $TrpE^-$ streaks, the growth of the latter was more luxuriant than that of the former, indicating that the $TrpD^-$ bacteria excrete a product (anthranilic acid) which the $TrpE^-$ bacteria can convert into tryptophan. Hence the metabolic block of $TrpD^-$ is at a later stage of the sequence than that of $TrpE^-$. At the points of contact between the $TrpD^-$ and $TrpB^-$, as well as between $TrpE^-$ and $TrpB^-$, the growth of $TrpD^-$ and $TrpE^-$ was more luxuriant than that of $TrpB^-$, indicating that $TrpB^-$ bacteria excrete a product (indole) which both $TrpE^-$ and $TrpD^-$ can convert into tryptophan. Hence the metabolic block of $TrpB^-$ is at a later stage of the sequence than the blocks of $TrpE^-$ and $TrpD^-$. This simple experiment shows, therefore, that the sequential order of the metabolic blocks is

$$TrpE^- \longrightarrow TrpD^- \longrightarrow TrpB^-$$

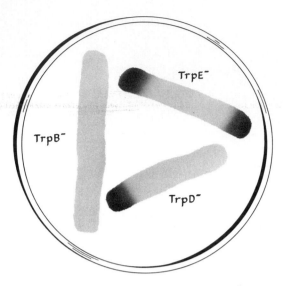

FIGURE 5-7.   Diagrammatic presentation of the result of a cross-feeding test of three different Trp⁻ auxotrophs of *E. coli*. Samples of cultures of TrpB⁻, TrpD⁻, and TrpE⁻ mutants were streaked in a triangular pattern across a plate of minimal medium agar containing a small amount of tryptophan. The degree of shading of the streak indicates the luxuriance of the growth.

The phosphorylated accumulation products IGP and CDRP of TrpC⁻ and TrpA⁻ auxotrophs cannot, however, participate in such cross-feeding experiments, since they cannot enter the recipient cells that might otherwise be able to convert them into tryptophan.

GENE-PROTEIN RELATIONS

Just as Garrod's investigations on the human disease alkaptonuria gave the first inkling that genes control the formation of enzymes, so did studies of another human hereditary disease, sickle-cell anemia, forty years later provide the first indication that genes exert this control by directing the primary structure of polypeptide chains. The red blood cells of persons afflicted with sickle-cell anemia assume an abnormal, or sickle-shape at low oxygen pressure, a condition which may lead to serious circulatory difficulties (Figure 5-8). Analyses of the incidence of this disease in family pedigrees had shown that it is the result of a single recessive mutant gene in the human genome. In 1949 Linus Pauling and his colleagues H. Itano and J. Singer demonstrated

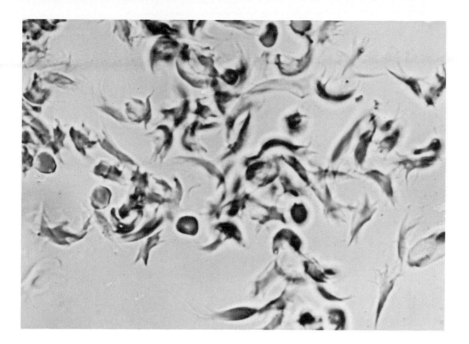

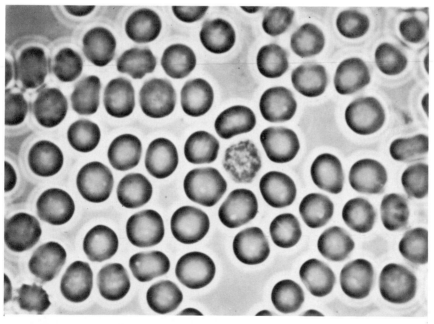

FIGURE 5-8. Sickle-cell anemia in man. The upper photomicrograph shows red blood cells (under low oxygen tension) of individuals carrying the sickle-cell trait. The lower photomicrograph shows red blood cells of normal individuals under similar conditions. [These photographs, by Anthony C. Allison, appeared in "How Do Genes Act?," by V. M. Ingram, *Scientific American*, January 1958.]

that sickle-cell anemia of homozygous carriers of the sickle-cell mutant gene is attributable to the presence of an abnormal hemoglobin molecule in the red blood cells. In alkaline solution, these sickle-cell hemoglobin molecules were found to carry a lower negative charge than the normal human hemoglobin. Furthermore, the hemoglobin of persons suspected to be *heterozygous* carriers of the recessive sickle-cell mutant gene (and hence not afflicted with the disease) was found to be a mixture of normal and abnormal hemoglobin molecules. Pauling and his colleagues concluded from this finding that sickle-cell anemia is a gene-controlled "molecular disease." At the time of this work Sanger was just developing the chemical tools for the analysis of primary protein structure, and once the success of his procedures was established, V. Ingram applied them to the sickle-cell hemoglobin. In 1956 Ingram announced that the difference between normal and sickle-cell hemoglobin is that the mutant hemoglobin carries a *valine* residue at a site on its polypeptide chain at which the normal hemoglobin carries a *glutamic acid* residue. (This explained the observed difference in negative charge, since one ionizable —COOH group is lost upon replacement of the glutamic acid side chain by the valine side chain. See Figure 2-2.) Later work was to show that this amino acid replacement takes place at the sixth residue from the amino terminus of the β-polypeptide chain, two of which make up the hemoglobin molecule in consort with a pair of α-polypeptide chains (see Figure 4-13).

By 1956, the enzyme-cannot-make-enzyme paradox was widely appreciated, and the idea that genes direct the assembly of amino acids into polypeptides had become commonplace. But Ingram's demonstration that a gene mutation engenders the replacement of one polypeptide amino acid residue by another was the first direct proof that this idea is a description of the real world.

Bibliography

PATOOMB

George W. Beadle. Biochemical Genetics: Some recollections.

HAYES

Chapters 5 and 6.

ORIGINAL RESEARCH PAPERS

Beadle, G. W., and E. L. Tatum. Genetic control of biochemical reactions in Neurospora. *Proc. Natl. Acad. Sci.*, **27**, 499 (1941).

Beadle, G. W., and B. Ephrussi. The differentiation of eye pigments in Drosophila as studied by transplantation. *Genetics*, **21**, 225 (1936).

SPECIALIZED TEXTS, MONOGRAPHS, AND REVIEWS

Wagner, R. P., and H. K. Mitchell. *Genetics and Metabolism* (2nd ed.). Wiley, New York, 1964.

Yanofsky, C. Gene-enzyme relationships. *In* I. C. Gunsalus and R. Y. Stanier (eds.), *The Bacteria.* Academic Press, New York, 1964. Vol. 5, p. 373.

6. Mutation

Without much ado, the discussions of the preceding chapter applied the fundamental Mendelian concept of the gene—and the notion of gene mutation—to a bacterium such as *E. coli*. But nothing could resemble less the actual development of bacterial genetics than this free-and-easy way of extending to the realm of lowly prokaryota the hereditary principles worked out by classical geneticists for eukaryota: until the 1940's few bacteriologists thought that bacteria even *have* any heredity. Since, as was mentioned in Chapter 2, the prokaryota neither contain a true cell nucleus nor possess cytologically recognizable chromosomes, it was generally held that bacteria represent a more anarchic form of life, one not reigned over by genes from a nuclear throne. It was only with the rise of molecular genetics after World War II that bacteria finally became the objects of intensive genetic studies.

PLEOMORPHISM, MONOMORPHISM, AND DISSOCIATION

The Founding Fathers of bacteriology encountered a bewildering diversity of bacterial types in the latter part of the nineteenth century. This diversity of types, and the endless metamorphoses of which bacteria seemed capable, was first explained by the doctrine of *pleomorphism*. This doctrine held that a bacterium is an organism of enormous biologic plasticity, and that it can assume, according to need, any one of a variety of morphological forms and physiological functions. But the development in the 1870's by Robert Koch and by Ferdinand Cohn of techniques capable of culturing the descendants of a single bacterium soon revealed that all earlier bacteriologists, not excepting Pasteur, had invariably worked with *mixed* bacterial cultures. Once a *pure* culture derived from a single bacterium had been secured, the plasticity all but vanished and the culture bred true. Pleomorphism then gave way to *monomorphism*, the counter-doctrine, which held that every bacterial type represents an immutable species unto its own: the descendants of any given bacterium forever manifest one and only one morphology and physiology. Monomorphism held sway for the next fifty years, although during all of this time many examples of bacterial variation were reported in which a new bacterial type appeared to have arisen within a pure culture supposedly descended from a single parent cell. The adherents of monomorphism at first attributed these observations to contamination, or other faulty laboratory manipulations. But because such sudden or gradual variations in bacterial properties were found more and more often by bacteriologists working in research laboratories, as well as in fermentation industry and medicine, it was finally realized that even pure bacterial cultures could undergo variation of their properties. For instance, it was often noted that some particular type of pathogenic bacterium that grows well on one species of animal but grows poorly on another can be "adapted" to better growth on that second host species by repeated "passages" through it. Moreover, it was found that bacteria initially sensitive to being killed by some drug may develop resistance to that drug. Many of these changes gave the appearance of being reversible, in that the original character of the bacteria could be restored by "readapting" them to their erstwhile conditions. A particularly clearly analyzed case of bacterial variation, and one that will be of importance in the next chapter, concerned the morphology of the colonies formed by a pure strain of the pneumococcus (*Streptococcus pneumoniae*) on an agar plate: although nearly all of the bacteria isolated from a normal, or *smooth*, pneumococcus colony (one that has a smooth surface and glistens) form smooth colonies upon being replated, a very few form an abnormal, or *rough*, colony (i.e., one that has an irregular surface and is wrinkled and dull) (Figure 6-1). In contrast, nearly all the pneumococci picked from an abnormal rough colony form rough colonies, but a very few give rise to the normal smooth colony. This interconversion between smooth and rough colony formers was called *dissociation* by P. de Kruif in 1921. To reconcile the existence of dissociation with monomorphism, the *cyclogenic* or *ontogenic* theory was advanced in the late 1920's, which envisioned that smooth and rough forms of the pneumococcus represent different phases of its life-cycle. This notion was drawn from the known "life-cycles" of some protozoa, in which

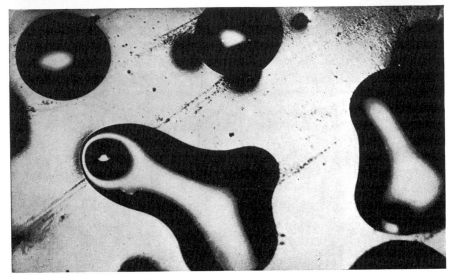

FIGURE 6-1.   Abnormal, or *rough* (above), and normal, or *smooth*, colonies of
Streptococcus pneumoniae (pneumococcus). [Photograph by Harriett Ephrussi-Taylor.]

one and the same organism does assume radically different forms depending
on its particular environment.

Possibly the first person to recognize that bacterial variation might reflect
the occurrence of *gene mutation* was M. W. Beijerinck. He proposed this
idea in the first decade of this century, soon after de Vries (Mendel's redis-
coverer and Beijerinck's colleague) had coined the term mutation to describe
the appearance of hereditary sports in higher forms. (Recent historical
research has revealed that Beijerinck owned a reprint of Mendel's paper and
that he may have drawn de Vries' attention to it.) But whereas the genetics
of higher forms took a tremendous upswing during the first four decades
of this century, the genetics of bacteria remained in the doldrums all the
while. During those years there seemed to be few people capable of, or
interested in, doing with bacteria the kind of incisive quantitative genetic
experiments by means of which the mechanics of inheritance were then being

worked out for eukaryotes. Thus it came to pass that after biologists had fully appreciated for several decades the role of gene mutations as the evolutionary source of diversity of all living forms, mutation and Darwinian selection still were not generally accepted as being at the root of bacterial variation. As a testimonial to this situation it may be noted that William Bulloch's definitive *History of Bacteriology* fails to mention any genetic concepts, especially mutation, even though it was written in 1938, when classical genetics was reaching its apogee.

An important reason for the long gestation period of bacterial genetics was that two features made its study more difficult conceptually than the study of the genetics of multicellular animals and plants. First, whereas it was the character differences between *individual* organisms which were usually examined in genetic experimentation on higher forms, the character differences observed in bacterial variation generally pertain to *populations*. Thus, such properties as pathogenic virulence or fermentative capacities of a bacterium are seen only as behavioral manifestations of a whole culture. Even such relatively clear-cut characters as rough and smooth colony formation are, in the last analysis, aspects presented by the million or so bacteria that make up a macroscopic colony rather than by the founder bacterium from which the members of the colony descended. Second, there exist, in fact, two radically different ways in which the character of a "pure" bacterial culture can change, of which only one actually derives from gene mutation. The other, nonhereditary process, for which the term *adaptation* is reserved nowadays, involves a direct response of *all* the members of a bacterial culture to a change in their environment. Such *adaptive* changes are merely temporary, so that the culture regains its initial character once the environment has returned to its initial condition. The hereditary process of gene mutation, however, brings about *permanent* changes, which persist even after the agent or condition that evoked them has vanished. These hereditary changes generally occur in only a small minority of the bacterial culture—namely, in mutant bacteria that had been *selected* because their fitness for the new environment exceeds that of their unmutated bacterial sib. Though bacterial genetics could not get under way until the distinction between adaptation and mutation was finally sorted out in the 1940's, it is now known that this distinction had been based on false criteria. Meanwhile, there have been discovered examples of nonhereditary, and hence adaptive, changes, which are nevertheless permanent, and examples of hereditary changes that can be induced by environmental conditions in nearly all of the bacteria of a population. Indeed, it is now clear that, vitally necessary though it was to make that distinction, it would have been *impossible* to establish valid operational criteria for distinguishing adaptative from hereditary changes before the molecular basis of heredity was finally understood. Small wonder that bacterial genetics had such a hard time getting under way.

ORIGIN OF PHAGE-RESISTANT VARIANTS

Just as the birth of genetics is considered to have taken place in 1865 upon
the appearance of Mendel's paper, so the birth of bacterial genetics can be
dated 1943, when S. E. Luria and Delbrück (Figure 6-2) published a paper
entitled "Mutations of Bacteria from Virus Resistance to Virus Sensitivity."
Luria and Delbrück were by no means the first to study bacterial mutation

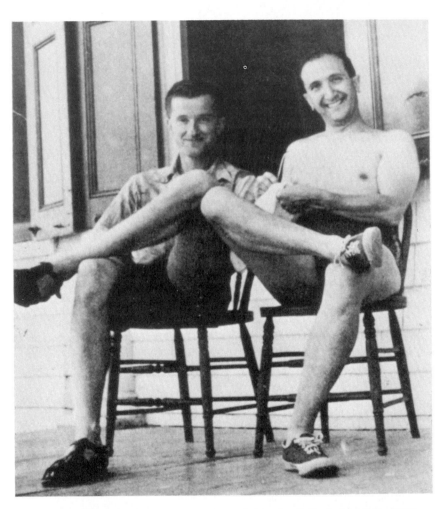

FIGURE 6-2.   Max Delbrück (left) (b. 1906) and Salvador E. Luria (right) (b. 1912)
at Cold Spring Harbor. [From J. Cairns, G. S. Stent, and J. D. Watson (eds.),
Phage and the Origins of Molecular Biology. Cold Spring Harbor Laboratory of
Quantitative Biology, New York, 1966.]

(as was seen in Chapter 5, Massini's work with *E. coli mutabile* preceded theirs by 35 years), no more than Mendel was the first to cross plants for the study of heredity. But with their paper, Luria and Delbrück did for bacterial genetics what Mendel had done for general genetics—namely, to show for the first time what kind of experimental arrangements, what kind of data treatment, and, above all, what kind of sophistication is required for obtaining meaningful and unambiguous results.

Before considering the experiments presented in that classic paper, it is necessary to make a preliminary, brief mention of *bacteriophages*, or *phages*, as they are called for short nowadays. Phages are subcellular living parasites that infect, multiply within, and kill bacteria. They are to be introduced in depth in Chapter 11 and will loom large in all subsequent chapters. For the present purpose it must suffice, therefore, to describe just one particular type of phage, namely the T1 phage active on *E. coli*, with which Luria and Delbrück were working. Figure 6-3 shows an electron micrograph of T1 phage particles. The tiny T1 phage, which is too small to be seen in ordinary microscopes, has a head and a tail and occupies about one-thousandth the volume of its *E. coli* host. Upon collision of such a T1 phage particle with an *E. coli* cell, the particle is fixed to the cell surface by interacting with *T1-phage receptor spots*, which form part of the *E. coli* cell wall. The interaction of a phage particle with its phage receptor spot has the same high degree of specificity as the interaction of the active site of an enzyme with its substrate. Once fixed to the receptor spot, the T1 phage particle invades the *E. coli* cell and destroys it. Let us now imagine an experiment in which about 10^5 *E. coli* cells are spread on the surface of a nutrient agar plate containing 10^{10} T1 phage particles. Upon incubation of that plate, the agar

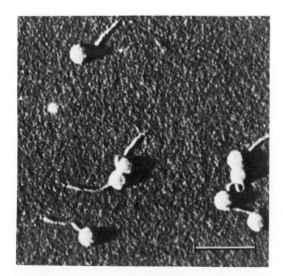

FIGURE 6-3. Electron micrograph of the bacteriophage T1 active on *E. coli*. The scale bar marked on the photograph represents 200 mμ. [From "Morphology of the Seven T-bacteriophages," by R. C. Williams and D. Fraser. *J. Bacteriol.*, **66**: 458 (1953).]

surface will most likely remain completely blank: not a single *E. coli* colony can be expected to appear on that plate, since every bacterium of the inoculum will be infected, and hence killed, by one or more of the 10^{10} phage particles on the plate.

The point of departure of Luria and Delbrück's paper was the observation that upon spreading about 10^9, rather than only about 10^5, *E. coli* cells upon agar containing an excess of T1 phage particles, the chances are rather good that a few *E. coli* colonies *will* appear on the agar surface. If one of these few surviving colonies is picked, and a sample of the cells making up that colony replated on agar containing T1 phage, it will be found that *all* of these cells grow into colonies. That is to say, *all* of the bacteria in one of the few surviving colonies are *T1-resistant*, or Ton^r. (By way of another genetic nomenclatural digression it must be noted that in the description of bacterial phenotypes pertaining to resistance and sensitivity to antibacterial agents, a capitalized, three-letter symbol such as "Ton" designates the agent in question, and the superscripts r and s designate "resistance" or "sensitivity," respectively.) The T1-resistant bacteria retain their Ton^r character also upon further cultivation in the *absence* of any T1 phage, as can be demonstrated by spreading samples of the Ton^r culture growing in T1-free medium on T1-containing agar. Thus the Ton^r bacteria perpetuate and pass on to their descendants the property of resistance to the phage, in contrast to the T1-sensitivity passed on by the normal Ton^s *E. coli* cells. The physiological basis of the T1 resistance resides in the structure of the bacterial cell wall, in that the cell wall of Ton^r bacteria does not feature the T1 receptor spots, to which the phage particles attach before they infect and kill the Ton^s cell. Hence the T1 phage particles cannot attach to, and *a fortiori*, cannot kill the Ton^r cells. Since the few Ton^r cells isolated by plating the original *E. coli* culture on T1-containing agar have clearly descended from the Ton^s ancestors that make up the bulk of the population, they must represent stable variants of the normal Ton^s type. That is, an element of the bacterium that controls the synthesis of the T1 receptor spots in the cell wall of the normal Ton^s cell has changed in some way in the Ton^r variant so that these receptor spots are no longer formed.

Such instances of the appearance of stable bacterial variants resistant to one or another antibacterial agent had been well-known to bacteriologists for many years when Luria and Delbrück designed an experiment that was to enable them to decide between the following two fundamentally different views of the origin of stable Ton^r variants in cultures of Ton^s *E. coli*:

1. The Ton^r character is *induced* as a consequence of the exposure of the Ton^s bacterial culture to T1 phage. That is, one may imagine that upon being attacked by a T1 phage particle, each Ton^s bacterium has a small but finite chance, say 10^{-7}, of surviving this attack. Having survived, the metabolism of the cell is now so altered that this cell and all of its descendants

henceforth possess the Ton^r character. This view of bacterial variation is strongly reminiscent of the inheritance of acquired characters envisaged in the late eighteenth century by J. B. Lamarck, and since it was this view that was long favored by the general run of bacteriologists to whom genetic thought was foreign, bacteriology remained according to Luria "the last stronghold of Lamarckism."

2. The Ton^r character *pre-exists* in a few cells *before* exposure of the bacterial culture to T1 phage. That is, one may imagine here that a small fraction of the cells of the Ton^s culture, say 10^{-7}, happen to be Ton^r *mutants*. These mutants, or their ancestors, would have arisen spontaneously during growth and division of the Ton^s culture, by sudden *mutation* of the bacterial *ton* gene controlling the enzymes responsible for synthesis of the T1 phage receptor spots. Such Ton^r mutants are then simply *selected* from among the vast majority of their Ton^s sib, because they and only they can survive on agar containing T1 phage. Since this view of bacterial variation accorded very well with the tenets of neo-Darwinism, it held more intrinsic appeal for anyone of genetic bent.

It had occurred to Luria that these two alternative views of the origin of Ton^r bacterial variants lead to different statistical predictions concerning the observed dynamics of their appearance. Thus, according to the first view of resistance induced by exposure of the bacteria to T1 phage, the *fraction* of cells of a growing culture of Ton^s bacteria that appear as Ton^r variants upon plating on an agar containing the phage should be constant at all stages of growth. For here each Ton^s cell is presumed to have a fixed probability of survival and induction of the Ton^r character. But under the second view, that of spontaneous mutation, the fraction of Ton^r variants in the culture should rise progressively with growth. For here the rare occurrence of the Ton$^s \rightarrow$ Ton^r mutation would steadily create new Ton^r mutant lines, or *clones*, which, once present, grow and divide at the same rate as the Ton^s bacterial population at large. These considerations may be readily stated analytically. Let us suppose, in accord with the hypothesis of induced resistance on the one hand, that contact with T1 phage converts each of a total of N Ton^s bacteria in the culture with probability a to the Ton^r form. Then the number n of individuals giving rise to Ton^r colonies on agar containing T1 phage is $n = a\,N$, and the fraction of such individuals in the culture would be constant for all values of N; that is,

$$n/N = a \qquad (6\text{-}1)$$

Let us suppose that, in accord with the hypothesis of gene mutation on the other hand, each Ton^s bacterium mutates spontaneously with probability a per cell per generation to the Ton^r type. After a culture derived from a single Ton^s cell has grown for g generations, the total number of cells will have reached the number $N = 2^g$. Then in the course of this growth, a total number of $a2^g$ spontaneous mutations will have taken place, of which $a2^i$ mutations occurred in the ith generation (where $i < g$); the progeny of bacteria that mutated in the ith generation contribute $a2^i2^{g-i} = a2^g$ mutant cells to the final mutant yield; and hence the mutations having taken place in all g generations contribute $n = ga2^g$ Ton^r mutant cells to the final mutant yield. It follows, therefore, that the fraction of Ton^r mutants in the culture would be

$$n/N = ga2^g/2^g = ga \qquad (6\text{-}2)$$

That is, here the fraction of Ton^r variants in the culture would rise continuously with the number of bacterial generations.

THE FLUCTUATION TEST

When Luria tried to carry out experiments to test whether the fraction of Ton^r variants in an *E. coli* population does or does not rise with growth of a culture started from a small number of Ton^s wild-type cells, he found to his surprise and annoyance that the fraction of Ton^r individuals present at the start of his experiments was subject to such great day-to-day fluctuations that he could not plan the experiments properly. Before long Luria, and Delbrück, whom Luria had consulted about these difficulties, realized that these annoying fluctuations were, in fact, of themselves a reflection of the spontaneous origin of Ton^r mutants. To reconstruct their reasoning, let us consider an experiment in which c identical *E. coli* cultures, each started from a single Ton^s bacterium, are grown for g generations to yield a final number $N = 2^g$ bacteria. The contents of each of the c cultures are then spread on agar containing T1 phage in order to determine the number n of Ton^r bacteria that give rise to a colony on this agar. Let the number of Ton^r bacteria found in the jth culture be n_j, then the *mean* number, $\bar{n}$, of Ton^r bacteria per culture is

$$\bar{n} = \left(\sum_{j=1}^{c} n_j\right)/c \qquad (6\text{-}3)$$

We may now consider the manner of distribution of the actual values n_j about their mean value $\bar{n}$. This distribution might be *narrow*, meaning that the number of Ton^r bacteria per culture is always rather close to the mean value, or it might be *broad*, meaning some cultures contain very many fewer and other cultures very many more Ton^r bacteria than the mean value. A convenient statistical measure of the degree of fluctuation of an observed parameter about its mean value is the *variance* of that parameter, which, in terms of the symbols defined here, is given by the formula

$$\text{variance}_n = \left[\sum_{j=1}^{c} (\bar{n} - n_j)^2 \right] /c \qquad (6\text{-}4)$$

The variance has a small value for a narrow distribution and a large value for a broad one, because the magnitude of the $(\bar{n} - n_j)^2$ difference term grows rapidly with departures from the mean.

Luria and Delbrück concluded that the two alternative views of the origin of Ton^r variants—induced resistance or spontaneous mutation—should lead to very different values of variance$_n$ in the outcome of the experiment designed to determine the number of Ton^r bacteria present in c Ton^s cultures. The reasoning underlying this conclusion is depicted diagrammatically in Figure 6-4. Both parts A and B of this figure present the hypothetical example of four cultures of bacteria, each of which has grown from a single Ton^s founder bacterium to a final number of 16 bacteria, making a grand total of $4 \times 16 = 64$ cells. Upon spreading each culture on agar containing T1 phage, a total of 10 Ton^r colonies is found. Part A of Figure 6-4 presents the situation that is to be expected if the Ton^r character were *induced* in a fraction $a = 10/64 = 0.15$ of the Ton^s bacteria upon contact with T1 phage on the agar surface. Here, the 10 Ton^r colonies should be distributed at *random* over the four agar plates if all Ton^s bacteria are equally eligible for acquiring the Ton^r character upon encountering the phage. Hence the variation in the number n_j about the mean value $\bar{n} = 10/4 = 2.5$ Ton^r cells per culture should be small, or no greater than that characteristic of random sampling processes. According to the statistics of random sampling,

$$\text{variance}_n/\bar{n} = 1$$

For the hypothetical example of part A of this figure, we may reckon that

$$\text{variance}_n/\bar{n} = [(2.5 - 3)^2 + (2.5 - 1)^2 + (2.5 - 5)^2 + (2.5 - 1)^2]/4/2.5 = 1.1$$

or a value close to the theoretical prediction of 1.

Part B of Figure 6-4 presents the situation that is to be expected if the Ton^r character arises by *spontaneous mutation* during the growth of Ton^s bacteria *before* exposure to any T1 phage. In our hypothetical example, a total of two Ton$^s \to$ Ton^r mutations have taken place during the $64 - 4 = 60$ cell divisions that produced the 64 bacteria in the four cultures from the four founder cells, corresponding to a probability $a = 2/60 = 0.033$ Ton^r mutations per cell per generation. One of these two mutations happened to have occurred in the last generation of growth of culture 1, giving rise to two Ton^r mutant bacteria, and the other mutation happened to have occurred during the first generation of growth of culture 3, giving rise finally to eight Ton^r bacteria. Cultures 2 and 4 produced no Ton$^s \to$ Ton^r mutation. These hypothetical results lead to the same total number of 10 Ton^r bacteria as

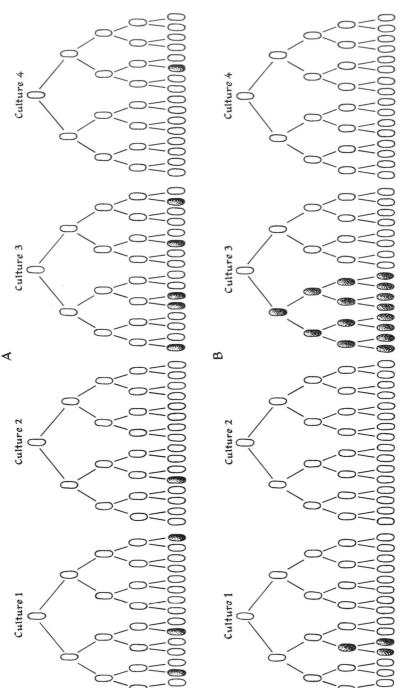

FIGURE 6-4. Schematic representation of the appearance of ten Ton^r mutant *E. coli* in four parallel cultures of Ton^s *E. coli*. (A) Sample distribution to be expected if the Ton^r character were *induced* in 0.15 of the Ton^s bacteria upon their contact with T1 phage. (B) Sample distribution to be expected if the Ton^r character appeared as a consequence of a *spontaneous mutation* with probability 0.033 per Ton^s cell per generation.

in part A of Figure 6-4, and hence to the same mean value $\bar{n} = (2 + 0 + 8 + 0)/$
$4 = 2.5$ Ton^r bacteria per culture. Here, however,

$$\text{variance}_n/\bar{n} = [(2.5 - 2)^2 + (2.5 - 0)^2 + (2.5 - 8)^2 + (2.5 - 0)^2]/4/2.5 = 4.3$$

Hence variance$_n/\bar{n}$ is much greater for spontaneous mutation than for induced phage resistance. Moreover, the difference in the expected variance$_n/\bar{n}$ ratio between induced resistance and spontaneous mutation grows progressively greater the more bacterial multiplication has occurred in the parallel cultures and the more opportunity there is for the appearance of large mutant subclones in a few of these cultures. In other words, the occurrence of spontaneous mutations, and hence of clones of mutant cells at various stages of growth of individual cultures, would result in much greater fluctuations in the observed number of Ton^r colonies per culture than the random induction of resistance following contact with the phage on the assay plate.

In order to examine whether the appearance of rare Ton^r variants in a culture of Ton^s *E. coli* manifests the fluctuations predicted by the spontaneous mutation hypothesis, Luria and Delbrück carried out the following experiment. Twenty 0.2-ml volumes ("individual cultures") and one 10-ml volume ("bulk culture") of nutrient medium were inoculated with about 10^3 *E. coli* per ml, and incubated until the bacteria in all cultures had grown for about $g = 17$ generations and thus reached a density of about 10^8 cells per ml. The content of each 0.2-ml individual culture was then spread on a single nutrient agar plate containing a very high concentration of T1 phage. At the same time ten 0.2-ml samples of the 10-ml bulk culture were similarly spread on agar containing T1 phage. After incubating these 30 plates overnight the number of Ton^r colonies shown in Table 6-1 were observed from these two sets of platings. It can be seen, first of all, that a mean number of 16.7 Ton^r colonies appeared per 0.2-ml sample of the bulk culture, indicating a mean frequency of $n/N = 16.7/(0.2 \times 10^8) = 8 \times 10^{-7}$ Ton^r variants per cell in that culture. Secondly, it is evident that the number of Ton^r colonies per plate arising from each 0.2-ml sample of the bulk culture is not subject to great variation, the ratio of variance to mean being close to one. This result is, of course, compatible with either induced resistance or spontaneous mutation hypothesis, and means only that the Ton^r bacteria found represent a random sample of the bulk culture. A very different situation is evident, however, for the number of Ton^r colonies per plate arising from each of the individual cultures. Although here the mean number of 11.3 Ton^r colonies per plate is similar to that found for the samples of the bulk culture, there exist extremely wide fluctuations about that mean of the individual numbers, so that the ratio of variance to mean has the value 61. This finding thus argued strongly in favor of the spontaneous mutation hypothesis, where such great fluctuations are to be expected, and argues strongly against

TABLE 6-1.   The Fluctuation Test of the Spontaneous
Origin of T1 Phage-Resistant *E. coli* Mutants

Individual cultures		Samples from bulk culture	
Culture no.	Ton^r bacteria found	Sample no.	Ton^r bacteria found
1	1	1	14
2	0	2	15
3	3	3	13
4	0	4	21
5	0	5	15
6	5	6	14
7	0	7	26
8	5	8	16
9	0	9	20
10	6	10	13
11	107		
12	0		
13	0		
14	0		
15	1		
16	0		
17	0		
18	64		
19	0		
20	35		
Mean ($\bar{n}$)	11.3		16.7
Variance$_n$	694		15
Variance$_n/\bar{n}$	61		0.9

From S. E. Luria and M. Delbrück, *Genetics*, **28**, 491 (1943).

the induced resistance hypothesis, where the variations should be random, and hence small.

Having thus demonstrated the spontaneous nature of the $Ton^s \rightarrow Ton^r$ mutation, Luria and Delbrück showed how the data of the fluctuation test can be used to determine an accurate value of a, the spontaneous *mutation rate* per cell per generation. Since $(N - N_0)$ cell generations occur when a culture increases from an initial number N_0 to a final number N of bacteria, it follows that the average number of mutations that has taken place per individual culture of the fluctuation test, where $N \gg N_0$ is aN. Since the occurrence of these mutations can be expected to be distributed at random over the c cultures, it follows from the Poisson law* that the fraction of p_0 of cultures in which *no* mutations happened to have occurred at all (i.e., in which no Ton^r mutants appeared on the assay plate) is

$$p_0 = e^{-aN} \tag{6-5}$$

Equation (6-5) may be rewritten as

$$a = (-\ln p_0)/N \tag{6-6}$$

Inspection of the individual culture data of Table 6-1 reveals that 11 of the 20 cultures contained no Ton^r mutant bacteria, i.e., did not produce the $Ton^s \rightarrow Ton^r$ mutation. Hence $p_0 = 11/20 = 0.55$, and since $N = 0.2 \times 10^8$ bacteria, it follows from equation (6-6) that $a = (-\ln 0.55)/0.2 \times 10^8 = 3 \times 10^{-8}$ mutations per cell per generation.

Luria and Delbrück's statistical proof of the spontaneous nature of bacterial mutation and measurement of mutation rate represents not only the beginning of bacterial genetics, but also the first of several fortunate choices of experimental material that were to aid the further development of this field. Luria and Delbrück's finding of spontaneous mutation to phage resistance turns out to have depended on their use of the T1 phage, a phage that, as will be seen later, is a "virulent" bacteriophage. Had they happened to pick one of the phage types that came to be known as "temperate" bacteriophages, Luria and Delbrück would have had to conclude that the bacterial variants acquire their resistant character by contact with the antibacterial agent on the test agar plate, and thus would have contributed, willy nilly, to the fortification of the last stronghold of Lamarckism.

---

* The Poisson law is a statistical distribution which states that

$$p_r = \frac{x^r}{r!} e^{-x},$$

where p_r is the fraction of a large number of boxes that will contain r objects each if an average of x objects per box is distributed *at random* over the ensemble of boxes.

THE RESPREADING TEST

A few years after the publication of the fluctuation test, H. Newcombe devised a second proof of the spontaneous origin of Ton^r mutants. The basic idea of Newcombe's proof is essentially the same as that of the fluctuation test, but it is simpler, both experimentally and conceptually, than its ideological progenitor. Newcombe spread about 5×10^4 Ton^s *E. coli* on each of twelve agar plates. These plates were incubated for five hours, during which time every bacterium produced some twelve generations of descendants and thus multiplied to form a microcolony of about 5,000 cells. At this point, the microcolonies present on six of the twelve plates were respread in order to distribute their bacteria uniformly over the agar surface. An excess of T1 phage was then sprayed as a fine mist onto both the respread and the undisturbed plates, and the number of Ton^r colonies appearing on the two sets of plates was compared after overnight incubation. Some simple considerations show that the two hypotheses, induced resistance and spontaneous mutation, lead to different predictions for the relative number of Ton^r colonies that should be found on undisturbed and respread plates. Since the theory of induced resistance envisages a small, constant probability per Ton^s cell of resistance induction upon contact with the phage, the final number n of Ton^r colonies appearing on any plate should depend only on N, the total number of cells present at the time that the phage was added. And since N is the same for undisturbed and respread plates, both kinds of plates should develop *equal* numbers of Ton^r colonies. The theory of spontaneous mutation, however, envisages the occurrence of clones of Ton^r mutants before addition of the phage. On the undisturbed plate each such Ton^r clone, no matter how many members it comprises at the time the phage is added, will give rise to one and only one Ton^r colony, since all the Ton^r clone members are localized in the same microcolony in which the clone arose. On the respread plate, in contrast, the Ton^r clone members have been separated, and each Ton^r mutant cell present on the plate at the time the phage is added will give rise to a separate resistant colony. Hence, respread plates should contain many more Ton^r colonies than the undisturbed plates if the Ton^r character is the consequence of spontaneous mutation.

Table 6-2 presents a result of Newcombe's test. It is immediately apparent here that the respread plates contain many more Ton^r colonies than the undisturbed plates, in full accord with the spontaneous mutation hypothesis, and in strong disagreement with the induced resistance hypothesis. This result is, furthermore, in good quantitative agreement with the statistical expectations for spontaneous mutation. First of all, the data of Table 6-2 permit another accurate estimate of a, the spontaneous mutation rate. Each Ton^r colony appearing on the *undisturbed* plates of the respreading experiment evidently represents the occurrence of *one* Ton$^s \rightarrow$ Ton^r mutation during the growth of the bacteria before spraying with T1 phage. Since the total number of bacterial divisions that took place on all six plates in the 5-hour incubation period before spraying was $6 \times (2.6 \times 10^8 - 5.1 \times 10^4)$ and the total number of mutations was 28, the mutation rate can be estimated as

$$a = \frac{28}{6(2.6 \times 10^8 - 5.1 \times 10^4)} = 1.8 \times 10^{-8} \text{ mutations per cell per generation.}$$

In view of the differences in experimental procedure employed, this value of the mutation rate can be considered to be in good agreement with that previously inferred from the results of the Luria-Delbrück fluctuation test. Moreover, since the total number of mutations is aN, it follows from equation (6-2) that the ratio of the number of *mutants* to the number of *mutations* that gave rise to these mutants ought to be equal to the number of generations g. As can be seen in Table 6-2, the ratio of resistant colonies found on respread and undisturbed plates, and hence the ratio of mutants to mutations, is $353/28 = 12.6$, in good agreement with the $g = 12.3$ generations during which the average microcolony grew to a clone of 5100 cells.

TABLE 6-2.   The Respreading Test of the Spontaneous Origin of T1 Phage-Resistant *E. coli* Mutants

Initial number of bacteria per plate	5.1×10^4
Hours of incubation before spraying with T1 phage	5
Number of bacteria per plate at time of spraying with T1 phage	2.6×10^8
Number of cells in average microcolony	5.1×10^3

Ton^r colonies on parallel plates

undisturbed	respread
5	194
3	14
4	16
8	13
2	4
6	112
28	353

From H. B. Newcombe, *Nature*, **164**, 150 (1949).

THE REPLICA-PLATING TEST

Though simpler conceptually than the fluctuation test, the interpretation of the respreading test still depends on statistical arguments. But in 1952 Joshua and Esther Lederberg finally demonstrated the spontaneous nature of bacterial mutation by a method that is independent of any statistical considerations. The Lederbergs made the very simple but technologically extremely useful discovery that it is possible to make a *replica* of bacterial colonies grown on an agar plate (Figure 6-5). For this purpose, the agar surface of the *master* plate containing the colonies is touched with a piece of velvet mounted on a circular wooden block whose diameter is equal to that of the petri plate. The velvet is then used to imprint a virgin agar surface, and

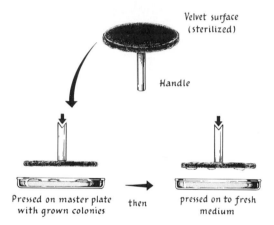

FIGURE 6-5. The replica plating technique. Diagram showing the method of mounting a sterile piece of velvet on a cylindrical wooden block and of touching the velvet with the master plate containing bacterial colonies. After lifting the master plate off the velvet surface, a sterile replica agar plate is inverted and allowed to rest on the velvet surface. [After W. Hayes, *The Genetics of Bacteria and their Viruses.* Wiley, New York, 1968.]

incubation of the second plate thus produces a *replica* of the original: on the replica plate a colony grows at every spot where there had been a colony on the master plate. To demonstrate the spontaneous origin of Ton^r mutants, the Lederbergs incubated for several hours a master plate seeded with about 10^7 Ton^s *E. coli* cells to allow each of the bacteria of the inoculum to form a clone of descendants. This master plate was then touched with the velvet, and imprints were made onto several replica plates containing a high concentration of T1 phage. Upon incubation there appeared a few Ton^r colonies on each replica plate, *the spatial pattern of distribution of these colonies being the same on every replica plate* (Figure 6-6). Now if the Ton^r character of these colonies were induced in the bacteria imprinted onto the replica plates only after their contact with the phage, then this random process could not produce the topographic correspondence of the Ton^r colonies on parallel replica plates. Thus the finding that a Ton^r colony appears in exactly the corresponding site on several replica plates can mean only that there existed a clone of Ton^r mutants on the master plate from which the velvet imprint

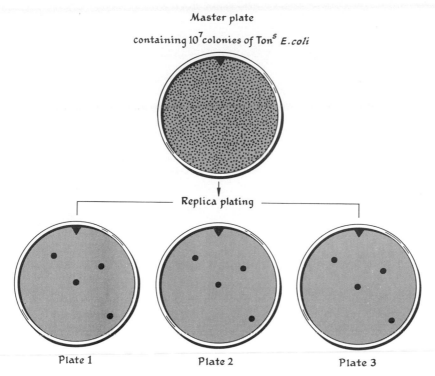

Master plate
containing 10^7 colonies of Ton^s E.coli

Replica plating

Plate 1   Plate 2   Plate 3

Series of replica plates containing high concentration
of T1 phage and four Ton^r colonies

FIGURE 6-6.   Proof of the spontaneous nature of the Ton^s → Ton^r mutation by replica plating. The fact that the spatial pattern of distribution of the four Ton^r colonies is the same on each of the three replica plates shows that there exist four Ton^r colonies on the master plate that have never been exposed to T1 phage and which therefore must have arisen spontaneously from Ton^s bacteria.

had been taken. Since that pre-existing clone had never seen any T1 phage, the origin of its Ton^r character must clearly be spontaneous.

Whereas the fluctuation test and the respreading test derive their present significance largely from having proved in their time a fundamental point, the replica plating test has remained one of the most important techniques in the practice of bacterial genetics. The replica procedure has found wide application in the isolation of auxotrophic mutants, such as those described in Chapter 5. In order to isolate auxotrophs by means of this method a series of nutrient-broth, or *complete*-agar, master plates is prepared, each plate carrying several hundred colonies of prototrophic bacteria. The master plates are then replicated onto *minimal*-agar plates. Any rare auxotrophic

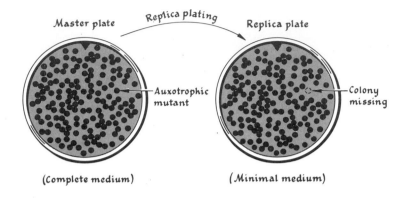

Master plate — Replica plating — Replica plate

Auxotrophic mutant

Colony missing

(Complete medium)   (Minimal medium)

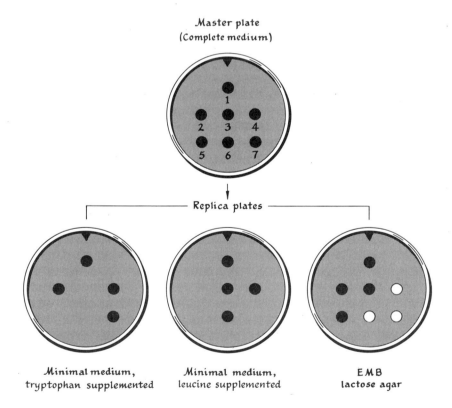

Master plate
(Complete medium)

Replica plates

Minimal medium,
tryptophan supplemented

Minimal medium,
leucine supplemented

EMB
lactose agar

mutant colony present on the complete-agar master plate will then make its presence known by failing to form a colony on the minimal-agar replica plate, in contrast to the prototrophic colonies, which all reappear on the replica plate (Figure 6-7). The auxotrophic mutant clone thus located on the master plate can then be picked and tested for its growth requirements. The replica plating method has been of even greater utility in the testing of these growth requirements, as well as of other phenotypic characters. Suppose it is desired to establish the character of a set of different *E. coli* mutants. A complete-agar master plate is prepared, and on its surface a small drop of each mutant culture to be tested is placed at a certain position with reference to a coordinate grid. Upon incubation, the master plate will display individual patches of bacterial growth in a regular array. The master plate is then replicated repeatedly onto a series of minimal-agar plates containing different putative growth factors, or onto EMB sugar-fermentation indicator plates, or onto complete-medium plates containing an antibacterial agent such as T1 phage. By noting on which replica plate any given patch of the master plate has or has not grown or turned the EMB indicator red or white, the phenotype of the unknown bacterial mutant present in that patch can be scored at a glance (Figure 6-7).

CONTINUOUS CULTURE

The difficulties that Luria had encountered in his initial attempts to ascertain whether the fraction of Ton^r mutants in a growing culture of Ton^s *E. coli*

FIGURE 6-7.   Use of replica plating in the isolation and classification of auxotrophs. (Top) Several hundred *E. coli* prototrophs are spread on a complete medium agar. The resulting colonies are replica plated on minimal agar. Comparison of master and replica plates reveals that one master plate colony failed to appear on the replica plate. The clone representing that auxotroph is then picked from the master plate colony. (Bottom) Samples from seven different *E. coli* cultures are spread on parts of the surface of a complete-medium agar. After incubation, the resulting patches of bacterial growth are replica-plated onto tryptophan-supplemented or leucine-supplemented minimal agar, or onto EMB lactose indicator agar (on which Lac$^+$ bacteria form red and Lac$^-$ bacteria form white colonies).

The growth pattern shown on the replica plates indicates the following phenotypic characters for the seven cultures:

Patch	Trp	Leu	Lac
1	+	+	+
2	−	+	+
3	+	−	+
4	+	+	−
5	?	?	+
6	+	−	−
7	−	+	−

remains constant, according to the prediction of equation (6-1), or increases continuously with the number of generations produced, according to the prediction of equation (6-2), were eventually overcome by growing the *same* culture of bacteria continuously for hundreds of generations and sampling its mutant content from time to time. At first, such continuous culture was achieved by periodic transfer of samples of the culture to fresh growth medium. But in 1949, a device was invented independently by Jacques Monod and by Novick and Szilard which achieves an automatic continuous culture. This device, the *chemostat* (Figure 6-8), consists of a culture vessel into which fresh medium drips continuously from a reservoir. The vessel has an overflow spout from which excess culture fluid drips out into a collecting vessel, so that the total fluid volume of the culture remains constant.

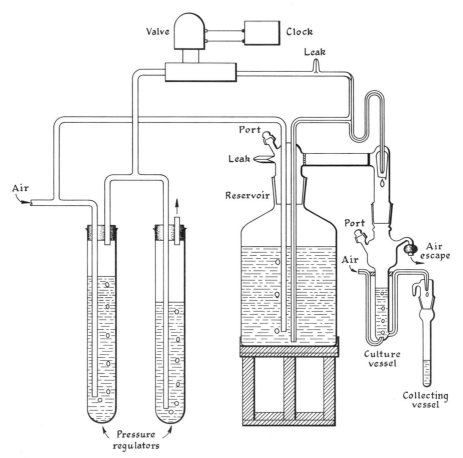

FIGURE 6-8. The chemostat device for continuous bacterial culture. [After A. Novick and L. Szilard, *Cold Spring Harbor Symp. Quant. Biol.* **16**, 338 (1951).]

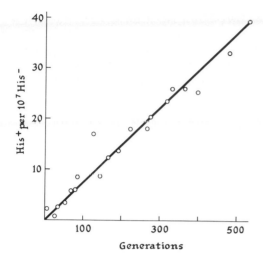

FIGURE 6-9. The accumulation of spontaneous His⁺ mutants during continuous culture of His⁻ *E. coli* in complete medium. The ordinate indicates the number of bacteria per 10^7 cells of culture that are capable of forming a colony on histidine-free minimal agar. The abscissa shows the number of generations of growth that have taken place. [After F. J. Ryan and L. K. Wainwright, *J. Gen. Microbiol.*, **11**, 376 (1954).]

In this way the bacterial culture can be maintained and observed under constant physiological conditions of growth for hundreds, or thousands, of generation periods.

Figure 6-9 presents the result of a continuous culture experiment in which F. Ryan grew a culture of a His⁻ auxotroph of *E. coli* for more than 200 generations in a medium supplemented with the required histidine growth factor. Samples of the culture were spread periodically on minimal, *histidine-free* agar in order to test how many His⁺ prototrophic mutants generated by His⁻ → His⁺ mutations are present after various generation times. As can be seen from these data, the fraction of His⁺ prototrophs in the culture, or n/N, does rise proportionally with the number of generations, g, which is in agreement with equation (6-2) and the spontaneous occurrence of mutations. Equation (6-2) may be restated as

$$a = n/Ng \qquad (6\text{-}2)$$

and hence it follows that the slope of the straight line connecting the experimental points of Figure 6-9 gives the probability a of the His⁻ → His⁺ mutation per cell per generation, or the mutation rate. That slope is 7.5×10^{-9} His⁺ mutations per His⁻ cell per generation, or a rate whose order of magnitude is comparable to that of the Ton^s → Ton^r mutation.

DRUG-RESISTANT MUTANTS

As soon as drug treatment of bacterial diseases came into use at the turn of this century it was noted that exposure of a bacterial culture to a drug often results in the conversion of the culture from a drug-sensitive to a drug-resistant form. That is, the bacteria in the resistant culture are no longer affected by an amount of the drug that would have killed the original, sensitive bacteria. With the advent of the widespread use of antibiotics, first of sulfanilamide in the 1930's and then of penicillin and streptomycin in the 1940's, the phenomenon of bacterial development of drug resistance became evermore commonplace and began to be (and still remains) a problem of the utmost practical concern. As far as an explanation of the origin of bacterial drug resistance was concerned, the general view had been that the bacteria acquire their resistance only upon exposure to the drug. One of the most eminent champions of this view was Cyril Hinshelwood, who developed a nongenetic theory of drug adaptation in his book *The Chemical Kinetics of the Bacterial Cell.* Hinshelwood envisaged that in the few resistant bacteria that survive exposure to a drug, the normal steady-state balance of metabolic reactions is shifted under the influence of the drug to a new steady-state balance that is less susceptible to interference by the drug. Hinshelwood's book appeared in 1946, or three years after Luria and Delbrück had already demonstrated the spontaneous origin of phage-resistant bacterial mutants. It would have seemed natural, therefore, to entertain the idea that the change from drug sensitivity to drug resistance also arises from spontaneous mutation of a small fraction of the drug-sensitive bacterial population. In the presence of the drug, there would be a strong selection for such mutants, because *they* can grow under these conditions, whereas all the sensitive wild types are killed. But Hinshelwood was not impressed by the fluctuation test and published several specious criticisms of its interpretations; he was so convinced of the justice of his kinetic theory of adaptation that he wrote, as late as 1953, that "adaptive changes should so easily occur in ways generally similar to those suggested by the [kinetic] models which have been studied that if they do not, then it is hard to evade the question why not." Hinshelwood's esteem as a chemical kineticist lent authority to his views, and in consequence probably delayed the development of bacterial genetics in his own country, Great Britain, by some years.

At the base of Hinshelwood's argument was one fact that did seem to argue in favor of the theory of induced resistance rather than of the theory of spontaneous mutation: the development of resistance to many drugs does resemble a "training" process, in that the resistant culture resists only that concentration of the drug to which it had been previously exposed, but remains sensitive to higher concentrations of the same drug.

Penicillin induces such apparent "training." Addition of 500 μg/ml of this antibiotic (Figure 6-10) to a culture of *E. coli* will cause the death of almost all of the bacteria. As was noted in Chapter 5, penicillin stops the formation of an essential component of the bacterial cell wall without interfering with the general biosynthetic processes. Thus, in the presence of

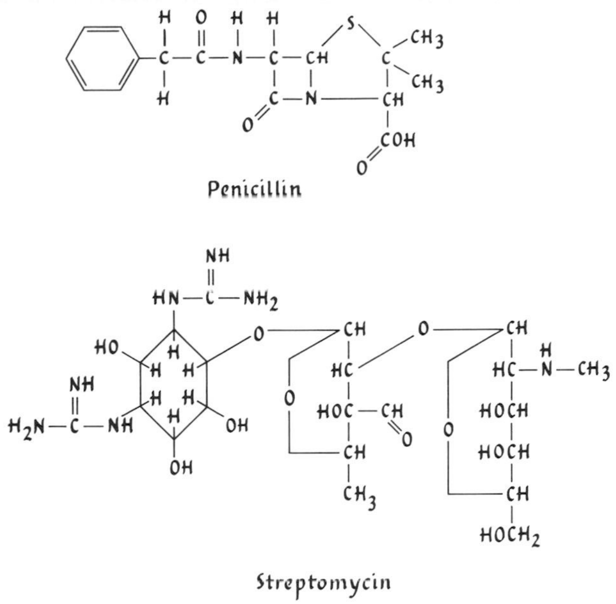

Penicillin

Streptomycin

FIGURE 6-10. The chemical formulas of the two antibiotics penicillin and streptomycin.

penicillin, bacteria burst because they outgrow their integuments. If samples of an *E. coli* culture are spread on a series of agar plates containing various concentrations of penicillin, fewer and fewer colonies survive higher and higher drug concentrations, until at 500 μg/ml penicillin less than 1 in 10^8 bacteria plated survive to form a colony. If one of the last few surviving colonies from the 500 μg/ml penicillin plate is picked, and the plating experiment repeated, it is found that the bacteria of this "first-step isolate" are more resistant to penicillin than the parent culture. Many more survivors are now found at any given penicillin concentration, though 1250 μg/ml penicillin leaves viable less than 1 in 10^8 of the bacteria of the "first-step isolate." If one of the last few surviving colonies from the 1250 μg/ml penicillin plate is picked and the plating experiment repeated once more, it is found that the bacteria of the "second-step isolate" have attained an even higher degree of penicillin resistance. This picking and replating may be repeated for a few more steps, each step resulting in a higher degree of resistance, until a bacterial population develops that is completely unaffected by any penicillin concentration to which it is exposed.

Even though this stepwise development of penicillin-resistance suggested the operation of a penicillin-induced "training" process, it was presently shown by means of fluctuation, respreading, and replica-plating tests that penicillin resistance does, after all, arise by spontaneous mutation, just as does resistance to T1 phage. And in 1948, M. L. Demerec explained how—in view of the spontaneous origin of its mutant character, and the failure of penicillin to play any role in inducing penicillin resistance—the resistant mutant bacterium can possibly "remember" the penicillin concentration at which it was selected. Demerec showed that the penicillin-resistance character, or Pen^r, is controlled by several equipotent bacterial *pen* genes, each of which can mutate spontaneously with a frequency of about 10^{-7} per cell per generation to confer "one-step" resistance on the bacterium. If only one *pen* gene has mutated, the bacterium manifests the lowest grade of Pen^r character. As soon as two *pen* genes have mutated, the physiological effect of their joint presence stands in a geometric (rather than arithmetic) relation to the effect produced by either *pen* mutant gene alone, and the bacterium attains a much higher, "two-step" resistance. Mutation of a third *pen* gene results in a further multiplication of the Pen^r level, and so on. Now the chance that in any one bacterium mutations happen in two *pen* genes simultaneously is only $10^{-7} \times 10^{-7} = 10^{-14}$ per cell per generation; that is, in the bacterial population, first-step resistance mutants are 10^7 times more abundant than second-step resistance mutants. Hence at any penicillin concentration that allows growth of first-step mutants, chance selection of any second-step mutant is extremely improbable. But once a first-step mutant *has* been selected and allowed to grow into a large population, second-step mutants become as frequent among its descendants as first-step mutants were among the original sensitive bacteria present.

The physiological basis of the Pen^r character is not yet fully understood. It is known, however, that mutation of some of the relevant *pen* genes confers penicillin resistance on the bacterium by preventing the uptake of penicillin into the cell, by altering the enzymatic ensemble that governs cell-wall synthesis, and by causing a chemical modification of penicillin itself, so as to detoxify it.

Demerec also studied the mechanism by which bacterial mutants become resistant to *streptomycin* (Figure 6-10). He could show by means of the fluctuation test that streptomycin-resistant, or Str^r, mutants of *E. coli* arise spontaneously. In contrast to penicillin, however, streptomycin does not produce the stepwise "training" phenomenon. Thus, if a streptomycin sensitive, or Str^s, *E. coli* culture is spread on an agar plate containing 100 μg/ml streptomycin, only about 1 in 10^9 of the bacteria can form colonies. If, now, several of the few surviving colonies are picked and the bacteria of these "first-step" isolates are replated on a series of agar plates containing increasing amounts of streptomycin, it is found that there will appear among the bacteria at least three different classes of Str^r mutants: (a) mutants with *slight* resistance that can withstand only the original concentration of 100 μg/ml streptomycin, but are killed by higher concentrations; (b) mutants with *intermediate* resistance that can withstand up to 500 μg/ml streptomycin, and (c) mutants with *high* resistance that can withstand even the highest streptomycin concentrations put on the plate. Since all three types of mutants had appeared in roughly equal frequency in a single selective step at 100 μg/ml streptomycin, Demerec concluded that the Str^r character is controlled by three or more bacterial *str* genes of *unequal potency*, each gene capable of mutating with roughly equal mutation frequency of 10^{-10} mutations per cell per generation to produce that level of resistance (slight, intermediate, or high) which it can confer on the bacterium. Later studies showed that streptomycin kills Str^s bacteria by interfering with the mechanics of bacterial protein synthesis and that Str^r mutants owe their resistance to a genetically controlled alteration of structural and catalytic members of the cellular engine for protein synthesis.

But even these earliest observations of Demerec taught important lessons for the clinical application of antibiotics such as penicillin and streptomycin. If a patient afflicted with a bacterial infection is to be treated with pencillin, the physician must employ the highest possible concentration of the drug at the very outset; for in that way the chance is maximized that all the pathogenic bacteria are killed and the chance that high-level resistant mutants are selected is minimized. If, in contrast, the patient is first given a low dose of antibiotic, and then, if he does not respond to this treatment, a somewhat higher dose is administered, and if *that* does not work, an even higher dose is given, one risks by this escalation of drug levels the selection of successively higher resistance mutants. The avoidance of such stepwise programs is less important in streptomycin treatment, since mutants with high resistance to that drug are selected even in the presence of low concentrations, and little can be done to avoid their appearance. Fortunately, most of the mutations to antibiotic resistance confer resistance against only one kind of antibiotic and leave the bacterium sensitive to many other kinds. Hence, the surest way to avoid clinical appearance of drug-resistant strains is to administer simultaneously two or more kinds of antibiotics, since the chance that a multiple-resistance mutant will appear is given by the product of the individual mutation rates. Alas, a new phenomenon of bacterial acquisition of antibiotic resistance has made its appearance among bacteria in recent years that defeats even this stratagem. Consideration of this ominous development, apparently brought about by the indiscriminate broadcast of antibiotics in the 1950's, must await Chapter 10.

FORWARD, REVERSE, AND SUPPRESSOR MUTATIONS

Comparison of the rates of various spontaneous mutations in *E. coli* reveals that these rates span a more than ten-thousandfold range. Thus the frequency of the mutation Str^s → Str^r was seen in the preceding section to have the very low value of about 10^{-10} per cell per generation, whereas the frequencies of His$^-$ → His$^+$ and Ton^s → Ton^r were seen to have the higher values of 7.5×10^{-9} and about 2×10^{-8} per cell per generation. Other studies not described here have shown that the frequency of the mutation His$^+$ → His$^-$ is higher yet at 2×10^{-6} per cell per generation, and that the mutations Lac$^+$ → Lac$^-$ and Trp$^+$ → Trp$^-$ occur at similarly high frequencies. We may now attempt a preliminary explanation of the meaning of these great differences in mutation rates from the one-gene-one-polypeptide-primary-structure viewpoint set forth in the preceding chapter. For this purpose, it is convenient to categorize the nature of the changes in the cellular protein ensemble that one might imagine are responsible for changes in cell phenotype.

LOSS OF FUNCTION

The most readily explainable type of mutation is exemplified by Lac$^+$ → Lac$^-$, or Trp$^+$ → Trp$^-$, or His$^+$ → His$^-$. Here the mutant phenotype is obviously attributable to the loss of catalytic function of an enzyme, which loss is in turn attributable to mutation of the gene that controls the primary structure of the corresponding polypeptide chain. This kind of mutation can be expected to occur with high frequency. First, since a replacement of one amino acid by another at any one of many sites on the polypeptide chain is likely to distort the tertiary and quaternary structure so as to abolish catalytic function, it can be anticipated that any one of many different possible mutational alterations of the corresponding gene engenders the functionally defective mutant phenotype. Second, since the prototrophic, or sugar-fermenting wild type depends on the tandem action of several enzymes, the mutant auxotrophic, or nonfermenting, phenotype can be produced by mutation of any one of several genes. The nature of the Ton^s → Ton^r mutation is less evident, since the mechanism of synthesis of the T1 phage receptor spots in the *E. coli* cell wall is still poorly understood. Nevertheless, indirect evidence makes it appear that the absence of the T1 receptor spots from the Ton^r mutant cell is attributable to the loss of a functional protein ordinarily present in the Ton^s cell. But why, in view of this conclusion, does the Ton^s → Ton^r mutation occur a hundred times less frequently than the auxotrophic and sugar-fermentation loss mutations? A factor of about five of this hundredfold discrepancy can be explained by invoking the idea that the Ton^r phenotype depends on the function of only one protein, and hence requires mutation of the particular *ton* gene controlling that protein. But another reason must be sought to account for the remaining twentyfold discrepancy in mutation rates. The following considerations provide for such a reason. We may imagine that the protein controlled by the *ton* gene does not act as a catalyst of a chemical reaction but is itself a structural element of the cell wall. In particular, its role might be that of a *spacer* that holds other cell-wall elements in their proper topographic relation. In the absence of that spacer protein, the portion of the mosaic of cell-wall elements that serve as T1 phage receptor spots is distorted, so that the phage particle is incapable of making its stereospecific attachment. Hence the bacterium is of Ton^r phenotype. But if the *ton* gene protein is merely a spacer, then it is plausible to suppose that many mutational alterations of its primary polypeptide structure—mutations of the kind that would surely abolish the catalytic activity of an enzyme—do not, in fact, alter tertiary and quaternary structure enough to eliminate the spacer function. That is to say, the low frequency of the Ton^s → Ton^r mutation may be attributable to the possibility that most mutations of the *ton* gene are *silent mutations*, in that they do not alter the structure of the corresponding protein enough to evoke the Ton^r phenotype. Only those mutations of the *ton* gene that cause a really drastic alteration of tertiary structure, or which lead to a complete absence of the spacer protein, would be registered.

Although this explanation of the comparatively low frequency of the Ton^s → Ton^r mutation is still largely speculative, there is now not only ample evidence that silent mutations occur but also that they may comprise the great majority of all mutations, even in genes that control enzyme proteins with exigent conformational requirements. But it is true that the fraction of silent mutations in such enzyme proteins is lower than in proteins that play a merely structural, rather than catalytic, role. We shall consider the direct evidence for such silent mutations in Chapter 14, and merely conclude at this point that the observed mutation rate of any character from wild to mutant phenotype diminishes with the relative preponderance of silent mutations in the pertinent gene or genes.

CHANGE OF FUNCTION

The reasons for the very low frequency of the Str^s → Str^r mutation are to be sought in an explanation that is the antithesis of the silent-mutation concept. It was previously stated that the genes which mutate to produce the Str^r phenotype control the formation of components of the cellular machinery for protein synthesis, and hence control an "indispensable" function in the sense of the discussions of the preceding chapter. Now it can be readily appreciated that any mutation that leads to a *loss* of the indispensable function is *lethal*: a cell that cannot carry out the normal process of polypeptide chain assembly is irrevocably dead, beyond rescue by any growth factors that can be added to the medium. Acquisition of the Str^r mutant character demands, therefore, not loss but *change* of function of the gene-controlled protein; this change must preserve the indispensable function yet render it immune from the interference by streptomycin to which the wild-type function is subject. Since only a very limited number of the total possible changes in primary polypeptide structure are likely to engender the requisite changes in tertiary and quaternary structure that satisfy such a stringent functional criterion, it is not surprising that the frequency of change-of-function mutations is very much lower than that of loss-of-function mutations.

Another case of a change-of-function mutation concerns the loss of sensitivity to end-product feedback inhibition of enzymes in biosynthetic pathways. In this connection, we may consider the example of the mutation of *E. coli* from sensitivity to resistance to the tryptophan analog 5-methyltryptophan

The *E. coli* wild type cannot grow in minimal media containing 10 μg/ml 5-methyltryptophan. The physiological basis of this toxic effect has been traced to the false feedback inhibition by 5-methyltryptophan of *anthranilate synthetase*, the first enzyme of the terminal stage of tryptophan biosynthesis (see Figure 3-11). As was stated in Chapter 4, the catalytic activity of anthranilate synthetase decreases as the intracellular level of the pathway end-product tryptophan rises. This inhibition is achieved by combination of tryptophan with an allosteric site of the enzyme protein, and 5-methyltryptophan resembles tryptophan sufficiently to be able to combine with that site as well. Thus in the presence of 5-methyltryptophan the pathway of tryptophan biosynthesis is shut down, and the cells cannot grow for lack of tryptophan. But upon spreading 10^{10} or more wild-type *E. coli* on minimal agar containing 5-methyltryptophan, an occasional colony does grow, whose member bacteria represent a clone of 5-methyltryptophan-resistant mutants. These mutants owe their resistance to a mutation in the gene that controls the primary structure of anthranilate synthetase. This mutation leaves intact the conformation of the active site, and hence the capacity of the protein to catalyze the reaction of glutamine and chorismic acid to yield anthranilic acid, but alters the conformation of the allosteric regulatory site so as to abolish the sensitivity to feedback inhibition. The anthranilate synthetase enzyme of such mutants is not only insensitive to 5-methyltryptophan but also to its natural inhibitor, tryptophan. Accordingly, such mutant *E. coli* overproduce tryptophan, which they excrete into the medium in very great amounts. It stands to reason that the mutation that results in loss of feedback inhibition occurs with very low frequency, ce only those alterations in primary protein structure produce the feedback-insensitive phenotype that preserve the active site while destroying the allosteric regulatory site.

The loss-of-function and change-of-function mutations fall into the class of *forward* mutations, in that they engender an abnormal phenotype that differs from the wild type. But as was illustrated by the example of the His⁻→His⁺ mutation studied in the experiment of Figure 6-9 there occur also *reverse* mutations that restore the wild phenotype to a mutant organism. It is fairly obvious that a reverse mutation, such as $His^- \rightarrow His^+$, which restores a function, ought to occur at a very much lower frequency than the $His^+ \rightarrow His^-$ forward mutation by which that function is lost. Although a forward mutation may abolish function of any one of several wild-type proteins in any one of very many different ways, the reverse mutation must repair that and only that one particular lesion with which the forward mutation has encumbered the particular mutant protein. The data presented here are in accord with this reasoning, since the $His^+ \rightarrow His^-$ forward mutation proceeds at a rate about 300 times higher than the $His^- \rightarrow His^+$ reverse mutation.

The analysis of reverse mutations, however, is actually much more complicated than this brief discussion might suggest. Indeed, it is nearly without meaning to speak of a rate for *the* $His^- \rightarrow His^+$ reverse mutation, since, as will be seen in Chapter 13, the rate of reverse mutation depends very strongly on the kind of molecular change the gene had undergone in the forward mutation whose reversal is sought. Thus different His⁻ mutants revert to the His⁺ wild type at widely different rates, the rate of 7.5×10^{-9} per cell per generation inferred from the data of Figure 6-9 having relevance only for the particular auxotroph that Ryan chose for his study. Indeed, the very concept of a reverse mutation is not as straightforward as one might think. As was first set forth by A. H. Sturtevant in 1920, some mutations restore the wild-type character to a mutant phenotype without having actually restored the mutant gene to its pristine wild-type state. Sturtevant gave the name *suppressor* to such mutations. Suppressor mutations were later recognized to encompass a great diversity of physiologico-genetic phenomena, in many of which it is much more likely that a mutant recovers the wild phenotype through a suppressor mutation than through a "true" reverse mutation. Hence measurements of reverse mutation rates more often than not reveal the frequency with which diverse suppressor mutations occur.

Until the end of the 1950's, two general classes of suppressor activities had been recognized: the *indirect, intergenic* and the *direct, intragenic*. The indirect, intergenic suppressor class circumvents rather than repairs the primary genetic lesion and results from mutations in genes other than the one that sustained the forward mutation. By way of an example, we may consider the case of a TrpE⁻ auxotroph (see Table 5-1) that is unable to synthesize tryptophan because it carries a mutant anthranilate synthetase enzyme whose affinity for its chorismic acid substrate is far lower than normal. This mutant could revert to the Trp⁺ prototroph phenotype by a suppressor mutation that reduces the sensitivity to feedback inhibition of an enzyme in the biosynthetic pathway leading to chorismic acid. Such a mutation would result in an overproduction of chorismic acid. The resulting higher intracellular concentration of chorismic acid would offset the effect of the reduced affinity of the mutant anthranilate synthetase enzyme for its substrate and thus restore to the mutant cell the capacity to synthesize tryptophan.

The direct, intragenic suppressor class resembles "true" reverse mutations in that it does repair the primary genetic lesion and does represent another mutation of the original mutant gene. But unlike the "true" reverse mutation, which restores the original primary structure to the polypeptide chain, the intragenic suppressor mutation repairs the primary protein structure by introducing into it a second alteration that *compensates* for the first and allows for formation of a functional tertiary and quaternary structure. An instance of such direct, intragenic suppressors will be of importance in Chapter 13. And a third class of suppressors, which was discovered only in 1960, will figure prominently in Chapter 18.

Bibliography

PATOOMB

S. E. Luria. Mutations of bacteria and of bacteriophage.

HAYES

Chapters 9 and 10.

ORIGINAL RESEARCH PAPERS

Lederberg, J., and E. M. Lederberg. Replica plating and indirect selection of bacterial mutants. *J. Bacteriol.* **63**, 399 (1952).

Luria, S. E., and M. Delbrück. Mutations of bacteria from virus sensitivity to virus resistance. *Genetics*, **28**, 491 (1943).

Newcombe, H. B. Origin of bacterial variants. *Nature*, **164**, 150 (1949).

Novick, A., and L. Szilard. Experiments with the chemostat on spontaneous mutations in bacteria. *Proc. Natl. Acad. Sci. Wash.*, **36**, 708 (1950).

7. Transformation

After having discussed the role of genes in determining enzyme primary structure and examined the mutation of bacterial genes during cell growth and reproduction, it is now high time that we consider the molecular nature of these hereditary elements.

THE TRANSFORMING PRINCIPLE

The story of the discovery of the chemical identity of genes begins in 1928, when F. Griffith was carrying out experiments on the infection of mice with *Streptococcus pneumoniae*, or "pneumococcus." The pneumococcus is one of the causative agents of human pneumonia and is extremely pathogenic for the mouse. Injection into a mouse of the sputum of a patient suffering from pneumococcal pneumonia causes death of the mouse within 24 hours, by which time vast numbers of the pneumococcus can be found in the heart blood of the animal. The pneumococcus owes its pathogenicity to a polysaccharide capsule, which it forms outside its cell wall proper. An important constituent of the capsular polysaccharide is the polymer composed of glucose and glucuronic acid, shown in Figure 7-1. The capsule protects the bacterium from destruction by the ordinary defense mechanisms of the infected animal. It is to the presence of this capsule that colonies of the

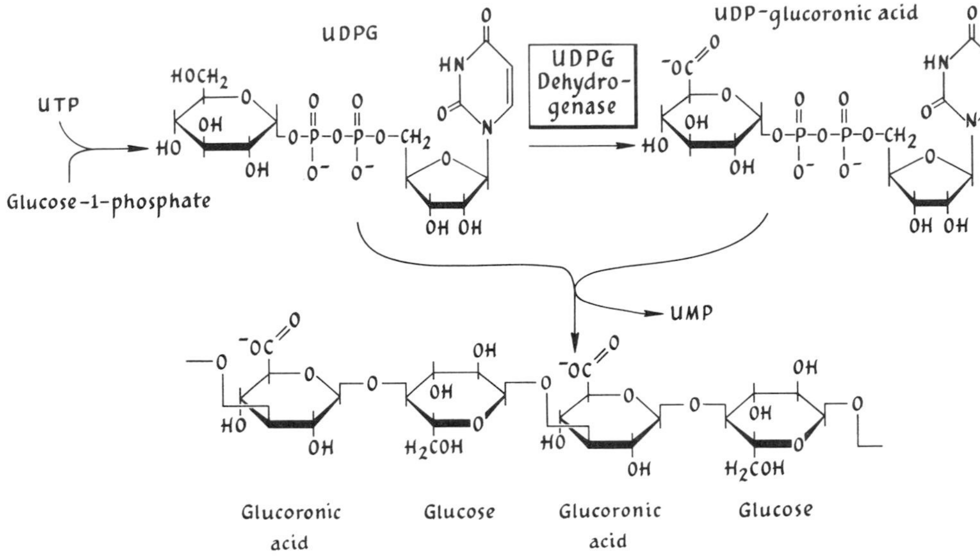

FIGURE 7-1.   Synthesis and structure of the capsular polysaccharide of the pneumococcus.

pneumococcus owe their glistening, smooth, or *S*, character. As was stated in Chapter 6, the normal *S* type "dissociates" into rough, or *R*, variants of different colonial morphology (see Figure 6-1). These *R* variants are, in fact, pneumococcal *mutants* that have permanently lost the capacity to synthesize their protective capsule, and are therefore no longer pathogenic to man or mouse. At least one class of *R* mutants owes its failure to synthesize the capsule to mutation in a gene that controls the enzyme UDPG-dehydrogenase. This enzyme converts UDPG into UDP-glucuronic acid, and thus its mutational loss prevents formation of one of the two basic building blocks of the capsular polysaccharide (Figure 7-1). Griffith had isolated one of these noncapsulated *R* mutants and found that it would no longer kill mice. He discovered to his great surprise, however, that injection of this nonpathogenic *R* mutant in a mixture with a *heat-killed*, or "dead," sample of the normal, pathogenic *S* strain did kill his mice (Figure 7-2). Griffith then isolated and characterized the bacteria present in the heart blood of the dead mice and concluded that the presence of the heat-killed *S* bacteria must have caused a *transformation* of the living *R* bacteria, so as to restore to them the capacity for capsule formation they had earlier lost by gene mutation. Three years later it was shown that the mouse had been merely the detector for but not an essential component of this transformation, which could be produced also *in vitro* merely by growing a culture of *R* mutants in the presence of heat-killed *S* bacteria. Within another two years it was shown that the transformation *R* → *S* would also proceed if no more than a cell-free extract of *S* bacteria were added to a growing culture of *R* mutants (Figure 7-3).

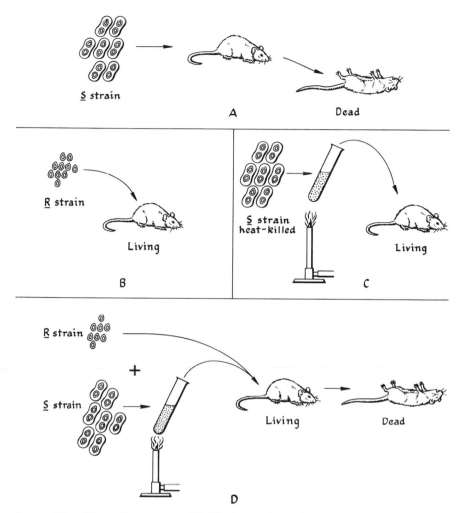

FIGURE 7-2.   Schematic summary of Griffith's transformation experiment. (A) Mice injected with a culture of the pathogenic, encapsulated S strain of the pneumococcus die. (B) Mice injected with a culture of the nonpathogenic, nonencapsulated R mutant of the normal S strain do not die. (C) Mice injected with a heat-killed culture of the S strain do not die. (D) Mice injected with a mixture of a living culture of the R mutant and a heat-killed culture of the normal S strain die, and living S bacteria are present in their heart-blood. [Adapted from R. Sager and F. J. Ryan, *Cell Heredity*, Wiley, New York, 1961.]

The stage was now set for the work of O. T. Avery (Figure 7-4) and his colleagues C. M. MacLeod and M. J. McCarty, who went about fractionating the cell-free extract of S bacteria containing the $R \rightarrow S$ *transforming principle*, in order to identify its chemical nature. They found that they could remove proteins, lipids, polysaccharides, and ribonucleic acid from the extract by a

Propagation of $\underline{S}$ strain:

Preparation of transforming principle:

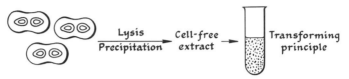

Transformation of $\underline{R}$ strain:

FIGURE 7-3.   Transformation of R mutants of the pneumococcus into S types by a cell-free extract (transforming principle) of S donor bacteria.

FIGURE 7-4.   Oswald T. Avery (1877–1955). [From *Biographical Memoirs of Fellows of the Royal Society* **2**, 35 (1956).]

variety of chemical and enzymatic methods without seriously diminishing its power to transform R mutants into the S wild type. Finally, after further purification of the extract, Avery, MacLeod, and McCarty concluded that the transforming principle is DNA: addition of as little as one part in 6×10^8 of the purified DNA from the S cell extract to a culture of R cells was still capable of effecting the $R \to S$ transformation. Furthermore, DNA extracted from a culture of S bacteria descended from a bacterium that had itself been transformed earlier from an R cell was also capable of causing the $R \to S$ transformation. It followed, therefore, that the S-DNA extracted from the *donor* bacterium not only permanently restores to the R *recipient* bacterium the capacity to form the enzyme UDPG-dehydrogenase, and hence UDP-glucuronic acid, and hence the capsule polysaccharide, but also replicates itself within the recipient, thus endowing the progeny of the *transformant* bacterium with the capacity to serve as a source for further transformations. In other words, the bacterial DNA appears to be the carrier of bacterial heredity, the substance into which is inscribed the information carried by the gene that controls formation of UDPG-dehydrogenase.

DNA as a Carrier of Hereditary Information

Publication of Avery, MacLeod, and McCarty's conclusions in 1944 met with great surprise and disbelief, because hardly anyone had previously considered such an informational role for DNA. Admittedly, DNA had long been suspected of exerting *some* function in hereditary processes, particularly after R. Feulgen had shown in 1924 that DNA is a major component of the chromosomes. But the then current views of the molecular nature of DNA made it well-nigh inconceivable that DNA *could* be a carrier of hereditary information. First of all, until well into the 1930's DNA was generally thought to be merely a *tetranucleotide* composed of one residue each of adenylic, guanylic, thymidylic, and cytidylic acid, as shown in Figure 7-5. Secondly, even when it was finally realized by the early 1940's that the molecular weight of DNA is actually much higher than that demanded by the tetranucleotide theory, it was still widely believed that the tetranucleotide is the basic repeating unit of the larger DNA polymer, in which the four purine and pyrimidine bases recur in regular sequence. DNA was therefore viewed as a monotonously uniform macromolecule that, like other monotonous polymers such as starch (see Chapter 2), is always the same no matter what its biological source. The ubiquitous presence of DNA in the chromosomes was therefore generally explained in purely physiological or structural terms. Instead, it was usually to the chromosomal *protein* that the informational role of the genes had been assigned, since the great differences in the specificity of structure that exist between heterologous proteins in the same organism, or between homologous proteins in different organisms, had been appreciated since the beginning of the century. The conceptual difficulty of assigning the genetic role to DNA had

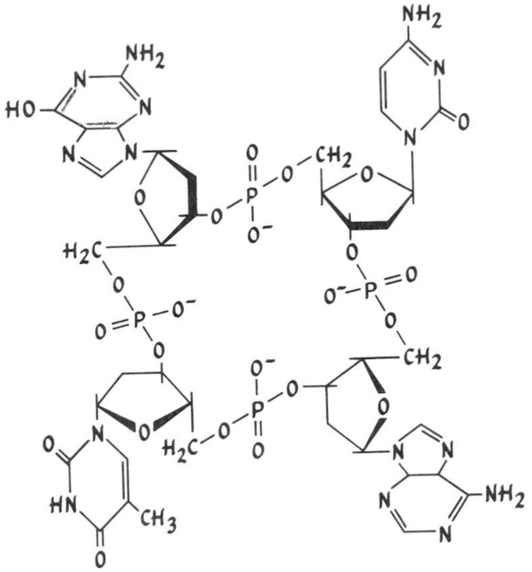

FIGURE 7-5. A version of the "tetranucleotide" structure of DNA proposed by H. Takahashi in the 1930's.

by no means escaped the notice of Avery, MacLeod, and McCarty, for the conclusion of their paper contained the statement that "if the results of the present study on the nature of the transforming principle are confirmed, then nucleic acids must be regarded as possessing biological specificity the chemical basis of which is as yet undetermined."

Thus rather than leading to the immediate acceptance of DNA as the genetic material in 1944, the identification of the pneumococcal transforming principle met with much conservative skepticism. The first and natural objection raised against that identification was the proposal that however pure the transforming DNA might have been by the chemical and physical criteria employed, it might still harbor a tiny trace of contaminating protein, which might actually be the active principle. This criticism was motivated in part by recollection of the famous mistake made twenty years earlier by Willstätter, who claimed to have prepared protein-free enzyme preparations. Willstätter, as was mentioned in Chapter 4, had concluded that enzymes are not proteins, and, because of his renown as a biochemist, had thus put back the study of enzyme by some ten years. Now the ghost of Willstätter's error was to put back the study of genes by another ten years.

By way of example of the thinking among contemporary expert students of these matters, we may consider an essay written by A. E. Mirsky in 1950, six years after the publication of Avery's discovery. Only two years earlier, both Mirsky and H. Ris, as well as A. Boivin and R. Vendrely, had shown that in the cells of different tissues of the same organism, the amount of DNA per haploid chromosome set is constant. This, as Mirsky pointed out, is what *ought* to be the case if DNA is the genetic material. But Mirsky was then

willing to conclude only that "if this component [i.e., DNA] of the chromosomes is indeed present in constant amount in the different somatic cells of an
organism and in one-half this amount in the germ-cells, then it may be said
that DNA is *part* of the gene substance." As far as the pneumococcus transformation work was concerned, Mirsky thought that "it is quite possible that
DNA, and nothing else, is responsible for the transforming activity, but this
has not been demonstrated. In purification of the active principle more and
more of the protein attached to DNA is removed, as indeed in the preparation
of DNA from any other source. It is difficult to eliminate the possibility that
the minute quantities of protein that probably remain attached to DNA,
though undectable by the tests applied, are necessary for activity—itself an
exceedingly sensitive test . . . . There is, accordingly, some doubt whether
DNA is itself the transforming agent, although it can be regarded as established that DNA is at least part of the active principle."

But Avery, MacLeod, and McCarty had already gone a long way toward
meeting the criticism of having repeated Willstätter's error. They had shown
that extensive treatment of the pneumococcus transforming DNA with a
variety of protein-splitting enzymes had no effect on its biological potency,
whereas even a very brief exposure to *deoxyribonuclease*, or *DNase*, a highly
specific enzyme whose hydrolytic action shown in Figure 7-6 is restricted to
the DNA polynucleotide, immediately destroyed all transforming activity.
Thus rather far-fetched *ad hoc* explanations were required to explain how the
putative active protein contaminant could be resistant to protein-splitting
enzymes while being sensitive to the DNA-specific DNase. In any case,

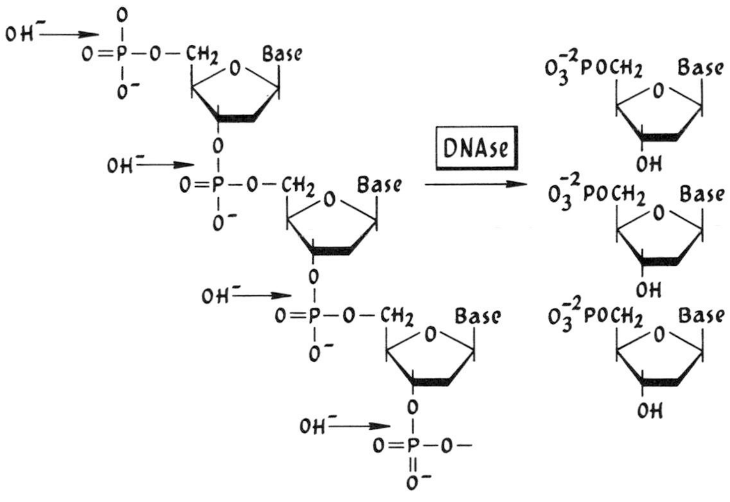

FIGURE 7-6.   The enzymic hydrolysis of DNA by deoxyribonuclease.

FIGURE 7-7. Rollin D. Hotchkiss (b. 1911). [Courtesy R. D. Hotchkiss.]

Rollin D. Hotchkiss (Figure 7-7) continued the chemical fractionation of the transforming principle, and by 1949 he had reduced the maximum amount of protein contamination of the active DNA to 0.02 %. But even this prodigious feat of purification did not convince everyone then that DNA was responsible for the hereditary changes attributed to it; and when in the mid-1950's the DNA nature of the genetic material finally had become a commonplace article of faith, it was not for the reason that anyone had produced any better direct proof than that already delivered by Hotchkiss. Rather, by the mid-1950's enough had become known about the nature of DNA to make its informational role so readily comprehensible that no further chemical proof of that proposition seemed necessary.

A second objection raised against inferring from the findings of Avery, MacLeod, and McCarty that DNA is the genetic material was that even if the active transforming principle really were DNA, it might act merely by exerting a direct chemical effect on capsule formation, rather than by being a carrier of hereditary information. Thus it was proposed that the uptake of donor DNA might displace some kind of biosynthetic equilibrium in the pathway of polysaccharide synthesis. In other words, the whole bacterial transformation phenomenon might be physiologic rather than genetic. In 1949, however, Harriet Taylor was able to adduce additional evidence that the pneumococcal transformation seems to operate on a genetic level. She had isolated from the rough R mutant strain a new mutant ER, or "extremely rough," which produced colonies even rougher, or more irregular than those of R, and found that DNA extracted from the R strain could accomplish the transformation ER → R. (These one- and two-letter symbols assigned to the pneumococcus capsular phenotype came into use long before adoption of the three-letter standard nomenclature of bacterial genetics introduced in Chapter 6.) Hence Taylor showed that the recipient strain R of the earlier experiments carries a transforming principle in its own right that can confer what little residual capsule-synthesizing capacity that strain R still possesses on the even more capsule-deficient ER strain. Furthermore, Taylor showed that addition of DNA extracted from the fully encapsulated S wild-type strain to the ER mutant transforms the recipient cells into R types. Re-exposure of a culture derived from one of these first-cycle R transformants to more S-donor DNA resulted in the production of second-cycle S transformants. Thus it could be concluded that the DNA extracted from S-donor bacteria contains at least two specific components, one capable of carrying out the transformation ER → R and the other R → S. The DNA extracted from R donor bacteria naturally contains only the first component. These findings would be rather hard to explain if in causing transformation the DNA merely intervened as an exogenous chemical agent in the synthesis of the poly-saccharide capsule.

All grounds for special pleading in regard to some peculiarity of the relation of DNA to capsule synthesis were removed in the same year, when Hotchkiss showed that transformation works also for bacterial characters that have nothing to do with capsule formation. He had isolated a penicillin-resistant, or Pen^r, mutant of the normal S-type pneumococcus, extracted its DNA, and then added the transforming principle to a culture of penicillin-sensitive, or Pen^s, R mutants. Hotchkiss found that as a result of this treatment, some of the Pen^s-R recipient bacteria had been transformed into Pen^s-S types, others into Pen^r-R types, and, finally, a very small fraction into the Pen^r-S donor type. Hence it could be concluded that the pneumococcal DNA carries information not only for capsule formation but also for the formation of those cellular structures that pertain to penicillin resistance. Capsule formation and penicillin resistance, furthermore, appear to be controlled by different DNA molecules, since the transformations that produce one or the other of these characters are evidently independent events. Soon thereafter, entirely analogous transformation was obtained with DNA extracted from streptomycin-resistant, or Str^r, mutants of the S wild type. Indeed, Hotchkiss' results on Pen^r and Str^r transformation turned out to be in excellent agreement with Demerec's explanation of the manner in which bacteria acquire high-grade resistance to these two antibiotics in the first place. For Hotchkiss had found that treatment of Pen^s recipients with DNA extracted from a Pen^r

donor possessing high-grade penicillin resistance, necessarily acquired in several successive selective steps, results mainly in the appearance of Pen^r transformants of *low-grade* resistance, generation of high-grade Pen^r transformants requiring several successive transformation cycles. Thus each of the equipotent genes that cooperate in their mutant form to confer high-grade penicillin resistance on the bacterium appears to reside in a different DNA molecule, so that transformation to high-grade resistance requires the introduction of at least one of each of the relevant donor DNA molecules into one and the same recipient bacterium. In contrast, treatment of Str^s recipients with DNA extracted from a Str^r donor possessing high-grade streptomycin resistance acquired in a single selective step results in the appearance of Str^r transformants of high-grade resistance in the first cycle. Thus the single gene that can mutate to confer high-grade streptomycin resistance on the bacterium resides in a single DNA molecule whose introduction into the recipient bacterium suffices to produce transformation to high-grade resistance.

POLYNUCLEOTIDE AND POLYPEPTIDE SEQUENCE

The tetranucleotide theory of DNA structure, which regarded that poly-nucleotide as a monotonous macromolecule ill-suited as an information carrier, was finally disposed of in 1950. The origin of that theory had been the finding of Levene and others in the 1920's that DNA contains approximately equal molar proportions of the four bases adenine, guanine, cytosine, and thymine (see Figure 2-5). But by 1948, both Hotchkiss and Erwin Chargaff had adapted the then recently invented methods of *paper chromatography* to the separation and quantitative estimation of the constituents of nucleic acid. This made possible more refined compositional analyses of the nucleotide bases liberated upon complete hydrolysis of DNA samples. These analyses showed that, contrary to the demands of the tetranucleotide theory, the four nucleotide bases are *not* necessarily present in DNA in *exactly* equal proportions. Thus Chargaff reported that the DNA extracted from calf thymus nuclei contains the four bases in the following molar proportions: 28% adenine [A]; 24% guanine [G]; 20% cytosine [C]; and 28% thymine [T]. When Chargaff proceeded to analyze a variety of DNA samples obtained from different organisms he found that the exact base composition of DNA differs according to its biological source. That is, he found that the actual molar proportions of the bases can vary within wide limits, suggesting that DNA may not be a monotonous polymer after all. If it is *not* monotonous, then its base composition bids fair to be the reflection of its biological specificity.

The way was now clear to formulate a theory of how DNA can act as the carrier of genetic information in the transformation experiment. It seems impossible today to establish who was actually responsible for originating these notions. The theory suddenly seemed to be in the air after 1950 and had come to be embraced as dogmatic belief by many molecular geneticists by 1952. The key proposition of this theory is that if the DNA molecule contains genetic information, then that information cannot be carried in any way other than as the *specific sequence of the four nucleotide bases* along the polynucleotide chain. That is to say, the DNA molecule is Schrödinger's "aperiodic crystal," and the four bases are the small number of "isomeric elements" whose exact succession represents the "hereditary code" (see Chapter 1). But since, as the considerations of Chapter 5 showed, the informational content of genes must specify the amino acid sequence of polypeptide chains, one arrives at the basic insight that the *meaning* of the sequence of the four nucleotide bases in a DNA molecule that makes up a gene is none other than the exact sequence of the amino acids of the protein molecule over whose synthesis that gene presides. From this insight follows effortlessly the molecular explanation of mutation as a *change in DNA nucleotide base sequence*. For such a change in base sequence would alter permanently the information content of the gene and thus cause appearance in the mutant cell of a protein of altered amino acid sequence and hence of altered function.

From this viewpoint, the transformation of an *R*-recipient pneumococcus by DNA extracted from an *S* donor cell can be interpreted in the following terms. There exists an alteration in base sequence in that sector of the DNA polynucleotide chain of the *R* mutant bacterium in which the amino acid sequence of the UDPG-dehydrogenase enzyme protein is specified. This alteration in base sequence causes failure of the mutant bacterium to synthesize the active enzyme, and hence is responsible for the absence of the polysaccharide capsule. Uptake of a molecule of *S*-type donor DNA introduces into the recipient *R* cell the normal nucleotide base sequence that specifies the normal amino acid sequence of the UDPG-dehydrogenase enzyme protein. In this way, the transformed cell regains the capacity to synthesize the active enzyme, and hence, to form the capsule.

The pregnant ideas that were merely adumbrated in the preceding paragraphs will be elaborated in much more detail in the following chapters. Indeed, these ideas are altogether central in the discussions of most of the remainder of this text.

EFFICIENCY OF TRANSFORMATION

Hotchkiss's discovery of the transformation of drug-sensitive to drug-resistant pneumococci made it feasible to study the quantitative aspects of the transformation reaction. Thus, in order to assay the number of Str^s recipient cells transformed upon their exposure to transforming DNA extracted from Str^r donor cells, it suffices to plate the recipient culture on streptomycin-containing agar, on which only the Str^r transformants can grow into colonies.

The results of an experiment in which the *efficiency* of transformation was ascertained in just this way are shown in Figure 7-8. A series of culture tubes was set up, each containing per ml 10^9 Str^s pneumococci, to which were added progressively greater concentrations of DNA extracted from Str^r donor cells. After contact of donor DNA and recipient cells had been allowed for 5 minutes, samples from each tube were plated for colony count on agar containing 100 μg/ml streptomycin. It can be seen in Figure 7-8, first of all, that at the low concentrations of donor DNA the number of Str^r transformants produced is directly proportional to the concentration of donor DNA added. This finding shows that each transformation event is the consequence of the interaction of a recipient cell with a single molecule of donor DNA. For if interaction with two, or more, molecules of donor DNA were required, the number of transformants would be proportional to the second, or higher, power of the donor DNA concentration. We can now reckon the efficiency of transformation as follows: each pneumococcus contains about 5×10^{-12} mg DNA per "nucleus," so that 10^{-6} mg transforming DNA represents the genetic material of about 2×10^5 donor cells. And since in the range of proportional response, 10^{-6} mg of donor DNA can be seen to transform about 10^4 recipient cells, the efficiency of the transformation is evidently $10^4/(2 \times 10^5)$ = 0.05. That is to say, about one in twenty of the DNA molecules carrying the

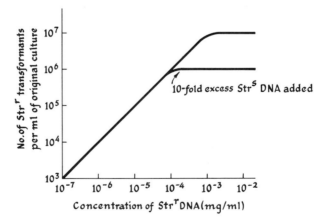

FIGURE 7-8.   Efficiency of the Str^s → Str^r transformation. The ordinate shows the number of Str^r colonies appearing on agar containing 100 μg/ml streptomycin upon plating samples from cultures of 10^9 Str^s pneumococci/ml to which the amount of DNA extracted from Str^r donor pneumococci shown on the abscissa had been added. The curve exhibiting the lower plateau represents the result of a transformation set in which the Str^r donor DNA was diluted with a tenfold excess of DNA extracted from Str^s-donor pneumococci.

Str^r gene of the donor bacteria succeeds in imposing its hereditary character on a recipient cell. Second, it can be seen in Figure 7-8 that at donor DNA concentrations exceeding 10^{-3} mg/ml a final plateau is reached at about 10^7 transformants/ml (corresponding to 1% of the Str^s recipient bacteria) which is not surpassed even at very much higher donor DNA inputs.

Hotchkiss discovered that one of the reasons for the existence of this final plateau is that at high DNA concentrations an *auto-inhibition* of DNA by DNA comes into play. This can be shown by repeating the preceding DNA-dose-response experiment but employing Str^r donor DNA to which has been added a tenfold excess of extraneous DNA extracted from Str^s bacteria. The result of such an experiment is shown also in Figure 7-8, where it can be seen that at low DNA inputs the presence of extraneous Str^s DNA does not affect the transformation efficiency. At high DNA inputs, however, the presence of excess Str^s DNA evidently causes the final plateau of Str^r transformants to be reached at a tenfold lower concentration of Str^r DNA and to be at a tenfold lower level. It follows from this finding that the pneumococci have a limited number of DNA *receptors*, for which all DNA molecules—both relevant and irrelevant for the particular transformation under study—compete. Since even in the "pure" Str^r-transforming DNA only a minor fraction of all DNA molecules could possibly be concerned with the streptomycin-resistant character for which transformation of the Str^s recipients is sought, the DNA molecules that bear other genes will start to compete for the limited number of receptors at high DNA inputs. And once all the receptors have been saturated, no further increase in transformants can be achieved by raising the transforming DNA input to even higher levels.

COMPETENCE

Another reason for the failure to transform *all* of the Str^s bacteria to Str^r types at the highest DNA inputs is that only a minority of the recipient cells are *competent* at any one time to be transformed at all. For, as was first shown by Hotchkiss, the pneumococcus bacterium passes through phases of competence and noncompetence in its division cycle. These cyclical changes became manifest when Hotchkiss managed to *synchronize* the growth and division of pneumococcal cultures by a program of temperature shifts. An ordinary bacterial culture presents a random sample of bacteria at all stages of their division cycle, so that at any moment some of the bacteria are just about to divide and others may still have half the generation time to go before their division. When Hotchkiss incubated a growing culture of pneumococci alternately at 25°C and at 37°C he found the result shown in Figure 7-9. It is apparent that almost none of the bacteria divide during the 25°C episode and that nearly all of the bacteria divide at the very beginning of the 37°C episode. This pattern is no doubt the consequence of the differential effect of temperature

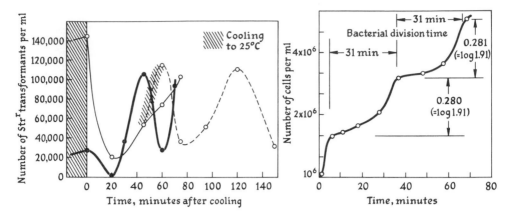

FIGURE 7-9.   Transformability of synchronized pneumococcus cultures. (Left) Cycles of transformability following maintenance of a Str^s culture at 25°C for 15 minutes and then returning it to 37°C. The heavy curve shows the maximum number of Str^r transformants obtainable by addition of excess Str^r donor DNA at various times after the cooling period. The light curve shows the results obtained with another, similarly treated culture, of which a sample (broken curve) was subjected to a second 15-minute cooling period 45 minutes after the first one. (Right) Synchronized growth of the pneumococcus culture at 37°C following the 15 minute cooling period at 25°C. The ordinate shows the number of Str^s colony-forming cells (estimated error ± 10%) in the culture at the time after its return to 37°C shown on the abscissa. [Redrawn from R. D. Hotchkiss, *Proc. Natl. Acad. Sci. Wash.* **40**, 53 (1954).]

on the rates of the various metabolic reactions on which the division process depends. That is, upon being placed at 25°C the bacteria quickly run through those steps whose rates are little effected by a lowering of temperature and tend to be arrested at that stage where the temperature dependence is the greatest. Upon being returned to 37°C, all the bacteria of the culture start off more or less synchronously from that same stage at which they had all been arrested and ultimately divide all at the same moment. Hotchkiss then added a saturating amount of Str^r DNA to a Str^s recipient culture at various stages of this synchronized division cycle and scored the number of Str^r transformants produced. The result of this experiment is shown also in Figure 7-9, where it can be seen that the fraction of all cells transformed by an excess amount of donor DNA varies cyclically over a three- to tenfold range, its value depending on the precise stage of the division cycle at which transformation occurs. It thus appears that the pneumococcus enters a competent phase shortly after its division and then returns to a relatively non-competent phase until the next division.

Competence represents the capacity of the bacterium to take up DNA from its surroundings. This was shown by preparing transforming DNA labeled with radiophosphorous ^{32}P (in its phosphate diester bridges linking successive nucleotides) from Str^r donor pneumococci that were grown in a ^{32}P-labeled medium. Saturation amounts of such ^{32}P-labeled DNA were added to a non-synchronized, nonlabeled Str^s recipient culture. The recipient bacteria, after having had sufficiently long contact with donor DNA, were then centrifuged into a pellet and freed of any contaminating free DNA. Finally both the amount of ^{32}P-label fixed in the bacterial pellet and the number of Str^r transformants produced were assayed. Two variations of this experiment were also carried out. One consisted in using ^{32}P-labeled Str^r donor DNA to which a tenfold excess of nonlabeled extraneous Str^s DNA had been added; the other, in using Str^s recipient bacteria from cultures in synchronized growth in phases of both maximum and minimum competence. The result of the three experiments can be summarized as follows: The amount of ^{32}P-labeled DNA taken up *per recipient cell* in the culture was higher when pure ^{32}P-labeled donor DNA was used than when the donor DNA had been mixed with nonlabeled Str^s DNA. That amount, furthermore, was higher in the synchronized recipient culture in its phase of maximum competence than in the nonsynchronized culture, whose uptake in turn was higher than that of the synchronized recipient culture in its phase of minimum competence. From these results it follows that the limited number of bacterial receptors for which the DNA molecules compete at high DNA inputs pertain to DNA uptake, and that competent cells take up more DNA molecules than non-competent cells. But when the amount of ^{32}P-labeled DNA taken up was compared with the number of cells transformed from the Str^s to the Str^r state in all three experiments, it could be seen that in each one the amount of ^{32}P-labeled DNA taken up *per transformed cell* was the same. Thus once a donor DNA molecule *has* been taken up by a recipient cell, the chance of its actually causing transformation depends neither on the presence of other, extraneous DNA molecules in the preparation nor on the state of competence of the recipient cell. These data allowed a more refined calculation of the efficiency of the transformation process than that previously made merely on the basis of the DNA input into the culture. For the probability of trans-formation per Str^r-donor-gene-bearing DNA molecule taken up is obtained by dividing the constant amount of ^{32}-P labeled DNA taken up per trans-formed cell by the amount of ^{32}P-labeled DNA extracted per donor bacterium. That probability turns out to be about 0.3. Thus it follows that once a donor DNA molecule bearing the relevant gene has managed to get into the recipient cell, it causes transformation with a rather high efficiency.

INTEGRATION

We may now consider how, once taken up from the medium, the donor DNA molecule manages to impose its nucleotide sequence on the recipient bacterium. Although we are not yet ready at this stage to discuss the molecular processes responsible for this event, we can assert that the event itself is an example of *genetic recombination*. That is, the exogenous DNA molecule bearing the S or Str^r genes of the donor bacterium manages to find its *homologous*, endogenous DNA molecule bearing the R or Str^s genes in the recipient cell. From the theory of the informational essence of DNA set forth in this chapter it follows that the homology of exogenous S or Str^r donor genes with the corresponding endogenous R or Str^r recipient genes represents a correspondence in the nucleotide base sequence of each pair of DNA molecules. That is to say, the base sequence of homologous donor and recipient DNA molecules is exactly the same, except for that limited domain where mutation had changed the base sequence of one of the two homologs and thus engendered the expression of a protein of altered amino acid sequence. Once the exogenous DNA molecule has found its endogenous homolog, the two molecules engage in an act of genetic exchange. This exchange results in integration of the exogenous and expulsion of the homologous endogenous DNA molecule from the gene ensemble of the recipient bacterium, and hence in its genetic transformation from R to S or from Str^s to Str^r type. Once integrated into the recipient genetic structures, the exogenous DNA molecule is replicated together with all the other DNA molecules of the transformant, and thus transmitted to all of the descendent bacteria, from which it can then be extracted for further transformation reactions.

GENE LINKAGE

Does each of the molecules present in the preparation of transforming DNA represent a single gene, or can several genes form part of the same polynucleotide continuity? The answer to this question was provided in 1954 by Hotchkiss and J. Marmur, after they had managed to achieve transformation of a number of other mutant characters of the pneumococcus. As we have seen, the genes responsible for the transformations $ER \rightarrow R$, $R \rightarrow S$, and $Str^s \rightarrow Str^r$, as well as the equipotent mutant genes that cooperate in producing high-grade penicillin resistance, all appear to reside in separate DNA molecules, since transformation for each of these characters is evidently an independent event. Though Hotchkiss and Marmur found a similar independence of transformation for several other characters as well, they discovered that transformation involving the *mtl* gene responsible for mannitol fermentation and the *str* gene are strongly correlated. Thus by adding transforming DNA extracted from a Str^r Mtl^+ donor pneumococcus that can ferment mannitol to Str^s Mtl^- mutant recipients that cannot ferment mannitol, they obtained a good yield of all three transformant types: Str^r Mtl^- and Str^s Mtl^+ transformants, each of which had received one of the donor characters, and Str^r Mtl^+ transformants, which had received *both* donor characters (Table 7-1). The finding of a good yield of Str^r Mtl^+ double transformants was thus in strong contrast to the previous general experience of finding only single-character transformation. Hotchkiss and Marmur next showed that this double transformant arises only if the two donor characters are extracted from one and the same donor cell, since addition to Str^s Mtl^- recipients of a *mixture* of transforming DNA extracted from Str^r Mtl^- and Str^s Mtl^+ bacteria resulted in production of only the Str^r Mtl^- and Str^s Mtl^+ single transformants (Table 7-1). The absence of any appreciable number of Str^r Mtl^+ double transformants in this control experiment indicates that their presence in the main experiment reflects the coexistence of *str* and *mtl* genes on the same DNA molecule in the extract obtained from Str^r Mtl^+ donor cells. Thus one DNA molecule can evidently carry more than one gene.

TABLE 7-1.   Demonstration of Linkage of *str* and *mtl* Genes
on Pneumococcal Transforming DNA

Donor DNA	Percent of Str^s Mtl^- recipients found transformed to		
	Str^r Mtl^-	Str^s Mtl^+	Str^r Mtl^+
Str^r Mtl^+	4.3	0.40	0.17
Str^r Mtl^- and Str^s Mtl^+	2.8	0.85	0.0066

From R. D. Hotchkiss and J. Marmur, *Proc. Natl. Acad. Sci. Wash.* **40**, 55 (1954).

Two reasons account for the finding that not *all* of the transformants of the main experiment were of the Str^r Mtl^+ double-transformant type. First, the native DNA molecules are adventitiously broken into the smaller fragments in the extraction process, so that in Str^r Mtl^+ donor extract not all of the *str* and *mtl* genes still reside in the same DNA molecule. Second, even though two genes are *linked* on the same DNA molecule, they are not necessarily *integrated* together in the genetic exchange process that finally gives rise to the transformant bacterium. As we shall see in our later considerations, the probability of joint integration of two linked genes is, in fact, an excellent measure of the physical distance between them in the DNA molecule where their information is inscribed.

In 1958, J. Spizizen discovered that mutants of the soil bacterium *Bacillus subtilis* can be transformed by addition of DNA extracted from genetically differentiated donor bacteria. The great interest of that discovery lay not so much in providing yet another instance of transformation but in the great superiority of *B. subtilis* over the pneumococcus as experimental material. For whereas the pneumococcus must be cultured in rather complex media, tends to grow in chains or clusters that hinder quantification of cell numbers, and is generally temperamental, *B. subtilis* is satisfied with the simple synthetic medium specified in Table 2-1, grows as individual cells, and is easy to manipulate. These properties of *B. subtilis* allowed not only the isolation of many more biochemical mutant types than had been available in the pneumococcus, but also provided for a rapid expansion of the use of transformation as an experimental tool among molecular geneticists, many of whom had earlier been loath to learn the black art of pneumococcus manipulation. One consequence of this development has been the identification of many more linked bacterial genes which, like *str* and *mtl* in the pneumococcus, are transferred from donor to recipient *B. subtilis* on the same DNA molecule. We shall have occasion to refer to some of these *B. subtilis* transformation experiments in later chapters.

Bibliography

PATOOMB

Rollin D. Hotchkiss. Gene, transforming principle and DNA.

HAYES

Chapter 20.

ORIGINAL RESEARCH PAPERS

Avery, O. T., C. M. MacLeod, and M. McCarty. Studies on the chemical nature of the substance inducing transformation of pneumococcal types. Induction of transformation by a desoxyribonucleic acid fraction isolated from pneumococcus type III. *J. Exp. Med.*, **79**, 137 (1944).

Hotchkiss, R. D. Criteria for quantitative genetic transformation of bacteria. *In* W. D. McElroy and B. Glass (eds.), The Chemical Basis of Heredity. The Johns Hopkins Press, Baltimore, 1957, p. 321.

Hotchkiss, R. D. Cyclical behavior in pneumococcal growth and transformability occasioned by environmental changes. *Proc. Natl. Acad. Sci. Wash.* **40**, 49 (1949).

Hotchkiss, R. D. and J. Marmur. Double marker transformations as evidence of linked factors in desoxyribonucleate transforming agents. *Proc. Natl. Acad. Sci., Wash.* **40**, 55 (1954).

SPECIALIZED TEXTS, MONOGRAPHS, AND REVIEWS

Ravin, A. W. The genetics of transformation. *Advances in Genetics,* **10**, 61 (1961).

Schaefer, P. Transformation. *In* I. C. Gunsalus and R. Y. Stanier (eds.), *The Bacteria.* Academic Press, New York, 1964. Vol. 5, p. 87.

Spizizen, J., B. E. Reilly, and A. H. Evans. Microbial transformation and transfection. *Ann. Rev. Microbiol.,* **20**, 371 (1966).

8. DNA Structure

As was set forth briefly in the preceding chapter, Chargaff's analyses of the nucleotide base composition of DNA samples obtained from different organisms had shown that the molar proportion of the bases adenine [A], guanine [G], cytosine [C], and thymine [T] varies within wide limits. Hence the requirement of the tetranucleotide theory of DNA structure that [A] = [G] = [C] = [T] was found not to obtain, and the idea slowly gained currency that DNA, rather than being a monotonous polymer, carries genetic information in the form of specific nucleotide base sequences.

THE EQUIVALENCE RULE

Chargaff published the data of his compositional studies of DNA in 1950 in a paper in which the following statement may be found: "The results serve to disprove the tetranucleotide hypothesis. It is, however, noteworthy—whether this is more than accidental, cannot yet be said—that in all deoxyribose nucleic acids examined thus far the molar ratios of total purines to total pyrimidines, and also of adenine to thymine and of guanine to cytosine, were not far from 1." This statement was the first enunciation of an important structural feature of DNA: despite the rather wide compositional variations exhibited by different types of DNA, the molar proportion of A is always very nearly equal to that of T, and the molar proportion of G is always very nearly equal to that of C, or

$$[A] = [T] \quad \text{and} \quad [G] = [C]$$

That Chargaff found this nucleotide base equivalence "noteworthy" suggests that he suspected it to be the reflection of some significant aspect of the structure of DNA. But his prudent reluctance to say that the equivalence rule is more than accidental suggests also that he did not yet suspect just what that aspect might be. Indeed, it would have been most astounding if, on the basis of the information available to him in 1950, Chargaff could have guessed the meaning of the equivalence rule he had discovered.

Subsequent nucleotide base composition analyses conducted during the next 15 years on DNA extracted from the most diverse organisms, ranging from viruses through bacteria, protozoa, fungi, algae, higher plants, invertebrates, and on to the vertebrates almost always confirmed the $[A] = [T]$ and $[G] = [C]$ equivalence rule. And those rare instances in which this rule did *not* seem to apply were, as we shall see later, the exceptions that prove the rule. The near-universality of the equivalence rule now allowed description of the relative base composition of different DNA's in terms of a *single* parameter, rather than in terms of the three parameters that would have been required if the four bases varied independently of each other. For specifying the mole fraction, X of $[G] + [C]$ in any type of DNA evidently specifies the relative abundance of all four bases, since if $[A] = [T]$ and $[G] = [C]$, it follows that $[G] = X/2$, $[C] = X/2$, $[A] = (1 - X)/2$, and $[T] = (1 - X)/2$. The greatest variation in the value of X was encountered in the base composition analysis of DNA extracted from different bacterial species, where, as can be seen in Table 8-1, the mole fraction $[G] + [C]$ ranges from a low of 26% in *Welchia perfringens* to a high of 74% in *Streptococcus griseus*. The DNA base composition of other bacteria represents a well-nigh continuous spectrum between these two extremes. *E. coli* DNA at 52% happens to be near the $[A] = [T] = [G] = [C]$ equivalence point demanded by the defunct tetranucleotide hypothesis, whereas *Streptococcus pneumoniae* DNA at 38% places the pneumococcus in the category of "A-T-rich" bacteria. The DNA of viruses and monocellular eukaryotes such as protozoa spans a range of base compositions nearly as wide as that of the bacteria, but the DNA of the higher plants and animals spans the much narrower A-T-rich range, from about 35% to 45% $[G] + [C]$. Most strikingly, from an anthropocentric point of view, the DNA of all vertebrates, from fishes through amphibia, birds, rodents to man, is confined to the narrow band from 40% to 44%, within which the differences come close to the limits of accuracy of the actual base analyses.

Though we are not yet ready at this stage of our discussion to consider in detail the meaning of the differences in base composition of the DNA of different organisms, we may note already that they undoubtedly reflect differences in genetic information carried by these hereditary molecules. Hence knowledge of the [G] + [C] content is of patent value for evolutionary and taxonomic deliberations. For instance, the much wider range of DNA base compositions among bacteria than among vertebrates suggests that the extant ensemble of bacteria represents the branches of a phylogenetic tree vastly more ramified than that of the look-alike vertebrates. Moreover, the near-equality of the DNA base composition of *E. coli, Salmonella typhimurium,*

TABLE 8-1. Nucleotide Base Compositions of a Variety of Bacterial Species

Mole fraction [G] + [C] (in percent)	Bacteria	Vertebrates
26	*Welchia perfringens*	
30	*Micrococcus pyogenes*	
34	*Bacillus cereus*	
36	*Proteus vulgaris*	
38	*Streptococcus pneumoniae*	
40	*Hemophilus influenzae*	man
42	*Bacillus subtilis*	salmon, frog
44	*Vibrio cholerae*	mouse, hen
48	*Corynebacterium acnes*	
50	*Salmonella typhimurium*	
52	*Escherichia coli*	
54	*Shigella dysenteriae*	
56	*Aerobacter aerogenes*	
58	*Serratia marcescens*	
60	*Pseudomonas fragii*	
64	*Pseudomonas fluorescens*	
66	*Mycobacterium phlei*	
68	*Mycobacterium tuberculosis*	
72	*Micrococcus lysodeikticus*	
74	*Streptococcus griseus*	

After N. Sueoka, *in* I. C. Gunsalus and R. Y. Stanier, eds., *The Bacteria*, Academic Press, Vol. 5, p. 422 (1964).

and *Shigella dysenteriae* confirms earlier inferences about the close taxonomic relation of these three intestinal bacteria. Indeed, an appreciation
of the value of DNA base composition data has made possible the hitherto
extremely difficult task of constructing a rational taxonomy of bacteria. For
example, Table 8-1 shows that *Streptococcus griseus* and *Streptococcus
pneumoniae*, which weigh in at nearly opposite ends of the compositional
scale, are obviously not close relatives at all and cannot, despite their common
generic name, belong to the same genus, whatever "genus" might actually
mean for bacterial taxonomy.

THE WATSON-CRICK STRUCTURE

A key development that led to the unraveling of the structure of DNA was
the successful application of X-ray crystallography to biological macromolecules. As was set forth in Chapter 4, the determination of molecular
structure by analysis of the diffraction pattern of X-rays impingent upon
crystals had begun in 1912 with study of simple inorganic salts and was
gradually applied to ever-more complicated organic molecules. An important
milestone in this development was passed during World War II, at about the
time that DNA was identified as the transforming principle, when Dorothy
Crowfoot and her colleagues managed to work out the complete three-
dimensional structure of the penicillin molecule by X-ray crystallographic
techniques before organic chemists had even determined its primary chemical
structure, which is shown in Figure 6-10. One of the first persons to give any
thought to the three-dimensional structure of DNA was W. T. Astbury,
who, as was mentioned in Chapter 4, had been one of the pioneers in the
X-ray crystallographic study of proteins. (Astbury, it should be noted, invented
the term "molecular biology" in the 1940's.) Astbury concluded, first of all,
that the density of dry samples of DNA is so high that the distance separating
successive nucleotides in the polynucleotide chain must be very small. Hence
he proposed that the DNA polymer is a column of nucleotides stacked on
top of each other. The individual nucleotides he imagined to be flat plates,
with the plane of the deoxyribose sugar ring being parallel to that of the
purine or pyrimidine ring, all perpendicular to the long axis of the molecule,
as shown in Figure 8-1. In the 1940's Astbury took some X-ray diffraction
photographs of DNA, which, though of insufficient quality to reveal much
detail of fine structure, did allow him to confirm his previous inference that
the polynucleotide is a stack of flat nucleotides. His measurements showed
that nucleotide residues, each oriented perpendicular to the long axis of the
molecule, were situated every 3.4 Å along the stack.

Three teams of research workers took up the X-ray crystallographic
analysis of DNA where Astbury had left it at the beginning of the 1950's.

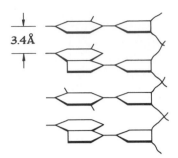

FIGURE 8-1.   The base-stacking structure of DNA proposed by
W. T. Astbury in 1945. In this structure the deoxyribose ring is
coplanar with its attached purine or pyrimidine ring, both sugar
and base being perpendicular to the long axis of the
polynucleotide chain molecule. [After Le Gros Clark and P. B.
Medawar (eds), *Essays in Growth and Form*, Oxford Univ.
Press, 1945.]

One team consisted of Pauling and his colleagues, whose then recent phenome-
nal success in solving the secondary structure of protein encouraged them to
apply the same techniques to DNA. Pauling's efforts did not meet with
success, however, for early in 1953 he proposed a DNA structure that was
proven wrong as soon as it was published. The second team, working under
the leadership of M. H. Wilkins, achieved an important technical advance:
they managed to prepare highly oriented DNA fibers that allowed them to
obtain an X-ray diffraction photograph showing a wealth of detail not
previously seen (Figure 8-2). In this superior picture, taken by Rosalind
Franklin, there were features that made certain facts immediately evident to
the trained eye of a crystallographer; in particular, one feature clearly confirmed Astbury's earlier inference of the 3.4 Å internucleotide distance. In
the winter 1952-1953 Wilkins and his collaborators were still trying to convert
their data into a reasonable molecular structure when Franklin's picture
(Figure 8-2) was seen by James Watson and Francis Crick (Figure 8-3), the
third team then in quest of the structure of DNA. Watson and Crick had
already considered a variety of possible structures, but because of the poor
quality of the X-ray photographs from which they had been working, they
had been unable to reach any definitive conclusions. But Franklin's picture
told them what they needed to know: within a few weeks they had worked
out the structure of DNA. In April 1953 Watson and Crick published their
conclusions about the structure of DNA in the same issue of *Nature* in which
Wilkins and his colleagues presented the X-ray evidence for that structure.

   Watson and Crick had concluded directly from the photograph shown in
Figure 8-2 that (1) the DNA polynucleotide chain has the form of a regular
helix, (2) the helix has a diameter of about 20 Å and, (3) the helix makes one

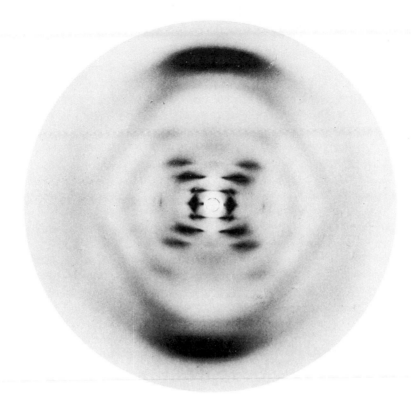

FIGURE 8-2.   X-ray diffraction photograph of DNA taken by Rosalind
Franklin in the laboratory of M. H. Wilkins. It was in this photograph
that Watson and Crick found the clues to their double helical structure.
The helical form is indicated by the dark areas that form a cross pattern
in the center of the picture. The large black areas at the top and bottom
of the picture indicate the 3.4 Å internucleotide distance in the base stack.
[This picture was first published by R. E. Franklin and R. Gosling in
Nature **171**, 740 (1953). Original photograph by courtesy of M. H. F.
Wilkins.]

complete turn every 34 Å along its length, and hence, since the internucleotide
distance is 3.4 Å, contains a stack of ten nucleotides per turn. Considering
the known density of the DNA molecule, Watson and Crick next concluded
that the helix must contain *two* polynucleotide chains, or two stacks of ten
nucleotides each per turn, since the density of a cylinder 20 Å in diameter and
34 Å long would be too low if it contained but a single stack of ten, and too
high if it contained three or more stacks of ten nucleotides each. Before
trying to arrange these two polynucleotide chains into a regular helix of the
required dimensions, however, Watson and Crick placed a further restriction
on their model—a restriction that derived from their knowledge that DNA is,

FIGURE 8-3.   (Left) Francis Crick (b. 1916). (Right) James Watson (b. 1928). [From
J. D. Watson, *The Double Helix*, Atheneum, New York, 1968.]

after all, the genetic material. If DNA is to contain hereditary information,
so they reasoned, and if that information is inscribed as a specific sequence of
the four bases along the polynucleotide chain, then the molecular structure of
DNA must be able to accommodate *any arbitrary sequence* of bases along its
polynucleotide chains. Otherwise the capacity of DNA as an information
carrier would be too severely limited. Hence Watson and Crick addressed
themselves to the problem of building a regular helix that, though composed
of two polynucleotide chains containing an arbitrary sequence of nucleotide
bases every 3.4 Å along their length, would nevertheless have a constant
diameter of 20 Å. Since the dimension of the purine ring is greater than that
of the pyrimidine ring, Watson and Crick hit upon the idea that the two-chain
helix could have a constant diameter if there existed a *complementary relation*
between the two nucleotide stacks, so that at every level one stack harbors
a purine base and the other a pyrimidine base. Finally, to endow the helix
with thermodynamic stability, the structure would have to allow ample
opportunities for the formation of hydrogen bonds between amino- or
hydroxyl-hydrogens and keto-oxygens or immino-nitrogens of the purine and
pyrimidine bases.

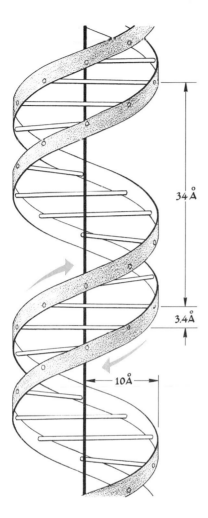

34 Å

3.4 Å

10 Å

FIGURE 8-4.   A schematic representation of the double-helical DNA molecule. The two ribbons symbolize the two deoxyribose-phosphate diester chains, and the horizontal rods the pairs of hydrogen-bonded bases that hold the chains together. The vertical line marks the fiber axis of the molecule.

These considerations led to the model shown in Figure 8-4, according to which the helix consists of two intertwined polynucleotide chains. These two chains are *antiparallel*—that is, have opposite directions. That a polynucleotide chain *has* a direction can be seen by inspecting the chemical formula of Figure 2-7 or the abbreviated, symbolic form shown in Figure 8-5. Moving along the chain from top to bottom is evidently different from moving along it from bottom to top. As in Astbury's model of the polynucleotide chain of Figure 8-1, the purine and pyrimidine rings are stacked as flat plates perpendicular to the long axis of the molecule. Contrary to Astbury's model, however, the plane of the deoxyribose sugar, forming with its esterified phosphate the backbone of the polynucleotide chain, is parallel to the main axis, and hence perpendicular to the plane of the base rings. The bases point inward,

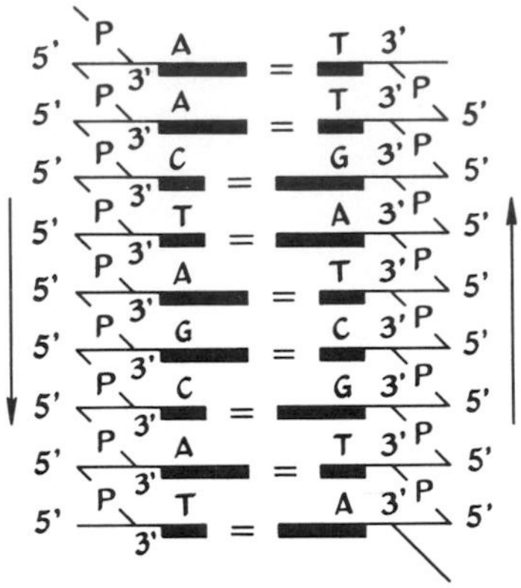

FIGURE 8-5. Another schematic representation of the double-helical DNA molecule. The polynucleotide chains are shown here in a symbolic form in which the symbol P represents the phosphate diester link between the 3'-OH of one deoxyribose residue (shown here as a thin horizontal line) and the 5'-OH of the next. The heavy horizontal lines represent the bases (A=adenine; G=guanine; T=thymine and C=cytosine) and the thin parallel lines the hydrogen bonds formed between complementary A-T and G-C base pairs. It is evident that the two polynucleotide chains run in opposite directions.

and at each nucleotide residue the two polynucleotide chains are held together by hydrogen bonds formed between a purine on one chain and a pyrimidine on the other chain. Probably the single most important feature of this structure is the manner in which these hydrogen bonds are formed (Figure 8-6): opposite every adenine base on one chain there is a thymine base at the corresponding position on the other chain: and a similar complementary relation holds between the guanine and cytosine bases of the two chains. In this way, and only in this way, is it possible to form the hydrogen bonds and at the same time maintain the constant diameter of the double helix. Though a good pair of hydrogen bonds could be formed between adenine and guanine, on the one hand, and between thymine and cytosine on the other hand, apposition of the two purines or of the two pyrimidines would occupy either too much or too little space to allow any regular double helix of constant diameter. Moreover, although the apposition of adenine and cytosine or of guanine and thymine would satisfy the spatial requirements, neither would allow formation of the hydrogen bonds. A photograph of the Watson-Crick structure of DNA built from space-filling atomic models is shown in Figure 8-7.

Though Watson and Crick had deduced the complementary relation of the four bases in DNA from their imposition of the boundary condition of arbitrary base sequence in the face of structural regularity, they were not slow to realize that the obligatory adenine-thymine and guanine-cytosine base

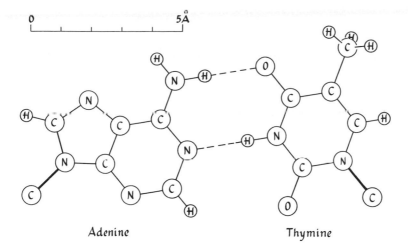

Adenine   Thymine

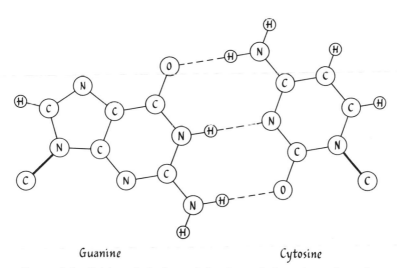

Guanine   Cytosine

FIGURE 8-6.   Pairing of adenine and thymine and of guanine and cytosine in the double-helical DNA molecule. Hydrogen bonds are shown as broken lines. The 1'-carbon of the deoxyribose residues is also shown. [From *Molecular Biology of Bacterial Viruses* by G. S. Stent. W. H. Freeman and Company. Copyright © 1963.]

pairing both provided the explanation for, and derived unexpected support from, Chargaff's previously mysterious equivalence rule. But the absolutely capital importance of the discovery of base pairing for the subsequent course of molecular genetics lay not in explaining these curious analytical data, but

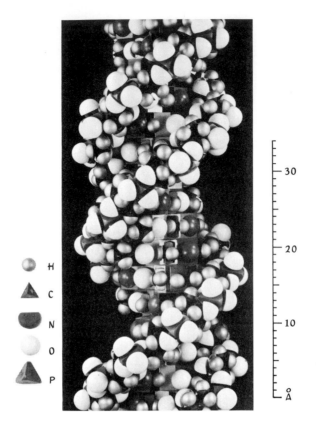

FIGURE 8-7.   A space-filling model of the double-helical DNA molecule. The sizes of the spheres that represent different atoms correspond to their relative radii. [Courtesy of M. H. F. Wilkins.]

in leading to the recognition that the whole DNA molecule is self-complementary: if the hereditary information is inscribed into the polynucleotide chain as a specific sequence of the four bases, then every DNA molecule carries *two* complete sets of that information, albeit written in complementary notation. By way of comment on this fact, Watson and Crick concluded their first letter to *Nature* describing the double helix with what can surely lay claim to being one of the most coy statements in the literature of science: "It has not escaped our notice that the specific pairing we have postulated immediately suggests a possible copying mechanism for the genetic material."

DENATURATION OF THE DOUBLE HELIX

Additional, more extensive X-ray crystallographic analyses by Wilkins and his colleagues soon confirmed the general validity of the Watson-Crick

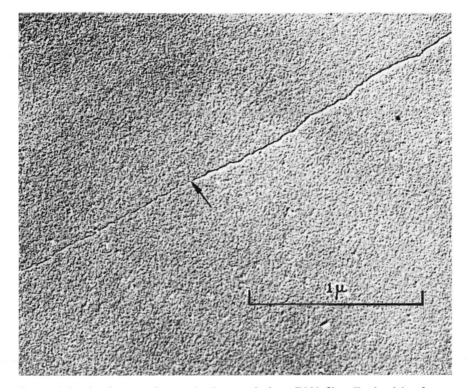

FIGURE 8-8.   An electron micrograph of a stretched-out DNA fiber. To the right of the arrow, the molecule consists of the double-stranded helix, which here can be seen directly to have the 20 Å diameter first inferred from X-ray crystallographic analyses. To the left of the arrow, the molecule consists of only a single polynucleotide strand, one of the two complementary strands having been removed by a special technique before the photograph was taken. The diameter of the single polynucleotide strand is clearly less than that of the double helix. [Unpublished photograph taken by L. A. MacHattie and C. A. Thomas, Jr.]

structure, though they led to some modifications of the spatial coordinates originally assigned to the atoms of the double helix. Physico-chemical studies of the behavior of DNA molecules in solution and direct electron-microscopical visualization of DNA (Figure 8-8) soon left little doubt that the double helix does describe correctly the three-dimensional structure of DNA. One particularly significant later development was the demonstration that the double helix collapses upon rupture of the purine-pyrimidine hydrogen bonds that hold the two chains together. For instance, rupture of the hydrogen bonds by heating a DNA solution to the boiling point and then cooling it rapidly causes the two complementary strands of the double helix to separate. Such separation engenders some drastic changes in the properties of the DNA molecule, since the two separated polynucleotide chains no longer maintain the stiff, rod-like helix but instead assume the configuration of a *random coil*.

One obvious consequence of the helix-coil transition is that the *viscosity* of the DNA solution is much reduced after the heat treatment. Another consequence of the collapse of the double helix is that the amino groups of the purine and pyrimidine bases, which are rather unreactive to formaldehyde in their hydrogen-bonded base-pairing configuration of the double helix, become exposed and readily react with formaldehyde to produce the corresponding hydroxymethyl derivatives:

$$NH_2 \qquad + CH_2O \longrightarrow \qquad NHCH_2OH$$

adenine   formaldehyde   6-hydroxymethylaminopurine

A third consequence is that the absorbance of ultraviolet light at the wavelength of 260 mμ by the DNA solution increases by about 40%. The disruption of the double helix and the dispersal of the bases from their stack causes an immediate increase in their absorbance of UV light because the close stacking of the bases in the intact double helix causes a reduction of their intrinsic absorbance below that manifested by the free bases. Figure 8-9 shows the result of an experiment in which the UV absorbance of the DNA of four different bacterial species was measured at various temperatures. As can be seen, each of these DNA species shows a sudden increase in light absorbance once a certain temperature, its so-called "melting point," has been reached. The melting point of each DNA species, however, is distinctly different, ranging from about 86°C for *Diplococcus pneumoniae* through 90°C for *E. coli* and 94°C for *Serratia marcescens* to 97°C for *Mycobacterium phlei*. Reference to Table 8-1 shows that this order of increasing melting points corresponds exactly to the order of increasing [G] + [C] contents of these same bacterial DNA species. This correspondence is to be explained by the relatively greater thermal stability of the three hydrogen bonds formed by the G-C base pair, compared to the two hydrogen bonds formed by the A-T pair. That is, the higher the [G] + [C] content of any DNA, the higher the temperature necessary to melt its hydrogen bonds. This inference finds most dramatic support from the melting-point determination of a fifth material shown in Figure 8-8, an artificially synthesized "DNA" containing *only* A-T base pairs. As is evident, the pure A-T "DNA" melts at a temperature as low as 65°C, the lowest melting point by far. Thus the melting temperature of any DNA can serve as an index of its base composition.

The activity of DNA molecules in the bacterial transformation phenomenon discussed in Chapter 7 makes it possible to follow the denaturation of the double helix also by means of a biological assay. For this purpose a solution of transforming DNA extracted from a genetically marked donor strain of

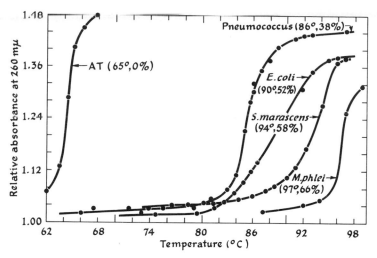

FIGURE 8-9.   The helix-coil transition, or "melting," of double-helical
DNA molecules extracted from four different bacterial species and of
the artificially synthesized "DNA" containing an alternating sequence
of A-T base pairs. The ordinate shows the relative absorbance of UV
light (260 mμ wavelength) of the various DNA solutions (containing
0.15 M NaCl and 0.015 M sodium citrate) at the temperatures indicated
on the abscissa. The figures in parentheses attached to each curve
indicate the "melting point," or the temperature at which the steeply
rising part of each curve reaches half the ultimate increase in UV
absorbance, and the [G] + [C] content of the sample. [After J. Marmur
and P. Doty, *Nature* **183**, 1427 (1959).]

D. pneumoniae is slowly raised to higher and higher temperatures. Samples
of this solution are taken from time to time and rapidly cooled in ice. These
samples are then added to a culture of recipient pneumococci differing in
genetic character from the donor strain, and the number of transformant
cells that have acquired the donor alleles is then scored. This experiment
shows that there occurs a precipitous decrease in transforming activity as
soon as the 86°C melting point of pneumococcal DNA is reached. Use of
[32]P-labeled transforming DNA in such heat-denaturation experiments shows
that the loss of transforming activity upon melting of the double helix is
attributable to failure of the recipient pneumococci to take up the single-
stranded polynucleotide chains.

RENATURATION OF THE DOUBLE HELIX

Whereas the finding of denaturation and strand separation of the DNA molecule at elevated temperatures did not cause much surprise once the Watson-
Crick structure had been put forward, the discovery made by J. Marmur in

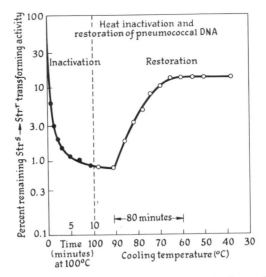

FIGURE 8-10.   Heat inactivation and restoration of the Str^s → Str^r transforming activity of DNA extracted from Str^r pneumococci. A DNA solution at 20 μg/ml in 0.15 M NaCl and 0.015 M sodium citrate was placed in a boiling water bath. At the times indicated on the abscissa, samples of the hot DNA solution were diluted into an ice-cold solution of 1.5 M NaCl and 0.15 M sodium citrate. One sample of the hot DNA solution was diluted into a hot solution of 1.5 M NaCl and 0.15 M sodium citrate at the 10th minute of heat treatment, placed in a large water bath, and allowed to cool slowly. When the temperature of the large bath reached the value shown on the abscissa just to the right of the broken vertical line, the samples were then chilled in ice. The ordinate indicates the residual capacity of the various heat-treated and then chilled DNA samples to transform Str^s-recipient bacteria to the Str^r state. After J. Marmur, C. L. Schildkraut, and P. Doty. In *The Molecular Basis of Neoplasia*, Univ. Texas Press, Austin, 1962.]

1960 that the double helix can be *reconstituted* from complementary single polynucleotide strands in solution was rather unexpected. One of Marmur's experiments is presented in Figure 8-10. The left hand part of this figure presents the kinetics of inactivation of the transforming activity of Str^r pneumococcus DNA at 100°C. As can be seen, within 10 minutes the transforming activity fell to less than 1 % of its initial value, as a result of denaturation and unwinding of the DNA molecules. The right-hand part of this figure presents the transforming activity of samples of the denatured DNA assayed while the 100°C water bath was cooling slowly to 60°C over a period of 80 minutes. It is apparent that during the cooling there occurs a gradual restoration of activity, and hence renaturation, of the double-helical DNA molecules, until 15 % of the initial activity is finally reached.

The renaturation of denatured DNA molecules can be followed also by means of the ultraviolet light absorbance of the solution. That is, upon incubating at 65°C a sample of DNA that has been denatured by heating to a temperature above its melting point, the ultraviolet absorbance of the solution gradually falls to a lower value, as the double helices reform and the nucleotide bases are restacked. The rate at which the single strands reform double helices is, however, strongly dependent on the type of DNA whose renaturation is being examined. For example, the absorbance of denatured pneumococcal DNA falls from the initial value of 1.4 to a value 1.2 times the corresponding absorbance of native DNA within an hour's incubation at 65°C. It can be concluded, therefore, that about half the double helices are reformed during that time. Denatured calf-thymus DNA responds differently: incubation at 65°C for one hour produces practically no renaturation; in contrast, renaturation of denatured bacteriophage DNA is nearly complete within a few minutes (Figure 8-11).

Marmur explained the variation in the renaturation behavior of different types of DNA in the following terms. Since the reformation of the double helices demands the collision in free solution of two polynucleotide chains of complementary base sequence, it stands to reason that the rate of renaturation rises with the probability that any given single polynucleotide chain will collide with a chain of complementary sequence. Among the three types of DNA considered here this chance is maximal for the bacteriophage DNA, inasmuch as the DNA complement of one phage particle embodies two complementary base sequences comprising only 10^5 nucleotides each. The DNA complement of the pneumococcus embodies two complementary base sequences comprising about 3×10^6 nucleotides each; hence the chance that complementary base sequences of denatured pneumococcal DNA would encounter one another is obviously less than for denatured phage DNA. Finally, the DNA complement of the mammalian chromosome set comprises 5000 times more nucleotides than that of the pneumococcus, making the collision of polynucleotides of denatured calf thymus DNA with complementary base sequences even less likely.

The technique of melting the DNA double helix and then allowing its reformation from complementary single-strand polynucleotide chains has found one of its most interesting applications in the systematics, or study of relatedness, of higher organisms. The basic idea underlying this approach is that the more genes two organisms share in common, and hence the more DNA polynucleotide base sequences they share in common, the more closely they are related. Thus in order to establish the degree of relatedness between organism A and organism B it is necessary only to extract the DNA from their cells, heat and anneal a mixture of DNA from their cells, and ascertain the extent to which *hybrid* double helices are reformed that carry one polynucleotide strand derived from A and one derived from B. In order to carry out such experiments, E. T. Bolton and B. J. McCarthy developed an easy method for the detection and quantitative estimation of hybrid DNA double helices. For this purpose, the DNA extracted from organism A is heated to

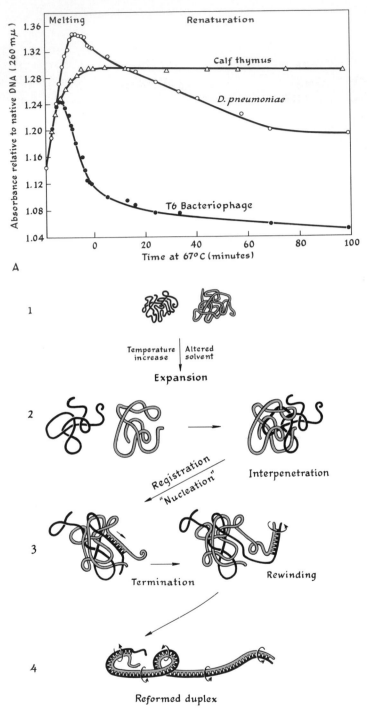

A

(Graph labels)
Absorbance relative to native DNA (260 mμ)
Melting | Renaturation
Calf thymus
D. pneumoniae
T6 Bacteriophage
Time at 67° C (minutes)

B

1

Temperature increase | Altered solvent

Expansion

2

Registration
"Nucleation"

Interpenetration

3

Termination   Rewinding

4

Reformed duplex

100°C and cooled quickly, in order to separate the native DNA molecules into single polynucleotide strands. These single strands are added to a hot solution of melted agar, which is then quickly cooled. Upon solidification of the agar, the single DNA strands become embedded and immobilized in the agar gel. Cells of organism B have meanwhile been grown in the presence of a radioactive DNA precursor, such as $^{32}PO_4^{-3}$ or ^{14}C-thymine. The radioactive DNA is then extracted from the B cells, broken by mechanical shear into relatively short polynucleotide fragments about a thousand nucleotides in length, heated and quickly cooled to separate the double helices into single strands, and then added to the agar in which the single strands of the DNA from organism A are already embedded. The agar is then heated to 60°C overnight, at which temperature double helices composed of single polynucleotide strands from organism A and single polynucleotide strands from organism B begin to reform. A salt solution is then passed through the agar to wash out all type-B polynucleotide strands that did not succeed in reforming a double helix with embedded type-A polynucleotide strands, and hence did not become trapped in the agar. Measurement of the trapped, radioactively labeled type-B strands gives the fraction of the labeled DNA of organism B that can reform double helices, and hence has nucleotide sequences in common with the nonlabeled DNA of organism A.

Table 8-2 presents the result of a survey carried out by means of this method on some vertebrate animals. It will be recalled from the discussion of the data of Table 8-1 that the base composition of the DNA of the vertebrates spans

FIGURE 8-11.   Renaturation of denatured DNA. **A.** Effect of the biological provenance of the denatured DNA sample on the kinetics of renaturation. DNA extracted from calf thymus, pneumococci, and T6 bacteriophage was dissolved to a concentration of 20 $\mu g/ml$ in 0.3 M NaCl and 0.03 M sodium citrate. The solutions were heated to 100°C for 10 minutes then quickly cooled in ice and placed in a spectrophotometer chamber maintained at 67°C in which the ultraviolet light absorbance (shown on the ordinate) could be measured continuously. The left-hand part of the graph shows the increase in UV absorbance ("melting") as the initially cold, denatured DNA samples warm up to the 65°C temperature of the UV absorption chamber. The right-hand of the graph shows the gradual *decrease* in UV absorbance (renaturation) of the DNA samples maintained at various times after having reached 65°C. [After J. Marmur, C. L. Schildkraut, and P. Doty. In *The Molecular Basis of Neoplasia*, Univ. Texas Press, Austin, 1962.] **B.** Schematic presentation of the renaturation of denatured DNA molecules. (1) Tightly coiled single polynucleotide chains (at low temperatures and high ionic solution strengths) are nonpenetrating and hence weakly interacting. (2) The single-strand coils expand with increase in temperature and decrease in ionic strength, allowing their interpenetration. (3) Nucleation of an incipient renatured double-stranded DNA molecule is accomplished by formation of hydrogen bonds between complementary base sequence regions that are in register. (4) The remainder of the molecule renatures rapidly by zipper-like action while the two polynucleotide chains rewind. [After C. A. Thomas, Jr. *In* G. C. Quarton, T. Melnechuck, and F. O. Schmitt (eds.), *The Neurosciences*, The Rockefeller Univ. Press, New York, 1967].

the rather narrow range of 40 to 44% [G] + [C], but, as the data of Table 8-2 show, there do exist considerable differences in nucleotide base sequences among the different members of this taxonomic phylum. In this survey, DNA from a variety of vertebrates (as well as from *E. coli*) was embedded in agar, and the trapping of radioactively labeled DNA fragments from either mouse or human cells was tested. Three important results can be noted. First, it can be seen that only 18% of the added human DNA fragments are trapped by human DNA. Moreover, only 22% of the added mouse DNA fragments are similarly trapped by mouse DNA. The failure to achieve 100% trapping of the labeled DNA from two homologous organisms is attributable to the fact that reformation of double helices occurs more frequently between the added polynucleotide chain fragments themselves than between them and the embedded DNA. Hence the figure of about 20% trapping can be taken as the upper limit to be expected between completely homologous embedded and added test DNA species. Second, it can be seen that 6% of the added human DNA is trapped by the mouse DNA and that 5% of the added mouse DNA is trapped by the human DNA. These figures indicate that human and mouse DNA have about $6/22 = 0.27$ or $5/18 = 0.27$ of their polynucleotide sequences in common. Third, it can be seen that, in number of nucleotide sequences, the DNA of the rhesus monkey is closer to that of the human, and the DNA of the rat and hamster are closer to that of the mouse, all of which is fully

TABLE 8-2. Hybrid Double-helix Formation Between DNA Molecules Extracted from Various Vertebrate Animals

Source of DNA embedded in agar	Percentage of labeled DNA fragments trapped in agar	
	Human DNA	Mouse DNA
Man	18	5
Mouse	6	22
Rhesus monkey	14	8
Rat	3	14
Guinea pig	3	3
Salmon	1.5	1.5
Escherichia coli	0.4	0.4
No DNA in agar	0.4	0.4

From B. H. Hoyer, B. J. McCarthy, and E. T. Bolton, *Science*, **144**, 959 (1964).

consonant with old-fashioned taxonomic criteria. Guinea pig and rabbit DNA are as equally remote from human DNA as are rat and hamster DNA. Salmon DNA is even more remote from human and mouse DNA than is the DNA of their fellow mammals. Finally, no more *E. coli* DNA is trapped by agar containing human or mouse DNA than is trapped by agar into which no DNA at all has been embedded, leading to the conclusion that man and mouse have practically no genes in common with *E. coli.*

ADUMBRATION OF THE GENETIC CODE

Just as the 1946 Cold Spring Harbor Symposium had been dominated by discussions of the ascendant one-gene–one-enzyme theory, so was the 1953 Cold Spring Harbor Symposium dominated by discussions of the implications of the then three-month-old Watson-Crick structure of DNA. No one who listened to Watson's lecture at that meeting (Figure 8-12) needed much imagination to realize that with the discovery of the double helix, the understanding of the gene was about to reach a higher plane. A new era was obviously dawning for genetics. One of the most pregnant issues of these

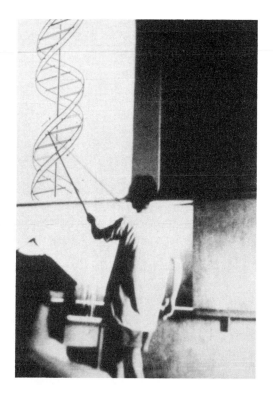

FIGURE 8-12.   James Watson presenting the Watson-Crick structure of DNA at the 1953 Cold Spring Harbor Symposium. [From J. Cairns, G. S. Stent, and J. D. Watson (eds.), *Phage and the Origins of Molecular Biology.* Cold Spring Harbor Laboratory of Quantitative Biology, New York, 1966.]

discussions was a general formulation of the problem of the *genetic code*, which was immanent in the two recently promulgated fundamental dogmas outlined briefly in Chapter 7. The first of these dogmas held that it is the exact sequence of nucleotide bases in the DNA polynucleotide chains that represents the genetic information of the gene. That is, the long DNA double helix was to be thought of as a tape on which information is inscribed in a language employing the four-letter alphabet, A, G, C, and T. This information is, as has been noted in this chapter, inscribed *twice* in every DNA molecule, and hence twice in every gene, since any base on one chain of the two helically inter-twined polynucleotides determines its complementary base on the other chain. From this it followed that though the same four-letter alphabet is used for inscribing the hereditary information on both chains, the two chains do not record this information in the same language. The second of these dogmas, the latter-day version of the one-gene–one-enzyme theory brought about by considerations of protein structure, held that the information content of any gene, and hence the actual *meaning* of the four-letter language recorded in the DNA, could not be anything but a representation of the primary structure of a given polypeptide. Taken together, both dogmas implied that the meaning of any particular purine-pyrimidine base sequence of the DNA polynucleo-tide chains that constitute a gene is nothing other than the specification of the *amino acid sequence* of some particular polypeptide chain. So, in the summer of 1953 the idea sprung into being that there must exist a *code* that relates nucleotide base sequences in polynucleotides to amino acid sequences in polypeptides. Some simple considerations quickly revealed the minimum complexity of this code, since for each amino acid residue in the polypeptide chain information must be provided as to just which of the twenty standard amino acids is to be present there. Certainly there cannot exist a one-to-one correspondence between purine and pyrimidine bases in the DNA and amino acids in the polypeptide, because the four bases A, G, C, and T taken one at a time could specify only one out of four, and not one out of twenty, kinds of amino acids. Nor would it suffice that *two* adjacent base pairs of the DNA double helix specify one amino acid residue in the polypeptide, since this would mean that no more than $4 \times 4 = 16$ kinds of different amino acids could be coded by the four bases. Hence the code must involve the specifica-tion of one amino acid residue by at least *three*, probably successive, bases in the DNA polynucleotide chains. Four kinds of bases taken three at a time could specify more than enough, or $4 \times 4 \times 4 = 64$ different kinds of amino acids.

The first to publish a formal scheme for a genetic code was the physicist-cosmologist George Gamow, who proposed in 1954 that each particular constellation of four contiguous purine and pyrimidine residues of the DNA double helix—two on one and two on the other of the two complementary polynucleotide chains—generates one of twenty specific kinds of "cavities" in the surface of the DNA macromolecule into which the side-chain of one and only one of the twenty protein amino acids can fit. In this way, Gamow thought, the DNA can function as the direct template for the ordered assembly of the specific polypeptides from their amino acid building blocks. When Gamow found that this scheme leads to internal contradictions and his attention was drawn to the fact that proteins are probably not synthesized *directly* on the DNA anyway, he generalized his proposal to take account of the possibility that not the DNA but some other substance is the immediate template of protein synthesis. The purely informational essence of Gamow's revised scheme in its most general form can be summarized as follows:

1. Three successive nucleotide base pairs on the DNA polynucleotide chains code for one amino acid. There are, therefore, 4^3 or 64 possible triplet words for the standard set of 20 amino acids.

2. Each nucleotide pair participates in the coding of three amino acids (Figure 8-13). The code is, therefore *overlapping*.

3. An amino acid can be represented by more than one kind of nucleotide pair triplet, or word. The code, therefore, contains synonyms.

Such an overlapping code imposes serious restrictions on the possible sequences of amino acids that could occur in nature. For instance, in any polypeptide an amino acid whose code word is ATT could not be the neighbor of an amino acid whose code word is ACC. If this type of code were actually in use, it should be possible to "break" it through purely cryptographic methods by looking for the occurrence of neighbor restrictions among amino acids found in proteins. A survey made by S. Brenner of the protein amino acid sequences known by 1957 revealed, however, that there do not seem to exist any such neighbor restrictions; hence there was no hope of breaking the code in this way. Brenner even demonstrated that the actual structure of proteins appears to rule out all overlapping codes of the type proposed by Gamow.

Crick and two colleagues then devised a nonoverlapping code. Under this code each triplet of nucleotide base pairs in the DNA double helix specifies one amino acid, and each base pair participates in the specification of only one amino acid (Figure 8-13). With such a code a new difficulty had to be faced: How is the "reader" supposed to know which of the triplets to "read"? For example, in the sequence of nucleotides . . . , ATT, GCA, TCG, TGG, . . . shown in Figure 8-13, where ATT represents one amino acid, GCA another, and so on, how could this message be read correctly if the commas were removed? To resolve this difficulty, an additional restriction was placed on this code; each code word triplet, such as ATT, GCA, TCG, TGG, can be automatically recognized as a word by means of a *dictionary of sense words*. The set of sense words would have the property that no sense word can be

A

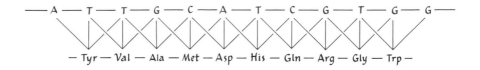

B

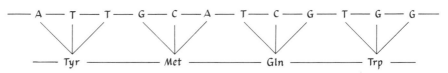

FIGURE 8-13.   Two general types of codes relating nucleotide sequence in DNA to amino acid sequence in polypeptides. In both codes, one amino acid is specified by three successive nucleotides. The assignment made here of any amino acid to a given nucleotide triplet is entirely hypothetical (and, as will be seen in Table 18-3, wrong). **A.** An *overlapping* code in which each nucleotide partakes in the specification of *three* amino acids. The *coding ratio*, or the ratio of the number of nucleotides in a segment of DNA to the number of amino acids in the polypeptide chain coded by this segment, is 1:1. **B.** A *nonoverlapping* code, in which each nucleotide partakes in the specification of one amino acid. Here the coding ratio is 3:1. [From *Molecular Biology of Bacterial Viruses* by G. S. Stent. W. H. Freeman and Company. Copyright © 1963.]

formed by misreading parts of any two adjacent sense triplets. Only sense words of nucleotide triplets correspond to amino acids in the dictionary; all the other triplets—those that might be formed by misreading adjacent sense triplets—do not code for any amino acid at all and are, therefore, *nonsense.* For example, if ATT and GCA are sense words in the dictionary then TTG, TGC, CAA, and AAT, which might be misread from the fragments ATTGCA and GCAATT, are necessarily nonsense words not to be found in the dictionary. In this way, a *comma-less* code can be generated, which assures correct reading of the message without any necessity for starting the reading at the beginning of the polynucleotide chain or having commas between adjacent sense triplets. Such comma-less codes are not only formally possible but in any such code no more than 20 of the 64 possible triplets can make sense, and at least 44 of the remaining triplets are necessarily nonsense. Thus this kind of comma-less code has no synonyms—only one code word per amino acid—and marvelously explains the apparent excess of possible triplets over the number of amino acid types. The restricted nature of the dictionary allows the automatic recognition of the beginning and end of each code word and this eliminates the need for a comma or spacer symbol.

As it turned out, nature does not seem to avail herself of this clever system of information storage and retrieval, perhaps because it is *too* clever by half! The comma-less code without synonyms probably lacks evolutionary flexibility; and, to be workable, the whole system, like Athena, would have to have sprung full-blown from Zeus' head. Instead, as experiments to be presented in later chapters were to show, the code is of the simplest type. It employs nucleotide triplets; each represents an amino acid, and the triplets are read simply, three by three from the beginning of the message to the end.

Taking the existence of such a nucleotide *triplet code* for granted for the time being, a rough estimate can be made of the total number of different proteins that are likely to be coded by the *E. coli* DNA. As was stated in Chapter 2, the DNA complement of an *E. coli* nucleus comprises about 3×10^6 nucleotide base pairs. A reasonable guess of the number of amino acids in the average *E. coli* protein might be 330. This means that under the triplet code $3 \times 330 = 1000$ nucleotides in one of the two DNA chains, or 1000 nucleotide base pairs of the double helix, would be involved in coding for the average protein. Hence the number of different average-sized proteins coded by the *E. coli* cell would be $3 \times 10^6/1000 = 3000$. Since each enzyme corresponds to at least one particular polypeptide chain, the figure of 3000 would also represent an upper limit to the number of specifically catalyzed biochemical reactions that *E. coli* could carry out. That figure exceeds by a factor of only two or three the number of reactions already known to the students of intermediary metabolism. And hence the prospect of a total bio-chemical description of the *E. coli* cell seems not far out of reach.

Bibliography

PATOOMB

J. D. Watson. Growing up in the phage group.

HAYES

Chapter 11.

ORIGINAL RESEARCH PAPERS

Chargaff, E. Chemical specificity of the nucleic acids and mechanism of their enzymatic degradation. *Experientia*, **6**, 201 (1950).

Hoyer, B. H., B. J. McCarthy, and E. T. Bolton. A molecular approach in the systematics of higher organisms. *Science*, **144**, 959 (1964).

Marmur, J., R. Rownd, and C. L. Schildkraut. Denaturation and renaturation of deoxyribonucleic acid. *Progress in Nucleic Acid Research*, **1**, 231–300 (1963).

Watson, J. D., and F. H. C. Crick. A structure for deoxyribosenucleic acid. *Nature*, **171**, 737 (1953).

SPECIALIZED TEXTS, MONOGRAPHS, AND REVIEWS

Davidson, J. N. *The Biochemistry of the Nucleic Acids* (5th ed.). Methuen, London; Wiley, New York, 1965.

Olby, R. "Francis Crick." *Daedalus,* **99,** 938 (1970). (A biography.)

Watson, J. D. *The Double Helix.* Atheneum, New York, 1968; Mentor Books, New York, 1969. (An autobiographical account of the discovery of the DNA structure.)

9. DNA Replication

At the beginning of Chapter 5, the basic problem of cellular self-reproduction was restated in terms of the question of how the entire apparatus of cellular proteins doubles in the course of the cell-generation period. And this question was, in turn, reduced at once to asking how the particular sequence of the twenty protein amino acids that makes up the primary structure of any given protein species is assembled into its unique order. Having seen meanwhile that the genes, or more specifically, the nucleotide sequences of the DNA are the self-reproducing informational elements which govern this assembly process, we are now ready to restate the basic problem in terms of two conceptually distinct functions of DNA, namely the *heterocatalytic* and the *autocatalytic* functions.

SECOND STATEMENT OF THE BASIC PROBLEM

For the purpose of this restatement, we ask two questions. First, how does the bacterial DNA manage to control, or *heterocatalytically* preside over, the synthesis of the polypeptides whose primary structure is inscribed into it, so as to provide for the daughter bacterium the enzymatic ensemble to which it owes its character? Or, more specifically, how is the purine-pyrimidine base sequence of the hereditary polynucleotide *translated* into the amino acid sequences that it specifies via the genetic code? Second, how does the bacterial DNA manage to reproduce itself *autocatalytically*, so as to regenerate one exact copy of every bacterial gene with which the daughter bacterium is to be endowed? Or, more specifically, how is a parental DNA molecule replicated to give rise to two DNA molecules of identical purine-pyrimidine base sequence? The ultimate solution of the problem of the heterocatalytic function of DNA—that is, of the molecular mechanism of protein synthesis—was not found until the 1960's, and its discussion must await later chapters. The autocatalytic function of DNA, however, happened to be the very thing that had not escaped Watson and Crick's notice. In a second letter to *Nature*, which appeared a few weeks after the one in which they announced the structure of the double helix, they clarified the meaning of the coy sentence with which they had concluded their first letter.

THE WATSON-CRICK REPLICATION MECHANISM

What the complementary pairing of purine and pyrimidine bases of the two polynucleotide chains of the double helix had suggested to Watson and Crick was that the DNA molecule could replicate itself directly by having each chain serve as a *template* for the formation of its own complementary chain. In this act of replication, they thought, the two strands of the double helix separate, and each purine and pyrimidine base attracts a complementary free nucleotide available for polymerization in the cell and holds it in place by means of the specific hydrogen bonds shown in Figure 8-5. Once held in place on the parental template chain, the free nucleotides are sewn together by formation of the phosphate diester bonds that link adjacent deoxyribose residues, forming a new polynucleotide molecule of predetermined base sequence (Figure 9-1). Thus after growth of complementary replica chains has taken place along the full length of both parental polynucleotide chains, two DNA molecules have been generated that have identical sequences of the four bases, and hence have the same information content as the parental DNA double helix. At this point, the autocatalytic function has been achieved. When the time comes for the next replication cycle in the two daughter cells that were endowed with the two daughter DNA molecules generated by the first cycle, the four polynucleotide chains of both daughter DNA molecules again separate. Each chain then acts as a template for the growth of further polynucleotide chains, thereby generating four granddaughter DNA double helices, each of which is identical in base sequence to the grandparental double helix and provides the genetic endowment for four granddaughter cells.

No sooner had the Watson-Crick replication mechanism been proposed than it was realized that its occurrence would be reflected in a singular distribution

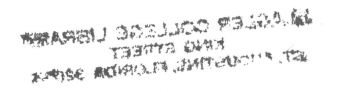

of the *substance* of the parental DNA molecule over the daughter molecules. This manner of distribution, it is fair to say, would not have been imagined by anyone before the discovery of the complementary double helix. Since in the act of serving as templates for the growth of new replica chains the two parental polynucleotide chains separate, the atoms of the parental

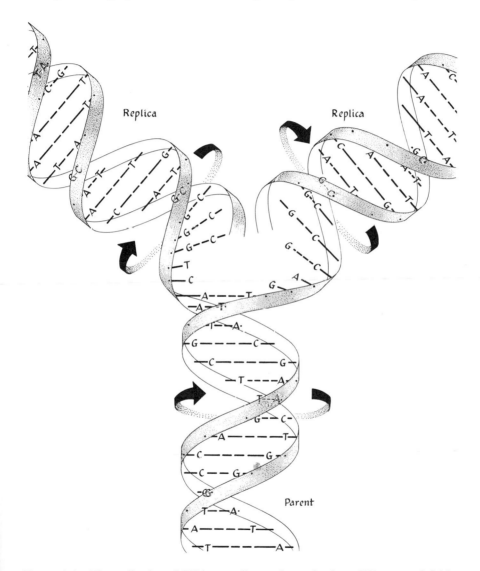

FIGURE 9-1.   The replication of DNA according to the mechanism of Watson and Crick. [From *Molecular Biology of Bacterial Viruses*, by G. S. Stent. W. H. Freeman and Company. Copyright © 1963.]

double helix become equally distributed over the two daughter molecules of the first replication cycle. In subsequent replication cycles, however, no further dispersal of the original parental DNA atoms should occur, since the individual polynucleotide chains of the parental DNA molecules remain intact. For instance, among the four daughter double helices generated by the second replication cycle, two contain one of the parental polynucleotide chains and one de novo chain, and the other two contain only nonparental substance. More precisely, in every one of the 2^g daughter double helices generated by g synchronous replication cycles of a parental DNA molecule, one chain was synthesized in the very last cycle, and its complementary chain was synthesized with probability 2^{-i} in the ith cycle before the last. This mode of distribution of the parental atoms has been called *semiconservative*, in contradistinction to a *conservative* distribution, which conserves the integrity of the whole parental double helix in the replication process. Under conservative replication, there is among the daughter DNA molecules generated by g replication cycles always one double helix whose atoms are entirely parental, whereas the substance of all other helices is entirely de novo (Figure 9-2).

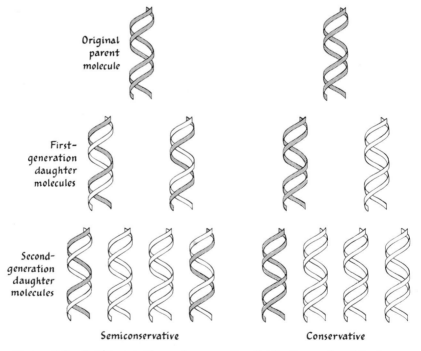

Original
parent
molecule

First-
generation
daughter
molecules

Second-
generation
daughter
molecules

Semiconservative   Conservative

FIGURE 9-2. Semiconservative and conservative distribution modes of the two parental polynucleotide chains as possible alternatives in the replication of DNA. [From *Molecular Biology of Bacterial Viruses*, by G. S. Stent. W. H. Freeman and Company. Copyright © 1963.]

Thus, in the years following the discovery of the DNA double helix, tests of the prediction of semiconservative replication seemed an urgent matter of business. For it did not follow that just because Watson and Crick had been right about the *structure* of DNA, they were also right about the manner of its *replication*. Indeed, Watson and Crick themselves had pointed out in their lecture delivered at the 1953 Cold Spring Harbor Symposium some difficulties inherent in their replication scheme, and further possible objections—some real, some imagined—were soon proposed by others. Indeed, it then still seemed conceivable to some students of this problem that DNA does not replicate itself directly at all, but replicates by transfer of the information in its nucleotide sequence to a non-DNA intermediary which then, in turn, serves as the template for synthesis of the replica nucleotide chains. Such schemes of indirect replication, for which there seemed to be some experimental support, led to the counter-proposal of conservative replication.

THE MESELSON-STAHL EXPERIMENT

Efforts to determine experimentally the fate and manner of distribution of the atoms of replicating DNA molecules had actually begun three years before Watson and Crick proposed their mechanism. Watson, in fact, had published a paper on this subject as early as 1951. In these *transfer experiments*, the parental DNA molecules were labeled with radiophosphorous, ^{32}P, or radiocarbon, ^{14}C, replication was allowed to proceed in a nonradioactive growth medium, and then the content and relative distribution of the parental radioisotopes in the daughter DNA molecules were ascertained. Once the singular prediction of semiconservative distribution had been made such efforts were intensified, since their outcome promised to deliver proof or disproof of the Watson-Crick proposal. Technically, however, these experiments turned out to be quite difficult, since they required the development of methods for measuring the tiny radioactive content of individual DNA molecules. Although some of the radioisotope transfer experiments did elicit some interesting physiological and genetic information, they failed to settle the main point for which they had been designed. An experimental breakthrough finally came in 1957, when M. Meselson and F. W. Stahl (Figure 9-3) decided to carry out a transfer experiment in which the heavy, stable isotope of nitrogen, ^{15}N, was used as a DNA label, instead of the radioisotopes ^{32}P or ^{14}C. This procedure became feasible after Meselson, Stahl, and J. Vinograd showed that the replacement of the ordinary, *light* isotope ^{14}N in purine and pyrimidine nitrogen atoms by the *heavy* isotope ^{15}N causes an appreciable increase in the *density* of the DNA molecule, as measured by *density-gradient equilibrium sedimentation*. For the purpose of this measurement, the DNA is

FIGURE 9-3.   (Left) Matthew Meselson (b. 1930). (Right) Franklin W. Stahl (b. 1929).
[Courtesy of M. Meselson.]

dissolved in a solution of $6M$ CsCl, whose density is about 1.7 g/cm^3, or very close to that of the DNA itself. This solution is then centrifuged at a velocity of about 50,000 rpm, which subjects the molecules in the solution to centrifugal forces about 10^5 times gravity. Under the influence of these high centrifugal forces, the Cs$^+$ and Cl$^-$ ions tend to settle to the bottom of the centrifuge tube, but this tendency is opposed by the rapid back diffusion, or Brownian movement, of these ions. Finally, after the centrifuge has been running for a few hours, the ions reach an equilibrium distribution—that is, a continuous concentration gradient of CsCl, and hence a density gradient, exists in the tube, the density being greatest at the bottom (Figure 9-4). Meanwhile, the DNA molecules in this solution have also been subject to the centrifugal force. Those DNA molecules that were initially in the lower part of the CsCl gradient, where the density of the solution now *exceeds* their own density, migrate upward, and the DNA molecules that were initially in the upper part of the gradient, where the density of the solution is *less* than their own density, migrate downward. Both upward and downward migration of DNA molecules finally stop in that zone of the density gradient where the density of the solution happens to be exactly that of the DNA molecules themselves. Thus the dissolved DNA molecules are centrifuged into a narrow

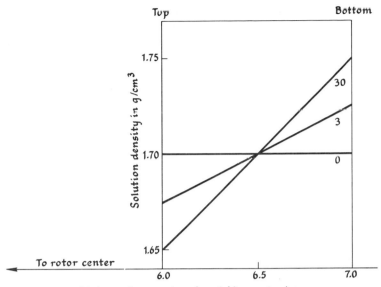

FIGURE 9-4.   Establishment of a CsCl density gradient by high-speed
centrifugation. The ordinates, whose positions mark the top and bottom
of a 1-cm-high centrifuge tube filled with a $6M$ CsCl solution, register the
solution density at the elevation indicated on the abscissa. The three lines
on this graph show the density profile of the solution after the centrifuge
has run for 0, 3, and 30 hours, respectively. The 30-hour profile is, for
all practical purposes, the equilibrium density gradient. DNA molecules
of density 1.70 g/cm^3 would be compressed into a band in the exact
middle of the centrifuge tube.

band, whose position in the tube is a precise measure of the density of the
molecules in that band.

As a preliminary to their transfer experiment, Meselson and Stahl grew a
culture of *E. coli* in a synthetic glucose-salts medium similar to that specified
in Table 2-1, whose source of nitrogen, the NH_4Cl, contained heavy ^{15}N,
instead of ordinary, light ^{14}N. They then extracted the DNA from these
"heavy" bacteria, mixed it with DNA extracted from ordinary, or "light,"
^{14}N-grown bacteria, and centrifuged this DNA mixture in a CsCl solution.
They found two distinct DNA bands in this way, whose difference in position
corresponded to the expected density difference of 0.014 g/cm^3. Thus "heavy"
and "light" DNA molecules can be easily resolved by this method.

To carry out their transfer experiment, Meselson and Stahl added a great
excess of $^{14}NH_4Cl$ and some ^{14}N-containing organic nitrogen sources to a
"heavy" culture of *E. coli* previously grown in a synthetic medium containing
$^{15}NH_4Cl$. This addition represented the density transfer point, after which all
nitrogen assimilated by the bacteria, and hence every newly synthesized

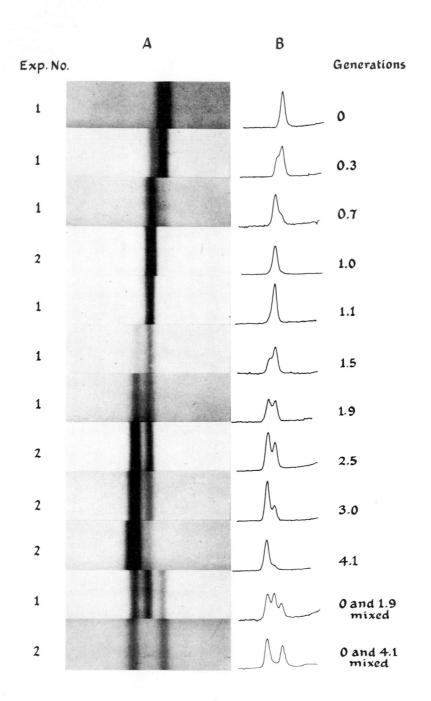

polynucleotide chain was of the light ^{14}N kind. At various times after the density transfer, DNA was extracted from a part of the growing bacterial culture and subjected to CsCl density-gradient equilibrium sedimentation. The result of this experiment, shown in Figure 9-5, was that after exactly one generation of growth—when the DNA present in the culture at the density transfer point had doubled in amount and every parental DNA molecule must have replicated itself in the meantime—the DNA is found to be present as a ^{15}N-^{14}N *hybrid* that sediments into a density band characteristic of molecules containing equal proportions of the two nitrogen isotopes. Furthermore, after two generation times of growth, when the parental DNA had undergone two replication cycles, half of the DNA can be seen to be in the hybrid band and half to be in a band characteristic of fully light, ^{14}N-DNA. Thus Meselson and Stahl showed that the distribution of the parental ^{15}N atoms of *E. coli* DNA actually follows the semiconservative mode demanded by the Watson-Crick replication mechanism, according to the schema of Figure 9-2. Thus it appeared that Watson and Crick had been right from the very first about both the structure of the hereditary polynucleotide *and* its manner of replication.

DENSITY DETERMINATIONS

The following objection can be raised against the inference just drawn that the result of the Meselson and Stahl experiment validates the Watson-Crick replication mechanism: whereas this experiment certainly shows that there is some elementary DNA unit that replicates semiconservatively, the data

FIGURE 9-5. The demonstration of semiconservative replication of *E. coli* DNA.
A. Ultraviolet absorption photographs of a centrifuge cell, in which DNA molecules are banded in a density gradient established in a concentrated cesium chloride solution. Each frame represents the bands formed by DNA extracted from an initially ^{15}N-labeled *E. coli* culture growth in a ^{14}N-labeled medium for the number of generations indicated. The density of the cesium chloride solution increases to the right, regions of equal density occupying the same horizontal position on each photograph. **B.** Densitometer tracings of the photographs at left; the height of the densitometer curve is proportional to the concentration of DNA. The relative content of ^{15}N and ^{14}N isotopes of a molecular species is indicated by the position of its band in relation to the bands of fully ^{15}N- and fully ^{14}N-labeled DNA shown in the lowest frame. It can be seen that after 1.0 generation of growth, all the DNA is found in a *hybrid* band, characteristic of the density of molecules containing equal proportions of the two nitrogen isotopes. After 1.9 generations, half of the DNA is present in a hybrid band and half is present in a band characteristic of molecules containing only ^{14}N. A test of the inference that the DNA in the band of intermediate density contains just equal proportions of the two nitrogen isotopes is provided by the frame showing the bands obtained by centrifugation of a mixture of DNA extracted from bacteria after 0 and 1.9 generations. [From M. Meselson and F. W. Stahl, *Proc. Natl. Acad. Sci. Wash.* **43**, 581 (1958).]

Semiconservative   Dispersive   End-to-end conservative

FIGURE 9-6.   Dispersive and end-to-end conservative distribution modes of the
two parental DNA chains, which, like the semiconservative mode, are also
compatible with the finding of the Meselson-Stahl experiment of Figure 9-5
that the first-generation replicas are half-heavy, half-light hybrid molecules.
But upon denaturation and strand separation of the first-generation hybrid DNA
molecules, only the semiconservative mode would lead to fully heavy and fully
light single strands, whereas the two other modes would both lead to half-heavy,
half-light hybrid single strands.

presented do not prove that this unit is actually the DNA double helix. For instance, it is conceivable that the first-generation daughter DNA molecules owe their ^{14}N-^{15}N hybrid density not to the semiconservative mode indicated in Figure 9-2 but instead to one of two alternative modes of distribution outlined in Figure 9-6. One of these can be referred to as *dispersive* and the other as *end-to-end conservative*. An experimental distinction between these alternative modes of distribution can be made upon denaturing the first-generation hybrid daughter DNA molecules and ascertaining the density of the resulting single polynucleotide chains. As can be seen in the schema of Figure 9-6, denaturation of ^{14}N-^{15}N hybrid DNA double helices carrying the two isotopes in either dispersive or end-to-end conservative modes would give rise to single polynucleotide strands of hybrid density, whereas denaturation of the true semiconservatively labeled hybrid double helix would give rise to two classes of single strands, one class carrying only ^{14}N and the other only ^{15}N atoms.

The result of an experiment that demonstrates which mode prevails is shown in Figure 9-7. Before interpreting the meaning of this result it should be mentioned that Meselson discovered that the buoyant density in CsCl solutions of *single* DNA strands is higher than that of the DNA double helix, from which they are released by denaturation. Thus the single strands of *E. coli* DNA show a density of 1.725 g/cm^3, compared to the density of 1.710 g/cm^3 of their double helix. As can be seen in Figure 9-7, denaturation of an ^{14}N-^{15}N first-generation hybrid double-helical molecule of density 1.717 g/cm^3 gives rise to two distinct species of single polynucleotide chains, one of density 1.725 g/cm^3, corresponding to ordinary ^{14}N-labeled single DNA strands, and the other of density 1.740 g/cm^3, corresponding to fully ^{15}N-labeled single DNA strands. Thus it can be concluded that the first-generation daughter molecules of the Meselson-Stahl experiment do owe their hybrid density to the distribution of the parental nitrogen atoms demanded by the semiconservative Watson-Crick replication mechanism, rather than to a dispersive or end-to-end conservative distribution.

The proof of semiconservative DNA replication was only the first, albeit the most important, application of CsCl density-gradient equilibrium sedimentation. This technique of detecting minute density differences was to become one of the main working tools of molecular biology, as the repeated mention of it throughout the remainder of this text will attest. One further application is manifest in an additional datum contained in Figure 9-7. Native, undenatured DNA extracted from *D. pneumoniae* was present in the CsCl solutions of that experiment as a density reference marker, and as can be seen it has a density of 1.700 g/cm^3, which is 0.010 g/cm^3 less than the density of native *E. coli* DNA. This density difference reflects the fact that the buoyant density of DNA molecules in CsCl increases with the [G] + [C]

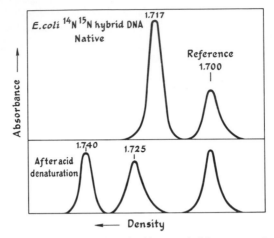

FIGURE 9-7.   The density of native, double-stranded first generation ^{14}N-^{15}N hybrid DNA molecules and of single polynucleotide strands derived from them. One part of a solution of hybrid DNA extracted from ^{15}N-labeled *E. coli* bacteria grown for one generation in ^{14}N-labeled medium was denatured so as to allow separation of the two complementary polynucleotide strands. Reference DNA of density 1.700 g/cm^3 extracted from a culture of pneumococci was then added to both denatured (lower panel) and undenatured (upper panel) *E. coli* hybrid DNA, and the two DNA samples were banded by centrifugation in a cesium chloride density gradient. The ordinate indicates the UV-absorbance and hence the concentration of DNA in the density gradient. [Redrawn from J. Marmur, C. L. Schildkraut, and P. Doty. *In the Molecular Basis of Neoplasia*, University of Texas Press, Austin, 1962.]

content, as is shown by the representative data of Table 9-1. More extensive density surveys of DNA obtained from many different organisms, whose

TABLE 9-1.   Density of DNA as a Function of [G] + [C]

DNA source	Density (g/cm^3)	[G] + [C] (in percent)
A-T polymer	1.679	0
Pneumococcus	1.700	42
E. coli	1.710	51
Serratia marcescens	1.718	59
Mycobacterium phlei	1.732	73

From L. C. Schildkraut, J. Marmur, and P. Doty,
J. Mol. Biol., **4**, 430 (1962).

DNA nucleotide base composition spans the wide range listed in Table 8-1, showed that the dependence of the buoyant density, ρ, of the mole percent [G] + [C] content of native DNA can be described by the empirical relation

$$\rho = 1.660 + 0.00098 \; \{[G] + [C]\} \qquad (9\text{-}1)$$

Thus by putting trust in the validity of this empirical formula it became possible to establish the nucleotide base composition of an unknown DNA sample by means of determining its density ρ through density-gradient equilibrium sedimentation. This procedure not only requires much less labor than direct chemical analysis of the [G] + [C] content, but also can be carried out on very small and chemically impure samples of DNA.

The density-gradient equilibrium-sedimentation technique has been applied also to the study of the *renaturation* of denatured double-helical DNA molecules. As was set forth in Chapter 8, such renaturation can be achieved by maintaining the complementary single strands at 65°C for a few hours. The

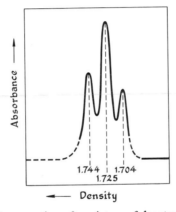

FIGURE 9-8.   Renaturation of a mixture of denatured fully "light" and fully "heavy" *B. subtilis* DNA. A solution of equal proportions of DNA extracted from *B. subtilis* cultures grown in either $^{14}NH_4Cl\text{-}H_2O$ or $^{15}NH_4Cl\text{-}D_2O$ medium was heated above the DNA "melting point" and then cooled slowly to 65°C to allow renaturation of the double helices. The solution was then treated with phosphodiesterase (an enzyme that digests single-stranded but not double-stranded DNA polynucleotide chains) and subjected to cesium chloride density-gradient sedimentation. Ordinate and abscissa as in Figure 9-7. Since on this graph the area of each band is proportional to the amount of DNA it contains, it can be seen that approximately twice as much hybrid DNA has been formed by renaturation as fully heavy or fully light DNA. [Redrawn from J. Marmur, C. L. Schildkraut, and P. Doty. *In The Molecular Basis of Neoplasia*, University of Texas Press, Austin, 1962].

experimental data presented in Figure 9-8 illustrate such an application. DNA was extracted from two cultures of *Bacillus subtilis*, one grown in an ordinary, "light" synthetic medium and the other grown in a "heavy" synthetic medium containing $^{15}NH_4Cl$ and D_2O (instead of H_2O). The native, light ^{14}N-^1H *B. subtilis* DNA has a density of 1.704 g/cm^3, and its heavy ^{15}N-^2D version has a density of 1.744 g/cm^3. A mixture of equal proportions of light and heavy DNA molecules was heated above the DNA "melting point" (see Chapter 8), slowly cooled to 65°C, and maintained at that temperature to allow renaturation of the double helices. The solution was finally treated with a special enzyme that digests all single-stranded DNA molecules that fail to renature but leaves intact all renatured double helices. As can be seen in Figure 9-8, renaturation generates three classes of DNA molecules: fully light molecules and fully heavy molecules, which have drawn complementary polynucleotide chains of the same isotope content, and hybrid molecules that have drawn one light and one heavy strand. This result proves, as the experiments cited in Chapter 8 did not, that renaturation of double-helical DNA molecules from complementary polynucleotide chains not previously paired together does in fact occur.

REPLICATION ORDER

Measurement of the size of the DNA molecules present in the extract used by Meselson and Stahl in their density-gradient analysis showed these molecules to have a molecular weight amounting to about 1% of the total mass of a single bacterial "nucleus." Thus semiconservative replication by strand separation could be inferred to occur in long double helices composed of polynucleotide chains that each contain tens of thousands of nucleotide residues. Besides leading to this most fundamental inference, however, the Meselson-Stahl experiment demonstrated one further point: individual sectors of the entire bacterial DNA complement *are replicated in the same order in successive cell generations*, since, as can be seen in Figure 9-5, after one generation time of growth, *all* of the DNA molecules in the culture are of hybrid density. If during this one generation time some sectors of the DNA *had* replicated twice and others not at all, then some light and some heavy DNA molecules should have been found in addition to the hybrid molecules. But as the data of Figure 9-5 show, at no stage of the transfer experiment is any joint presence of light and heavy molecules manifest.

Beyond this conclusion, however, the Meselson-Stahl experiment provides no detailed insights into the order in which the 1% sectors of the bacterial nuclear DNA complement are replicated during the generation time. In particular it leaves open the question of how many of these sectors actually replicate simultaneously, or, to restate the question, how many replicating Y-forks of the type shown in Figure 9-1 are simultaneously at work in a single *E. coli* nucleus. This question was settled in 1963 by the results of three completely different experiments, one of which, designed by F. Bonhoeffer and A. Gierer, was simply a refinement of the Meselson-Stahl experiment. The principle of Bonhoeffer and Gierer's experiment is that if there are *n* sectors of the DNA complement replicating simultaneously, or *n* replicating Y-forks per bacterial nucleus, then it ought to take a fraction $f = nx$ of the generation time to replicate a DNA molecule that represents a fraction *x* of the total DNA of the nucleus. Hence $n = f/x$. In order to measure the length of time required for replicating a given length of DNA, a transfer experiment was carried out in which thymine and its analog, 5-bromouracil (BU) (Figure 9-9), were used as density markers, instead of the two isotopes of nitrogen. This is possible because if thymine-requiring (Thy⁻) auxotrophs of *E. coli* ⸱are given BU instead of thymine in their growth medium, they will readily incorporate BU into their DNA in place of thymine, and DNA molecules

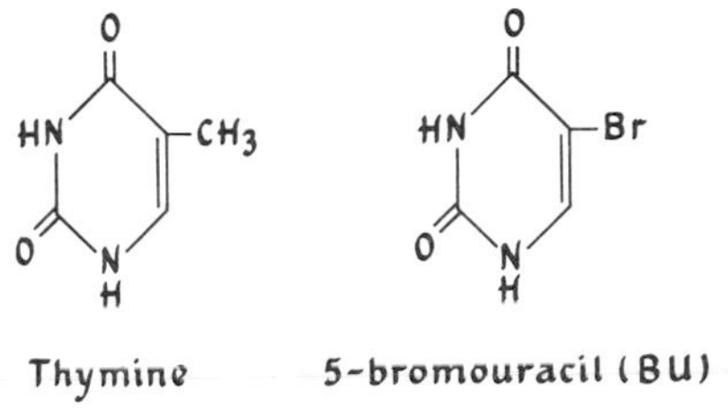

FIGURE 9-9.   The structures of thymine and its brominated analog, 5-bromouracil.

that contain this brominated pyrimidine are much denser than normal, thymine-containing polynucleotide molecules. Hence, for the purpose of this experiment, three samples were taken from a culture of Thy⁻ *E. coli* that was growing very slowly in a thymine-supplemented synthetic medium at the abnormally low temperature of 23°C, which extended the generation time to 1000 minutes. These samples were fed ¹⁴C-labeled BU instead of thymine for 2.5, 5, and 10 minutes, respectively. The DNA was then extracted from the bacteria of each sample and mixed with two kinds of ³H-thymine-labeled reference DNA: ordinary, or light, DNA and BU-thymine hybrid DNA extracted from ³H-thymine-labeled Thy⁻ *E. coli* grown in the presence of BU for exactly one generation time. The density distribution of both the ¹⁴C-BU DNA molecules and the ³H-labeled reference DNA molecules was determined by means of a modification of the original CsCl density-gradient equilibrium-sedimentation method. The principle of this modification (which had been invented by J. J. Weigle in connection with another investigation to be considered later) is illustrated in Figure 9-10.

The results of Bonhoeffer and Gierer's transfer experiment are shown in Figure 9-11. First it can be seen, that the two kinds of ³H-labeled reference DNA, light and hybrid, are well-separated in the CsCl density gradient.

Second, it can be seen that during the short ^{14}C-BU assimilation times of 2.5 and 5 minutes, which represent, respectively, 0.0025 and 0.005 of the 1000-minute generation time at 23°C, the DNA molecules containing ^{14}C-BU have attained densities intermediate between those of light and hybrid DNA molecules. In other words, these molecules are composed of one strand that

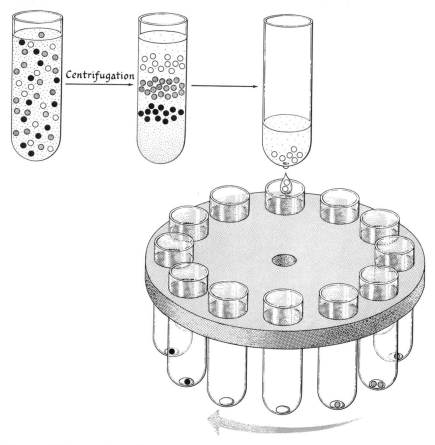

FIGURE 9-10.   Analysis of bands formed in cesium chloride density-gradient sedimentation. A mixture of heavy (solid dots), hybrid (stippled dots), and light (open dots) DNA molecules is dissolved in a concentrated solution of cesium chloride (tiny dots). Upon centrifugation the cesium chloride forms a density gradient, the salt being concentrated toward the bottom of the tube; each species of DNA forms a band at that level at which the cesium chloride solution matches its own density. When the sedimentation process has reached near-equilibrium, the centrifuge is stopped, a hole is punched through the bottom of the centrifuge tube, and the contents of the tube collected drop by drop in fraction collector. Each drop is then analyzed for its content of radioactivity in a radiation counter and for its absorption of ultraviolet light in a spectrophotometer. [After "Single-Stranded DNA," by R. L. Sinsheimer, *Scientific American*, Copyright © 1962 by Scientific American Inc. All rights reserved.]

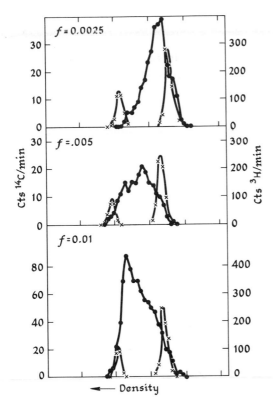

FIGURE 9-11.   Density distribution of *E. coli* DNA labeled with ^{14}C-labeled BU for three different, small fractions f of the generation time. Before cesium chloride density-gradient centrifugation, a mixture of ^{3}H-labeled fully light and half-heavy, half-light, reference DNA molecules were added to each of the three ^{14}C-labeled DNA extracts. The ordinates indicate the amounts of [14]C (black dots) and [3]H (crosses) radioactivity found per drop recovered after centrifugation and collection by the method shown in Figure 9-10. The abscissa shows the density of the cesium chloride solution in the drops. [After F. Bonhoeffer and A. Gierer, *J. Mol. Biol.* **7**, 534 (1963).]

contains only thymine and one that contains both thymine and BU along its length. After assimilation of ^{14}C-BU for 10 minutes, however, the DNA molecules have attained nearly the density of the fully hybrid marker DNA. In other words, after assimilation of BU during $f = 0.01$ of the generation time, the double helices have one strand that contains only thymine and the other mainly BU; that is, the DNA molecules are nearly fully replicated. Since the average molecular weight of the DNA molecules in these extracts was about 1 % of the mass of the bacterial nucleus ($x = 0.01$), it follows that $n = f/x = 0.01/0.01 = 1$. In other words, there appears to be only one sector of the bacterial DNA complement replicating at a time, or only one replicating Y-fork per nucleus.

The second experiment that helped to settle the question of replication order of the sectors of the *E. coli* DNA complement was carried out by J. Cairns. By the time Cairns designed this experiment it had become clear that however large the DNA molecules might have been that anyone had ever extracted from bacteria, and however uniform their size distribution, these molecules were undoubtedly artifactual random fragments of still larger intracellular DNA molecules. For the enormous axial ratio of long polynucleotide content—a double helix of the nucleotide content of the whole DNA complement of the *E. coli* nucleus would be a stiff cable about 600,000 times as long as it is thick—renders them extremely fragile, subject to breakage by even the slightest shearing forces that arise in any solution involved in ordinary laboratory manipulations. Cairns therefore tried to develop a method that would liberate the bacterial DNA under conditions that would reduce manipulations to an absolute minimum and hence might reveal the true molecular size of the intracellular DNA. For this purpose, he allowed a Thy⁻ strain of *E. coli* growing in a synthetic medium to incorporate ³H-labeled thymine into its DNA for slightly less than two generation times. The bacteria were then lysed by digesting part of their cell wall with lysozyme, and the products of lysis were allowed to settle slowly onto a piece of filter paper. This filter paper was then mounted on a microscope slide, overlaid with a photographic film, and stored for about two months. At the end of this time the film was developed. The resulting *autoradiograph* was then examined under the microscope to reveal the grains of the photographic emulsion that had been blackened by the emission of electrons from the radioactive ³H atoms residing in the thymine residues of the bacterial DNA. A microphotograph of one such preparation, and its interpretative cartoon, is shown in Figure 9-12.

The continuous line of blackened photographic grains in this autoradiograph shows, first of all, that Cairns did manage to liberate intact the entire DNA complement of an *E. coli* nucleus. Second, it can be concluded that the *E. coli* DNA is a closed, or circular, structure, thus confirming an earlier inference of the circular arrangement of the *E. coli* genes, which, as will be discussed in Chapter 10, was first made by Jacob and Wollman in 1957 on entirely different grounds. Third, it is evident that this is a picture of a circular structure which, at the time of its extraction, had been about two-thirds replicated, since the contour length of the sector C (representing the as-yet unreplicated part) is about one-half of the contour lengths of sectors A and B (representing the already replicated parts). One of the two branch points, p_1 or p_2, must correspond to a replicating Y-fork, and the other to the point at which replication of the circular structure had begun. Fourth, the sum of the contour lengths of sectors A + C or B + C, corresponding to the length of one whole unreplicated *E. coli* nucleus, is seen to be 1100 μ (or about 500 times as long as the *E. coli* cell). Since the DNA double helix contains ten nucleotide base pairs per 34 Å, or 3.4×10^{-3} μ, along its length (Chapter 8), we can now reckon that the DNA complement, or genome, of *E. coli* comprises

$$1100 \times 10/(3.4 \times 10^{-3}) = 3.2 \times 10^6$$

nucleotide base pairs. This molecular chain length turns out to be about a thousand times longer than what had been thought to be the length of DNA molecules at the time of the discovery of the Watson-Crick structure.

In addition to making these topological and topometric inferences, Cairns also counted the density of photographic grains per micron of contour length in all the sectors of this autoradiograph. For the grain density is

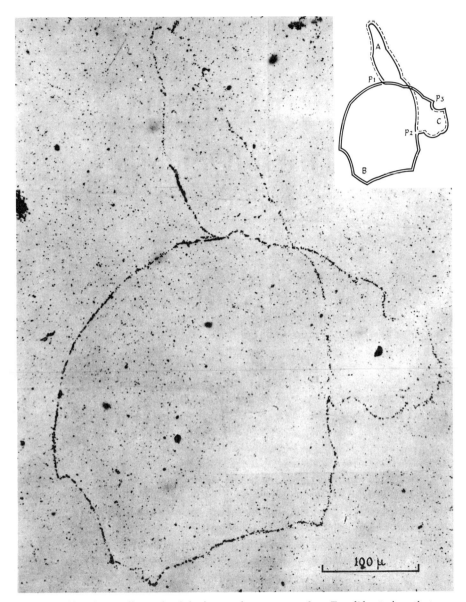

FIGURE 9-12. Autoradiograph of the intact chromosome of an *E. coli* bacterium that has been allowed to incorporate ^{3}H-thymine into its DNA for slightly less than two generation periods. The continuous lines of dark grains were produced by electrons emitted during a two-month storage period by decaying ^{3}H-thymine atoms in the DNA molecule. The scattered grains were produced by background radiation. The interpretative cartoon shows the portions of double- and single-stranded labeling that can be inferred to be present in the three sectors A (p_1 to p_2), B (p_1 to p_2), and C (p_1 to p_3 and p_3 to p_2) of the branched chromosome on the basis of the grain density per micron of contour length in the autoradiograph.

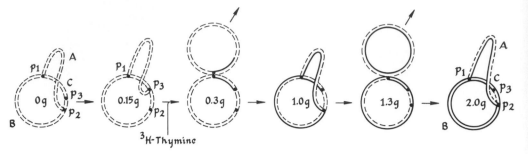

FIGURE 9-13. Diagrammatic history of the *E. coli* chromosome seen in the autoradiograph of Figure 9-12. The double lines represent the two DNA polynucleotide strands of the chromosomal double helix; the broken lines indicate unlabeled strands, and the solid lines, labeled strands. Exactly two generation periods before extraction of the chromosome ($0g$) the entirely unlabeled DNA of the grandparental bacterium was at the same stage of two-thirds completed replication as its granddaughter nucleus ($2.0g$) of Figure 9-12. Let us suppose that p_1 represents the point at which replication started and p_2 the position of the replicating Y-fork. A short while later ($0.15g$) the replicating Y-fork had moved to p_3; ^{3}H-thymine was then added, and all newly synthesized DNA strands contained ^{3}H label. One-sixth of a generation later ($0.3g$) replication of the chromosome was complete, since the replicating Y-fork had reached p_1. The two daughter chromosomes contained half-labeled double helices from p_3 to p_1, but were otherwise entirely unlabeled. We now follow one of the two daughter chromosomes in which the replicating fork had reached p_2 two-thirds of a generation later ($1.0g$). At this stage, sectors A and B, and sector C from p_3 to p_1 contained half-labeled double helices; sector C from p_2 to p_3 was still entirely unlabeled. One-third of a generation later ($1.3g$) the replicating Y fork was once more at p_1. One of the two granddaughter chromosomes contained fully labeled double helices from p_3 to p_1 and half-labeled double helices in the rest. We now follow one of the two granddaughter chromosomes in which the replicating fork has reached p_2 two-thirds of a generation later ($2.0g$), and thus given rise to a chromosome whose labeling pattern is precisely like that actually observed in the autoradiograph of Figure 9-12.

evidently proportional to the number of ^{3}H atoms that decayed during the two-month exposure of the film, and hence to the amount of ^{3}H-thymine present in the sector of the DNA responsible for any part of this picture. These grain counts revealed that per micron of contour length, sector B and the part of sector C from p_1 to p_3 contained twice as much ^{3}H-thymine as the part of sector C from p_3 to p_2 and sector A. As outlined in the diagrammatic history of the chromosome (Figure 9-13), this particular distribution of radioactive label means that p_1 represents the point at which replication of the circular chromosome begins, p_2 the position of the Y-fork at which semiconservative replication proceeds, and p_3 the position of the Y-fork at the time ^{3}H-thymine had first been presented to the bacterium from which that chromosome was later extracted. Thus Cairns was able to infer from this autoradiograph that the replication of *E. coli* DNA occurs in an orderly manner. A single replicating Y-fork proceeds along the giant circular molecule from a fixed starting point and generates two circular daughter

FIGURE 9-14. Diagramatic demonstration of the inference that a gene G_1 inscribed in a sector of the chromosome close to the starting point of replication p_1, and hence replicated early in the growth cycle, is present, on the average, in more copies than a gene G_2 distant from p_1, and hence replicated late in the growth cycle.

nuclei by the semiconservative replication mechanism of Watson and Crick.

The third experiment designed to help determine the order of replication of the bacterial nucleus was carried out by H. Yoshikawa and N. Sueoka with *Bacillus subtilis*. They reasoned that if the DNA of the bacterial chromosome is replicated at a single replicating Y-fork in an orderly manner, from a fixed starting point p_1, then in an exponentially and asynchronously growing culture of bacteria, whose cells represent a random sample of all stages of the growth cycle, a gene G_1 inscribed in a sector of the chromosome close to p_1 and hence replicated early in the growth cycle, would be present in more copies than a gene G_2 distant from p_1, and hence replicated late in the growth cycle (Fig. 9-14). More precisely, in a random population of growing bacteria the number of copies per nucleus $g(x)$ of a gene replicated at a stage of the growth cycle when a fraction x of the chromosome has replicated would be

$$g(x) = 2^{1-x} \qquad (9-2)$$

Evidently, the values of the function $g(x)$ range from 2 for genes replicated very early to 1 for genes replicated very late in the growth cycle. But in a population of "stationary phase" bacteria, or in a population of bacterial spores, when replication of the chromosome might be presumed to be arrested in all cells at the starting point p_1, *every* gene would be present in a *single* copy, so that $g(x) = 1$ for all values of x.

In order to determine whether the relative frequency of different *B. subtilis* genes actually corresponds to that expected from the hypothesis of orderly replication by a single replicating Y-fork from a fixed starting point, Yoshikawa and Sueoka extracted transforming DNA from a prototrophic wild-type strain of *B. subtilis* in both exponential and stationary phases and then carried out quantitative transformation experiments with auxotrophic recipient strains carrying eleven different mutant genes. These mutant strains required for growth on synthetic medium one or more of the following supplements: adenine (Ade), histidine (His), isoleucine (Ile), indole (Ind), leucine (Leu), lysine (Lys), methionine (Met), phenylalanine (Phe), threonine (Thr), or tyrosine (Tyr). In order to determine the relative frequency of the eleven wild-type genes in exponentially growing and stationary donor cultures, equivalent samples of the preparations of transforming DNA were added to competent cultures of each of the auxotrophs and the number of prototrophic transformants scored for each mutant gene. That number was then normalized by dividing it by the number of Met^+ transformants produced by addition of the same DNA preparation to Met^- recipient cells, and the ratio obtained in this way for the DNA extracted from the exponential donor culture was divided by the corresponding ratio obtained for the stationary donor culture DNA. A typical result is presented in Table 9-2, where it can be seen that in

TABLE 9-2. Relative Frequencies of Gene Transformation by DNA Extracted from Exponentially Growing (exp) and Stationary (sta) *B. Subtilis* Cultures

$(ade/met)_{exp}/(ade/met)_{sta}$ =	1.92
$(thr/met)_{exp}/(thr/met)_{sta}$ =	1.67
$(his/met)_{exp}/(his/met)_{sta}$ =	1.28
$(leu/met)_{exp}/(leu/met)_{sta}$ =	1.24
$(ile/met)_{exp}/(ile/met)_{sta}$ =	1.04
$(met/met)_{exp}/(met/met)_{sta}$ =	(1.00)

From H. Yoshikawa and N. Sueoka, *Proc. Natl. Acad. Sci. Wash.*, **49**, 559 (1963).

agreement with the prediction of equation (9-2) these normalized ratios cover the range from 2 to 1—namely, from a maximum of 1.92 for the *ade* gene to a minimum of 1.04 for the *ile* gene. Hence Yoshikawa and Sueoka concluded that the *B. subtilis* chromosome is replicated in an orderly manner from a fixed starting point, the DNA sector that carries the *ade* gene being one of the first and the DNA sector that carries the *met* gene being one of the last to be replicated. On the basis of such data and equation (9-2), they then constructed the gene map of *B. subtilis* shown in Figure 9-15. Strong support for the validity of this map is to be found in the results of transformation experiments similar to the one presented in Table 7-1. In these experiments it was found that such genes as *tyr* and *his*, or *met* and *ile* (shown here to be in close temporal linkage), actually reside on the same DNA fragment—that is, they are linked in transformation.

We can now calculate the *rate of chain growth* of the bacterial DNA during its replication. Since according to our previous calculation the *E. coli* nucleus consists of $1100 \times 10/(3.4 \times 10^{-3})$ nucleotide base pairs, and since its replication is achieved by a single replicating Y-fork in a generation time of about 40 minutes, or 2400 seconds, the rate of chain growth of the replica nucleotide chains is evidently $1100 \times 10/3.4 \times 10^{-3} \times 2400 = 1400$ nucleotides/second. That this is a very high rate of chain growth indeed can perhaps

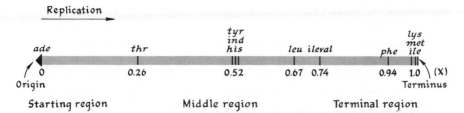

FIGURE 9-15.   A genetic map of *B. subtilis* inferred from the relative frequencies of gene transformation shown in Table 9-2. The numbers give the value of the parameter *x* of equation (9-2) for each gene—that is, the fraction of the generation time that has elapsed when the Y-replicating fork passes that gene during the replication of the circular chromosome. [From N. Sueoka and H. Yoshikawa, *Cold Spring Harbor Symp. Quant. Biol.* **28**, 47 (1963).]

be visualized best by considering that in order to replicate itself in the semi-conservative manner shown in Figure 9-1, the parental double helix must rotate to allow separation of its complementary nucleotide strands. Since one complete rotation is required for every ten nucleotides added to the growing replica chain, the replicating bacterial DNA must then turn at $140 \times 60 = 8400$ rpm, or as fast as a high-speed centrifuge.

DNA SEGREGATION

We may now inquire into the mechanism of segregation of the bacterial DNA during bacterial cell division. That is, we may ask how the replica DNA molecules manage to be distributed over the daughter cells, so as to insure that each of the daughters created by binary fission of the parent bacterium is endowed with a complete set of genetic information. The solution of this problem obviously required an analysis of the relation of the replication cycle of the circular bacterial DNA molecule to the overall division of the cell. One of the first insights into the nature of this relation was obtained in 1959, when it was found that more than 70% of the cells of a growing *E. coli* culture incorporate $^{32}PO_4^{-3}$ into their DNA within no more than 11% of the generation time. Hence it could be concluded that bacterial DNA replication proceeds well-nigh continuously throughout the whole generation period. The result of Bonhoeffer and Gierer's transfer experiment (that $n = 1$ replicating Y-fork per nucleus) confirmed this earlier inference, of course, in that their measurement of the average chain length of the DNA molecules replicated in a given short fraction of the generation time would have indicated that $n < 1$ replicating Y-fork per nucleus if the replication of the entire circular DNA molecule took only a *part* of the generation time. A second important insight was attained in 1965, when it was found that the rate of DNA synthesis in individual *E. coli* cells doubles about *midway* through their generation time. Hence it could be concluded that replication of the circular parental DNA molecule is complete, and that replication of the two daughter molecules begins, long before fission of the parent cell. Taken together, these two insights imply that at the moment the daughter cells are born by fission of the parent cell at the end of the generation period, each daughter contains an already partially replicated genome. But how do these two partially replicated parental genomes manage to move toward the two domains of the parent cell which, upon fission, are to form the central, or nuclear, regions of the two daughter cells? To answer this question, F. Jacob, S. Brenner, and F. Cuzin then postulated that the bacterial DNA molecule is attached to the cell membrane and that the synthesis of new cell membrane material during elongation of the rod-shaped bacterium occurs in a narrow zone of growth that is situated between the points of attachment of the two partially replicated DNA molecules. Thus growth of the membrane between these points of attachment would cause the continuous separation of the two future daughter genomes. The overall process is illustrated schematically in Figure 9-16, where the additional postulate has been introduced that the point of membrane attachment of the bacterial DNA molecule is the replicating Y-fork p_2 of Figure 9-13. Thus, for its replication, the long parental DNA molecule would be continuously paid out of the nuclear region to allow it to run through the membrane-attached replicating machinery, whence its two replicas would return to the nuclear region. As can be seen in Figure 9-16, in the first 20 minutes that elapse after birth of the daughter bacterium, replication of the partially replicated circular DNA molecule has been completed. Meanwhile, growth of new membrane material between the point of DNA attachment and the septum, whose growth in the preceding generation period had fissioned the parent bacterium, has moved the point of attachment to the center of the daughter cell. The two just-completed replica circles now acquire separate points of attachment (that is, separate replicating Y-forks), which straddle the narrow growing zone of the cell membrane. After the lapse of another 15 minutes, the bacterium is nearing the end of its generation period. Both of the replica DNA circles are already partially replicated, their points of attachment having moved apart because of growth of new cell membrane material between them. Meanwhile, the septum has begun to form between the attachment points. Finally, at the end of the 40-minute generation period, the bacterium has fissioned, giving rise to two daughter cells, each endowed with a partially replicated circular DNA molecule.

Jacob and his collaborators subsequently demonstrated the verity of this hypothesis. First, by sectioning bacteria longitudinally into six or seven slices, and then examining these slices serially under the electron microscope they found that the bacterial DNA is indeed attached to the bacterial membrane. Figure 2-10 is an electron micrograph of such a slice of an *E. coli* bacterium, in which the cell membrane can be seen to have been pulled toward the nuclear region at the point of attachment of the DNA molecule. Second, by depositing a uniform layer of tellurium crystals on the bacterial membrane

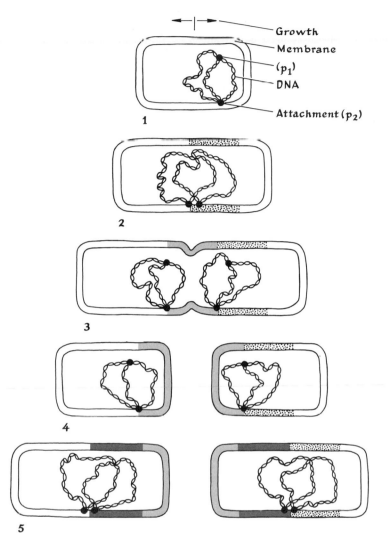

FIGURE 9-16. Mechanism of distribution of the bacterial chromosome. (1) Daughter bacterium has just been born. It contains a partially replicated circular DNA molecule attached to the bacterial membrane at the site of replication (p_2) of the circular structure. Replication started at p_1 in the parent cell. (2) Bacterium is midway through generation period. Replication of the circle has just been completed, and the two replica circles have separate points of attachment. The sector of the bacterial membrane that has been formed meanwhile is lightly stippled. (3) Bacterium is nearing end of generation period. Both replica circles are already partially replicated. Their points of attachment have moved apart because of growth of the sector of the bacterial membrane shown in light gray. (4) Bacterium has divided in two between the points of attachment of the two partially replicated genomes. (5) Both daughter bacteria are midway through the next generation period. Replication of their circular chromosomes is now complete, whose points of attachment have moved to the center of the cell as a result of growth of the new membrane material, shown in dark gray, since the moment of cell division.

(by treating growing cells with potassium tellurite), allowing the bacteria to grow for somewhat less than a generation period, and then examining the distribution of these crystals under the electron microscope, they found that in cells nearing the end of their generation period the membrane region between two points of DNA attachment was generally free of crystals, the crystals being present mainly outside of that region. In other words, in agreement with the hypothesis, growth of new membrane—formed since uniform deposition of the crystals—had occurred mainly *between* the attachment points.

ENZYMATIC DNA SYNTHESIS

While efforts were underway in the mid-1950's to study DNA replication through transfer experiments, Arthur Kornberg (Figure 9-17) and his collaborators attacked this problem from an entirely different direction. Kornberg had decided that growth of the replica polynucleotide chains must be catalyzed by an enzyme, and he proposed to isolate this enzyme and study the mechanism of its action. For this purpose Kornberg made protein extracts from *E. coli* bacteria and added these extracts to a reaction mixture containing ^{14}C- or ^{32}P-labeled deoxynucleoside triphosphates of the type

carrying the four DNA bases adenine, guanine, cytosine, and thymine (dATP, dGTP, dCTP, and dTTP), Mg^{++} and template DNA molecules. These mixtures, he hoped, would produce enzymic conversion of the low-molecular-weight radioactivity residing in the deoxyribonucleoside triphosphates into high-molecular-weight polynucleotide material. His efforts soon met with success, since an enzyme-catalyzed polymerization of the labeled deoxynucleoside triphosphates into high-molecular-weight polynucleotide chains was obviously taking place in his mixtures. This polymerization, furthermore, seemed to bear some resemblance to DNA synthesis, since it would proceed only if, in addition to the protein extract, *all four* deoxynucleoside triphosphates were present, as well as a template DNA. The availability of an assay of its activity thus allowed purification of the enzyme, which Kornberg called *DNA polymerase*, and after some eight years of diligent preparative work involving enormous quantities of bacteria, *E. coli* DNA polymerase was finally obtained in pure form. The purified enzyme was found to consist of a single polypeptide chain about 1000 amino acids in length, and the amount of enzyme recoverable from a bacterial culture

FIGURE 9-17. Arthur Kornberg
(b. 1918). [Courtesy Richard Calendar].

indicated that there are about 300 DNA polymerase molecules per *E. coli* cell.

The enzyme-catalyzed polynucleotide synthesis liberates one mole of pyrophosphate (P-P) per mole of nucleoside triphosphate polymerized. Hence the overall reaction can be written as

$$
\begin{matrix} n_1 dATP \\ + \\ n_2 dGTP \\ + \\ n_1 dTTP \\ + \\ n_2 dCTP \end{matrix}
\quad \xrightarrow[\substack{\text{DNA template} \\ \text{DNA polymerase}}]{\text{Mg}^{++}} \quad
\begin{bmatrix} dAMP \\ dGMP \\ dTMP \\ dCMP \end{bmatrix}_{2n_1 + 2n_2}
+ 2(n_1 + n_2) \text{ P-P}
$$

The most important evidence in favor of the inference that the *in vitro* reaction catalyzed by DNA polymerase is not merely a random polymerization of nucleotides, but does amount to DNA replication, came from the demonstration that the DNA template added to the reaction mixture is not only required for polymerization to proceed but actually determines the character of the polynucleotide formed. Thus it can be seen from the results presented in Table 9-3 that the nucleotide base compositions of the enzymatic reaction products formed in the presence of a variety of template DNA species, ranging in [G] + [C] content from 44% to 68%, bears a strong resemblance to the base composition of their templates. Subsequent, more-refined analysis of the enzymatically synthesized polynucleotides showed,

TABLE 9-3. Nucleotide Base Compositions (in percent) of Enzymatically
Synthesized DNA Products and Their DNA Templates

Source	[A]	[T]	[G]	[C]	[G] + [C]
Calf					
Template	29	27	23	21	44
Product	30	30	20	20	40
Escherichia coli					
Template	25	24	25	26	51
Product	26	25	24	25	49
Aerobacter aerogenes					
Template	22	22	28	28	56
Product	25	25	25	25	50
Mycobacterium phlei					
Template	16	16	34	34	68
Product	16	20	29	35	64

After I. R. Lehmann, M. J. Bessman, E. S. Simms, and A. Kornberg, *J. Biol. Chem.* **233,** 163 (1958).

furthermore, that the DNA template governs not only the overall base composition of the product but also the relative frequencies with which any two bases occur as nearest neighbors in the replica polynucleotide chains.

Thus the manner in which *in vitro* polynucleotide synthesis was catalyzed by the purified *E. coli* DNA polymerase indicated that, as in Watson and Crick's proposed mechanism for the autocatalytic function of DNA, the DNA acts directly as a template for the orderly copolymerization of its replicas, without requiring synthesis of any non-DNA intermediaries. But it is still not known whether the *in vivo* replication of the *E. coli* DNA is actually catalyzed by the DNA polymerase isolated by Kornberg and his collegues. For, as will be shown later, the orderly replication of the bacterial nucleus is not the only DNA synthesis that takes place within the *E. coli* cell: there also occur DNA *repair* processes that result in extracurricular polynucleotide synthesis. Moreover, because each bacterium contains some 400 molecules of DNA polymerase, and because replication of its DNA molecule proceeds at a single replicating Y-fork, it is unlikely that all of these polymerase molecules are active in the autocatalytic function of DNA. Indeed, it is not even clear how the DNA polymerase would catalyze *both* of the chemical reactions, which, as is shown in Figure 9-18, would be expected to proceed at the replicating Y-fork. Study of the mechanism of the *in vitro* polynucleotide chain growth

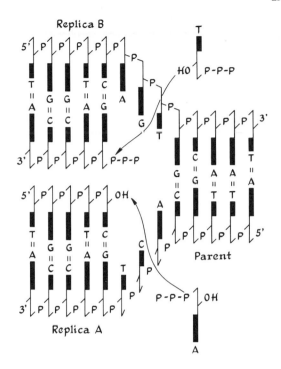

FIGURE 9-18. Schematic representation of the two chemical reactions that might be expected to proceed at the Y-fork of DNA replication. Here the nascent polynucleotide chain of replica A is growing in the direction $5' \to 3'$ and that of replica B in the direction $3' \to 5'$.

catalyzed by DNA polymerase has so far revealed only the reaction shown for growth replica A in Figure 9-18, in which the 3'-OH group of the last nucleotide of the nascent chain condenses with the 5'-α-phosphate of the monomeric nucleoside triphosphate to form the diester bond that joins the next nucleotide residue to the chain. It has so far been impossible to demonstrate *in vitro* the occurrence of the complementary reaction shown for growth of replica B, in which the 5'-α- phosphate of the triphosphate of the last nucleotide of the nascent chain condenses with the 3'-OH group of the monomeric nucleoside triphosphate. Thus it is not excluded at present that an as-yet undiscovered enzyme complex capable of catalyzing the simultaneous growth of the two antiparallel replica polynucleotide chains is at work at the replicating Y-fork, and that the principal physiologic task of the DNA polymerase is not the autocatalytic function of DNA after all.

Another possibility which must be considered is that the schema of Figure 9-18 does not give a true representation of the detailed events that occur at the replicating Y-fork. In fact, two discoveries made in 1967 provided a new direction for thought about those events. One of these discoveries was the nearly simultaneous finding by several different workers of the enzyme *DNA ligase* in *E. coli*. As is shown in Figure 9-19, DNA ligase catalyzes the formation of an internucleotide phosphate diester bond between 5'-phosphate and 3'-hydroxyl of two adjacent nucleotides belonging to an interrupted polynucleotide forming part of a double-stranded DNA helix. Formation of

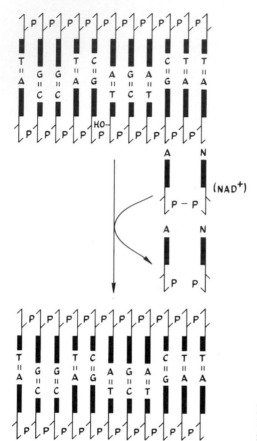

FIGURE 9-19. The reaction catalyzed by the DNA ligase enzyme. (N= nicotinamide; see Figure 3-3.)

this diester bond is coupled to and driven by the hydrolysis of NAD^+ (see Figure 3-3) to yield AMP and nicotinamide mononucleotide. It is evident that DNA ligase is capable of *repairing* single-strand interruptions of the DNA double helix, and we will consider the significance of this repair function in later chapters. That DNA ligase may also play an important role in DNA replication, however, was suggested by the other of the two discoveries— namely, the finding by R. Okazaki that in *E. coli* the most recently synthesized DNA polynucleotide chains are present as rather short molecular fragments. This conclusion followed from experiments in which *E. coli* growing very slowly at 20°C with a generation time of many hours were labeled with ^{3}H-thymine for various short periods, ranging from about 0.02 to 2% of the generation time. Total DNA was extracted from the bacteria and denatured in order to separate the two complementary polynucleotide strands. The chain length of the newly synthesized strands containing the recently assimi-lated ^{3}H label was then determined by following their rate of sedimentation

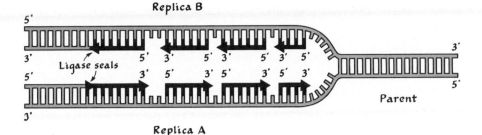

FIGURE 9-20.   Structure of the Y-fork under the proposed mechanism of *discontinuous* DNA replication. Here the nascent polynucleotide chains of both replica A and replica B are growing in the direction $5' \rightarrow 3'$. Several short nascent chain fragments are simultaneously growing in tandem on both parental template chains, to be joined eventually into a continuous replica chain by DNA ligase action.

in a centrifugal field. The result of this experiment was that the ^{3}H label was found to be incorporated into polynucleotide chains no more than about 1000 to 2000 nucleotides in length when assimilation of ^{3}H-thymine was permitted for less than 0.1 of the generation time. This short chain length is to be contrasted with the approximately 10,000-nucleotide-long single strand fragments of the total (or "old") *E. coli* DNA present in the same extract in unlabeled form. Upon assimilation of ^{3}H-thymine for longer and longer periods, however, an ever-increasing fraction of the [3]II label was found to appear in long polynucleotide chains having the same size as the strand fragments of the total (or "old") DNA.

Okazaki concluded from this finding that DNA replication within the Y-fork proceeds by assembly of short polynucleotide fragments, which are only later linked into the covalent continuity of the nascent replica strands by action of DNA ligase. As Figure 9-20 shows, this proposal can resolve the dilemma of the simultaneous growth of the two antiparallel replica polynucleotide chains. If DNA replication is a microscopically discontinuous process, as Okazaki proposed, then the synthesis of *both* antiparallel replica chains A and B can be envisaged to proceed by the one chemical reaction known to be catalyzed by DNA polymerase—namely, condensation of the 3'-OH group of the most recent nucleotide to have been incorporated into the nascent chain with the 5'-α-phosphate of the next nucleoside triphosphate to be added to the chain. Thus all short fragments of the nascent replica A strand would grow in the same direction as the overall movement of the replicating Y-fork, and all short fragments of the nascent replica B strand would grow in the opposite direction.

Bibliography

PATOOMB

Matthew Meselson and Franklin W. Stahl. Demonstration of the semiconservative mode of DNA duplication.
John Cairns. The autoradiography of DNA.
Ole Maaløe. The relation between nuclear and cellular division in *Escherichia coli*.

ORIGINAL RESEARCH PAPERS

Bonhoeffer, F., and A. Gierer. On the growth mechanism of the bacterial chromosome. *J. Mol. Biol., 7*, 534–540 (1963).

Cairns, J. The chromosome of *Escherichia coli. Cold Spring Harbor Symp. Quant. Biol., 28*, 43–45 (1963).

Replication of DNA in Microorganisms. XXXIII Cold Spring Harbor Symposium on Quantitative Biology. Cold Spring Harbor, New York, 1968.

Kornberg, A. Biological synthesis of deoxyribonucleic acid. *Science, 131*, 1503 (1960).

Meselson, M., and F. W. Stahl. The replication of DNA in *E. coli. Proc. Nat. Acad. Sci. Wash., 44*, 671 (1958).

Okazaki, R. T., K. Okazaki, K. Sakabe, K. Sugimoto, and A. Sugino. Mechanism of DNA chain growth. I. Possible discontinuity and unusual secondary structure of newly synthesized chains. *Proc. Natl. Acad. Sci., Wash., 59*, 598 (1968).

Watson, J. D., and F. H. C. Crick. Genetical implications of the structure of deoxyribonucleic acid. *Nature, 171*, 964 (1953).

Yoshikawa, H. and N. Sueoka. Sequential replication of *Bacillus subtilis* chromosome. I. Comparison of marker frequencies in exponential and stationary growth phases. *Proc. Natl. Acad. Sci., Wash., 49*, 559–566 (1963).

SPECIALIZED TEXTS, REVIEWS, AND MONOGRAPHS

Delbrück, M., and G. S. Stent. On the mechanism of DNA replication. *In* (W. D. McElroy and B. Glass, (eds.), *The Chemical Basis of Heredity*. The Johns Hopkins Press, Baltimore, 1957, p. 699.

10. Conjugation

During the long pre-1943 gestation period of bacterial genetics there had been repeated searches for sexual phenomena among bacteria. Some of these efforts claimed success, but since they amounted to no more than direct microscopic observation of conjugation or fusion of bacterial cells, and were never coupled with any sort of genetic experiments, it was impossible to assess their true meaning. Other efforts, which did involve the search for genetic recombination of parental heredity characters, reported negative results, but the nature of the genetic characters employed in those experiments would have required the phenomenon of genetic recombination to occur with very high frequency in order to be detected. But after Luria and Delbrück had established the spontaneous nature of bacterial mutation and developed the sort of quantitative reasoning necessary for obtaining meaningful results in bacterial genetics, Joshua Lederberg (Figure 10-1) realized what kind of experimental design would provide a critical test of the existence of genetic recombination among bacteria.

THE CLASSICAL CROSS

Lederberg thought that a favorable procedure would be to "cross" two different auxotrophic mutant bacteria, by growing them in joint culture in a complete medium and then plating samples of the culture on *minimal*

FIGURE 10-1.   Joshua Lederberg
(b. 1925). [Courtesy News and
Publication Service, Stanford
University.]

medium to select any rare prototrophic recombinants that might arise. But Lederberg realized also that if recombination were so rare an event that less than one in 10^6 parent bacteria are turned into prototrophic recombinants, then *reverse mutation* to prototrophy of one or both of the two auxotrophic parents (a mutation that could be expected to occur at a comparable frequency) might frustrate the detection of true recombinants. He therefore conceived the idea that crosses between *multiple auxotrophs*, in which each parent strain carries two or more mutant genes, had to be set up, since the probability would be vanishingly low that two or more genes might back-mutate in the same parent individual to reconstitute a prototroph. In 1946 Lederberg decided to carry out such crosses, but rather than going through the very great labor that would then have been necessary to prepare multiple auxotrophs himself, he proposed to E. L. Tatum that they join forces and search for genetic recombinants among the biochemical mutants of *E. coli* that, as was described in Chapter 5, Tatum had then recently isolated. This collaboration turned out to be one of the fortunate accidents in the history of science, for Tatum happened to have selected his mutants in the *E. coli* strain K12. His choice of wild-type strain had been simply a matter of convenience, because the K12 strain had been used for many years in bacteriology courses at Stanford University, where Beadle and Tatum were then working.

Lederberg and Tatum's search for recombination in strain K12 was almost immediately successful; but had Tatum selected his mutants in almost any other of the then popular but now known to be sexually infertile *E. coli* strains, such as the one with which Luria and Delbrück were working, Lederberg and Tatum's experiment would almost certainly have failed. They announced their success that same year at the 11th Cold Spring Harbor Symposium, which was referred to in Chapter 5.

Lederberg and Tatum's classical experiment was this (Figure 10-2): two auxotrophic *E. coli* mutant strains, one (which we will call strain A) Met⁻ Bio⁻ and the other (which we will call strain B) Thr⁻ Leu⁻ Thi⁻, were grown together overnight in a complete medium. The mixed culture was then centrifuged, washed free of the complete medium, and plated on minimal agar. The result of this experiment was that one Met⁺ Bio⁺ Thr⁺ Leu⁺ Thi⁺ prototrophic colony appeared on the minimal agar for about every 10^7 parent cells plated. Since control platings on minimal agar of either of the two parent strains did not lead to the appearance of *any* prototrophic colonies, it could be concluded that the prototrophs recovered from the mixed culture were *genetic recombinants* of some kind. That is, these prototrophs must have incorporated into their genome the *met bio* genes of strain B and the *thr leu thi* genes of strain A.

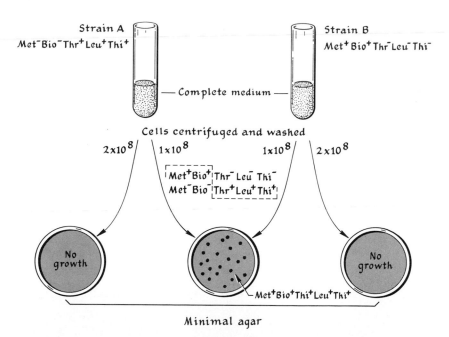

FIGURE 10-2. Schematic summary of Lederberg and Tatum's classical genetic cross of two auxotrophic mutant strains of *E. coli* K12.

Since genetic transformation of the pneumococcus and its mediation by bacterial DNA was already known at the time of this experiment, it seemed possible on first sight that Lederberg and Tatum had merely observed another instance of such transformation. For example, during joint growth of the two auxotrophic parent strains, DNA molecules of the Met^+ Bio^+ strain B bearing its *met bio* genes might have been liberated by spontaneous lysis of a few B cells and taken up by competent Thr^+ Leu^+ Thi^+ A cells to produce prototrophic transformants. It became clear very soon, however, that transformation could hardly account for the origin of these recombinants, since *direct contact between the bacteria of strains A and B was necessary for their appearance*. For Lederberg and Tatum had shown that no prototrophs were produced when sterile filtrates of media in which one of the two strains had grown were added to cultures of the other strain, even when the cells of the first strain had been broken or lysed just before filtration of their medium. In 1950, B. D. Davis added support to this demonstration by devising his **U**-tube experiment (Fig. 10-3). The two auxotrophic strains A and B were inoculated into opposite arms of the **U**, which were separated at the bottom by a sintered glass filter impervious to *E. coli* cells but allowing passage of particles smaller than about 0.1 μ, and hence of free DNA molecules. The two bacterial cultures were then allowed to grow to saturation while the culture fluid was slowly pushed back and forth from one arm to the other through the filter by alternate suction and pressure on one end of the **U**. In this way the two auxotrophic strains shared the same growth medium without their cells having come into contact. The result of this experiment was that no prototrophs appeared in either arm of the **U**, showing that formation of the recombinant bacterial genomes requires contact between the two conjugating parental cells.

From the very start, Lederberg and Tatum interpreted their discovery in terms of conventional eukaryotic sexuality. That is, they reasoned that cell

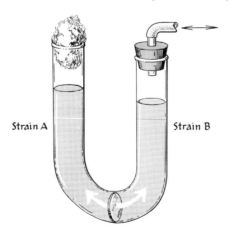

Strain A   Strain B

FIGURE 10-3.. The Davis **U**-tube. The two arms of the **U** are separated by a sintered glass filter impervious to bacterial cells but allowing free passage of the growth medium. The two auxotrophic parent strains are inoculated into opposite arms, and alternating pressure and suction is applied to one arm of the **U**. [From *Molecular Biology of Bacterial Viruses*, by G. S. Stent. W. H. Freeman and Company. Copyright © 1963.]

contact must be necessary for the appearance of recombinant bacterial genotypes because they believed that recombination in *E. coli* proceeded according to the following sequence of events: (1) cell fusion, (2) fusion of the nuclei of the parent cells to form a diploid *zygote*, and (3) "meiotic" reduction division of the zygote with attendant crossing over to give rise to segregant bacteria carrying recombinant genomes. Both conjugating haploid cells were regarded as playing equivalent roles in this conjugal process—that is the sexual system of *E. coli* was thought to be *homothallic*.

A further and even more convincing demonstration that the genetic recombination he and Tatum had discovered in *E. coli* was not just another mode of DNA-mediated transformation came from Lederberg's analysis in 1947 of the segregation of *unselected* characters in these crosses. This analysis made it clear that, in agreement with the hypothetical sequence of cellular and nuclear fusion, a rather considerable portion of the whole genome of both parents participates in the elementary recombination act. To make this analysis Lederberg repeated the classical cross (A) Bio$^-$Met$^-$ × (B) Thr$^-$Leu$^-$ Thi$^-$, but plated the mixed culture on four different kinds of minimal agar, each supplemented with *one* of the growth factors required by either parent strain. In this way, the gene controlling the synthesis of a particular growth factor was not selected for, and any recombinant would grow on the selective agar as long as it was prototrophic for the *other* characters. He then isolated many of the recombinant colonies that appeared on each of the selective agars and, by plating the isolates on minimal agar, determined how many colonies in each recombinant class had drawn the wild type (+) gene and how many had drawn the mutant (−) gene of the unselected character. The data of this experiment are presented in Table 10-1. They show first of all that the chance that any unselected character will appear among the recombinants is neither

TABLE 10-1.   Relative Frequency of Unselected Characters in the Cross (A) Bio$^-$Met$^-$Thr$^+$Leu$^+$Thi$^+$ × (B) Bio$^+$ Met$^+$Thr$^-$Leu$^-$Thi$^-$

Agar supplemented with (or unselected character)	Ratio of unselected character ($-/+$) among recombinants
Bio	0.17
Thr	0.24
Leu	0.096
Thi	9.88

From J. Lederberg, *Genetics* 32, 505–527 (1947).

extremely high nor extremely low, leading to the inference that an appreciable portion of the genome of both parents participates in the conjugal act. Second, they show that *genetic linkage* can be detected between these genes. For example, *bio* appears to be linked to *met*, since the unselected Bio$^+$ character is found more often in the prototrophs than Bio$^-$. Since Bio$^+$ was carried by the Met$^+$ parent, and since the prototrophs were selected for being Met$^+$, it follows that *bio* and *met* have a tendency to go together, that is, *bio* is linked to *met*. Similarly, since the unselected Thi$^+$ character is found more *rarely* in the prototrophs than Thi$^-$, and since Thi$^-$ was carried by the Met$^+$ Bio$^+$ parent, it follows that *thi* has a tendency to go together with *met* and *bio*—that is, *thi* is linked to *met* and *bio*. Analogous reasoning leads to the conclusion, finally, that *thr* and *leu* are linked.

Lederberg then introduced three further genetic characters, Tsx^r, Ton^r, and Lac$^-$, into strain B and by analyzing the frequency of unselected characters appearing in crosses of these highly marked B strains to strain A, was able to construct the first genetic map of *E. coli*, shown in Figure 10-4.

In the next four years, Lederberg, and others who had taken up the study of his and Tatum's important discovery, continued the exploration of the *E. coli* genetic map. For this purpose more and more genetic characters were introduced by isolating further mutants in both strains A and B and analyzing the segration pattern of unselected characters in relevant crosses. This work soon led to serious interpretative difficulties, since the data seemed to make it impossible to place the known genes on a simple linear map. For instance, the pairs of linked genes *str* and *mal*, *mtl* and *xyl*, and *tsx* and *lac* all appeared to be linked to *met*, without the pairs showing any linkage between them. In order to explain this paradoxical result, Lederberg proposed in 1951, probably as a last resort, that the *E. coli* chromosome might have a triple *branch point* near the *met* gene, as shown in Figure 10-4.

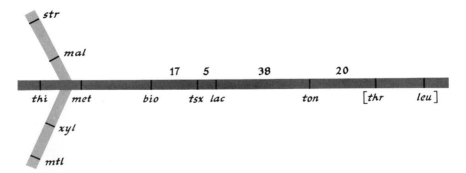

FIGURE 10-4.   The first genetic map of genes on the *E. coli* chromosome. *Dark gray line:* Lederberg's 1947 map. The numbers indicate the percent recombination observed between adjacent gene pairs in crosses in which the characters determined by these genes were not used for recombinant selection. *Light gray line:* The arms of a triple branch point added by Lederberg in 1951.

ONE-WAY TRANSFER AND THE F SEX FACTOR

Just when the attempts to establish a linear genetic map of E. *coli* began to encounter some difficulties, an unexpected finding by William Hayes (Figure 10-5) presaged that the previously held homothallic view of bacterial conjugation, under which both parent bacteria are supposed to play an equivalent role in the recombination process, was probably wrong. In 1952 Hayes discovered to his surprise that the number of prototrophic recombinants produced by Lederberg and Tatum's cross (A) Met$^-$ × (B) Thr$^-$ Leu$^-$ Thi$^-$ is not greatly diminished if, before the cross is carried out, strain A is sterilized by treatment with high doses of streptomycin. He also found, however, that similar streptomycin treatment of strain B prior to the cross completely prevents recombination. To Hayes these findings implied that genetic recombination in E. *coli* is the result of a unidirectional process, in which strain A acts as a *donor* of its genetic material, which enters the *recipient* strain B bacterium upon contact between the parental cells. Genetic recombination then ensues between donor and recipient genomes in the strain B recipient cell turned zygote. Sterilization by streptomycin of the donor strain would not necessarily impede its capacity to donate its genetic material, whereas sterilization of the recipient cell—the future home of the zygote—would

FIGURE 10-5.   William Hayes (left) (b. 1913) and Guiseppe Bertani (right) (b. 1923).

certainly abort all recombinants. Hayes concluded, therefore, that bacterial conjugation is a *heterothallic* process, in which both parents do *not* play an equivalent role in the recombination process. By analogy with the sexual roles of internally fertilizing, dioecious higher forms, the donor strain would correspond to the *male* and the recipient strain to the *female sex*.

Soon after reaching these conclusions, Hayes discovered, quite by chance, a variant of the putative donor strain A that had apparently lost the ability to donate its genetic material to strain B. This infertile variant had arisen spontaneously among the surviving bacteria of a culture of the original A strain after a year's storage in the refrigerator, and produced no prototrophic recombinants when crossed with the normal B strain. Hayes then isolated a Str^r mutant of this infertile variant of strain A and grew it in mixed culture with the original, fertile Str^s A strain. By plating the bacteria of this mixed culture on Streptomycin-containing agar, Hayes could reisolate a number of Str^r clones of the previously infertile A strain variant and make test-crosses of these isolates with the normal test strain B. The result of this experiment was that about one-third of the reisolated Str^r clones had *regained* their fertility by mixed growth with the fertile Str^s A strain. Soon after the completion of these experiments, Hayes learned that Lederberg, his wife Esther, and L. L. Cavalli had made entirely analogous discoveries with *their* strains A and B. An agreement was soon reached between them to place the following interpretation on these and other findings related to the fertility of crosses: Certain strains of *E. coli* (such as strain A), termed F$^+$, carry a *sex factor*, F, whereas other strains, such as strain B and the infertile variant of A, termed F$^-$, do not. For any cross to be fertile, the presence of the sex factor is required in at *least one* of the parents. Thus the crosses F$^+$ × F$^+$ and F$^+$ × F$^-$ are fertile, whereas the cross F$^-$ × F$^-$ is sterile. The sex factor is *transmissible,* from F$^+$ to F$^-$ bacteria, by means of a process for which cell contact is necessary, just as it is for genetic recombination. But unlike genetic recombination between F$^+$ and F$^-$ bacteria, which proceeds with an efficiency of about one recombinant per 10^7 parents, transfer of the F factor occurs with an efficiency higher than one F$^-$ → F$^+$ conversion per 10 parents. The sex factor can be lost spontaneously, as it was in the infertile variant of strain A, but it can never be regained except by transfer from an F$^+$ cell.

Further experiments by Hayes and by the Lederbergs and Cavalli soon showed that the genetic constitution of the recombinants of an F$^+$ × F$^-$ bacterial cross depends very strongly on *which* of the two parent strains carries the F factor: as the following two crosses show, *most of the unselected markers appearing in recombinants are derived from the F$^-$ parent.* In both crosses the Met$^-$ strain A was crossed with the Thr$^-$Leu$^-$ strain B, but in the first cross strain A carried the F factor and in the second cross strain B carried the F factor. Recombinants of Met$^+$ Thr$^+$ Leu$^+$ type were then selected on minimal agar and retested for the presence of four other genetic characters in which the two parent strains differed and for which no direct selection had been made. The results of these two crosses are presented in Table 10-2. It can be seen that among the Met$^+$ Thr$^+$ Leu$^+$ recombinants

TABLE 10-2. Influence of the F Factor on the Genetic Constitution of the Recombinants of an F$^+$ × F$^-$ Bacterial Cross

Parents	Strain A: Met$^-$ Str^r Strain B: Thr$^-$ Leu$^-$ Lac$^-$ Thi$^-$ Mal$^-$ Str^s			
Cross	Characters → ↓	Mal$^+$ Str^r	Mal$^-$ Str^s	Mal$^+$ Str^s or Mal$^-$ Str^r
AF$^+$ × BF$^-$	Lac$^+$ Thi$^+$	0	1.7	1.0
	Lac$^+$ Thi$^-$	0	23.0	
	Lac$^-$ Thi$^+$	0	3.7	
	Lac$^-$ Thi$^-$	0	70.6	
AF$^-$ × BF$^+$	Lac$^+$ Thi$^+$	35.2	2.0	4.9
	Lac$^+$ Thi$^-$	50.0	6.0	
	Lac$^-$ Thi$^+$	0.3	0	
	Lac$^-$ Thi$^-$	1.3	0.3	

The numbers indicate the percent of genetic recombinants selected for their Met$^+$ Thr$^+$ Leu$^+$ character which upon retesting carry various constellations of the unselected characters.
From W. Hayes, *Cold Spring Harbor Symp. Quant. Biol.* **18**, 75 (1953).

of the AF$^+$ × BF$^-$ cross, the most frequent type is the one that carries the Lac$^-$ Thi$^-$ Mal$^-$ Str^s characters of the BF$^-$ parent, the next most frequent being the one that carries all BF$^-$ characters except Lac$^-$. In contrast, none of the recombinants carry the Mal$^+$ Str^r characters of the AF$^+$ parent. Conversely, among the Met$^+$ Thr$^+$ Leu$^+$ recombinants of the AF$^-$ × BF$^+$ cross most of the recombinants carry the three Lac$^+$ Mal$^+$ Str^r characters of the AF$^-$ parent, whereas almost none carry the three Lac$^-$ Mal$^-$ Str^s characters of the BF$^+$ parent.

It soon became obvious that the previously unknown polarity introduced into bacterial crosses by the disposition of the F factor must have been at least partly responsible for the difficulties encountered in the attempts to construct a simple genetic map of the *E. coli* chromosome from gene linkage data. For the frequency of appearance of an unselected character in the recombinants of any cross depends not only on its linkage to a selected character but also on whether the F$^+$ or the F$^-$ parent had introduced it into the cross.

Though there was no disagreement on the facts concerning the evident sexual polarity they had discovered in bacterial recombination, the Lederbergs and Cavalli on the one hand, and Hayes on the other, made radically different interpretations of how these facts might explain the basic mechanism of the conjugation process. The Lederbergs and Cavalli continued to believe in the original notion of complete fusion of parent cells and nuclei, and attributed the polarity introduced into the cross by the F factor to a preferential *post-zygotic elimination* of most of the F⁺ parental genome from a true diploid zygote. They were to persist in this view until 1958, and abandoned it only when overwhelming evidence had finally been marshalled against their idea. Hayes, however, immediately drew what eventually turned out to be the more or less correct conclusion. On the basis of the differential effect of streptomycin treatment on the fertility of the cross, he proposed that the F⁺ cell, which he had already identified as a gene *donor* in the recombination process, transfers only *part* of its genome to the F⁻ recipient, so that the resulting zygote is not a complete but a partial zygote, or *merozygote*. This transfer is accomplished by means of the F factor itself, thought to be a self-reproducing "vector" capable of passing from donor to recipient cell with high efficiency, and carrying, with low probability, any given, selected-for sector of the donor genome. This view explained easily why F⁻ × F⁻ crosses are sterile, since neither parent carries the "vector" required for gene transfer; why, provided that function of the "vector" is not sensitive to streptomycin, treatment of the F⁺ parent with streptomycin does not abolish its fertility; and why most of the *unselected* markers in the recombinants are those of the F⁻ genotype.

Being a self-reproducing vector, and thus evidently containing some genetic material of its own, the F sex factor seemed likely to contain DNA. But for the next eight years this supposition proved quite difficult to confirm, since there seemed to be no way of isolating the F factor from the F⁺ cell and of determining its chemical composition. Finally, in 1961, indirect radiobiological studies of the rate of inactivation of the F factor by decay of incorporated radiophosphorous ³²P atoms strongly suggested that the F factor does contain DNA. One result of these studies was that the amount of F factor DNA complement was estimated to be about 2% of that of the *E. coli* chromosome, or $0.02 \times 10^6 = 6 \times 10^4$ nucleotide base pairs. Soon thereafter a *direct* method of isolating the F factor became available, when S. Falkow and L. S. Baron discovered that upon mixing an F⁺ strain of *E. coli* with a culture of *Serratia marcescens*, the *E. coli* F sex factor enters and perpetuates itself in some of the cells of this taxonomically very remote recipient bacterium. As was seen in Table 9-1, the purine-pyrimidine base compositions, and hence the buoyant densities of the DNA of *E. coli* and of *S. marcescens*, are significantly different. Consequently, when the DNA was extracted from

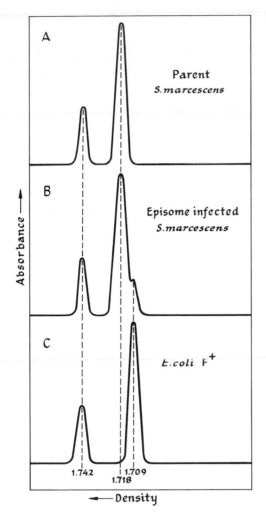

FIGURE 10-6. Isolation of the
E. coli F sex factor. DNA was
extracted from a culture of normal
S. marcescens (panel **A**), from a
culture of *S. marcescens* to which the
E. coli F factor had been transferred
(panel **B**), and from a culture of
F^+ *E. coli* carrying its F factor
(panel **C**). To these extracts was
added ^{15}N-labeled DNA from
Pseudomonas aeroginosa (density
1.742 gm/cm^3) as a density
reference marker. The DNA
extracts were then subjected to
cesium chloride density-gradient
centrifugation. The ordinates
indicate the amount of UV
absorbance in the solution (hence
DNA concentration) at the gradient
densities shown on the abscissa.
The satellite band at the density of
1.709 gm/cm^3 in panel **B** represents
the F factor DNA. [After
J. Marmur, R. Rownd, S. Falkow,
L. S. Brown, C. Schildkraut, and
P. Doty, *Proc. Natl. Acad. Sci.
Wash.* **47**, 974 (1961).]

a pure culture of F^+ *S. marcescens* carrying the *E. coli* F factor and the
extract subjected to CsCl density-gradient sedimentation, two distinct DNA
bands were seen (Figure 10-6). One, the *major band,* appears at the density
of 1.718 g/cm^3, corresponding to the *S. marcescens* chromosome, and the
other, the *satellite* band, appears at the density of 1.709 g/cm^3, characteristic
of *E. coli* DNA and hence corresponding to the exotic F factor. Thus the F
factor really did turn out to be DNA, and measurements of the amount of
DNA in the satellite band confirmed the indirect estimate of its chain length
being about 6×10^4 nucleotide base pairs.

HIGH-FREQUENCY, OR Hfr, DONOR STRAINS

The key discovery that ultimately led to an understanding of the true nature of bacterial conjugation had, however, already been made by Cavalli nearly three years before the recognition of the role of the F sex factor in this process. In 1950 Cavalli had treated a culture of the classical AF^+ Met^- strain with nitrogen mustard and isolated among the surviving bacteria a new strain that, in crosses with the standard BF^- Thr^- Leu^- strain, engendered protorophic recombinants with at least 10^3 times greater frequency than the original AF^+ strain in a standard $F^+ \times F^-$ cross. He called this strain *Hfr*, for *high frequency of recombination*; it now goes under the name *Hfr Cavalli*. Three years later Hayes isolated, also by chance, an Hfr strain from his AF^+ strain —the *Hfr Hayes*, which, like *Hfr Cavalli*, produces prototrophic recombinants at a frequency 10^3 times higher than its parent strain. Hayes showed that his Hfr strain behaves in two respects like an F^+ donor; (1) it can be sterilized by streptomycin treatment without seriously affecting the fertility of a cross with the normal BF^- parent, and (2) most of the unselected characters appearing among the recombinants are those of the F^- recipient. In one important respect, however, the Hfr strain was found to differ from an ordinary F^+ strain: in joint growth with an F^- strain it does *not* transfer the F sex factor to the F^- recipients, which become neither F^+ nor Hfr upon contact with the Hfr strain. It thus appeared that concomitant with the enormous gain in capacity to donate genes to the F^- recipient, the Hfr strain lost its capacity to donate with high efficiency its F sex factor.

Hayes's further study of his Hfr strain revealed one additional, most important attribute of the Hfr state: *the high-frequency donor character pertains only to a limited portion of the donor genome*—namely, to such genes as *thr*, *leu*, and *lac*, which are situated on the right half of the map shown in Figure 10-4. In crosses of the Hfr Hayes strain with BF^- recipients in which transfer of the *mal*, *str*, *xyl*, *mtl* or *thi* genes, which are situated on the left half of the map, of the donor is selected, recombinants are produced at the low frequency characteristic of an ordinary $F^+ \times F^-$ cross. Hayes interpreted these findings to mean that the $F^+ \rightarrow$ Hfr change was attributable to a permanent alteration of the F sex factor, which engendered at the same time a loss of its own transferability to an F^- strain and a very much greater transferability of a limited sector of the donor genome.

ORIENTED TRANSFER

Despite, or possibly because of, their evident simplicity and plausibility, the notions set forth by Hayes did not immediately gain general acceptance in the community of bacterial geneticists. Thus, for another few years genetic

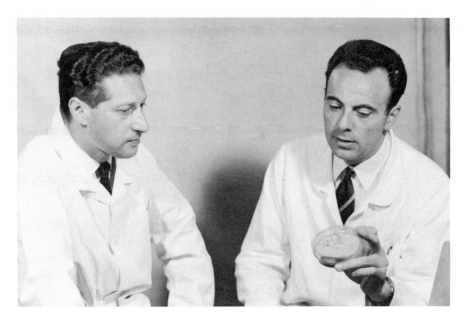

FIGURE 10-7.   Elie L. Wollman (left) (b. 1917) and Francois Jacob (right) (b. 1920).

recombination in *E. coli* remained a highly controversial subject of Byzantine complexity. Meanwhile, E. L. Wollman and F. Jacob (Figure 10-7), using as their point of departure both the bacterial strains and the working hypothesis provided them by Hayes, began a series of experiments that, by 1957, had finally explained the *E. coli* conjugation phenomenon discovered by Lederberg and Tatum a decade earlier. In one of these experiments, Wollman and Jacob studied the kinetics of formation of genetic recombinants in the cross

Hfr Hayes Met$^-$Azi^r Ton^r Str^s × BF$^-$ Thr$^-$ Leu$^-$Azi^s Ton^s Lac$^-$ Gal$^-$ Str^r

For this purpose they mixed cultures of Hfr and F$^-$ bacteria in the ratio of one Hfr cell per 10 F$^-$ cells (at a density of about 10^8 bacteria per ml) and incubated the mixture for various times. Samples of the mating mixture were then diluted and plated on minimal glucose agar containing methionine and streptomycin to select for Thr$^+$Leu$^+$Str^r recombinants, or on minimal galactose agar containing threonine, leucine, methionine, and streptomycin to select for Gal$^+$Str^r recombinants. They obtained the result shown by broken lines in Figure 10-8—namely, that the number of recombinants increases linearly with time from the moment the parental cultures are mixed until a plateau of 18 Thr$^+$Leu$^+$Str^r and 5 Gal$^+$Str^r recombinants per 100 Hfr cells is reached after about an hour. This plateau frequency of recombination evidently exceeds by another factor of 100 the first Hfr recombination values observed by Cavalli and by Hayes, and thus shows that more than

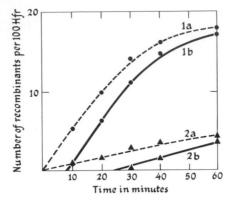

FIGURE 10-8.   Kinetics of gene transfer in an Hfr Hayes Azi^r Ton^r Str^s × F$^-$ Thr$^-$ Leu$^-$ Azi^s Ton^s Gal$^-$ Str^r cross. Cultures of Hfr and F$^-$ bacteria were mixed, and at the time shown on the abscissa samples were either plated directly on selective agar (broken curves, 1a and 2a) or agitated for 2 minutes in a Waring blendor and then plated on selective agar (solid curves, 1b and 2b). The 1a and 1b set of platings were made on minimal *glucose* agar containing methionine and streptomycin, and hence the ordinate indicates the frequency of Thr$^+$ Leu$^+$ Str^r recombinants recovered on the selective agar at various times. The 2a and 2b set of platings were made on minimal *galactose* agar containing threonine, leucine, methionine, and streptomycin, and hence the ordinate indicates the frequency of Gal$^+$ Str^r recombinants recovered on the selective agar at various times. [After F. Jacob and E. L. Wollman, *Sexuality and the Genetics of Bacteria*. Academic Press, New York, 1961.]

10% of the cells of any Hfr population can donate part of their genetic material to an F$^-$ recipient cell. The solid lines in Figure 10-8 represent a small but extremely important modification of this kinetic experiment in which the mating population of Hfr and F$^-$ cells was not only diluted at various times after the parent strains were mixed, but the diluted sample was also agitated in a Waring blendor for two minutes before being plated on the selective agar. The purpose of this procedure was to use the strong shear forces generated in the violently agitated suspending fluid to separate any Hfr and F$^-$ cells that might be stuck to each other. As can be seen, blending the mating mixture just before plating for recombinants leads to rather different kinetics. Here the number of recombinants recovered does *not* rise linearly from the moment of mixing the parental strains; instead *no* Thr$^+$ *Leu$^+$ Str^r recombinants at all appear if a sample of the mating mixture is blended during the first 8 minutes of cell contact*, and no Gal$^+$ Str^r recombinants appear upon blending during the first 25 minutes. Blending samples of the mating mixture at later times generates a linear increase in the number of recombinants, until finally the recombinant frequencies level off at the same plateaus attained in the first part of the experiment, which does not involve any blending.

   Wollman and Jacob interpreted these results in the following terms. In the first part of the experiment, which involved no blending, Hfr and F$^-$ cells that had come into stable contact and started their conjugation could continue to do so after dilution of the mixture, since they would remain together on the selective agar plate and there complete the act of recombination. The initial rise of the broken curves thus signals the rate at which the mating cells established contact. In the second part of the experiment, however, the blending has forcibly separated the two conjugating cells, and no further transfer of genetic material from donor to recipient can proceed on the selective agar plate. That the solid curve for Thr$^+$ Leu$^+$ Str^r recombinants begins to rise only after a latent period of 8 minutes means, therefore, that once contact between Hfr and F$^-$ cell *has* been established, a minimum of 8 minutes is required before the first *thr* and *leu* genes of the Hfr donor have been transferred to the F$^-$ recipient. Once this segment of the donor genome *has* entered the recipient cell, the two conjugating cells may be separated without prejudicing the ultimate appearance of a Thr$^+$ Leu$^+$ recombinant. The linear rise of the solid curve thus signals the rate of entry of the *thr-leu* donor segment into individual members of the F$^-$ recipient population. That the solid curve for Gal$^+$ Str^r recombinants begins to rise only after a 25-minute latent period means, furthermore, that it takes 17 minutes longer for the first *gal* donor gene than for the first *thr-leu* donor segment to enter the recipient cell. This finding suggested to Wollman and Jacob that the transfer of genetic material from Hfr donor to F$^-$ recipient proceeds in an *oriented* manner, transfer of the *thr-leu* segment always preceeding transfer of the *gal* gene.

Wollman and Jacob quickly confirmed this inference by examining the distribution of the unselected donor characters among the Thr$^+$ Leu$^+$ Str^r recombinants recovered after blending the mating mixture at various times. The result of this analysis is shown in Figure 10-9. As can be seen, about 10% of earliest recombinants that appear 10 minutes after the onset of conjugation contain also the donor *azi* gene, but none yet contain the *ton*, *lac*, or *gal* donor genes. After 15 minutes, 70% of the Thr$^+$ Leu$^+$ Str^r recombinants contain the *azi* donor gene, and 30% contain the *ton* gene, but none as yet contain either *lac* or *gal* donor genes. After 20 minutes, the recombinants start to show the *lac* donor gene, but only after 30 minutes do any of them show the *gal* donor gene. After 60 minutes when according to the kinetics of Figure 10-8, the number of Thr$^+$Leu$^+$Str^r recombinants has reached its final plateau, the percentage of unselected donor genes among them has also leveled off at values ranging from 90% for *azi* to 25% for *gal*. Thus, upon encountering an F$^-$ bacterium and forming a stable contact, every Hfr Hayes bacterium evidently begins transfer of its chromosome from a definite *origin*, or *O* locus, and continues transfer in the gene order *O-thr-leu-azi-ton-lac-gal*. Hence, assuming that the rate of chromosome migration is constant per unit length, the *time of entry* of each of these genes into the F$^-$ recipient cell provides a measure of their genetic distance—in other words, represents a genetic map.

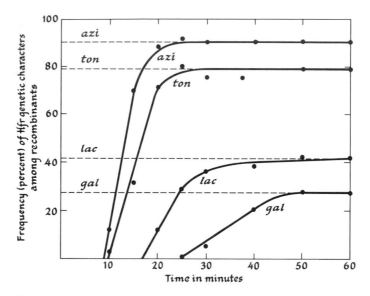

FIGURE 10-9.   Kinetics of transfer of genes carrying *unselected* characters in the Hfr × F⁻ cross of Figure 10-8. The Thr⁺ Leu⁺ Strʳ recombinants recovered at various times, either after direct plating (broken curves) or after blendor treatment (solid curves), were analyzed for the state of their unselected Azi, Ton, Lac, and Gal characters. The ordinate shows the relative frequency of the Hfr *azi, ton, lac,* and *gal* genes among the Thr⁺ Leu⁺ Strʳ recombinants sampled from the conjugation mixture at the time shown on the abscissa. [After F. Jacob and E. L. Wollman, *Sexuality and the Genetics of Bacteria.* Academic Press, New York, 1961.]

Once Wollman and Jacob reached these insights, it was not long before they realized why the high frequency character of the Hfr strain seemed to pertain only to a limited part of its genome—namely, to that part proximal to the *origin* of transfer. For the disruption of the mating bacterial couple, which can be induced artificially by the blendor treatment, also apparently occurs spontaneously by shearing forces extant in normal fluid mating mixtures, causing a spontaneous interruption of the transfer process. If, as seems likely, there exists a constant probability k per minute (and hence per unit length of donor chromosome transferred) that transfer is interrupted, then the probability p of transfer of a gene located at a distance from the origin O such that its transfer requires x minutes can be shown to be

$$p = e^{-kx} \tag{10-1}$$

Since, according to the data of Fig. 10-8, the ratio of p for *thr-leu* to that for *gal* is $0.18/0.05 = 3.6$, and the corresponding x values are 8 and 25 minutes, respectively, one may reckon from equation (10-1) that

$$k = \frac{\ln 3.6}{25 - 8} = 0.075/\text{minute}$$

Substituting this value of k into equation 10-1, it can be calculated that a donor gene that is, say, 3 times as far from O as *gal* and whose transfer would thus require 75 minutes, would be expected to appear in only about 0.02 as many recombinants as the donor *gal* gene, or with a frequency of about 10^{-3} per Hfr donor.

Wollman and Jacob then crossed a variety of auxotrophic and otherwise genetically marked F$^-$ strains with Hfr Hayes, in order to study by means of the interrupted mating technique the transfer of as many genes of the donor genome as possible. These experiments showed, first of all, that provided sufficient time is allowed for the conjugation process, this Hfr strain does eventually transfer all of its genome to the recipient cell, each gene having a characteristic time of entry. The last genes to be transferred turned out to be *met* and *thi*, whose time of entry appeared to be about 100 minutes. Second, the final number of recombinants formed requiring transfer of any given donor gene were found to decrease with the time of entry of that gene in the exponential manner predicted by equation (10-1). Wollman and Jacob were thus able to construct a genetic map of the chromosome of the Hfr Hayes strain in terms of *time*, a parameter never previously used in genetic mapping. This first temporal map is shown in Figure 10-10. A schematic

FIGURE 10-10. The first temporal map of a sector of the *E. coli* chromosome established by Jacob and Wollman in 1956. The number of minutes shown indicates the minimum time that must elapse between contact of Hfr Hayes donor and F$^-$ recipient cell and transfer of that gene from donor to recipient cell.

summary of the overall features of these general notions concerning the Hfr × F$^-$ conjugation process is shown in Figure 10-11.

NATURE OF THE F$^+$ × F$^-$ CROSS

Once they had understood the general nature of the Hfr × F$^-$ cross, Wollman and Jacob addressed themselves to the problem of how genetic recombinants arise in the "classical" F$^+$ × F$^-$ crosses. Is, they asked, the very low frequency of recombination of 10^{-7} per parent cell in these crosses the consequence of the capacity of *every* F$^+$ cell to transfer with small but equal probability *any* given sector of its genome to the F$^-$ cell it happens to encounter? Or is the F$^+$ cell altogether *incapable* of donating any genetic material other than

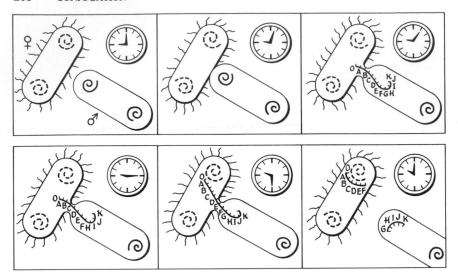

FIGURE 10-11. Wollman and Jacob's schematic representation of the conjugation process between Hfr and F⁻ bacteria. The F⁻ recipient cell is shown here with bristles, although there is no connection between these appendages and the F⁻ condition. Conjugation begins shortly (upper right) after the two bacteria come into contact, and is soon followed by the linear transfer of genetic material. The capital letters represent the location of various genes along the donor chromosomes. [After F. Jacob and E. L. Wollman, *Sexuality and the Genetics of Bacteria*. Academic Press, New York, 1961.]

its F sex factor, and are the recombinants actually formed by a small proportion of different Hfr mutants hidden in the F^+ population, each Hfr mutant capable of transferring with high efficiency some limited sector of the *E. coli* genome? By stating this problem in terms of these two antithetical possibilities it became formally analogous to the problem of the spontaneous versus induced origin of bacterial mutants discussed in Chapter 6, and hence capable of resolution by means of the Luria-Delbrück fluctuation test.

In order to carry out this test, a minimal medium containing a limiting amount of methionine was inoculated with a very small concentration (250 cells/ml) of the AF^+ Met⁻ strain. Half of this suspension was divided into 50 small portions, and both the small portions as well as the remaining bulk half of the culture were incubated until bacterial growth had exhausted the methionine supply and attained a density of about 2×10^8 cells/ml in all the cultures. The contents of each small culture, as well as each of 50 aliquots of the bulk culture, were then mixed with an excess of bacteria from a bulk culture of the BF^- Thr⁻ Leu⁻ Str^r strain and the number of Thr⁺ Leu⁺ Str^r recombinants produced in each of these 100 crosses scored on streptomycin-containing minimal agar supplemented with methionine. The summary result of this fluctuation test is presented in Table 10-3. It can be seen that

TABLE 10-3.   Fluctuation Test on the Number of Recombinants Formed by Independent Cultures of F^+ Met$^-$ Str^s Bacteria

	Number of Thr$^+$ Leu$^+$ Str^r recombinants formed in crosses with F$^-$ Thr$^-$ Leu$^-$ Str^r	
	F^+ donor from 50 individual cultures	F^+ donor from 50 aliquots of big culture
Minimum found	1	10
Maximum found	116	23
Mean	15.3	16.3
Variance	351.5	13

From F. Jacob and E. L. Wollman, *Sexuality and the Genetics of Bacteria*, Academic Press, New York, 1961.

the number of recombinants produced by crosses involving F^+ cells from the individual small cultures is subject to a very great fluctuation, whereas the number of recombinants produced by crosses involving F^+ cells from aliquots of the bulk culture manifests no greater than random variation. It can be concluded, therefore, that the formation of genetic recombinants in the F^+ × F^- cross does depend on the spontaneous appearance of Hfr "mutants" during the growth of the F^+ culture, since, according to the reasoning presented in Chapter 5, the number of such mutants in a series of parallel cultures, and hence the fertility of crosses carried out with such cultures, would be expected to manifest a large variance. If, on the contrary, *every* F^+ cell were potentially capable of giving rise to any type of recombinant upon mixing F^+ and F^- cultures, then the variance in the number of recombinants produced in crosses carried out with parallel cultures should have been no greater than the mean, as actually observed for the control crosses involving aliquots of the bulk culture.

Having demonstrated by means of the fluctuation test that the spontaneous occurrence of Hfr "mutants" within the F^+ population is responsible for the fertility of F^+ × F^- crosses, Wollman and Jacob retraced a further step of the earlier efforts on behalf of spontaneous mutation—namely, the Lederbergs' use of replica plating for direct isolation of mutants. For this purpose, a master plate containing complete agar was seeded with enough cells of an auxotrophic F^+ culture to produce growth of a confluent bacterial lawn. This lawn was then touched with the velvet block and replicated onto a plate containing a minimal, selective agar whose surface was densely covered with auxotrophic F^- bacteria unable to grow on it. The genetic constitution of F^+

and F⁻ strains was chosen in such a way that genetic recombinants *could* arise that would be able to grow on the selective agar of the replica plate. Wherever the velvet happened to have picked up part of a rare Hfr clone capable of producing the selected recombinant type among the F⁺ bacteria of the master plate and deposited it among the F⁻ bacteria of the replica plate, a recombinant colony arose on the replica plate. The exact location of that recombinant colony thus signaled an area on the master plate lawn of F⁺ bacteria where a clone of putative Hfr bacteria was to be found. That area of the lawn was picked, respread on a second master plate, and the replica plating procedure repeated once or twice in order to procure a selective enrichment of the Hfr bacteria. Finally, individual colonies of the final Hfr-enriched population of F⁺ cells were inspected for their fertility, and among them pure clones of Hfr individuals were found. By varying the F⁻ genotype and the selective conditions of the master plate, Wollman and Jacob succeeded in isolating from the same F⁺ parent strain a variety of Hfr strains, some of which, like Hfr Hayes, donate the *thr-leu* sector of their genome with high frequency, and some of which donate sectors other than *thr-leu* with high frequency. Thus, the inference that the fertility of F⁺ × F⁻ crosses is actually the work of a minority of diverse Hfr bacteria that arise spontaneously during the growth of the F⁺ culture was directly confirmed.

THE CIRCULAR CHROMOSOME

Wollman and Jacob next proceeded to characterize in more detail the nature of the various Hfr strains they had isolated. In particular they carried out interrupted mating experiments with each strain similar to that described for Hfr Hayes in Figures 10-8 and 10-9, in order to establish the gene order by means of which that strain transfers its chromosome to the F⁻ cell. The result of this survey (including also three Hfr strains subsequently isolated by other workers) is shown in Table 10-4. At first glance, the data of this table suggest that different Hfr strains differ completely in their order of gene transfer, any gene locus being apparently eligible for being next to the transfer origin, *O*. A second glance reveals, however, that there is a high degree of structure to the information contained in this table, in that any pair of neighboring genes in any one row is also neighboring in every other row. That is, the gene transfer sequence does appear to be exactly the same for all these Hfr strains, but what is different from row to row is the position of *O* in that sequence and the *direction* in which the genes are transferred. For instance, in comparing Hfr Hayes and Hfr 2 it becomes apparent that the former transfers the segment *thr-leu-azi-ton-pro* in that order, whereas the latter transfers the same segment in the reverse direction. Or, Hfr 2 and Hfr 3 evidently transfer their genes in the same direction, but in the latter the

TABLE 10-4.   Gene Order in Conjugational Transfer by an Ensemble of Different Hfr Strains

Hfr strain	Order of gene transfer
Hayes	O-thr-leu-azi-ton-pro-lac-pur-gal-trp-his-gly-str-mal-xyl-mtl-ile-met-thi
1	O-leu-thr-thi-met-ile-mtl-xyl-mal-str-gly-his-trp-gal-pur-lac-pro-ton-azi
2	O-pro-ton-azi-leu-thr-thi-met-ile-mtl-xyl-mal-str-gly-his-trp-gal-pur-lac
3	O-pur-lac-pro-ton-azi-leu-thr-thi-met-ile-mtl-xyl-mal-str-gly-his-trp-gal
4	O-thi-met-ile-mtl-xyl-mal-str-gly-his-trp-gal-pur-lac-pro-ton-azi-leu-thr
5	O-met-thi-thr-leu-azi-ton-pro-lac-pur-gal-trp-his-gly-str-mal-xyl-mtl-ile
6	O-ile-met-thi-thr-leu-azi-ton-pro-lac-pur-gal-trp-his-gly-str-mal-xyl-mtl
7	O-ton-azi-leu-thr-thi-met-ile-mtl-xyl-mal-str-gly-his-trp-gal-pur-lac-pro
AB311	O-his-trp-gal-pur-lac-pro-ton-azi-leu-thr-thi-met-ile-mtl-xyl-mal-str-gly
AB312	O-str-mal-xyl-mtl-ile-met-thi-thr-leu-azi-ton-pro-lac-pur-gal-trp-his-gly
AB313	O-mtl-xyl-mal-str-gly-his-trp-gal-pur-lac-pro-ton-azi-leu-thr-thi-met-ile

From F. Jacob and E. L. Wollman, *Sexuality and the Genetics of Bacteria*, Academic Press, New York, 1961.

segment *pur-lac* is interposed between *O* and *pro*. In order to interpret these surprising results, Wollman and Jacob made another proposal, which for the year 1957 was quite radical—namely, that the chromosome of the F^+ bacterium is *circular*, as shown in Figure 10-12. They then envisaged that the "mutation" $F^+ \rightarrow$ Hfr represents the spontaneous "breakage" of the circular chromosome and assignment of an origin *O* to one of the two chromosome ends created by that "breakage." This no longer circular but rectilinear Hfr chromosome was then supposed to be capable of being transferred to the F^- cell, beginning with the *O* locus, a process that the circular F^+ (and, presumably, also circular F^-) chromosome was imagined to be incapable of performing. Thus Cavalli's Hfr strain (similar to strain Hfr 3 of Table 10-4) would have arisen by a "breakage" of the circle between *gal* and *pur* genes, with assignment of *O* to the *pur* end, whereas Hayes Hfr strain would have resulted from a "break" between *thi* and *thr*, with assignment of *O* to the *thr* end.

As we saw already in Chapter 9, particularly in Figure 9-12, the circularity of the *E. coli* chromosome first postulated by Wollman and Jacob was directly confirmed by Cairns' later autoradiographic visualization of the bacterial DNA molecule. Furthermore, as the discussions of later chapters will show, the inference of circularity was soon to recur in the analysis of more and more self-replicating DNA structures. Nowadays, when DNA circularity is more

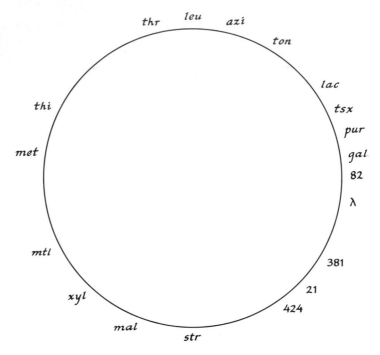

FIGURE 10-12.   The first circular map of the *E. coli* chromosome, as
inferred from a comparative study of the chromosome segments
transferred with high frequency by different Hfr donor strains. This
diagram intends to present only the sequence of the genes and not the
relative distances between them. The symbols 82, λ, 381, 21, and 424
refer to the chromosomal sites at which the prophages of various
temperate phages are attached, as will be discussed in Chapter 14.

or less taken for granted, it is hard to imagine the surprise and skepticism
with which this notion was first met in 1957. Indeed, circularity turned out
to be even more universal than Wollman and Jacob might have wished.
For it transpired that the chromosome of the vegetatively growing Hfr
bacterium is also circular. Hence the spontaneous $F^+ \rightarrow Hfr$ conversion
cannot, in fact, represent the postulated chromosomal "break." That is, the
rectilinear topology of the Hfr chromosome appears to pertain only to its
state during conjugational transfer from donor to recipient cell. The
"fertility mutation" must thus correspond to the creation of a *potential*
rather than actual break at some site of the F^+ chromosome, the potentiality
of the break being realized only upon encounter of a nubile F^- cell.

How *does* the presence of the F factor manage to cause the spontaneous
$F^+ \rightarrow Hfr$ conversion? Wollman and Jacob proposed that this event represents
an act of genetic recombination between genomes (and hence DNA molecules)
of sex factor and bacterial chromosome. As a consequence of this act, the

chromosome was thought to "break," and the sex factor was thought to be integrated into the chromosome at the site at which recombination happened to have taken place. Once it was realized, however, that the "break" is only potential rather than actual, the "fertility mutation" was explained by A. M. Campbell in the following terms (Figure 10-13). The F factor DNA,

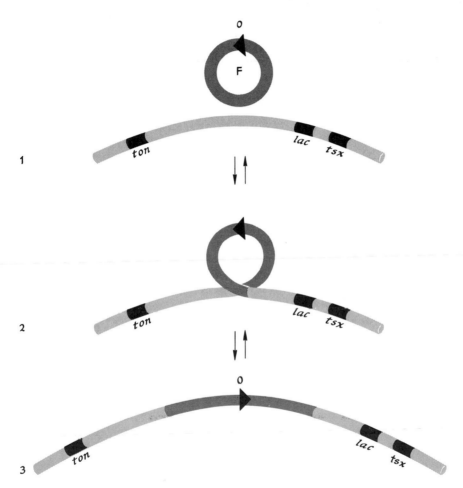

FIGURE 10-13.   The conversion of an F⁺ bacterium to the Hfr state. The circular F factor DNA, bearing among its genes *O*, the origin of oriented chromosome transfer from donor to recipient cell happens to pair with the DNA of the circular bacterial chromosome between the *ton* and *lac* genes. A reciprocal crossover between F factor and bacterial chromosomes results in the insertion of the F factor into the continuity of the chromosome. This act has created an Hfr strain that transfers *ton* as one of its first and *lac* as one of its last genes upon conjugation with an F⁻ cell. This sequence of events is reversible, so that the Hfr chromosome can loop out to reconstitute the circular F factor and revert to the F⁺ state by another reciprocal crossover.

like the bacterial chromosome, is circular and bears among its genes one that represents O, the origin of oriented conjugal chromosome transfer. Among its nucleotide sequences the F factor DNA contains one or more domains that are sufficiently "homologous" to a number of domains in various sectors of the chromosomal DNA, such that pairing, or synapsis, of F factor and chromosome can occur within one of these sectors of homology. A reciprocal crossover then occurs between F factor and chromosomal DNA which results in the insertion of the F factor, and hence of its O gene, into the continuity of the chromosome. Since the synapsis may occur in either one of two relative attitudes of the two paired DNA structures, depending on the nature of the homology, the sex factor, and hence its origin, may be inserted into the circle in either a "clockwise" or "counterclockwise" orientation. Thus the position and orientation of the integrated sex factor in the bacterial chromosome determines the gene-transfer character of the Hfr strain that results from the integration act. According to this picture, the transferred Hfr chromosome bears both at its origin as well as at its terminus part of the genes of the integrated sex factor. That this is actually true can be inferred from the following facts. Whereas the recombinants issuing from Hfr $\times$ F$^-$ crosses are almost always F$^-$ (an observation dating back to Hayes' first studies in 1953), Wollman and Jacob found that among the very rare recombinants of such crosses selected for transfer and incorporation of *both* an origin-proximal *and* a near-terminal gene of the donor chromosome, a high proportion *do* acquire the Hfr character of the donor parent. Thus an Hfr bacterium *can* transfer its sex factor after all, provided that conditions obtain that allow both parts of the F-factor genome, origin-proximal and terminal, to reach the F$^-$ recipient.

THE SEX FACTOR AS AN EPISOME

It thus appeared that the sex factor is a self-reproducing genetic element that can exist in two states, *autonomous* and *integrated*. In the autonomous state, that obtaining in an F$^+$ cell, the F factor does not form part of, and replicates independently of, though in synchrony with, the bacterial chromosome. The products of one or more genes carried by the F factor endow the F$^+$ bacterium with the capacity to transmit the F factor to an F$^-$ cell, where the F factor establishes itself and effects with high efficiency the F$^- \rightarrow$ F$^+$ conversion. In the integrated state—that obtaining in an Hfr cell—the F factor forms part of, and is replicated as part of the bacterial chromosome. The very act of F factor integration, according to Jacob and Wollman, is responsible for the rare spontaneous chromosomal modification that constitutes the F$^+ \rightarrow$ Hfr fertility "mutation." Once integrated into the chromosome the F factor is no longer independently transmissible, which explains why contact of Hfr cells with F^- cells does not convert the latter to F'. Even in the integrated state, however, the sex factor *does* promote its own transfer from donor to recipient cell, except that here the O-proximal sector of the F factor functions as a locomotive that pulls in its train the Hfr chromosome into the recipient cell.

On the basis of this view of the nature of the F sex factor, Jacob and Wollman defined a new class of genetic elements—the *episomes*, which are *added* to the genome of a cell, and hence may be either present or absent. When absent, an episome can be acquired only from an external source; it cannot simply appear by mutation, rearrangement, or alteration of the endogenous cellular genome. When present in the cell, an episome alternates between two mutually exclusive states, the autonomous and the integrated. The phenotypic characters acquired by the cell through possession of an episome are usually dispensable, since episomes are not necessary cellular constituents. Once a bacterium has been "cured" of an episome (such as in the change $F^+ \rightarrow F^-$), the phenotypic character associated with the episome is also irretrievably lost. Temperate bacteriophages, the subject matter of Chapter 15, will be seen to represent another important group of episomes, since such bacteriophages, as was first adumbrated by Hayes in 1952, share these fundamental properties with the sex factor.

Inspection of the schema of F sex factor integration outlined in Figure 10-13 suggests that the $F^+ \rightarrow Hfr$ "fertility-mutation" process ought to be reversible. For it can be readily imagined that the circular DNA genome of an Hfr bacterium containing the integrated episome in its continuity would "loop out" again by specific pairing of its homologous sectors and regenerate an autonomous circular F factor DNA by means of another reciprocal crossover. That is to say, the sequence of events depicted in Figure 10-13 should run both forward and backward. This expected occurrence of the $Hfr \rightarrow F^+$ reverse "mutation" is, in fact, met in practice, since there is present in most cultures of Hfr strains a small (and sometimes even a large) proportion of F^+ bacteria.

In 1959 E. H. Adelberg was intending to carry out conjugation experiments with one of Wollman and Jacob's then recently isolated Hfr strains, but he found to his annoyance that his culture of that strain had partially reverted to the F^+ state, and hence had lost its potency as a high-frequency donor strain. But when Adelberg examined the properties of the revertant F^+ strain he made a most important discovery: the F sex factor carried by the F^+ revertant bacteria had undergone a genetic modification. The modification of this sex factor, to which Adelberg assigned the symbol F', expressed itself in a great increase in the frequency of occurrence of both the $F^+ \rightarrow Hfr$ forward and the $Hfr \rightarrow F^+$ reverse integration processes. Furthermore, the chromosomal site at which the F' factor became attached upon its eventual reintegration was invariably the same as the site at which the original Hfr strain had carried its F sex factor. This preference for a unique attachment site of F′ is to be contrasted with the behavior of the normal F sex factor, whose integration proceeds at many different sites of the *E. coli* chromosome and hence gives rise to a diversity of Hfr strains. To account for his finding, Adelberg proposed that the F′ sex factor acquired part of the bacterial chromosome and hence retained a "memory" of the genetic site at which Wollman and Jacob had first found it to be integrated in their Hfr strain.

This proposal encouraged Jacob and Adelberg to search for the incorporation of identifiable bacterial genes into F′ sex factors in other F$^+$ revertants of Hfr strains. For this purpose, they mated an F$^+$ revertant culture derived from an Hfr strain (whose origin is between *ton* and *lac* and which transfers *lac* as one of the last genes) with an F$^-$ recipient strain carrying many mutant characters. As was to be expected, genetic recombinants for the mutant characters arose only in the low-frequency characteristic for F$^+$ × F$^-$ crosses. There was one apparent exception, however; the Lac$^+$ character of the F$^+$ revertant donor strain was transferred to the F$^-$ Lac$^-$ recipient bacteria with a very high frequency similar to that obtainable in the most fertile Hfr × F$^-$ crosses. But close examination of the Lac$^+$ recombinants formed as a result of this mating showed that they were not ordinary recombinants in which the donor wild-type *lac* gene had replaced its recipient mutant allele. Instead, the Lac$^+$ recombinants were found to carry *two* sets of *lac* genes, those of the Lac$^+$ F$^+$ parent and those of the Lac$^-$ F$^-$ parent. This could be inferred from the *instability* of the Lac$^+$ character of these bacteria, of whose descendants about 0.1 % lost the donor wild-type *lac* gene and reverted to the Lac$^-$ character of the mutant *lac* gene of the original F$^-$ recipient cell (Plate III). Furthermore, these Lac$^+$/Lac$^-$ *diploid* recombinants had gained from their donor parent not only the Lac$^+$ character but also the capacity to donate that character at a high frequency to other F$^-$ Lac$^-$ recipient cells. The explanation of this phenomenon is that the F$^+$ revertant donor strain isolated by Jacob and Adelberg carried an F′ sex factor, designated as F-*lac*, that upon its release from the Lac$^+$ Hfr genome by the Hfr → F$^+$ reverse integration process had managed to incorporate the *lac* genes of the bacterial genome. As shown in Figure 10-14, such incorporation can occur through an improper "looping out" in the Hfr → F$^+$ reversion process, leading to the generation of an oversize F′ factor. Transfer of this F′ factor to an F$^-$ cell then brings along and maintains in that cell not only the sex-factor genes but also whatever bacterial genes the episome happens to carry. Thus transfer of the wild-type *lac* gene of the Lac$^+$ Hfr genome carried by the F-*lac* sex factor into the F$^-$ Lac$^-$ cell carrying a nonfunctional mutant *lac* gene will evidently give rise to a Lac$^+$ bacterium. For the presence of the functional sex-factor *lac* gene, which is capable of directing the synthesis of one of the enzymes of lactose metabolism, bestows on the diploid cell the capacity to ferment lactose, even if there is also present in the bacterial genome a second, mutant allele of the *lac* gene that is unable to give rise to that enzyme. In genetic parlance, the wild-type *lac* gene is therefore said to be *dominant* over its mutant allele (and the mutant *lac* gene is said to be *recessive* to its wild type allele), since the overt character of the Lac⁺/Lac⁻ diploid bacterium is Lac⁺.

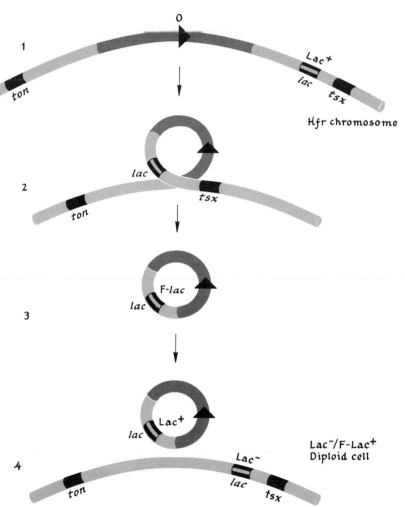

FIGURE 10-14.   The production of an F′ sex factor, bearing the a *lac* gene of the bacterial chromosome, or *F-lac*. The F factor integrated into the Hfr chromosome between the *ton* and *lac* genes loops out improperly, so that the chromosomal *lac* sector has been included in the looped-out portion bearing the sex-factor genes. A reciprocal crossover generates a circular F factor into which the *lac* gene has been incorporated and which can now be readily transferred from F⁺ donor to F⁻ recipient cell.

Since Jacob and Adelberg's discovery of F-*lac*, a great variety of F' sex factors have been found that carry different portions of the bacterial genome as part of their own episomal genome. Some of these F', like the original F-*lac*, carry only a very few bacterial genes, whereas others are monster-bastards that carry as much as a quarter of the whole bacterial genome. Most of these sex factors may be merely laboratory curiosities, of interest only to bacterial geneticists, but an ominous type of sex factor of the greatest clinical import and social portent was identified in Japan shortly after the discovery of F-*lac*. By that time, such antibiotic drugs as penicillin, streptomycin, chloramphenicol, and sulfonamide had been in widespread use in Japan for more than a decade, and drug-resistant mutant types, such as Pen^r and Str^r, of many pathogenic bacterial species had made their appearance. Bacterial strains isolated from dysentery patients during the late 1950's were found to be simultaneously resistant to three or four different commonly used antibiotics: even the stratagem of multiple antibiotic treatment, referred to in Chapter 6, was ineffective against these strains. Japanese bacteriologists then discovered that some of these chemotherapeutically invulnerable pathogenic bacteria can *transfer* their multiple drug resistance to other drug-sensitive bacteria, not only of their own species but to a broad spectrum of taxonomically diverse genera. Study of this alarming development was taken up by T. Watanabe and T. Fukasawa, who showed that transfer of multiple drug resistance is attributable to a sex-factor-like episome, to which was given the name *resistance-transfer factor*, or RTF. The RTF carries, in addition to the set of sex factor genes, a set of drug-resistance genes, each of which governs the synthesis of some protein product that confers resistance to a particular drug. It is obvious that with the appearance of RTF, drug resistance can spread like wild-fire through the bacterial flora. As Watanabe has pointed out, there was virtually no incidence of drug-resistant dysentery in Japan in 1956, but by 1964 about half of all the bacterial strains isolated from dysentery patients were simultaneously resistant to four commonly used and formerly highly effective antibiotics. In the meantime, RTF's have made their appearance all over the world, and multiple drug-resistant bacterial pathogens have become the order of the day in medical bacteriology. The indiscriminate broadcast use of antibiotics characteristic of the 1950's and 1960's will have to be halted, and the environmental conditions that favor the natural selection of RTF-carrying bacteria must be averted if drug therapy of bacterial diseases is not to suffer the fate of Chinese acupuncture as an effective medical tool.

THE CONJUGAL TUBE AND MOBILIZATION

As soon as the one-way transfer of genetic material from donor to recipient cell in bacterial conjugation had been established, it became a matter of great interest to know just *how* this transfer is effected once cell contact has been established. In 1957 T. F. Anderson succeeded in obtaining an electron micrograph (Figure 10-15) of what appear to be two conjugating bacteria. This picture was taken of a mating mixture in which the Hfr donor strain consisted of long, thin, rod-shaped cells, and the F⁻ recipient strain of short, fat, round cells. It can be seen in this now-famous picture that the two bacteria are connected by a narrow tube, through which, it seems reasonable enough to suppose, the genetic material passes from donor to recipient cell.

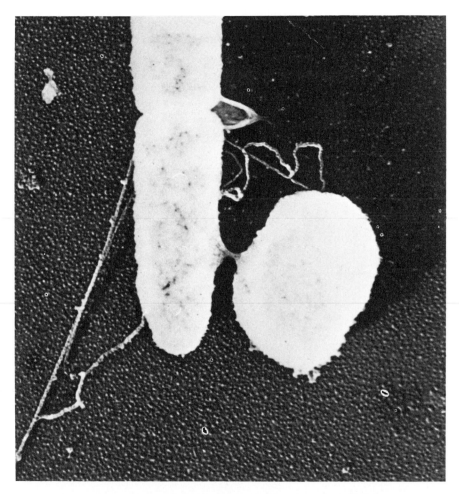

FIGURE 10-15.   Electron micrograph of two conjugating *E. coli* bacteria. A slender connecting bridge has been formed between an elongated Hfr donor cell and a rotund F⁻ recipient cell. [From E. L. Wollman, F. Jacob, and W. Hayes, *Cold Spring Harbor Symp. Quant. Biol.* **21**, 141 (1950). Photograph taken by T. F. Anderson.]

Another seven years were to elapse, however, before further insight was gained into the possible origin of this conjugal tube. Finally, in 1964, C. C. Brinton discovered that F$^+$ and Hfr bacteria possess a superficial filamentous appendage, the *F pilus*, which is absent from F$^-$ cells (Figure 10-16). Because Brinton found that removal of its pilus by physical or chemical means destroys the fertility of the donor cell until such time as a new pilus has been regenerated, he proposed that the F pilus serves as the conjugation tube. This seems like a very plausible proposal; furthermore, there can be little doubt that the F pilus is the product of one or more genes carried by the F sex factor and that, as befits a putative genetic conduit, it is actually hollow. At the time of this writing, however, there is still no direct evidence that the donor F pilus attaches itself to the surface of the recipient F$^-$ cell and that genetic material passes through it, as Brinton envisaged.

But how does the F factor *mobilize* the genome of which it forms part, so that it begins to move through the conjugation tube toward its new home once donor has made contact with recipient cell? The first coherent proposal for this process was made in 1963 by Jacob, S. Brenner, and F. Cuzin. The

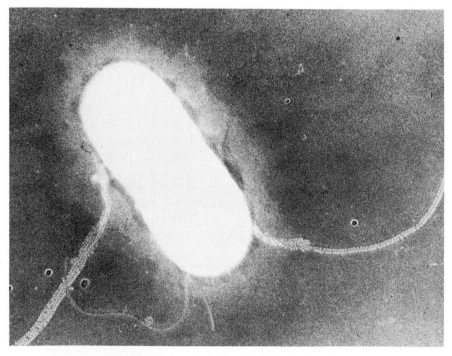

FIGURE 10-16.   Two sex pili of an Hfr *E. coli* cell. In this preparation, the sex pili are covered with male-specific phages (see Chapter 19). [Courtesy of R. C. Valentine and Alice Taylor. Magnification about 17,000 diameters.]

essence of that proposal was the assertion that the O locus of the F factor represents what they called a *replicator*, or site at which replication of the DNA molecule can be initiated by the DNA polymerase enzyme. Each replication cycle of the autonomous (i.e., nonintegrated) F factor present in an F^+ cell was thought to begin at that replicator site, in the same way in which replication of the bacterial chromosome was imagined to begin at its own replicators—that is, at the p_1 point of Figures 9-13 and 9-14—and then to proceed around the DNA circle by means of a replicating Y-fork. The F replicator at the O locus has one additional quality, however; it is "activated" upon contact of the donor with a recipient cell. This activation engenders a new round of semiconservative DNA replication of the sex factor DNA, in the course of which one of the two daughter duplexes is threaded, O-locus-first, into the conjugation tube and driven toward the F^- cell, while the other daughter duplex remains in the donor cell. As soon as the whole F genome has been replicated in this way, the F^- recipient is in possession of one copy of the episome and ready to change its sex.

The F replicator at the O locus of the integrated F factor of the Hfr cell is also "activated" upon contact with a recipient cell, but the semiconservative DNA replication "activated" in this manner does not remain confined to the sex factor genome but eventually involves also the chromosome (Figure 10-17). In this manner, one copy of the donor genome continues to be driven, O-locus-first, into the F^- cell, as long as the conjugation tube connecting the two bacteria remains intact. Thus it is activation of the F replicator and subsequent, continuous sequestration of one of the replica chromosomes that are responsible for the realization of the potentiality for chromosome breakage at the O locus and the conversion of the circular chromosome into a rectilinear

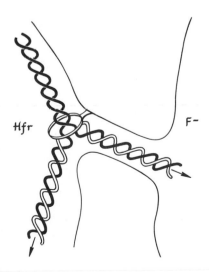

FIGURE 10-17.   The Jacob-Brenner-Cuzin model of chromosome transfer by replication during conjugation. In the donor Hfr bacterium, the F-factor replicator (F Rep) is assumed to be attached to the bacterial membrane in the vicinity of the point at which the conjugation tube (possibly the sex pilus) forms an intercellular bridge between Hfr donor and F^- recipient. Replication of the donor DNA is assumed to be initiated at F Rep once conjugal contact is established, one of the daughter DNA replicas being driven through the conjugation tube into the F^- cell, the other replica remaining in the Hfr donor cell. [Redrawn from F. Jacob, S. Brenner, and F. Cuzin, *Cold Spring Harbor Symp. Quant. Biol.* **28**, 329 (1963).]

Hfr

F^-

structure. The arrow affixed to the schematic drawing of the sex factor in Figure 10-13 thus symbolizes the *direction* of the replication initiated at the F replicator at the O locus upon its "activation" by conjugal contact, which, depending on the attitude of the integrated episome, leads to either "clockwise" or "counterclockwise" transfer of the Hfr chromosome.

This view of the nature of the sex-factor-mediated transfer mechanism has been the subject of some controversy since it was first proposed. For instance, there has been disagreement about one crucial matter of fact—namely, whether concomitant DNA synthesis by the donor cell is *required* for chromosome transfer, as is obviously demanded by the model of Jacob, Brenner, and Cuzin. Whereas the answer to this question still has not been established to everybody's satisfaction, partly because of technical and partly because of semantic difficulties, there is no longer any doubt that during conjugation under *normal* conditions, DNA replication *does* occur concomitantly with transfer of the donor genome. The experiments that established this most clearly were done by J. Gross and L. Caro, who carried out two sets of Hfr × F⁻ crosses with morphologically distinguishable donors and recipients, such as those shown in Figure 10-15. The donor DNA of one set was labeled with ³H-thymine *only before* conjugation, by first growing the Hfr donors in the presence of ³H-thymine and then mating them to nonlabeled F⁻ recipients in a nonlabeled medium. The donor DNA of the other set was labeled *only during* conjugation, by mating nonlabeled Hfr donors and F⁻ recipients in the presence of ³H-thymine under special experimental conditions that prevented ³H-thymine from directly entering the DNA of the F⁻ recipients. Autoradiograms of samples of these mating mixtures were then prepared after various times of conjugation, and the amount of ³H-thymine transferred to the recipients was determined by counting the number of photographic grains exposed by ³H-decay lying over bacteria of recognizable F⁻ morphology (Figure 10-18). The result of these determinations was that the donor DNA that is transferred to the recipient cells contained the same amount of ³H-thymine, whether the labeling was done only before or only during conjugation. This result shows that the DNA transferred in bacterial conjugation is half-new, half-old hybrid DNA, which is in agreement with the demands of the model of gene transfer by concomitant semiconservative DNA replication.

A further point of controversy has been the relation of the O-locus of the integrated F factor to the normal vegetative replicator of the bacterial chromosome. Whereas Jacob, Brenner, and Cuzin assumed that vegetative replication of the Hfr chromosome remains under the control of its normal replicator locus, a variety of experiments, the first of which was done soon after the model was proposed, have since suggested that once integrated, the F replicator serves also for the *vegetative* replication of the chromosome. Furthermore, and most surprisingly, the direction of this F-factor-governed vegetative DNA replication has been claimed to proceed in a direction

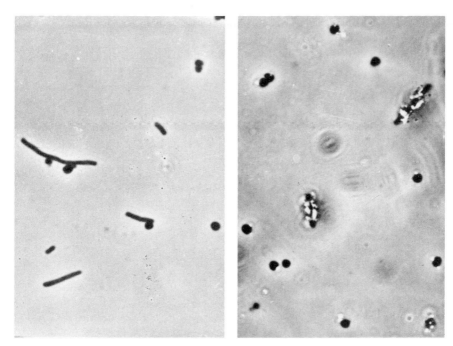

FIGURE 10-18.   Autoradiography of a conjugating mixture of ³H-thymine-labeled
Hfr and initially unlabeled F⁻ bacteria. (Left) Photomicrograph of the conjugating
mixture, showing two mating complexes formed between the long Hfr cells and the
round F⁻ cells. (Right) Autoradiograph of the mixture prepared after conjugation had
been allowed to proceed for 50 minutes. The grains formed in the autoradiographic
emulsion by the decaying ³H atoms during storage appear as bright dots. The two
elongated Hfr bacteria are densely covered by grains, indicative of their heavy ³H-thymine
label. Some of the round F⁻ bacteria have no grains; others have various numbers of
associated grains, indicating that they received some ³H-thymine labeled DNA from an
Hfr donor mate. [From J. D. Gross and L. G. Caro, *J. Mol. Biol.* **16**, 269 (1966).]

opposite from that in which chromosome transfer proceeds during conjugational DNA replication. In any case, it would seem that though the question
of whether the integrated F factor also plays a role in vegetative chromosome
replication of the Hfr cell is not devoid of intrinsic interest, it is not really the
touchstone of the model of conjugational chromosome transfer by initiation
of a special DNA replication cycle. It is to be noted, however, that if conjugational F-replicator-governed DNA replication does proceed *pari passu*
with chromosome transfer, then it must proceed at only one-third the rate of
vegetative DNA replication. For under growth conditions in which vegetative
replication of the whole *E. coli* chromosome requires 30 minutes (see Chapter
9), conjugational transfer of the whole chromosome requires 100 minutes.

The Jacob, Brenner, and Cuzin model is thus probably correct in its most
essential features, but the details of the F-factor mobilization processes still

remain to be worked out. Furthermore, the passive role assigned to the F⁻ cell in the transfer reaction under this model may also be in need of revision.

GENETIC RECOMBINATION IN THE MEROZYGOTE

We are now ready to consider the last stage of the conjugation process— namely, the production of a recombinant bacterial chromosome that contains some of the genome of the Hfr donor and some of the genome of the F⁻ recipient parent. This recombinant obviously arises in the F⁻ recipient cell turned merozygote, which, because of the spontaneous interruption of chromosome transfer, usually contains only a part of the genome of the Hfr donor in addition to its own chromosome. The merozygote is thus formally equivalent to the recipient cell in *bacterial transformation*, which cell, as was seen in Chapter 7, takes up an exogenous DNA molecule from the medium and thus similarly carries part of a donor genome in addition to its own entire chromosome. It is to be noted, of course, that the *extent* of diploidy is very much greater in conjugational merozygotes, which contain often a quarter, sometimes half, and, occasionally *all* of the donor genome, than in the transformational recipient cells, whose ingested exogenous DNA molecule can hardly amount to more than one percent of the donor genome. But just as in transformation, so in conjugation do the various sectors of the exogenous transferred donor DNA molecule find their homologous sectors of corre- sponding nucleotide base sequence of the endogenous recipient DNA molecule and then engage with them in acts of genetic exchange, as shown symbolically in Figure 7-9. This exchange results in the integration of parts of the trans- ferred Hfr donor DNA and in expulsion of the corresponding parts of the F⁻ recipient DNA. Once integrated into the recipient genome, the donor DNA sectors are replicated as part of the recombinant chromosome and transmitted to all of the descendant bacteria.

By comparing the frequency of recombination with that of *zygotic induction* —a process whose discussion must be deferred until Chapter 15 but which, we may now note, is a direct measure of donor gene *transfer*—Jacob and Wollman were able to estimate the *efficiency* of the integration process. They found in this way that once a donor gene *has* been transferred from Hfr to F⁻ cell, its chance of being integrated into a recombinant genome is about 0.5, an efficiency seen to be entirely comparable with that reckoned for gene integration in DNA transformation in Chapter 7.

The time course of this integration process was ascertained by J. Tomizawa in 1960 in the merozygotes formed in the cross Hfr Cavalli Lac⁺ Tsxˢ × F⁻ Lac⁻ Tsxʳ. Since, according to Table 10-4, the *lac* donor gene is one of the first genes to be transferred by the Hfr Cavalli strain, Tomizawa could inter- rupt the mating 30 minutes after the onset of conjugation by destroying the Tsxˢ Hfr cells with T6 phage, leaving the Tsxʳ merozygotes and F⁻ cells intact. He then examined the character of the very first Lac⁺/Lac⁻ merozygotes

formed at various later times by plating the mixed bacterial population on
EMB-lactose agar. On this agar the F⁻ Lac⁻ recipients form *white* colonies
(see Chapter 5). Nearly all of the Lac⁺/Lac⁻ merozygotes, however, at first
form red-white *variegated*, or Lacᵛ, colonies, since such merozygotes ultimately give rise to mixed colonies of both pure Lac⁺ and pure Lac⁻ descendants. Platings made at later times do give rise to nonvariegated, purely
red colonies, which signal the presence of stable Lac⁺ recombinant genomes
in the population of ex-conjugant bacteria. Tomizawa then determined time
dependence of the number of Lac⁺ and Lacᵛ colonies, the former being
evidently a measure of the degree to which integration of the donor *lac* genes
had proceeded, with the result shown in Figure 10-19. As can be seen, the
first pure Lac⁺ colonies appear some 50 minutes after onset of conjugation of
Hfr and F⁻ parents and then steadily mount in number. At the same time
there occurs a steady decrease in Lacᵛ colonies, so that the total number of
Lac⁺ and Lacᵛ colonies remains constant for about 100 minutes. After 130
minutes, by which time the number of Lacᵛ colonies has decreased more than
tenfold, multiplication of the Lac⁺ recombinant cells is underway. Tomizawa
concluded from these data that there is a one-to-one correspondence between
the appearance of pure red colonies and the disappearance of variegated
colonies—that is, each transferred donor *lac* gene gives rise to *only one* Lac⁺
recombinant genome. This means that the two complementary strands of the
transferred donor DNA molecule, both of which carry the Lac⁺ genetic

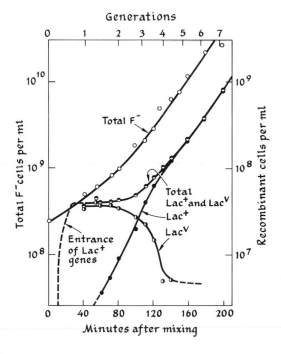

FIGURE 10-19.   Kinetics of
segregation of Lac⁺ recombinants
in the cross Hfr Cavalli
Lac⁺ Tsxˢ Strˢ × F⁻ Lac⁻ Tsxʳ
Strʳ. Thirty minutes after the
onset of conjugation, T6 phage
was added to the mating mixture
to destroy all Tsxˢ Hfr cells and
stop further conjugation. The
mating mixture was then diluted,
and samples were plated at the
times indicated on the abscissa
on EMB-lactose agar. The
curve labeled "total F⁻"
indicates the number of white, or
Lac⁻, colonies that appear on
the EMB-lactose agar. The
symbol Lacᵛ refers to the
red-white variegated colonies
that appear on the EMB-lactose
agar. [After J. Tomizawa,
Proc. Natl. Acad. Sci. **46**, 91
(1960).]

information, and hence each of which could theoretically engender its own Lac$^+$ recombinant, produce but a single recombinant genome. This fact will be of importance in later considerations of the molecular basis of genetic recombination. In any case, donor gene integration appears to be an "early" process, in that most of the recombinants that finally arise from the conjugational merozygote are seen to be formed within a time-span comparable to one bacterial generation period.

Earlier in this chapter it was shown that efforts to construct a genetic map of *E. coli* on the basis of joint inheritance of selected and unselected markers in F$^+$ × F$^-$ crosses had encountered serious difficulties, for the now obvious reason that the genetic contribution of the donor parent to the zygote is not complete. And thus a *necessary* condition for the joint appearance of any two donor genes in a recombinant is that they both be transferred to the F$^-$ cell before interruption of the conjugation process. But such joint transfer is by no means a *sufficient* condition, because even though both genes may be present in the merozygote, there is no guarantee that they will find joint integration in the ensuing recombination process. The chance that two genes *are* integrated together increases with the proximity of their respective nucleotide sequences on the transferred donor DNA molecule, or their *genetic linkage*. Once these general principles had been recognized, it became possible to construct a meaningful genetic map based also on relative recombination frequencies.

Thus Jacob and Wollman carried out the cross Hfr Hayes Str^s × F$^-$ Pro$^-$ Lac$^-$ Pur$^-$ Str^r, and selected Pur$^+$ Str^r recombinants, while treating Pro and Lac as unselected characters. Since, according to Table 10-4, the donor genes are transferred in the order *O-pro-lac-pur-str*, every Pur$^+$ recombinant they recovered must have descended from a merozygote that had contained also origin-proximal *pro* and *lac* genes, as shown in Figure 10-20. In order to generate a Pur$^+$ Str^r recombinant, two crossover events are evidently required, the first of which may occur in regions 1, 2 or 3 and the second of which *must* occur in region 4. By scoring the distribution of the two unselected donor characters among the recombinants it was found that 22% were Lac$^-$—that is, had *not* integrated the donor *lac* gene. This class of recombinants evidently arose by making the first crossover in region 3. Among the remaining, or Lac$^+$ recombinants, 20% turned out to be Pro$^-$—that is, they arose by making the first crossover in region 2. These respective crossover probabilities in regions 2 and 3 are, therefore, quantitative indications of the genetic linkage of *pro*, *lac*, and *pur* genes on the *E. coli* chromosome.

It is now possible to relate the genetic linkage inferred from the foregoing recombinant analysis to the temporal distances previously established on the basis of the transfer kinetics of Figure 10-8. In particular, we see that the two loci *lac* and *pur*, are separated, on the one hand, by a crossover probability of 22% and, on the other hand, by one minute of transfer time. Hence 22% linkage is equivalent to one minute of transfer. Since transfer of the entire

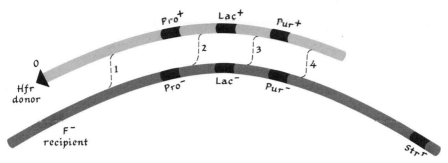

FIGURE 10-20.   Determination of genetic linkage by crossover frequency in bacterial conjugation. Hfr Hayes Pro⁺ Lac⁺ Pur⁺ Strˢ was mated to F⁻ Pro⁻ Lac⁻ Pur⁻ Strʳ, and Pur⁺ Strʳ recombinants were selected by plating the mating mixture on glucose minimal ᵃgar containing proline and streptomycin. Such recombinants can arise only from merozygotes that have received the O-*pur* segment of the donor chromosome and in which one crossover has taken place in region 4, situated between *pur* and *str* genes. The second crossover required for integrating the donor Pur⁺ character into the recombinant chromosome may occur in region 1, or in region 2, or in region 3, giving rise, respectively, to Pro⁺ Lac⁺, or to Pro⁻ Lac⁺, or to Pro⁻Lac⁻ recombinants. The relative frequency of these recombinant types thus provides a measure of the genetic distance of the *pro*, *lac*, and *pur* genes.

chromosome requires about 100 minutes, and since the *E. coli* genome is represented by a circular DNA molecule containing 3×10^6 nucleotide base pairs, we may reckon that the probability of crossover per nucleotide base pair is $22\% \times 100/(1 \times 3 \times 10^6) = 0.007\%$.

The method of mapping the *E. coli* chromosome by measuring joint appearance of nonselected markers that are origin-proximal to a selected marker in Hfr $\times$ F⁻ crosses turns out to be reliable only for rather short genetic distances, for reasons which are now understood but whose discussion would lead us too far astray here. Happily, it is precisely for longer distances that the method of mapping by determination of transfer kinetics has *its* greater reliability. Thus by combining both methods, the entire genome of *E. coli* was gradually charted. In 1967, with publication of the map shown in Figure 10-21, these efforts reached what is undoubtedly a rather close approximation to the true genetic topography. This composite map, which shows nearly 250 gene loci (whose symbolism is explained in Table 10-5), is based both on A. L. Taylor's own studies on the transfer kinetics of a variety of Hfr strains and on the data reported in more than 150 publications. The general features of this map are evidently similar to those of Wollman and Jacob's earlier map (Figure 10-12), but the relative size assigned to various sectors of the genome has undergone some change. Even Lederberg's first 1947 map (Figure 10-4) can be seen to have been ordinally congruent with the final map. It is to be noted also that the total temporal extent of the genome has been reduced from its earlier estimate of about 100 minutes to its present value of 90 minutes.

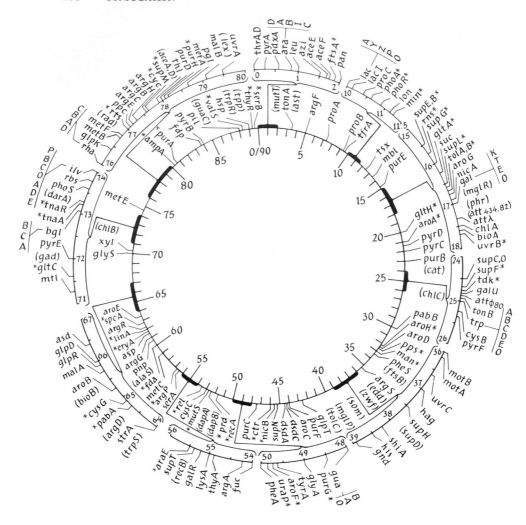

FIGURE 10-21. The definitive genetic map of the *E. coli* chromosome, drawn to scale. The inner circle with its associated time scale depicts the entire chromosome. The map is graduated in 1-minute intervals, beginning arbitrarily with zero at the *thr* gene. Selected portions of the map (e.g., the 0- to 2-minute segment) are displayed on arcs of the outer circle with a four-fold expanded time scale in order to accommodate crowded regions. Gene symbols are explained in Table 10-5. Genes in parentheses are only approximately mapped at the positions shown. Genes identified by asterisks have been mapped more precisely than the genes in parentheses, but their order relative to adjacent genes marked by asterisks has not yet been determined. [After A. L. Taylor and C. D. Trotter, *Bacteriol. Revs.* **31**, 332 (1967).]

TABLE 10 5.   Gene Catalog of *E. coli*

Gene symbol	Mnemonic	Map position (min)*	Other gene symbols; phenotypic trait affected
*ace*A	Acetate	(78)	Utilization of acetate; isocitrate lyase
*ace*D	Acetate	(78)	Regulation of the glyoxylate cycle
*ace*E	Acetate	2	*ace*E1; acetate requirement; pyruvate dehydrogenase (decarboxylase component)
*ace*F	Acetate	2	*ace*E2; acetate requirement; pyruvate dehydrogenase (lipoic reductase-transacetylase component)
*ala*S	Alanine	(60)	*ala-act*; alanyl tRNA synthetase
*amp*A	Ampicillin	80	Resistance or sensitivity to penicillin
*ara*A	Arabinose	1	L-Arabinose isomerase
*ara*B	Arabinose	1	L-Ribulokinase
*ara*C	Arabinose	1	Regulatory gene
*ara*D	Arabinose	1	L-Ribulose 5-phosphate 4-epimerase
*ara*E	Arabinose	56	L-Arabinose permease
*ara*I	Arabinose	1	Initiator locus
*arg*A	Arginine	54	*arg*B, $Arg1$, Arg_2 ; N-acetylglutamate synthetase
*arg*B	Arginine	77	*arg*C; α-N-acetyl-L-glutamate-5-phosphotransferase
*arg*C	Arginine	77	*arg*H, $Arg2$; N-acetylglutamic-γ-semialdehyde dehydrogenase
*arg*D	Arginine	(64)	*arg*G, Arg_1; acetylornithine-δ-transaminase
*arg*E	Arginine	77	*arg*A, $Arg4$; L-ornithine-N-acetylornithine lyase
*arg*F	Arginine	5	*arg*D, $Arg5$; ornithine transcarbamylase
*arg*G	Arginine	61	*arg*E, $Arg6$; argininosuccinic acid synthetase
*arg*H	Arginine	77	*arg*F, $Arg7$; L-argininosuccinate arginine lyase
*arg*P	Arginine	57	Arginine permease
*arg*R	Arginine	62	R*arg*; regulatory gene
*arg*S	Arginine	35	Arginyl tRNA synthetase
*aro*A	Aromatic	21	3-Enolpyruvylshikimate-5-phosphate synthetase
*aro*B	Aromatic	65	Dehydroquinate synthetase
*aro*C	Aromatic	44	Chorismic acid synthetase
*aro*D	Aromatic	32	Dehydroquinase
*aro*E	Aromatic	64	Dehydroshikimate reductase
*aro*F	Aromatic	50	3-Deoxy-D-arabinoheptulosonic acid-6-phosphate (DHAP) synthetase (tyrosine repressible isoenzyme)
*aro*G	Aromatic	17	DHAP synthetase (phenylalanine-repressible isoenzyme)

* Numbers refer to the time scale drawn in Figure 10-21. Parentheses indicate tentative map position.

TABLE 10-5—*continued*

Gene symbol	Mnemonic	Map position (min)*	Other gene symbols; phenotypic trait affected
*aro*H	Aromatic	32	DHAP synthetase (tryptophan-repressible isoenzyme)
asd	—	66	*dap* + *hom*; aspartic semialdehyde dehydrogenase
asp	Aspartate	62	Requirement
ast	Astasia	(4)	Generalized high mutability
*att*λ	Attachment	17	Integration site for prophage λ
*att*φ80	Attachment	25	Integration site for prophage φ80
*att*82	Attachment	17	Integration site for prophage 82
*att*434	Attachment	(17)	Integration site for prophage 434
azi	Azide	2	Resistance or sensitivity to sodium azide
*bgl*A	β-Glucoside	73	β-*gl*A; aryl β-glucosidase
*bgl*B	β-Glucoside	73	β-*gl*B; β-glucoside permease
*bgl*C	β-Glucoside	73	β-*gl*C; regulatory gene
*bio*A	Biotin	18	Requirement
*bio*B	Biotin	65	Requirement
cat	—	(25)	CR; catabolite repression
*chl*A	Chlorate	18	Pleiotropic mutations affecting nitrate-chlorate reductase and hydrogen lyase activity
*chl*B	Chlorate	(71)	Pleiotropic mutations affecting nitrate-chlorate reductase and hydrogen lyase activity
*chl*C	Chlorate	(25)	Nitrate-chlorate reductase
ctr	—	46	Pleiotropic mutations affecting uptake of carbohydrates
cyc	Cycloserine	78	Resistance or sensitivity to D-cycloserine
*cys*B	Cysteine	25	Reduction of thiosulfate or sulfite to sulfide
*cys*C	Cysteine	53	Reduction of sulfite to sulfide
*cys*G	Cysteine	65	Reduction of sulfite to sulfide
*dap*A	Diaminopimelate	(53)	Dihydropicolinic acid synthetase
*dap*B	Diaminopimelate	(53)	N-succinyl-diaminopimelic acid deacylase
*dar*A	Dark-repair	73	*dar*$_2$; repair of ultraviolet radiation damage to DNA
*dsd*A	D-Serine	45	D-Serine deaminase
*dsd*C	D-Serine	45	Regulatory gene
edd	—	(35)	Entner-Doudoroff dehydrase (gluconate-6-phosphate dehydrase)
*ery*A	Erythromycin	62	Resistance or sensitivity to erythromycin
fda	—	60	*ald*; fructose-1,6-diphosphate aldolase

* Numbers refer to the time scale drawn in Figure 10-21. Parentheses indicate tentative map position.

TABLE 10-5—*continued*

Gene symbol	Mnemonic	Map position (min)*	Other gene symbols; phenotypic trait affected
fdp	—	84	Fructose diphosphatase
*fts*Λ		2	*fts*-2,7; filamentous growth at 42°C
*fts*B	—	(35)	*fts*-9; filamentous growth and inhibition of nucleic acid synthesis at 42°C
fuc	Fucose	54	Utilization of L-fucose
gad	—	(72)	Glutamic acid decarboxylase
*gal*E	Galactose	17	*gal*D; uridinediphosphogalactose 4-epimerase
*gal*K	Galactose	17	*gal*A; galactokinase
*gal*O	Galactose	17	*gal*C; operator locus
*gal*T	Galactose	17	*gal*B; galactose 1-phosphate uridyl transferase
*gal*R	Galactose	55	R*gal*; regulatory gene
*gal*U	Galactose	25	UDPG; uridine diphosphoglucose pyrophosphorylase
*glp*D	Glycerol phosphate	66	*gly*D; L-α-glycerophosphate dehydrogenase
*glp*K	Glycerol phosphate	76	Glycerol kinase
*glp*T	Glycerol phosphate	43	L-α-Glycerophosphate transport system
*glp*R	Glycerol phosphate	66	Regulatory gene
*glt*A	Glutamate	16	*glut*; requirement for glutamate; citrate synthase
*glt*C	Glutamate	72	Structural or regulatory gene for glutamate permease
*glt*H	Glutamate	20	Requirement
*gly*A	Glycine	49	Serine hydroxymethyl transferase
*gly*S	Glycine	70	*gly-act*; glycyl tRNA synthetase
gnd	—	39	Gluconate-6-phosphate dehydrogenase
*gua*A	Guanine	48	*gua*$_b$; xanthosine-5′-monophosphate aminase
*gua*B	Guanine	48	*gua*$_a$; inosine-5′-monophosphate dehydrogenase
*gua*C	Guanine	(88)	Guanosine-5′-monophosphate reductase
*gua*O	Guanine	48	Operator locus
hag	H antigen	37	H; flagellar antigens (flagellin)
his	Histidine	39	Requirement
hsp	Host specificity	89	*hs, rm*; host restriction and modification of DNA
*ilv*A	Isoleucine-valine	74	*ile*; threonine deaminase
*ilv*B	Isoleucine-valine	74	Condensing enzyme (pyruvate + α-ketobutyrate)
*ilv*C	Isoleucine-valine	74	*ilv*A; α-hydroxy β-keto acid reductoisomerase
*ilv*D	Isoleucine-valine	74	*ilv*B; dehydrase

* Numbers refer to the time scale drawn in Figure 10-21. Parentheses indicate tentative map position.

TABLE 10-5—*continued*

Gene symbol	Mnemonic	Map position (min)*	Other gene symbols; phenotypic trait affected
*ilv*E	Isoleucine-valine	74	*ilv*C; transaminase B
*ilv*O	Isoleucine-valine	74	Operator locus for genes *ilv*A, D, E
*ilv*P	Isoleucine-valine	74	Operator locus for gene *ilv*B
*lac*A	Lactose	10	*a, lacAc*, thiogalactoside transacetylase
*lac*I	Lactose	10	*i*; regulatory gene
*lac*O	Lactose	10	*o*; operator locus
*lac*P	Lactose	10	*p*; promoter locus
*lac*Y	Lactose	10	*y*; galactoside permease (M protein)
*lac*Z	Lactose	10	*z*; β-galactosidase
leu	Leucine	1	Requirement
lex	—	(79)	Resistance or sensitivity to X rays and UV light
*lin*A	Lincomycin	62	Resistance or sensitivity to lincomycin
lon	Long form	11	*dir*; R1, *muc*; filamentous growth, slime formation, and radiation sensitivity
*lys*A	Lysine	55	Diaminopimelic acid decarboxylase
*mal*A	Maltose	66	Amylomaltase and maltodextrin phosphorylase
*mal*B	Maltose	79	*mal*-5; possibly maltose permease
man	Mannose	33	Utilization of D-mannose
mbl	Methylene blue	14	*Mb*; sensitivity to methylene blue and acridine dyes
*met*A	Methionine	78	met_3 ; homoserine O-transsuccinylase
*met*B	Methionine	77	*met-1*, met_1; cystathionine synthetase
*met*C	Methionine	59	Cystathionase
*met*E	Methionine	75	met-B_{12} ; N^5-methyltetrahydropteroyl triglutamate-homo-cysteine methylase
*met*F	Methionine	77	*met-2*, met_2 ; N^5,N^{10}-methyltetrahydrofolate reductase
*mgl*P	Methyl-galactoside	(40)	P-MG; methyl-galactoside permease
*mgl*R	Methyl-galactoside	(17)	R-MG; regulatory gene
min	Mini-cell	11	Formation of minute cells containing no DNA
*mot*A	Motility	36	Flagellar paralysis (complementation groups I and III)
*mot*B	Motility	36	Flagellar paralysis (complementation group II)
mtl	Mannitol	71	Utilization of D-mannitol
*mut*S	Mutator	53	Generalized high mutability
*mut*T	Mutator	(1)	Generalized high mutability

* Numbers refer to the time scale drawn in Figure 10-21. Parentheses indicate tentative map position.

TABLE 10-5—*continued*

Gene symbol	Mnemonic	Map position (min)*	Other gene symbols; phenotypic trait affected
*nic*A	Nicotinic acid	17	Requirement
*nic*B	Nicotinic acid	46	Requirement
*pab*A	*p*-Aminobenzoate	65	Requirement
*pab*B	*p*-Aminobenzoate	30	Requirement
pan	Pantothenic acid	2	Requirement
*pdx*A	Pyridoxine	1	Requirement
pgi	—	79	Phosphoglucoisomerase
*phe*A	Phenylalanine	50	Prephenic acid dehydratase
*phe*S	Phenylalanine	33	*phe-act*; phenylalanyl tRNA synthetase
*pho*A	Phosphatase	11	P; alkaline phosphatase
*pho*R	Phosphatase	11	R1 *pho*, R1; regulatory gene
*pho*S	Phosphatase	74	R2 *pho*, R2; regulatory gene
phr	Photoreactivation	(17)	Photoreactivation of UV-damaged DNA
pil	Pili	88	*fim*; presence or absence of pili (fimbriae)
pnp	—	61	Polynucleotide phosphorylase
ppc	—	77	*glu*, *asp*; succinate, aspartate, or glutamate requirement; phosphoenolpyruvate carboxylase
pps	—	33	Utilization of pyruvate or lactate; phosphopyruvate synthase
prd	Propanediol	53	1,2-Propanediol dehydrogenase
*pro*A	Proline	7	*pro₁*; block prior to L-glutamate semialdehyde
*pro*B	Proline	9	*pro₂*; block prior to L-glutamate semialdehyde
*pro*C	Proline	10	*pro₃*, Pro2; probably Δ-pyrroline-5-carboxylate reductase
*pur*A	Purine	81	ade_k, Ad_4; adenylosuccinic acid synthetase
*pur*B	Purine	23	ade_h; adenylosuccinase
*pur*C	Purine	48	ade_g; phosphoribosyl-aminoimidazole-succinocarboxamide synthetase
*pur*D	Purine	78	$adth_a$; phosphoribosylglycineamide synthetase
*pur*E	Purine	15	ade_3, ade_f; Pur_2; phosphoribosyl-aminoimidazole carboxylase
*pur*F	Purine	44	*pur*C, $ade_{a,b}$; phosphoribosyl-pyrophosphate amidotransferase
*pur*G	Purine	49	$adth_b$; phosphoribosylformylglycineamidine synthetase

* Numbers refer to the time scale drawn in Figure 10-21. Parentheses indicate tentative map position.

TABLE 10-5—*continued*

Gene symbol	Mnemonic	Map position (min)*	Other gene symbols; phenotypic trait affected
*pur*H	Purine	78	*ade*ᵢ; phosphoribosyl-aminoimidazole-carboxamide for-myltransferase
*pyr*A	Pyrimidine	0	*cap, arg* + *ura*; glutamino-carbamoyl phosphate synthetase
*pyr*B	Pyrimidine	84	Aspartate transcarbamylase
*pyr*C	Pyrimidine	22	Dihydroorotase
*pyr*D	Pyrimidine	21	Dihydroorotic acid dehydrogenase
*pyr*E	Pyrimidine	72	Orotidylic acid pyrophosphorylase
*pyr*F	Pyrimidine	25	Orotidylic acid decarboxylase
rad	Radiation	(77)	Resistance or sensitivity to X rays
rbs	Ribose	74	Utilization of D-ribose
*rec*A	Recombination	52	Ultraviolet sensitivity and competence for genetic recombination
*rec*B	Recombination	(55)	Ultraviolet sensitivity and competence for genetic recombination
rel	Relaxed	54	RC; regulation of RNA synthesis
*rha*A	Rhamnose	76	L-Rhamnose isomerase
*rha*B	Rhamnose	76	L-Rhamnulokinase
*rha*C	Rhamnose	76	Regulatory gene
*rha*D	Rhamnose	76	L-Rhamnulose-1-phosphate aldolase
rns	Ribonuclease	16	Ribonuclease I
rts	—	77	*ts*-9; altered electrophoretic mobility of 50S ribosomal subunit
*ser*A	Serine	57	3-Phosphoglyceric acid dehydrogenase
*ser*B	Serine	89	Phosphoserine phosphatase
*shi*A	Shikimic acid	38	Shikimate and dehydroshikimate permease
som	Somatic	(37)	*O*; somatic (O) antigens
*spc*A	Spectinomycin	63	Resistance or sensitivity to spectinomycin
*str*A	Streptomycin	64	Resistance, dependence, or sensitivity; "K-character" of the 30S ribosomal subunit
suc	Succinate	17	*lys* + *met*; aerobic requirement for succinate or lysine plus methionine; α-ketoglutarate dehydrogenase
*sup*B	Suppressor	16	*su*_B; suppressor of *ochre* mutations (not identical to *sup*L)

* Numbers refer to the time scale drawn in Figure 10-21. Parentheses indicate tentative map position.

TABLE 10-5—*continued*

Gene symbol	Mnemonic	Map position (min)*	Other gene symbols; phenotypic trait affected
*sup*C	Suppressor	25	*suc*, Su-4; suppressor of *ochre* mutations (possibly identical to *sup*O)
*sup*D	Suppressor	38	*su*₁, Su-1; suppressor of *amber* mutations
*sup*E	Suppressor	16	*su*ᵢᵢ, suppressor of *amber* mutations
*sup*F	Suppressor	25	*su*ᵢᵢᵢ, Su-3; suppressor of *amber* mutations
*sup*G	Suppressor	16	Su-5; suppressor of *ochre* mutations
*sup*H	Suppressor	38	
*sup*L	Suppressor	17	Suppressor of *ochre* mutations
*sup*M	Suppressor	78	Suppressor of *ochre* mutations
*sup*N	Suppressor	45	Suppressor of *ochre* mutations
*sup*O	Suppressor	25	Suppressor of *ochre* mutations (possibly identical to *sup*C)
*sup*T	Suppressor	55	
tdk	—	25	Deoxythymidine kinase
*tfr*A	T-four	9	ϕ^r; resistance or sensitivity to phages T4, T3, T7, and λ
thi	Thiamine	78	B₁; requirement
*thr*A	Threonine	0	Block between homoserine and threonine
*thr*D	Threonine	0	*HS*; aspartokinase I-homoserine dehydrogenase I complex
*thy*A	Thymine	55	Thymidylate synthetase
*thy*R	Thymine	89	Possibly deoxyriboaldolase
*tna*A	—	73	*ind*; tryptophanase
*tna*R	—	73	R_{tna}; regulatory gene
*tol*A	Tolerance	17	*cim*; *tol*-2; tolerance (immunity) to colicins E2, E3, A, and K
*tol*B	Tolerance	17	*tol*-3; tolerance (immunity) to colicins E1, E2, E3, A, and K
*tol*C	Tolerance	(41)	*colEl-i*; *tol*-8; tolerance (immunity) to colicin E1
*ton*A	T-one	4	T1, T5 *rec*; resistance or sensitivity to phages T1 and T5
*ton*B	T-one	25	T1 *rec*; resistance or sensitivity to phages T1, ϕ80 and colicins B, I, V
tpp	—	(89)	TP; thymidine phosphorylase
*trp*A	Tryptophan	25	*tryp* 2; tryptophan synthetase, A protein
*trp*B	Tryptophan	25	*tryp* 1; tryptophan synthetase, B protein
*trp*C	Tryptophan	25	*tryp* 3; indole-3-glycerol phosphate synthetase

* Numbers refer to the time scale drawn in Figure 10-21. Parentheses indicate tentative map position.

TABLE 10-5—*continued*

Gene symbol	Mnemonic	Map position (min)*	Other gene symbols; phenotypic trait affected
*trp*D	Tryptophan	25	*try*E; phosphoribosyl anthranilate transferase
*trp*E	Tryptophan	25	*try*D, *anth*, *tryp* 4; anthranilate synthetase
*trp*O	Tryptophan	25	Operator locus
*trp*R	Tryptophan	(89)	R*try*; regulatory gene
*trp*S	Tryptophan	(64)	Regulatory gene
tsx	T-six	13	T6 *rec*; resistance or sensitivity to phage T6 and colicin K
*tyr*A	Tyrosine	50	Prephenic acid dehydrogenase
*ura*P	Uracil	50	Uracil permease
*uvr*A	Ultraviolet	80	*dar* 3; repair of ultraviolet radiation damage to DNA
*uvr*B	Ultraviolet	18	*dar* 1, 6; repair of ultraviolet radiation damage to DNA
*uvr*C	Ultraviolet	37	*dar* 4, 5; repair of ultraviolet radiation damage to DNA
*val*S	Valine	88	*val-act*; valyl tRNA synthetase
xyl	Xylose	70	Utilization of D-xylose
zwf	Zwischenferment	(35)	Glucose-6-phosphate dehydrogenase

From A. L. Taylor and C. D. Trotter, *Bacterial Revs.*, **31**. 332 (1967).
* Numbers refer to the time scale drawn in Figure 10-21. Parentheses indicate tentative map position.

Bibliography

PATOOMB

William Hayes. Sexual differentiation in bacteria.
Elie L. Wollman. Bacterial conjugation.

HAYES

Chapters 3, 22, and 24

ORIGINAL RESEARCH PAPERS

Hayes, W. The mechanism of genetic recombination in *E. coli*. *Cold Spring Harbor Symp. Quant. Biol.*, **18**, 75 (1953).

Lederberg, J., and E. L. Tatum. Novel genotypes in mixed cultures of biochemical mutants of bacteria. *Cold Spring Harbor Symp. Quant. Biol.*, **11**, 113 (1946).

Wollman, E. L., F. Jacob, and W. Hayes. Conjugation and genetic recombination in *Escherichia coli* K-12. *Cold Spring Harbor Symp. Quant. Biol.*, **21**, 141 (1956).

SPECIALIZED TEXTS, MONOGRAPHS, AND REVIEWS

Campbell, A. M. *Episomes*. Harper & Row, New York, 1969.

Jacob, F., and E. L. Wollman. *Sexuality and the Genetics of Bacteria*. Academic Press, New York, 1961.

11. Phage Growth

Our discussions of the central biological problem of self-replication have so far focused on the reproduction of *cells*—in particular, on prokaryotic bacterial cells, whose level of organization is less complex than that of the eukaryotic cells from which the higher forms of life are constituted. In the preceding six chapters we have considered the role, mutation, structure, replication, and organization of the genetic material that presides over the reproduction of those cells. But there exist among living forms even simpler self-replicating biological entities than prokaryotic cells—namely, the *viruses*, which served as one of the main working tools in the molecular-genetic quest to fathom the mystery of the gene.

VIRUSES

Although for the past thirty years or more there has been tacit agreement among virologists as to the nature of the objects they have been studying, the term "virus" is not easy to define explicitly. The physicians of ancient Rome used "virus" to mean a *poison* of animal origin, and the diseases caused by such poisons were called "virulent." Thus the Romans considered rabies to be caused by a virus, an inference that, in view of the modern meaning of virus, was substantively wrong but terminologically correct. During the Renaissance it came to be realized that a distinction should be made between noncontagious diseases caused by poisons and contagious diseases caused by infectious agents, and the name virus was reserved for the latter. With the rise of microbiology in the nineteenth century, bacteria came to be recognized as the infectious agents responsible for many contagious diseases, and hence bacteria came to be referred to as the viruses of disease. The causative agents of some contagious diseases, however, could not be identified by the methods of nineteenth-century bacteriology. For instance, even though Pasteur was able to demonstrate that rabies is caused by a specific, transmissible living agent, he failed in his efforts to cultivate the agent on any bacteriological medium or to isolate any microscopically visible bacterium to which the responsibility for rabies could be attributed. At the close of the nineteenth century, it was found that these elusive, invisible infectious agents, like the one responsible for rabies, must be much smaller than bacteria, since they could be passed through filters with pores so fine that they retained all then-known types of bacteria. These small agents were then given the name "filtrable viruses." In the first few decades of this century, the use of the term "virus" in referring to *any* infectious pathogenic agent was gradually abandoned, instead, the term came to mean only what had previously been called "filtrable viruses." For during that period it was discovered that the tiny viruses are living objects whose organization is even less complex than that of bacteria. Viruses were recognized to be *subcellular entities capable of entering living cells and of reproducing only in such cells.* Viruses are therefore obligate intracellular parasites; but the special feature that sets them apart from other intracellular parasites (such as some species of bacteria that live parasitically within eukaryotic cells) is that the parasitism of viruses is at the *genetic level.* As the facts to be set forth in the remainder of this text will show, viruses arrogate the synthetic machinery of the cells that they enter by substituting their own genes for those of their host, and thus cause that machinery to produce viral rather than host gene products.

In 1915 F. W. Twort discovered that the parasitic life of viruses is not confined to eukaryotic cells, and that some viruses utilize even the lowly bacteria as their hosts. Twort had found a serially transmissible agent that destroys bacteria, and since this agent was capable of passing through filters so fine that all bacteria should have been retained, he proposed as one of three possible explanations that this agent might be a virus that grows on and kills the bacteria it infects. Twort published this finding in a brief note that remained unnoticed until, two years later, F. d'Hérelle announced *his*, probably independent, discovery of an entirely analogous filtrable agent. It was d'Hérelle who gave to this agent the name "bacteriophage" (from the Greek *phagein*, to devour). Later usage shortened this name to "phage," and it is in this vestigal form that it will be used most often in these pages. D'Hérelle's announcement caused an immediate sensation in the world of medical microbiology, because it was he who promulgated the ideas that phages are the chief agents of natural immunity against disease and that phages should be a panacea for infectious diseases. Despite holding to these in the event quite mistaken notions, d'Hérelle managed to recognize phages for what we now know them to be. From the very start of his studies, d'Hérelle thought of phages as particulate, invisible, filtrable, self-reproducing viruses that are obligate parasites of bacteria. Within two or three years of his original discovery, he had invented a method for the accurate assay, or titration, of the phages, and by 1923 he described their life cycle in these terms: the phage particle attaches itself to the surface of the bacterium and then enters the cell, where it reproduces itself to generate many progeny viruses. When the infected cell finally bursts open, or undergoes *lysis*, the progeny thus liberated are ready to infect other bacteria. Although in retrospect this description not only seems eminently plausible but also happens to be true, it was accepted by few of d'Hérelle's contemporaries. Especially the notion that the phage is a self-reproducing virus aroused widespread offense, for many of d'Hérelle's adversaries preferred to regard it as a self-stimulating enzyme endogenous to the bacterium. Nevertheless, as F. M. Burnet observed in 1934, "however agnostic they have been in regard to the nature of a phage, all workers manipulated and in practice thought of it as an extrinsic, virus-like agent."

Phage research made a Big Leap Forward with the appearance of M. Schlesinger on the bacterial virus scene. From about 1930 until his death in 1935, Schlesinger was the first to train on bacterial viruses the tools of what was to become molecular biology. Schlesinger showed by various indirect means, such as the adsorption capacity of the bacterial cell for phages and the sedimentation velocity of phages, that the phage particle has a maximum linear dimension of about 0.1 μ and a mass of about 4×10^{-16} g. Schlesinger also studied the attachment mechanism of the phage to its host cell and found that the kinetics of attachment imply that Brownian movement brings phage particles into random collisions with the bacterial surface. Possibly, Schlesinger's most important accomplishment was to purify a "weighable" amount of phage particles by differential centrifugation and graded filtration of crude phage suspensions; he then found by direct chemical analysis of the pure phage that it consists mainly of protein and DNA, in roughly equal proportions. Although phages are invisible under ordinary microscopes, Schlesinger was able to estimate directly the total number of particles in a purified phage preparation by counting the number of bright spots produced in a dark-field microscope. In this way he was able to establish that the number of physical phage particles is roughly equal to the infective titer estimated by d'Hérelle's assay method.

Schlesinger left no intellectual disciples, and "modern" phage research actually dates only from 1938, when M. Delbrück began to work with bacterial viruses. Delbrück became the focus of a new school of phage workers many of whom, like Delbrück himself, had been trained in the physical sciences. In a relatively short time, this group came to dominate not only phage research but also to exert a most important influence on the then nascent molecular biology. One reason for the rapid progress that was now made was that for more than 10 years the attention of most workers was confined to the seven strains of T phages active on *E. coli*, particularly to the T-even strains T2, T4 and T6. (The phage with which Schlesinger had worked is now known to have belonged to the T-even family, and the T1 member of the T phage group was discussed in Chapter 6 in relation to the spontaneity of the Ton^s → Ton^r mutation.) Thus the results obtained in different laboratories could be integrated much more readily than the efforts made during the earlier period of phage research, when every investigator seemed to take pains to develop his own phage-host system. But what set off the members of this new school most sharply from their predecessors was their single-minded interest in the central problem of biological replication. They mainly wanted to know what is going on inside the phage-infected bacterium while the parental phage particle manages to effect its own manifold replication. It was the desire to understand this process that motivated most of the latter-day phage workers in their work.

TITRATION

In his very earliest experiments on phage, d'Hérelle had observed "taches vierges" or *plaques* in areas of dense bacterial growth on nutrient agar surfaces. D'Hérelle found that these plaques are regions in which bacteria have been destroyed by the growth of phage colonies descended from parent phages present in the bacterial inoculum. D'Hérelle realized that the formation of these easily recognizable plaques could serve for the titration of the number of *infective units* contained in phage suspensions. He therefore developed the following *plaque assay* procedure, which is still in general use today: About 10^7 host bacteria and an appropriate dilution of the phage suspension to be titrated are spread over the surface of a solid layer of nutrient agar in a petri plate, and the plate is incubated. During the incubation, each bacterium of the inoculum grows into a tiny colony, so that finally 10^7 tiny colonies form a thick, turbid lawn of bacterial growth on the agar surface. At an early stage of the incubation, phage particles present on the plate infect bacteria in their immediate neighborhood and then grow in and lyse those bacteria to produce a crop of progeny phages. The progeny of each original parent phage then infect neighboring bacteria, which in turn are lysed upon the appearance of a second phage progeny generation. These progeny infect more neighborhood bacteria, and the process of phage reproduction and bacterial lysis continues on and on in each focus of infection. Thus as the bacterial lawn grows, holes or plaques,

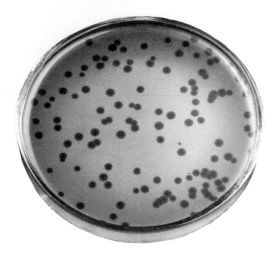

FIGURE 11-1. A petri plate
showing growth of a lawn of
E. coli bacteria on which T2
phages have formed plaques.
[From *Molecular Biology of
Bacterial Viruses*, by G. S. Stent.
W. H. Freeman and Company.
Copyright © 1963.]

develop in that lawn. The final diameter of the holes depends on the phage
type, the bacterial host strain, and the exact conditions of plating and incubation used, but it generally amounts to a few millimeters. Figure 11-1 shows a
photograph of an agar plate containing about 100 such phage plaques.

Each plaque is initiated by a single phage, and not, as would be theoretically conceivable, by the cooperation of two or more phage particles. This
follows from the observation that the number of plaques formed on any
plate is directly proportional to the volume of a given phage suspension
mixed with the bacterial inoculum. Figure 11-2 shows the result of an experiment in which equal volumes of increasing concentrations of the same phage
suspension were plated on a series of agar plates; it is evident that each twofold
increase in concentration produced a corresponding twofold increase in
average plaque number per plate—that is, plaque number and phage

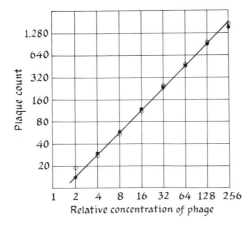

FIGURE 11-2. Proportionality of
phage concentration to plaque count.
Successive two-fold dilutions of a
phage lysate were plated in
duplicate on nutrient agar; 0.1 ml
on each plate. The plaque counts per
plate from two series of such
dilutions are plotted against the
relative phage concentration (the
reciprocal of the dilution) both on a
logarithmic scale. [After E. L. Ellis
and M. Delbrück, *J. Gen. Physiol.*
22, 365 (1939).]

concentration are directly proportional. If more than one phage particle were required to initiate a plaque, then the number of plaques in this experiment should have increased more rapidly than the phage concentration; that is, the chance that two or more plaque particles happen to fall close enough together on the agar surface to cooperate in the generation of an infection focus would be proportional to the second, or higher, power of the average concentration of phages per plate rather than to the first power. Thus, to reckon the infective titer of a phage suspension, it is only necessary to count the number of plaques formed on the assay plate and multiply that number by the dilution made before plating. For instance, if 0.1 ml of a 10^4-fold dilution of a phage suspension produced an average of 150 plaques per plate, one would calculate a titer of $(150/0.1) \times 10^4 = 1.5 \times 10^7$ infective units per milliliter for the suspension.

STRUCTURE AND COMPOSITION

The development of the electron microscope during the late 1930's finally made possible the direct visualization of the hitherto invisible phage particles. The first electron micrograph of a phage was taken by H. Ruska in 1940, and two years later Luria and T. F. Anderson obtained the first electron micrographs of the T phages. These fairly indistinct images revealed that phage particles are tadpole-shaped structures, which are endowed with a head and tail (Figure 11-3). Since the particles can be seen to be attached tail-first to the wall of their bacterial host cell, it follows that the phage tail is the organ of adsorption. These pictures confirmed also the earlier indirect particle size estimates of approximately 0.1 μ, and thus showed that the T phages occupy a volume about one-thousandth that of their *E. coli* host cell. Subsequent advances in the design of electron microscopes and in the techniques of preparing specimens for observation have greatly improved the quality of the electron micrographs of phages. Figure 11-4 presents a more highly resolved electron micrograph of a T-even phage particle, as well as an interpretative cartoon of what is to be seen in this picture. The phage tail is evidently a structure of great complexity, being constructed of at least four different components: a *sheath*, a *core*, a *base plate*, and six *tail fibers*. By breaking up the tail in various ways and testing the adsorbability of the different isolated components, it was found that the tail fibers are the actual adsorption organs by means of which the phage anchors itself to the surface of the host cell.

In the 1950's chemical analysis of highly purified T-even phage by means of modern techniques fully substantiated Schlesinger's by then 20-year-old conclusion that DNA and protein, in roughly equal proportions, constitute more than 90% of the dry weight of the particles. The total DNA complement of a single T-even phage was found to consist of 2×10^5 base pairs, or about

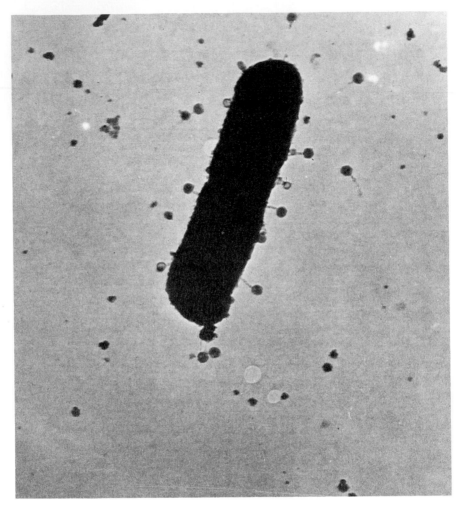

FIGURE 11-3. Electron micrograph of an *E. coli* cell to which T5 phages have attached themselves by their tail. [After T. F. Anderson, *Cold Spring Harbor Symp. Quant. Biol.* **18**, 197 (1953).]

6% that of the genome of the *E. coli* host cell. Chemical analyses of the total T-even phage protein did not bring to light any startling facts: its amino acid composition resembles more or less that of the global *E. coli* protein. Fractionation of the phage protein revealed that it is composed of at least five different types of polypeptides, of which the *head protein* makes up by far the major part. Each of the tail components, sheath, core, base plate, and tail fibers, is a specific polypeptide in its own right. Chemical analysis of the phage DNA did turn up two surprises, however. First, the T-even phage

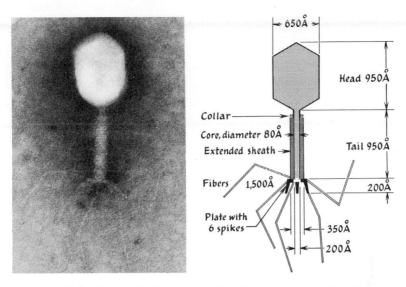

FIGURE 11-4.   The detailed structure of the T-even phage particle. [After E. Kellenberger, *Adv. Virus Research* **8**, 1 (1962).]

DNA, unlike all other types of DNA then known, was found to have no cytosine. Instead of cytosine, it contains the cytosine analog 5-hydroxymethyl-cytosine (HMC) (Figure 11-5), the relative base frequency being [A] = 32%, [T] = 33%, [G] = 18%, and [HMC] = 17%. Thus T-even phage DNA obeys the [A] = [T] and [G] = [C] equivalence relation demanded by the Watson-

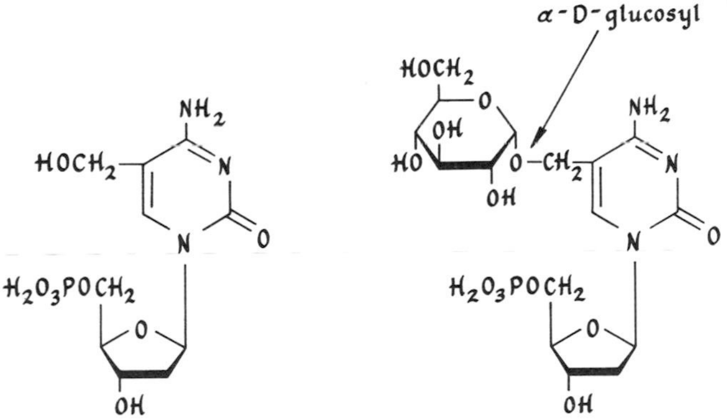

FIGURE 11-5.   The unusual deoxyribonucleotide containing 5-hydroxymethyl cytosine (HMC) found in the DNA of T-even phages (left) and the manner in which glucose is attached to some of the HMC bases (right). [From *Molecular Biology of Bacterial Viruses,* by G. S. Stent. W. H. Freeman and Company. Copyright © 1963.]

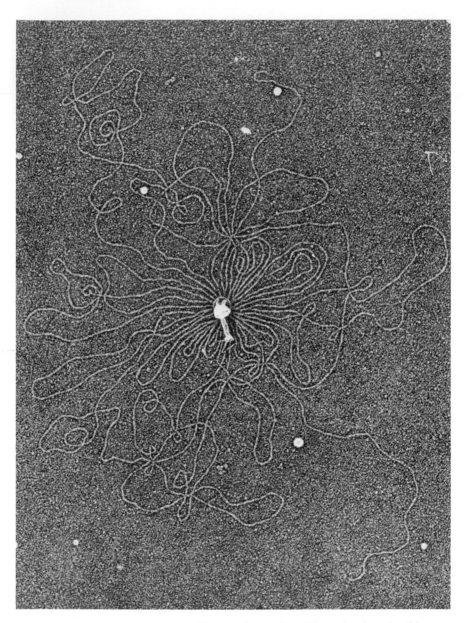

FIGURE 11-6.   The DNA molecule of T-even phage released from the phage head by osmotic shock. *Center:* The phage ghost. *Bottom right and top center:* The two ends of the DNA molecule. Approximate magnification of this electron micrograph 60,000. [After A. K. Kleinschmidt, D. Lang, D. Jacherts, and R. K. Zahn, *Biochim. Biophys. Acta.* **61,** 857 (1962).]

Crick structure, provided that HMC, which can form the same three hydrogen bonds with G as can C, replaces C in the double helix. The low [G] + [HMC] content of 35% places the T-even phage DNA in a compositional range quite remote from that of the DNA of its *E. coli* host, which as can be seen in Table 8-1, has a [G] + [C] content of 52%. Second, the T-even phage DNA was found to contain glucose attached to some of the hydroxymethyl groups of HMC (Figure 11-5). The biological function of these unusual features of the T-even phage DNA has not been explained to this day, but the features themselves have been of great practical help in the study of intracellular phage growth.

The key to the topological relation between the two principal phage components, phage DNA and phage protein, was provided in 1949, when T. F. Anderson found that T-even phages lose their infectivity when subjected to the osmotic shock attending a sudden dilution into distilled water from a suspension in concentrated salt solution. Electron micrographs showed that this inactivation derives from the rupture of the phage head and concomitant release of the phage DNA into the ambient medium. The effects of osmotic shock can be seen in Figure 11-6, which presents an electron micrograph of a T-even phage particle shocked *in situ* on the electron microscope specimen holder. In this picture the DNA released from the disrupted phage can be seen as one single, giant macromolecule. Contour measurements of this continuous fiber, from its one free end (lower right corner) to its other free end (top center), show that the phage DNA molecule has a length of about 50 μ, or 550 times that of the phage head which contains it. The occurrence of osmotic shock suggested, therefore, that the phage head consists of an outer semipermeable proteinaceous membrane and an inner core of phage DNA.

ONE-STEP GROWTH

Delbrück's first major contribution to phage research was his and E. Ellis' design of the *one-step growth experiment*, which demonstrated clearly that the progeny of the infecting phage particle appear only after a period of constant phage titer. The one-step growth experiment, which marked the beginning of modern phage work, is still today the basic procedure for studying phage multiplication: a dense suspension of growing bacteria is infected with a suitable number of phages, incubated for a few minutes to allow most of the phage particles to attach themselves to the bacteria, and then diluted with nutrient medium to a concentration that may range from one ten-thousandth to one millionth of that of the suspension. The diluted culture is then incubated further and samples plated on sensitive bacteria from time to time for plaque assay of the instantaneous number of infective units in the culture. The

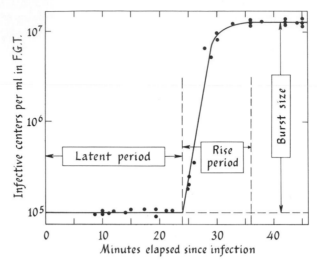

FIGURE 11-7. The one-step growth experiment of phage T4. Procedure: *E. coli* growing exponentially are concentrated by centrifugation to a density of 10^9 cells/ml and infected with an average of one T4 phage/cell. The mixture is incubated in an aerated medium for 2 minutes, during which time at least 80% of the phage input becomes fixed to the bacterial cells. The infected bacteria are then diluted 10,000-fold into fresh growth medium. The tube containing this dilution (*first growth tube*, or F.G.T.) and another tube containing a 20-fold further dilution of the first growth tube (*second growth tube*) are incubated, and samples from both tubes are plated periodically for plaque assay on sensitive indicator bacteria. During the latent period and early in the rise period, the titer of infective centers shown on the ordinate of the graph is estimated from plaque counts obtained by assay of the first-growth tube; thereafter, the titer is reckoned from plaque counts obtained by assay of the second growth tube. [After A. H. Doermann, *J. Gen. Physiol.* **35**, 645 (1952).]

protocol and the results of a typical one-step growth experiment are presented in Figure 11-7.

As can be seen, the number of plaque-forming units in the culture remains constant for the first 24 minutes after infection. This initial period, during which the infective titer shows no increase, is the *latent period*. After some 24 minutes have elapsed, the number of plaque-forming units in the culture begins to rise rapidly, until a final plateau is attained some ten minutes later, when no further increase in infectivity occurs. The period of time during which the number of plaque-forming units increases is the *rise-period*, and the ratio of the final titer of the plateau to the initial titer of phage-infected bacteria is the *burst size*. The latent period thus represents the time that elapses between the moment at which the bacterial culture is infected with a phage stock and the moment at which the first infected cells in the culture lyse, thereby liberating into the medium a litter of progeny phage particles. The rise period represents the time span during which more and more of the infected bacteria lyse, and the final plateau of infectivity is attained when all the infected bacteria that are going to lyse have done so; no further phage multiplication occurs after this stage, since progeny phage and residual uninfected bacteria in the culture have been separated from each other by the high dilution of the culture just after the initial infection. The burst size corresponds to the average number of progeny phage particles produced per infected bacterium, which in the experiment presented here amounts to about 100 phages per infected cell.

Let us now examine further the nature of the plaque-forming units that appear on the assay plates of the one-step growth experiment of Figure 11-7. Evidently, the plaques produced upon plating samples of the infected culture during the latent period derive not from individual free phage particles but, instead, from phage-infected bacteria that may already contain within them very many progeny phages. Nevertheless, the plating of each such phage-infected bacterium before the end of the latent period gives rise to only a single plaque, because the solid agar of the assay plate confines the hundred or so progeny phages ultimately liberated from that cell to a single focus of infection. Only after the conclusion of the latent period, when the intra-cellular phages have escaped from the host cell into the culture medium, can each progeny phage form its own focus of infection on the agar surface. The term infective center is used to describe the unit that forms a single plaque; thus an infective center may represent either a single, free phage particle or a phage-infected bacterium.

THE ECLIPSE

The one-step growth experiment demonstrated clearly the general nature and overall kinetics of the multiplication of bacterial viruses in their bacterial hosts. It thus brought into focus a question of fundamental biologic interest: what is taking place within the infected cell during the latent period while the parental phage particle is managing to cause its own several-hundredfold replication? With reference to this question let us first consider the manner in which the number of phage particles present within the infected cell increases from the moment of infection until the time of lysis. This information can be obtained by artificial lysis—breaking open the infected cells during the latent period and assaying for the infectivity of the material released. This experiment was first realized in 1948 by A. H. Doermann, with the result presented in Figure 11-8.

As can be seen from these data, Doermann's experiment led to a most surprising finding: the infectivity associated with the original parental phage is lost at the outset of the reproductive process, since no infective phages whatsoever were found in any of the infected bacteria lysed artificially within

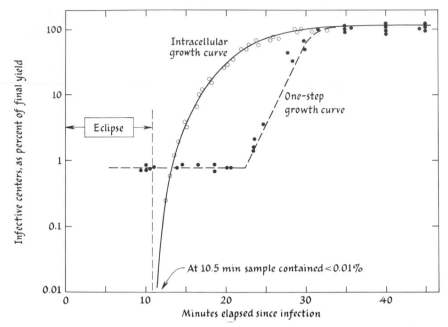

FIGURE 11-8.   The intracellular growth kinetics of phage T4, as first determined by Doermann. Procedure: The same as that described in the legend of Figure 11-7, except that here at various times after infection samples of the first growth tube are diluted 20-fold into a lysing medium containing 4×10^6 T6 phages/ml and 0.01 M KCN. After standing in the lysing medium for at least 30 minutes, samples of these dilutions are plated on agar with a Tsx^r strain of $E.\ coli$ on which only T4 but not T6 phages can form plaques. [After A. H. Doermann, $J.\ Gen.\ Physiol.$ **35**, 645 (1952).]

the first ten minutes following infection. After more than ten minutes has elapsed, however, an ever-increasing number of infective phages are seen to have made their intracellular appearance, until the same final litter of progeny has been produced which would have been released by the spontaneous lysis of all the infected bacteria at the end of the normal latent and rise periods. The time that elapses between infection and the first intracellular reappearance of infective phage particles—the stage of intracellular phage growth during which the infected host cell contains no material capable of initiating a plaque—is called the *eclipse*.

The discovery of the eclipse temporarily complicated the conclusions one might have hoped to draw from the intracellular growth curve, for it raised a new question: At what time during the latent period does the actual multiplication of the infecting phage particle take place? Does the eclipse period represent a waiting stage for the infecting phage particle, during which it is "masked," while the host cell undergoes some necessary renovations preliminary to the onset of intracellular phage growth? If this were so, the first infective phage particle present within the bacterium at the termination of the eclipse would be the original, once-more "unmasked" parental phage, which might then proceed to grow and divide as any other microbe. The daughter phages of the first division would in turn divide, until through successive division cycles the final crop of several hundred phages had been attained by the end of the latent period.

An entirely different explanation of the significance of the eclipse is that—far from being a stage of waiting—it is precisely that part of the latent period during which the substance of the phage progeny is constructed. If this is so, then the increase in intracellular progeny does not at all represent the multiplication of the parental virus but constitutes rather a terminal process of the completion of previously synthesized phage constituents as intact phage particles. The results of later investigations established that the latter point of view is the correct one; the infecting phage particle not only metamorphoses into a noninfective form at the outset of its infection of the host cell but also multiplies in this form to yield at first noninfective progeny structures, whose assembly, or maturation, into intact progeny phages signals the end of the eclipse. This noninfective form is called the *vegetative phage*, in contradistinction to the resting, mature, infective phage particle. The vegetative phage is thus the connecting link between parental and progeny phages, and the elucidation of its structure and function is the central problem of virus growth.

THE HERSHEY-CHASE EXPERIMENT

Since the reproduction of the infecting phage particle proceeds within the bacterium, it follows that after its adsorption to the bacterial surface the phage must somehow penetrate into the interior of the host cell, if it is to set in motion there the train of events that culminates in the construction of infective progeny phages. The "blendor" experiment that finally led to the understanding of just how phages manage to get into bacteria is one of the great milestones in the history of phage research. Two discoveries by T. F. Anderson, mentioned previously, were of undoubted heuristic value for the conception of this experiment: namely, that osmotic shock ruptures the phage and produces an empty-headed phage "ghost," and that phages attach to bacteria by their tail and thus form what ought to be a dynamically unstable union with the host cell surface. In support of the idea of dynamically unstable union, Anderson had found that violent agitation in a Waring blendor of mixed suspensions of phage and sensitive bacteria prevents infection. According to A. D. Hershey (Figure 11-9), it was an appreciation of these facts that "literally forced" him and Martha Chase to perform in 1952 the blendor experiment. This experiment showed that upon infection,

FIGURE 11-9.   Leo Szilard (left) (1898–1964) and Alfred D. Hershey (right) (b. 1908) at Cold Spring Harbor. [From J. Cairns, G. S. Stent, and J. D. Watson (eds.), *Phage and the Origins of Molecular Biology*. Cold Spring Harbor Laboratory of Quantitative Biology, New York, 1966.]

just as upon osmotic shock, the phage DNA leaves the phage head and is the only, or at least the principal, phage component that gains entrance into the host cell, the bulk of the phage protein remaining outside, beyond the wall. Before we can consider the experiment of Hershey and Chase in detail, however, a brief discussion of the use of radioactive tracers in phage research is required.

When a culture of *E. coli* is grown in a medium such as that specified in Table 2-1, containing phosphate and sulfate as the only sources of phosphorous and sulfur, then any radiophosphorus ^{32}P or radiosulfur ^{35}S added to the medium as $^{32}PO_4^{-3}$ or $^{35}SO_4^{-2}$ is incorporated into all of the phosphorylated or sulfurylated constituents of the cells. If, furthermore, such radioactive bacterial cultures are infected with phage, then the progeny grown on these bacteria also contain the radioisotope in their phosphorylated or sulfurylated components. Since the DNA contains practically all of the phosphorus of the phage (in polynucleotide phosphate diester bonds) and the protein all of its sulfur (in the two amino acids methionine and cysteine), a phage labeled only with ^{32}P carries all its radioactivity in DNA, whereas a phage labeled only with ^{35}S carries all its radioactivity in the protein.

Hershey and Chase thus infected different samples of an *E. coli* culture with purified stocks of ^{32}P- or ^{35}S-labeled T-even phages. After allowing a short time for host-cell attachment of most of the phage particles, the infected cells were separated from any unattached phages by low-speed centrifugation into a pellet and resuspension in fresh medium. This suspension was then agitated violently for various lengths of time in a Waring blendor to subject the phage-infected cells to the very strong shearing forces generated in the fluid by that device. After blendor treatment the capacity of the infected bacteria to yield progeny phage was determined by plaque assay of the infective centers. At the same time, the infected bacteria were centrifuged once more, and the fraction of the total radioactivity that remained in the supernatant fluid or sedimented into the bacterial pellet was determined in a radiation counter. It was possible, therefore, to measure by means of this procedure how much of either ^{32}P or ^{35}S label initially attached to the bacteria with the infecting phage particles is stripped off by violent agitation and how many of the infected cells can still continue to produce infective progeny phages after the blendor treatment. The result of this experiment is presented in Figure 11-10, and can be summarized as follows: (1) The capacity of the infected bacteria to yield progeny phage is not affected by the blendor treatment. (2) The shearing forces strip off 75–80% of the attached ^{35}S from the infected cells. (3) Only 20–35% of

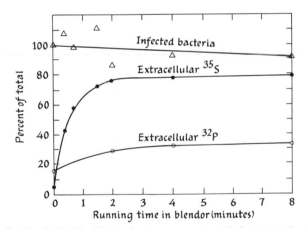

FIGURE 11-10.   The blendor experiment. The ordinate presents the percent of ^{35}S and ^{32}P removed from bacteria infected with radioisotope-labeled T2 phage, and the percent of infected bacteria surviving as infective centers, after the time of agitation in a Waring blendor, is shown on the abscissa. Procedure: E. coli are infected with ^{35}S- or ^{32}P-labeled T2 phages in buffered saline. After allowing a few minutes for phage adsorption, the bacteria are centrifuged and resuspended in water containing per liter 1 mM MgSO$_4$, 0.1 mM CaCl$_2$, and 0.1 g gelatin in order to separate them from any unadsorbed phages. This suspension is agitated in a Waring blendor at 10,000 rpm. The suspension is cooled briefly in ice water at the end of each 60-second running period. Samples are removed at intervals, assayed for plaque count to determine the number of bacteria capable of yielding phage, and centrifuged to measure the radioactivity in the supernatant fluid— that is, the proportion of initially adsorbed radioisotope released from the cells. [After A. D. Hershey and M. Chase. J. Gen. Physiol. **36**, 39 (1952).]

the attached ^{32}P is stripped from the bacteria; of that fraction, half has been already liberated into the medium before the blendor is turned on. Hence it may be inferred that most of the phage DNA enters the bacterium at the outset of infection, whereas most of the phage protein remains at the cell surface. The shearing forces of the blendor break the tail by which the phage "ghost" is still held to the cell wall and thus liberate into the medium only the empty head membrane, devoid of the phage DNA it once enclosed. Since the capacity to produce progeny phage survives the blendor treatment it may be inferred that the bulk of the phage protein has no further function in the intracellular reproduction process after the proteinaceous tail has anchored the phage particle to the bacterial surface and safely "injected" the phage DNA into the host cell. (Most of the 20% of the phage sulfur that is not removed from the bacteria by the blendor treatment consists of parts of the phage tail that adhere too firmly to the bacterial surface to be shaken loose.) Hershey and Chase further substantiated the inference that the phage DNA leaves the protective envelope of the phage head membrane at the moment of infection by showing that the phage DNA is rendered accessible to the hydrolytic action of DNase upon attachment of the phages to isolated bacterial cell walls or heat-killed bacteria, or upon freezing and thawing phage-infected bacteria. In the attachment of phages to isolated bacterial cell walls, the phage DNA is "injected" into the medium on the other side of the wall; in the infection of bacteria that have been either heat-killed before infection or frozen-thawed after infection, the bacterial cell membrane is rendered permeable to DNase enzyme molecules.

How does the single, enormous viral DNA macromolecule, seen in the electron micrograph of Figure 11-6, manage to travel from the phage head into the interior of the host cell? It seems likely that this is accomplished by mechanical penetration of the cell wall by the tail core upon contraction of the tail sheath. The tip of the phage tail must then become uncorked, probably by removal of the core, so that the DNA is free to pass through the channel provided by the sheath into the cytoplasm of the host cell.

The discovery that most or all of the DNA, and only very little of the protein, of the parental phage enters the host cell at the moment of infection thus showed that it must be the DNA which is the carrier of the genetic continuity of the phage; that is, its DNA is the germinal substance of the extracellular phage. (The 20–35% of attached phage ^{32}P liberated from infected bacteria in the blendor experiment probably represents the DNA complement of a minor fraction of the phage population that fails to infect, or properly inject its DNA into, the bacterial hosts.) The release of the phage DNA from its protein coat at the very moment of infection evidently accounts for the eclipse period that occurs at early stages of intracellular phage development, when, as Doermann's experiment on premature lysis showed, no infective phage particles can be recovered from the infected cell. For, having just been divested of its attachment and injection organs, the DNA of the parental phage is of course now unable to gain entrance into any bacteria to which it might be presented in the infectivity test after its release upon artificial lysis of the host cell.

Hershey and Chase's attribution of a germinal role to the phage DNA was thus in complete harmony with the earlier discovery by Avery, MacLeod, and McCarty, recounted in Chapter 7, that in bacterial transformation the genes of the donor bacterium are transferred to the recipient cell through the exclusive vehicle of bacterial DNA molecules. Now in retrospect it seems odd that, although in 1952 the genetic role of bacterial DNA had already been known for eight years, the notion that phage DNA is also the genetic material of bacterial viruses did not figure prominently in any of the numerous hypotheses on the nature of phage previously considered. Perhaps the idea that DNA is the germinal substance of the phage, if it *was* ever proposed, had been rejected as too hopelessly naive. In any case, it was Hershey and Chase's experiment that suddenly brought enlightenment to all but the most obdurate regarding the molecular nature of the gene. Learning of the results of this experiment was probably one of the most important incentives for Watson and Crick to work out the molecular structure of DNA in the following months.

SYNTHESIS OF PHAGE PRECURSORS

The divorce of parental nucleic acid and parental protein at the very outset of infection reflects a trait that distinguishes most clearly viral from cellular forms of life: in their life cycle viruses pass through a stage in which their genetic substance is the only material link connecting one generation with the next. Since the entire phage particle is thus reproduced only from its DNA, the two fundamental functions of DNA, *heterocatalytic* and *autocatalytic*, are manifest here even more clearly than in the case of the bacterium: once injected into the host cell, the parental phage DNA (1) presides over, or induces, the manufacture of several hundred replicas of the phage protein to provide the structural components of heads and tails for the somatic substance of the phage progeny and (2) achieves its own several-hundredfold replication to generate the genetic substance with which the progeny are to be endowed. In order to study these functions, experiments began in the 1950's in which T-even phage-infected bacteria were artificially lysed at various times during their latent period, and the lysates were analyzed for any materials whose properties rendered them likely to be precursors, or building blocks, of the progeny phages.

One line of search for intracellular phage *protein* precursors took advantage of the fact that T-even phages are excellent *antigens*, in that upon injection into a rabbit or a horse, phages elicit a specific immune response in the animal. That is, there appear in the blood serum of the immunized animal a class of globular proteins, or *antibodies*, that possess specific affinity for, and hence can combine specifically with, the phage proteins. (The process of formation of antibodies will be discussed in more detail in Chapter 21.) The specific combination of such antiphage antibodies present in samples of the immune serum with the homologous phage antigens that elicited their formation can be detected by the formation of enormous antigen-antibody complexes that precipitate out of solution. The T-even phage-protein antigens, furthermore, are sufficiently different in structure from all of the 3000-odd species of proteins present in *E. coli* so that the anti-phage antibodies do not *crossreact* with any of the nonphage proteins present in a lysate of a T-even infected culture. Such experiments revealed that significant amounts of antibody-precipitable phage protein appear in the T-even infected *E. coli* several minutes before any infective progeny phage are present. By the end of the eclipse period, when the first infective progeny make their intracellular appearance, enough phage protein is already present to provide the structural components for 10–20 progeny phage particles. For the remainder of the latent period, while the number of intracellular infective progeny rises steadily, synthesis of phage proteins continues at such a rate that there is maintained an *excess* of phage protein equivalent to the protein content of about 30–40 phage particles, over and above the number of infective phages present at any time. The fact that this protein both precedes the appearance and is later maintained in excess of the number of infective phages indicates that it is a developmental precursor of the mature progeny particles (Figure 11-11).

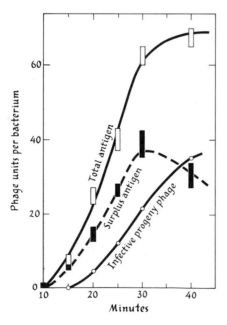

FIGURE 11-11.   Appearance of phage precursor protein in T2-infected bacteria. The curve labeled "total antigen" represents the amount of ^{35}S-labeled material specifically precipitable by anti-T2 serum that is liberated by premature lysis of *E. coli* infected with an average of five T2 particles per cell. (A "phage unit" is the amount of sulfur present in a single phage particle, or 1.5×10^{-12} μg.) The infected bacteria are growing in a minimal medium to which ^{35}SO$_4$$^{-3}$ is added at the moment of infection. The points on the curve labeled "surplus antigen" represent the difference between the number of phage units per cell of "total antigen" and the number of intracellular infective progeny per cell present at various times. [Data from I. Watanabe, *Biochim. Biophys. Acta*, **25**, 665 (1957).]

A second line of search for phage protein precursors consisted in direct electron microscopic comparison of premature lysates of normal and of phage-infected bacteria. It was hoped that, in addition to morphologically intact phage particles and structural components of the normal *E. coli* host cell, some *phage-related* objects might be discernible in the lysates of phage-infected cells. In this way it was found that although nothing is visible at the earliest stages of the eclipse that is not also seen in lysates of normal *E. coli*, about 3 minutes before the end of the eclipse some *empty phage heads* and isolated phage tail components begin to appear. After the first intact and infective phages have appeared, an average of about 35 empty heads per cell plus isolated tail components continue to remain throughout the rest of the latent period (Figure 11-12).

FIGURE 11-12.   Electron micrograph of a lysate of T2 phage-infected *E. coli*. Structurally intact phage particles, empty heads, and isolated tail components are visible. The tangled filaments in the upper part of the picture probably are DNA. [From E. Kellenberger and J. Séchaud, *Virology* **3**, 256 (1957).]

The discovery that the DNA of the T-even bacteriophages, unlike that of their *E. coli* host, contains 5-hydroxymethyl cytosine (HMC) in place of cytosine, provided the means of searching for an intracellular precursor state of the phage DNA. Thus analyses of infected *E. coli* for their content of HMC-containing DNA showed that synthesis of the progeny phage DNA begins half way through the eclipse and then proceeds so rapidly that at the end of the eclipse enough phage DNA is already present to provide the genetic material for about 80 progeny phages. The synthesis of phage DNA then continues at a more or less constant rate for the remainder of the latent period, so that there is always an excess of about 80 phage DNA complements over the total number of infective progeny particles accumulated at any time. Thus synthesis of both phage protein and phage DNA starts during the eclipse, well before the appearance of the first structurally intact infective progeny.

Once these findings had been made, it was evident that one could probe into the nature of the replication process responsible for the synthesis of the phage precursor DNA by following the fate of the parental DNA molecule injected into the host cell by the infecting phage. In particular, it seemed of greatest interest to ascertain whether the atoms of that parental molecule really become partitioned over its replicas in the semiconservative manner predicted by the mechanism of Watson and Crick. These first replication tests, like those used in the study of bacterial DNA, consisted of transfer experiments involving only radioactive labels, such as ^{32}P or ^{14}C, and did not lead to decisive results. But once Meselson and Stahl had invented the analysis of DNA *density* label distribution by equilibrium-density-gradient sedimentation, that method was quickly extended to the replication of the T-even phage DNA. In these transfer experiments *E. coli* growing in a "heavy" synthetic medium, containing either the rare isotopes ^{15}N and ^{13}C as the only nitrogen and carbon sources, or the thymine analog 5-bromouracil, were infected with ^{32}P-labeled "light" T-even phages containing the normal ^{14}N and ^{12}C isotopes, and thymine, in their DNA. From the progeny phage issuing from the "heavy" bacteria at the end of the normal latent period of this infection, the DNA was extracted and fragmented into molecules whose size was about 10% of that of the whole phage DNA complement. The density of the DNA molecules carrying any transferred parental ^{32}P molecules was then determined by the method previously shown in Figure 9-10. The result of this experiment was that most of the transferred parental ^{32}P atoms were found to reside in progeny phage DNA molecules of *hybrid density*, halfway between that of fully "light" parental DNA and that of fully "heavy" progeny DNA. Thus the atoms of the parental T-even phage DNA were found to be distributed semiconservatively over their vegetative replicas, so that it could be concluded that the Watson-Crick replication mechanism is also at work while the phage DNA, in its autocatalytic function, synthesizes the genetic patrimony for its progeny.

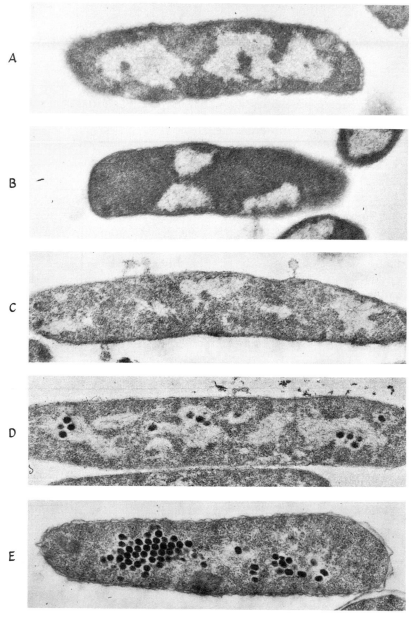

FIGURE 11-13.   Electron micrographs of ultrathin sections of T2 phage-infected *E. coli* at various stages of intracellular growth. **A.** At the moment of infection; the normal bacterial nuclei are visible as light areas. **B.** Two minutes after infection; the nuclei have changed their form and migrated toward the cell wall. **C.** Ten minutes after infection; the nuclei have disappeared and, instead, vacuoles filled with fibrillar material (phage DNA) have made their appearance. **D.** Fourteen minutes after infection; the first tail-less DNA-filled phage heads have been formed. **E.** Forty minutes after infection; many structurally intact, as well as incomplete, phages are present. [Micrographs courtesy of E. Kellenberger.]

It thus appeared that phage-precursor protein and phage-precursor DNA are first synthesized separately in the infected cell and ony later combined to constitute the mature infective progeny particles. In this *maturation process*, the phage head membrane arises by aggregation of a thousand or so identical head protein polypeptide subunits, which thus form their quaternary protein structure. The head membrane becomes filled with the very long and thin thread of phage-precursor DNA, which is thus packaged into a compact polyhedral body. Finally, the nascent DNA-filled phage head is endowed with a phage tail, and thus achieves the status of infectivity. This intracellular maturation sequence was first visualized by E. Kellenberger who, in 1958 developed a method for slicing bacteria into sections sufficiently thin to become transparent to the electron beam of the electron microscope. Figure 11-13 shows a series of Kellenberger's electron micrographs of *E. coli*, sliced and photographed at various times after their infection with T2 phage.

Since the phage tail was known to be an engine of considerable complexity its assembly from the diverse constituent protein subunits was thought to be so complicated a process that sorting out its details appeared beyond the reach of any experimental analysis. In 1965, however, R. S. Edgar and R. Wood made the quite unexpected discovery that assembly of the T-even phage tail and its attachment to a pre-existing, DNA-filled phage head can proceed spontaneously *in vitro*. For their experiments, Edgar and Wood obtained concentrated lysates of *E. coli* infected with a variety of genetically defective T-even phage *mutants* (whose nature will be the subject of later discussions). None of these lysates contained any infective phage particles, because under the conditions in which the infection was carried out each mutant phage lacks the capacity to synthesize some particular component of the phage tail. Mixing of certain of these lysates, however, resulted in a good yield of infective, structurally intact phage particles, evidently because each of the lysates provided the particular structural component of the tail that the other lacked. By making an extensive survey of which lysates could and which lysates could not complement each other in this manner, Edgar and Wood were able to provide the detailed morphogenetic sequence of the last episode of the phage maturation process, as shown in Figure 11-14.

FIGURE 11-14.   Morphogenesis of the T-even phage particle. **A.** The *in vitro* complementation test system. Two noninfective phage lysates are prepared by infecting *E. coli* with a pair of conditional lethal phage mutants (see Chapter 11) which, under the conditions of infection employed (but not under some other conditions), lack the capacity to form either tail fibers or heads. Upon incubating at 30°C a mixture of two lysates, one containing noninfective phage particles that lack tail fibers and another containing free tails and tail fibers, structurally intact, infective particles are formed by attachment of the free fibers to the fiberless particles (Figure continues on next page).

Mutant defective in tail fibers   Mutant defective in phage heads

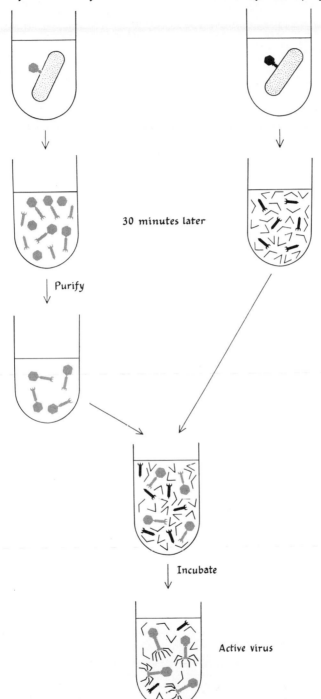

30 minutes later

Purify

Incubate

Active virus

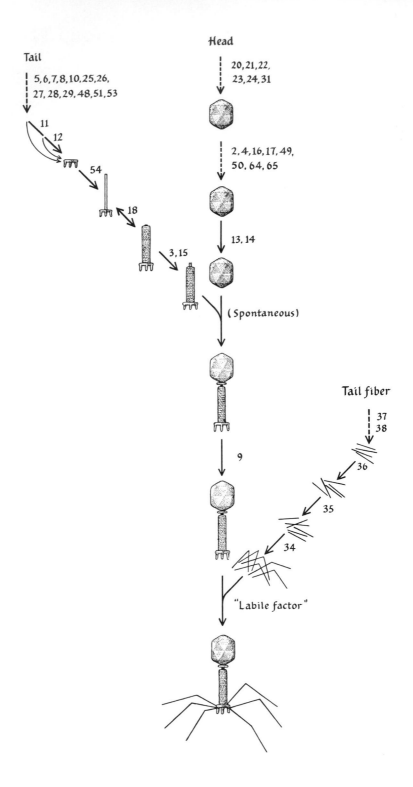

Tail

5,6,7,8,10,25,26,
27,28,29,48,51,53

11

12

54

18

3,15

Head

20,21,22,
23,24,31

2,4,16,17,49,
50,64,65

13,14

(Spontaneous)

9

Tail fiber

37

38

36

35

34

"Labile factor"

PHAGE-INDUCED ENZYMES

Further analyses of phage-infected cells eventually revealed that phage-precursor proteins are not by any means the *only* polypeptides whose synthesis within the infected host cell is presided over heterocatalytically by the parental DNA. Work by S. S. Cohen, by Kornberg, and by their colleagues showed, for instance, that at the outset of intracellular phage growth, the T-even phage DNA induces formation of an ensemble of "early" enzymes whose presence is required before replication of the phage DNA can begin. This ensemble, which is wholly foreign to the uninfected *E. coli*, includes enzymes that catalyze the synthesis and glucosylation of HMC, the synthesis of thymine by a new metabolic pathway, and the polymerization of the phage DNA from its nucleoside triphosphate building blocks. Synthesis of these "early" enzymes ceases near the end of the eclipse, by which time they are present in sufficient amount to sustain phage DNA replication for the remainder of the latent period. Another important enzyme formed in the phage-infected bacterium is the *phage lysozyme* that digests from within the structural members of the cell wall and thus prepares the bacterium for lysis and liberation of the intracellular phage progeny at the end of the latent period. In contrast to the "early" enzymes of DNA replication, the phage lysozyme is a "late" enzyme, in that it starts to make its intracellular appearance only half way through the latent period. No doubt the delayed start of synthesis of the phage lysozyme is as necessary for successful phage multiplication as the early start of the "early" enzymes related to DNA synthesis. For if the lysozyme were formed too soon after infection, it would commence erosion of the cell wall members too soon and thus engender lysis of the infected cell before any structurally intact, infective progeny had been produced.

It is important to realize, therefore, that the heterocatalytic function of the phage DNA is not confined solely to the construction of proteins that become incorporated in the mature phage particle. Using the same assumptions as those made in Chapter 8 for our earlier estimate of the total number of different proteins likely to be encoded in the *E. coli* DNA, we can now

FIGURE 11-14 (cont.)   **B.** The morphogenic pathway of phage maturation has three principal branches leading independently to formation of heads, tails, and tail fibers, which then combine to form complete phage particles. The numbers refer to the T-even phage genes, as listed in Table 12-1 and Figure 13-6, whose products are involved at each step. The solid arrows indicate the steps that have been shown to occur in extracts. Infection of *E. coli* with a phage carrying a mutation in any one of these genes leads to accumulation of the electron-microscopically visible structure shown immediately before the step in which that gene is involved, as well as of the last structure(s) of the other afferent pathway branch(es). [Parts A and B after " Building a Bacterial Virus," by W. B. Wood and R. S. Edgar, *Scientific American*, July 1967. Copyright © 1967 by Scientific American, Inc. All rights reserved.]

reckon that the sequence of nucleotide triplets of the T-even phage DNA molecule, made up of 2×10^5 base pairs, probably encodes something like $2 \times 10^5/(3 \times 330) = 200$ different proteins having an average of 330 amino acids in their chains. It seems most likely that much less than half of these 200 proteins are component parts of the infective T-even phage.

A MINUTE PHAGE

The facts set forth so far about the structure, composition, and intracellular growth of the T-even phages have rather general validity for a large variety of other phage types that infect not only *E. coli* but a wide spectrum of other bacterial species. Thus, though they may differ considerably in their detailed makeup—some phages have much smaller and less complicated tails than T-even phages, others have cylindrical or spherical heads rather than the polyhedral heads of T-even phages, yet others simply contain cytosine in their DNA instead of the glycosylated-HMC eccentricity of T-even phages—these other phage types all contain a double-stranded DNA molecule ranging from 10^4 to 3×10^5 nucleotide base pairs in length. This DNA molecule is always injected into the bacterial cell, where it replicates according to the Watson-Crick mechanism to provide the genetic material for hundreds of progeny phages.

By 1959, however, it had become permissible for disciples of Delbrück's school of phage workers to study phages other than the T series. One of these disciples, R. L. Sinsheimer, then began an intensive study of the *E. coli* phage ϕX174. This phage had then been known for about 25 years, and physical and chemical measurements of its particle size had long suggested that it must be a dwarf among bacterial viruses. Sinsheimer was able to show, first of all, that ϕX174 actually is as small as had been supposed. It is a tailless polyhedron only 250 Å in diameter, and hence of a volume about 1/40 that of the T-even phage head (Figure 11-15). It contains a single DNA molecule made up of 5000 nucleotides, or having only 1.3% the weight of the T-even DNA complement. But most importantly, the ϕX174 DNA does not manifest several of the properties characteristic of ordinary DNA. Unlike ordinary DNA, ϕX174 DNA does not obey the usual purine-pyrimidine base equivalence relation: 25% of its nucleotides contain A, 33% contain T, 24% contain G, and 18% contain C. Hence A and G cannot be paired with T and C in the ϕX174 DNA. Furthermore, unlike the relatively unreactive hydrogen-bonded amino groups of purines and pyrimidines of ordinary DNA, the amino groups of the ϕX174 DNA are readily accessible to formaldehyde reaction: apparently the nucleotide bases of that DNA do not take part in intramolecular hydrogen bonding. Finally, unlike the macromolecules of ordinary DNA, ϕX174 DNA molecules do not behave as rigid rods in solution. From these observations

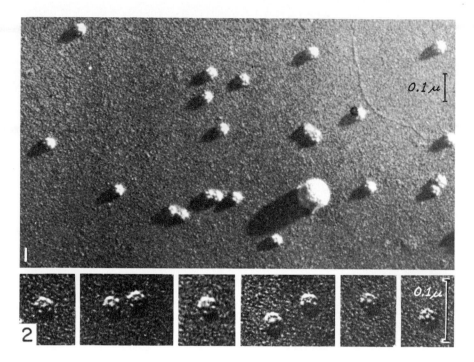

FIGURE 11-15.   Electron micrographs of phage ϕX174 [After C. E. Hall, E. C. Maclean, and I. Tessman, *J. Mol. Biol.* **1**, 192 (1959).]

Sinsheimer inferred that the DNA of the ϕX174 phage is not in the double-stranded, Watson-Crick configuration, but is, instead, composed of only a single polynucleotide strand. Later electron microscopic studies showed this single strand to be a circle of 1.7 μ contour length.

But if the DNA of the minute phage ϕX174 is a single-stranded rather than a double-stranded molecule, how then does its replication proceed in the infected cell? Does the semiconservative Watson-Crick replication process have any relevance here? It became possible to study this problem when it was found that *E. coli* bacteria can be infected with purified DNA extracted from ϕX174 phage particles, provided that the bacterial envelope is first rendered sufficiently permeable by partial lysozyme digestion to allow entrance of the phage DNA molecules. Thus, if phage DNA molecules are added to a suitably treated *E. coli* suspension, about 20–30% of the phage DNA molecules penetrate the cells successfully and give rise to infective centers that, at the conclusion of a latent period, yield several hundred normal, structurally intact ϕX174 particles. This experiment furnished, of course, an even more direct proof that DNA is the viral genetic material than Hershey and Chase's then eight-year-old blendor experiment with ^{32}P-labeled, T-even phage DNA. But more importantly, this procedure made it possible to assay the infectivity of the intracellular pool of vegetative phage DNA by presenting artificial lysates of ϕX174-infected cells to test suspensions of receptive bacteria.

In order to examine the intracellular fate of the single-stranded parental phage DNA, Sinsheimer infected *E. coli* bacteria growing in nonradioactive light (^{14}N-^{31}P) medium with heavy ^{15}N-^{32}P-labeled ϕX174 particles and lysed the infected bacteria at various times after infection. He then determined the density of the ^{32}P-labeled parental phage DNA present in the lysate by CsCl density-gradient-equilibrium sedimentation. The results of these analyses were that once intracellular phage growth is underway, the ^{32}P-labeled parental phage DNA extracted from the infected cell no longer bands in the CsCl density gradient at the position characteristic of the DNA extractable from the heavy ^{15}N-labeled ϕX174 parental phage. Instead, the ^{32}P-labeled DNA bands at a position whose density corresponds to that of a *double-stranded* ^{15}N-^{14}N hybrid DNA having the [G] + [C] content of ϕX174. Hence Sinsheimer inferred that the single-stranded DNA of the parent phage enters

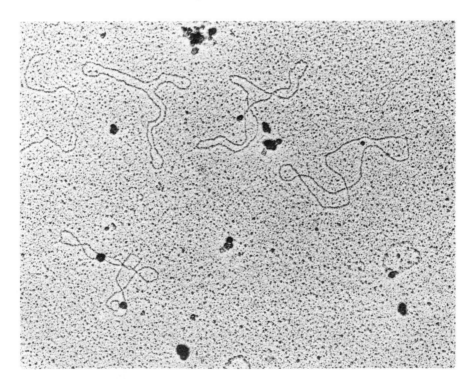

FIGURE 11-16.   Electron micrograph of the replicative form (RF) of phage ϕX174, extracted from infected *E. coli* cells. Four circular RF structures can be seen, whose contour lengths are all 1.80 ± 0.02 μ. This contour corresponds to double-stranded DNA molecules about 5000 base pairs in length. [Photograph courtesy of L. A. MacHattie.]

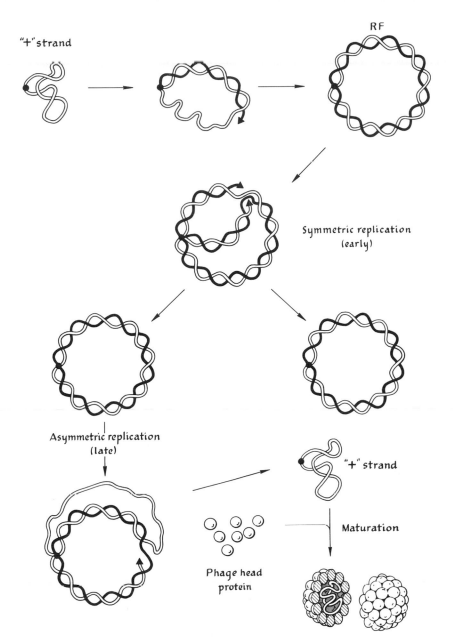

"+"strand

RF

Symmetric replication
(early)

Asymmetric replication
(late)

"+" strand

Maturation

Phage head
protein

FIGURE 11-17.   The general schema of intracellular growth of phage ϕX174. The single-stranded, circular "plus" strand of the infecting phage DNA serves as template for the synthesis of a complementary "minus" strand, yielding the double-stranded replicative form, or RF. At early stages of infection, RF replicates semiconservatively and symmetrically to yield daughter RF molecules. At late stages of infection, when phage head protein molecules are already present, an asymmetric replication of RF molecules sets in. Now only the minus strand of the RF serves as template for the synthesis of a daughter plus strand, expelling thereby the old plus strand from the RF molecule. The expelled plus strand is then encapsulated into a progeny phage head.

a double-stranded *replicative form* (RF) in the course of vegetative phage growth. Sinsheimer also examined the infectivity of artificial lysates of ϕX174-infected cells and found that the vegetative RF DNA is, like the DNA of the mature phage, infective. Electron micrographs of the purified RF soon showed that it is, in fact, a circular DNA molecule that has neither beginning nor end (Figure 11-16).

Studies on the intracellular growth of ϕX174 have led to the following picture of the replication of its genetic material (Figure 11-17). Upon its entrance into the host cell, the single-stranded, circular DNA molecule of the parent phage, or *plus strand*, first serves as the template for the synthesis of a complementary *minus strand,* to generate the double-stranded circular RF structure. Replication of the RF then proceeds by the ordinary semiconservative Watson-Crick process, so that the number of RF molecules per infected cell increases at early stages of the latent period. At later stages, however, when a pool of phage-head protein subunits is already present, an *asymmetric* DNA replication process comes into operation, and only the minus strand of the RF ring serves as template for the synthesis of a complementary plus strand. Growth of the new plus strand expels the old plus strand from the RF, and the expelled plus strand is encapsulated at once by the membrane of the head protein to become the single-stranded DNA genome of a mature, infective ϕX174 phage particle.

Bibliography

PATOOMB

Emory L. Ellis. Bacteriophage: one-step growth.

Thomas F. Anderson. Electron microscopy of phages.

A. H. Doermann. The eclipse in the bacteriophage life cycle.

A. D. Hershey. The injection of DNA into cells by phage.

Lloyd M. Kozloff. Transfer of parental material to progeny.

Edward Kellenberger. Electron microscopy of developing bacteriophage.

Robert L. Sinsheimer. ϕX: multum in parvo.

HAYES

Chapter 16.

MOBIBAV

Chapters 1, 3, 4, 5, 6, and 7.

ORIGINAL RESEARCH PAPERS

Anderson, T. F. The morphology and osmotic properties of bacteriophage systems. *Cold Spring Harbor Symp. Quant. Biology*, **18**, 197 (1953).

Doermann, A. H. The intracellular growth of bacteriophages. I. Liberation of intracellular bacteriophage T4 by premature lysis with another phage. *J. Gen. Physiol.*, **35**, 645 (1952).

Ellis, E. L., and M. Delbrück, The growth of bacteriophage. *J. Gen. Physiol.*, **22**, 365 (1939).

Hershey, A. D., and M. Chase. Independent functions of viral protein and nucleic acid in growth of bacteriophage. *J. Gen. Physiol.*, **36**, 39 (1952).

Koch, G., and A. D. Hershey. Synthesis of phage precursor protein in bacteria infected with T2. *J. Mol. Biol.*, **1**, 260 (1959).

Sinsheimer, R. L. A single-stranded deoxyribonucleic acid from bacteriophage ϕX174. *J. Mol. Biol.*, **1**, 43 (1959).

Sinsheimer, R. L., B. Starman., C. Nagler, and Guthric, S. The process of infection with bacteriophage ϕX174. I. Evidence for a "replicative" form. *J. Mol. Biol.*, **4**, 142 (1962).

SPECIALIZED TEXTS, MONOGRAPHS, AND REVIEWS

d'Hérelle, F. The bacteriophage. *Science News*, No. 14, p. 44. Penguin, Harmondsworth, 1949.

Twort, F. W. The discovery of the bacteriophage. *Science News*, No. 14, p. 33. Penguin, Harmondsworth, 1949.

12. Recombination

Study of the genetics of bacterial viruses *could* have gotten under way in 1936, when F. M. Burnet published a paper showing that phages can sport mutants whose plaques have a distinctly different appearance from those of the ordinary wild type. But although Burnet's paper clearly exposed the phenomenon of phage mutation and its attendant physiological consequences, it stimulated no further work with phages as objects of genetic study. And thus the development of phage genetics had to wait for another ten years, until the first mutants of T-even phages were isolated.

PHAGE MUTANTS

One of these first mutant types was discovered by Hershey, who noticed that one out of 10^3 or 10^4 plaques formed by T2 phages on agar plates seeded with *E. coli* looks rather different from the normal plaques. The normal plaque has a small, clear center and is surrounded by a turbid halo, whereas the rare variant plaque has a large, clear center and a sharp edge (Figure 12-1). When Hershey picked such a variant plaque from the agar and isolated and replated the phage particles that it contained, he found that all these phages themselves give rise to variant plaques identical in morphology to the variant plaque from which they came. In other words, the phages present in the original variant plaque *breed true*. Thus the variant plaque must reflect the presence in the phage inoculum of a T2 phage *mutant*, from which a line of mutant individuals descends during the growth of the plaque. This phage mutant possesses and passes on to its progeny the property of forming a plaque different in appearance from that formed by the normal or wild-type phage.

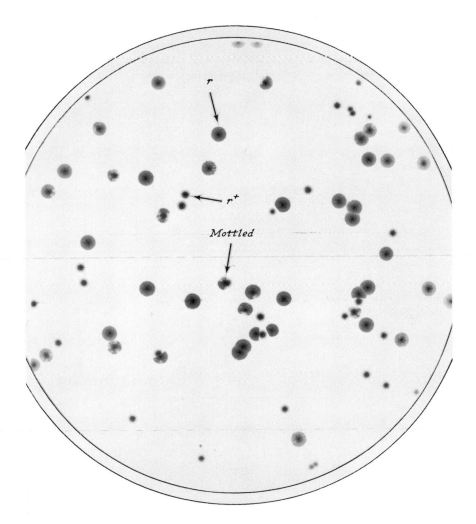

FIGURE 12-1.   Morphology of the normal T2r^+ (wild-type) plaque and of a variant T2r (rapid-lysis mutant) plaque. This plate also contains mottled plaques that owe their characteristic morphology to the joint growth of both r^+ and r phages in the same focus of infection. [From *Molecular Biology of Bacterial Viruses*, by G. S. Stent. W. H. Freeman and Company. Copyright © 1963.]

RAPID-LYSIS AND HOST-RANGE MUTANTS

Hershey gave this phage mutant the name rapid lysis, or r mutant. And in accordance with the nomenclature of classical genetics, Hershey represented the wild type—the normal, nonrapid-lysis T2 phage forming ordinary plaques—by the symbol r^+. Since later phage workers followed Hershey's lead, the superscript plus sign on the symbol of some mutant character now signifies in phage genetics, as in classical genetics, that the individual is "wild"—that is, does *not* possess the character in question. It is important to remember, however, that in bacterial genetics, which was developed at the same time and by some of the very same people as phage genetics, a different convention has been adopted: in bacterial genetics, different symbols are used to represent the gene that controls a certain character and the phenotype of that character: the superscript plus sign affixed to a phenotype symbol means that the individual *does* possess the character in question. Thus, as we saw in Chapter 5, Lac$^+$ signifies a bacterium that is able to ferment lactose, and a bacterium that is unable to do so is signified by Lac$^-$. Unfortunately, it is necessary to use both conventions in this text. In the remaining chapters the reader will thus have to be wary of the plus sign; when used to refer to phage genetic characters it will have a diametrically opposite meaning than it has when used to refer to bacterial genetic characters.

Though the r mutant character has been of capital importance in the development of phage genetics, its true physiological basis is still incompletely understood. It is known, however, that the turbid halo of the r^+ wild-type plaque constitutes a peripheral area of the bacterial lawn where there remain many unlysed bacteria that, though replete with progeny phage, do not lyse at the terminal phases of plaque growth. In the r mutant plaque, in contrast, bacterial lysis continues as long as there is any plaque growth, generating in the bacterial lawn a large, clear hole with a sharp edge.

The second type of phage mutant responsible for the rise of phage genetics was discovered by Luria, who, as we saw in Chapter 6, had studied the mutation of *E. coli* from Ton^s, or T1 phage-sensitivity to Ton^r, or T1 phage resistance at the dawn of bacterial genetics. Similar spontaneous bacterial mutations occur that convert the T2-sensitive, or Tto^s, *E. coli* wild type into the Tto^r form. These Tto^r bacterial mutants owe their resistance to structural modifications of their cell wall that prevent the stereospecific fixation of the attachment organs on the T2 phage tail to the T2 phage receptors on the bacterial cell wall, so that phage attachment, and, *a fortiori*, injection of the phage DNA into the interior of the host cell, can no longer take place. But why, if bacteria can mutate to phage resistance, are there still left in nature phage-sensitive bacterial strains? Why have sensitive strains not been replaced long ago by resistant forms through natural selection? Or why have bacterial viruses not yet been deprived of all suitable host organisms and thus vanished altogether? The answers to these questions, as to all evolutionary problems, are not immediately obvious, but one reason for the continued persistence of phage-sensitive bacterial strains is that the phages can, as Luria discovered in 1945, sport *host-range mutants*. Such host-range mutant phages are able to overcome the resistance of phage-resistant mutant bacteria because the structure of the attachment organs of the mutant phage differs in some subtle way from that of the wild-type phage. This structural difference permits the mutant attachment organs to make the necessary stereospecific reaction with the phage receptors on the wall of the phage-resistant mutant bacterium, in spite of the modification of the cell surface that thwarts attachment of the wild-type phage. With the appearance of host-range mutants, however, the phage has by no means put an end to the struggle for existence, since the bacterial strain that is both resistant to the wild-type phage and sensitive to the host-range mutant phage can sport superresistant bacterial mutants that are resistant to infection by both wild-type and host-range mutant phages. At this point the phage, not be to outdone, can respond to the appearance of a superresistant bacterial strain by sporting a super-host-range mutant. The coexistence in nature of bacteria and bacterial viruses is thus sustained by a delicate mutational equilibrium that saves both antagonists from total extinction.

If an agar plate seeded with about 10^7 Tto^r mutants of *E. coli* is inoculated with several hundred wild-type T2 particles, no plaques whatsoever will appear, since the phages are unable to multiply on the Tto^r bacteria presented to them. If, however, several million T2 phages are inoculated onto the Tto^r-seeded plate instead of several hundred, then a few plaques will usually appear. The phages present in one of these rare plaques formed on the Tto^r bacteria may then be picked and isolated. Upon replating these phages on plates seeded either with strain Tto^r or with strain Tto^s it will be found that the number of plaques formed on Tto^r is more or less equal to that formed on strain Tto^s; that is, the original plaque picked contained a clone of host-range mutant phage that was able to grow as well on the T2-resistant Tto^r strain as on the normal host strain Tto^s. Such host-range mutants are described by the symbol h: $T2h$ is a mutant of phage T2 that can infect and grow on strain Tto^r resistant to the $T2h^+$ wild type.

In genetic experiments with phages it is frequently necessary to determine the relative proportion of two types in a mixed population of phage particles. In a mixture of r and r^+ types, this can be done by simply scoring the number of r and r^+ plaques that are formed upon plating the phage mixture on wild-type *E. coli*. The relative proportion of h and h^+ types in a phage mixture can be estimated by two platings, one on agar seeded with the normal Tto^s bacteria and the other with the Tto^r mutant bacteria. The number of plaques that appear on the Tto^r-seeded plate indicates the number of h mutant type phages and the difference in the plaque count on plates seeded with strain Tto^s and with Tto^r indicates the number of h^+ wild-type phages. Fortunately, the technique of *mixed indicators*, invented by Delbrück in 1945, permits assay of the relative proportion of h and h^+ types in a single plating of the phage suspension. In this technique the phages are inoculated onto a plate seeded with a mixture of roughly equal parts of Tto^s and Tto^r bacteria; all h phage mutant types initially present on the plate will produce *normal* plaques, since they infect, grow on, and lyse equally well both bacterial indicator strains on this plate; all h^+ phage wild types, however, will produce *turbid* plaques, since of the two bacterial strains they can only lyse bacteria of strain Tto^s. In the area covered by each T2h^+ plaque growing on strain Tto^s there will remain intact all the T2h^+-resistant bacteria of strain Tto^r, which thus prevent the development of a clear plaque and give rise to a turbid plaque (Figure 12-2).

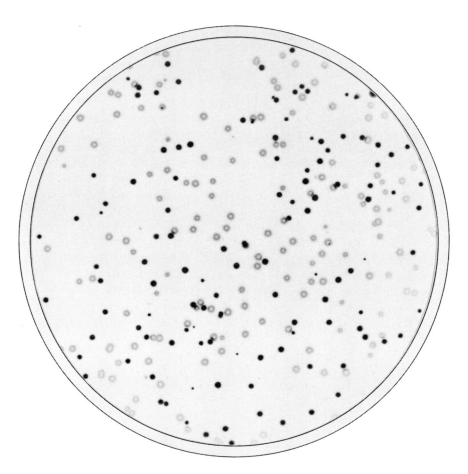

FIGURE 12-2.   Plaques formed by a mixture of T2h^+ wild type and T2h mutant phages on agar seeded with Tto^s + Tto^r *mixed-indicator* strains of *E. coli*. The h^+ wild type forms turbid plaques and the h mutant forms clear plaques. [From *Molecular Biology of Bacterial Viruses*, by G. S. Stent. W. H. Freeman and Company. Copyright © 1963.]

Not only is it possible to isolate phages carrying a single mutant character, but by two successive selective steps multiple mutants can also be obtained, which carry in their genome several mutant characters different from the wild type. For instance, starting with wild-type T2 phage one can first select an *h* mutant by plating a high concentration of wild-type phage on strain Tto^r and then select among the host-range mutant population an individual that produces an *r* plaque. Such a phage has the constitution T2*hr*, and thus differs in two of its characters from the T2h^+r^+ wild type. By means of the technique of plating on Tto^s + Tto^r *E. coli* mixed indicators, it is possible to recognize on a single plate (Figure 12-3) each of the four phage types (the

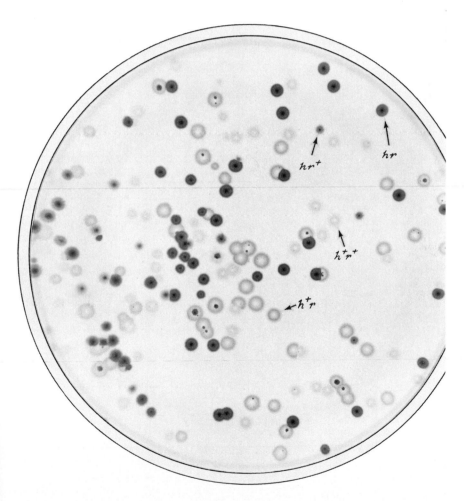

FIGURE 12-3.   The four types of plaques formed by T2h^+r^+, T2h^+r, T2*hr*$^+$, and T2*hr* on Tto^s + Tto^r mixed indicators. [From *Molecular Biology of Bacterial Viruses*, by G. S. Stent. W. H. Freeman and Company. Copyright © 1963.]

wild type, the double mutant, and the two single mutants) according to the following scheme:

Phage type	Plaque type on Tto^s + Tto^r E. coli
T2h^+r^+	turbid r^+
T2h^+r	turbid r
T2hr^+	clear r^+
T2hr	clear r

Conditional Lethal Mutants

In the 15 years following the first isolation of r and h mutant phages, a variety of other mutant types of T-even phages were identified. But though this ensemble of mutants was to provide the tools for carrying genetic analysis to so fine a level that the conceptual gap between DNA chemistry and gene structure could be closed, as will be set forth in Chapter 13, it became clear that all of these mutants pertained to only a relatively small fraction of the total genome of the phage. The reason for this was quite evident: most of the phage genes undoubtedly code for proteins that carry out *indispensable* functions, so that mutations in these genes would be irrevocably lethal. Despite this commonplace realization, however, for many years no one thought of extending to the T-even phages the astute method developed by Horowitz and Leupold for obtaining mutants of indispensable *E. coli* genes as *temperature-sensitive* phenotypes (See Chapter 5). Finally, in 1960, R. S. Edgar and R. H. Epstein got around to isolating *ts* mutants of T4, which do not form any plaques on agar incubated at 42°C but do form plaques at 25°C. The T4ts^+ wild-type phage, by contrast, is able to form plaques equally well at both incubation temperatures. Study of the physiology of growth of the *ts* mutants at the higher, *restrictive* temperature showed that different mutants appear to be blocked at different stages of the ontogenetic sequence of phage development. Thus one class of *ts* mutants is unable to initiate replication of the phage DNA at the restrictive temperature because one or another of its "early" enzymes involved in the metabolism of the nucleotide precursors cannot function at 42°C. Another class of *ts* mutants is able to initiate replication of the phage DNA but fails at some later stage of growth at the restrictive temperature. For instance, *ts* mutations occur in the gene coding for the phage lysozyme; bacteria infected with these mutants do not lyse at 42°C, though they contain infective progeny phage. Other *ts* mutations have been found in the many genes coding for the structural components of the phage; in bacteria infected with any of such mutants, no intact mature phages can be assembled at 42°C. High-temperature lysates of such mutants contain various kinds of incomplete phage components. In case the *ts* mutation occurs in a gene coding for the phage head protein, the lysate contains intact phage tails but no heads. In case the *ts* mutation occurs in a gene coding for the tail fibers, the nearly complete particles have heads and attached tails but lack the tail fibers necessary for attachment to the host cell.

The molecular explanation of the *ts* mutant phenotype is to be sought in an emendation of the fundamental principle of molecular biology introduced in Chapter 4, which stated that primary protein structure is solely responsible for secondary, tertiary, and quaternary protein structure. This emendation asserts that the particular secondary, tertiary, and quaternary structure actually assumed by a polypeptide chain of given primary structure depends on the environmental conditions, especially the *temperature*. Thus a given protein takes on its biologically functional tertiary and quaternary structure only within a limited "physiological" temperature range and takes on a nonfunctional, or denatured, structure outside that range. The primary structures of the proteins encoded in the genes of wild-type organisms are of such a character that they can take on their functional higher-level structures within the 25° to 42°C temperature range. But a gene carrying a *ts* mutation is one in which a change in nucleotide base sequence entails such a change in the primary structure of the polypeptide chain encoded into it that the resultant mutant protein, though still capable of taking on its functional higher-level structures at 25°C, takes on a nonfunctional denatured form at the restrictive temperature of 42°C.

At about the same time that the use of *ts* mutants was developed, another type of phage mutant was discovered that allows the recovery of T4 phages carrying mutations in genes coding for indispensable functions. This class of mutants was called *amber*, or *am*: T4*am* mutants cannot grow on most *E. coli* strains on which the T4*am*$^+$ wild type is able to grow, and these strains are said to be *nonpermissive*. (The name "amber" is not meant to be descriptive of any phenotypic aspect of these mutants; it first arose as a private laboratory joke and, eventually, became established general usage.) The *am* mutants *are* able to grow on some *E. coli* strains, however, and these strains are said to be *permissive*. Study of the physiology of growth of *am* mutants in the *nonpermissive E. coli* strains showed that *am* mutants manifest the same kinds of defects that the *ts* mutants manifest at the restrictive temperature. One class of *am* mutants is unable to initiate replication of the phage DNA; the abortive infective centers constituted by nonpermissive bacteria infected with mutants of this class do not lyse and do not contain any of the head or tail precursor proteins of the intact phage. Other classes of *am* mutants, though able to initiate the synthesis of phage DNA and cause lysis of the infected nonpermissive cell, yield in their lysates only various kinds of incomplete phage components but no structurally complete progeny phages. Finding the explanation of why an *am* mutation in a gene renders the phage unable to synthesize the corresponding polypeptide in functional form in one strain of host bacteria but leaves it still able to do so in another strain turned out to be a fascinating episode in the story of molecular genetics, and its account must await a later chapter.

Both *ts* and *am* mutants are thus members of the class of *conditional lethal mutants*, to which also belong the bacterial auxotrophs whose phenotype, as we saw in Chapter 5, is lethal in the absence of their required growth factors. Within a few years of their discovery the *ts* and *am* mutants made possible identification of almost all of the 200 or so genes of the T-even phage genome. Table 12-1 presents a list of the *am* and *ts* mutant types of T-even phages known by 1963.

TABLE 12-1.   Properties of *am* and *ts* Mutants of T4 Phage under Restrictive Growth Conditions

Gene	Mutant	Viable phage/cell	Lysis of host cell	Phage DNA	Tail fibers	Heads	Tails
—	Normal T4	200	+	+	+	0	+
1	*am*B24	0.03	0	0	0	0	0
	*am*A494	0.005	0	0	0	0	0
2	*am*N51	2	+	+	+	+	+
3	*ts*A2	0.01	+	+	+	+	+
4	*am*N112	2	+	+	+	+	+
5	*ts*A28	0.4	+	+	+	+	0
	*ts*B49	0.01	+	+	+	+	0
	*am*N135	0.008	+	+	+	+	0
	*am*B256	0.006	+	+	+	+	0
6	*ts*A25	0.01	+	+	+	+	0
	*am*N102	0.08	+	+	+	+	0
	*am*B251	0.002	+	+		+	0
	*am*B254	0.03	+			+	0
	*am*B274	0.02	+			+	0
7	*ts*B98	0.01	+	+	·+	+	0
	*am*B16	0.003	+	+	+	+	0
	*am*N115	0.009	+			+	0
	*am*B23	0.002	+	+		+	0
8	*ts*B25	0.01	+	+		+	0
	*am*N132	0.02	+	+	+	+	0

TABLE 12-1. *continued*

Gene	Mutant	Viable phage/cell	Lysis of host cell	Phage DNA	Tail fibers	Heads	Tails
9	tsN11	1.0	+	+	+	0	0
	tsL54	10				0	0
10	tsA10	0.5	+	+	+	+	0
	tsB64	0.02	+	+	+	+	0
	tsB12	0.02	+	+	+	+	0
	amB255	0.003	+	+	+	+	0
11	tsL140	<0.01	+	+		+	0
	amN93	0.07	+		+	+	0
	amN128	0.02	+	+	+	+	+
12	tsA13	0.02	+	+	+	+	+
	tsB60	0.05	+	+	+	+	+
	amN69	0.003	+	+	+	+	+
	amN104	0.01	+			+	+
	amN108	0.02	+			+	+
13	tsN49	1.0	+	+	+	+	+
14	amN71	0.10	+			+	+
	amB20	0.01	+	+	+	+	+
	amE351	0.1	+	+	+	+	+
15	tsN26	0.02	+	+	+	+	+
	amN133	1.0	+	+	+	+	+
16	amN66	0.01	+	+		+	+
	amN88	0.8	+	+	+	+	+
17	tsL51	2.0	+	+		+	+
	amN56	0.002	+		+	+	+
18	tsA38	0.4	+	+	+	+	+
19	tsN3	0.01	+	+	+	+	0
	tsB31	0.05	+	+		+	0
20	tsA23	0.01	+	+	+	0	+
	amB8	0.43	+			0	+
	amN83	0.002	+	+		0	+
	amN50	0.007	+	+	+	0	+
21	tsN8	0.03	+	+		0	+
	amN80	1.0	+	+		0	+
	amN121	0.3	+			0	+
	amN90	0.3	+	+	+	0	+
22	tsL147	0.5	+	+		0	+
	amB270	0.001	+	+	+	+	+

TABLE 12-1.—*continued*

Gene	Mutant	Viable phage/cell	Lysis of host cell	Phage DNA	Tail fibers	Heads	Tails
23	tsL65	0.1	+	+		+	+
	tsN37	1.0	+	+	+	+	+
	amB17	0.04	+	+	+	0	+
	amB272	0.006	+	+	+	0	+
24	tsN29	0.02	+	+	+	0	+
	amN65	0.01	+	+	+	0	+
	amB26	0.007	+	+	+	0	+
25	amN67	1.0	+		+		
	amN61	1.0	+	+	+	+	0
26	amN131	0.01	+	+	+	+	0
27	tsN34	0.5	+	+	+	+	0
	amN120	0.002	+	+	+	+	0
28	amA452	0.1	+	+	+	+	0
29	amB7	<0.001	+	+	+	+	0
	amN85	0.01	+	+	+	+	0
	tsL103	0.5	+	+		+	0
30	tsN7	0.5	+	+	+	0	+
	tsB20	0.01	+	+		0	+
31	amN54	0.02	+	+	+	0	+
	amN111	0.005	+	+	+	0	+
32	am453	<0.001	0	0	0	0	0
33	amN134	0.006	0	+	0	0	0
34	tsA20	0.1	+	+	+	0	0
	tsN1	0.04	+	+	+	0	0
	tsB3	0.03	+	+	+	0	0
	tsB22	0.3	+	+	+	0	0
	tsB57	0.01	+	+	+	0	0
	tsA44	0.02	+	+	+	0	0
	amN58	0.10	+	+	0	+	+
	amB25	0.04	+	+	0		
	amB258	0.07	+		0		
	amB288	0.03	+		0		
	amB265	0.03	+		0		
35	tsN30	0.1	+	+	+	0	0
	amB252	0.1	+	+	+	+	+
36	tsN41	1.0	+	+	+	0	0
	tsB6	1.0	+	+	+	0	0

TABLE 12-1.—*continued*

Gene	Mutants	Viable phage/cell	Lysis of host cell	Phage DNA	Tail fibers	Heads	Tails
37	*ts*B78	0.01	+	+	+	0	0
	*ts*B32	0.05	+	+	+	0	0
	*ts*N5	2.0	+	+	+	0	0
	*am*N52	0.03	+	+	0	+	+
	*am*N91	0.003	+		0	+	+
	*am*B280	0.06	+		0	+	+
38	*am*N62	1.2	+		0	+	+
	*am*B262	0.1	+	+	0	+	+
39	*ts*A41	∼50	+	+			
	*ts*G41	∼50	+	+	+		
	*am*N116	37	+	+	+	+	+
40	*ts*L177	0.01	+	+			
41	*ts*A14	0.01	0	0	0	0	0
	*am*N57	0.07	0	0	0	0	0
	*am*B15	0.05	0	0			
42	*ts*G25	5.0	0	0	0		
	*am*N122	0.007	0	0	0	0	0
43	*ts*L91	5.0	0	0	0		
	*ts*G37	10.0	0	0	0		
	*ts*L141	0.5	0	0	0		
	*ts*L56	0.4	0	0	0		
	*ts*L107	0.1	0	0	0	0	0
	*am*B22	0.08	0	0	0	0	0
	*am*N101	16	0	+	+	+	+
44	*am*N82	0.001	0	0	0	0	0
	*ts*B110	0.01	0	0			
45	*ts*L159	1.0	0	0			
46	*ts*L109	1.0	0	+			
	*am*N130	2.6	0	+	+	+	+
	*am*N94	0.9				+	+
	*ts*L166	0.01	0	+			
47	*ts*L86	1.0	0	+			
	*ts*B10	1.0	0	+			

From R. H. Epstein, A. Bolle, C. M. Steinberg, E. Kellenberger, E. Boy de laTour, R. Chevalley, R. S. Edgar, M. Sussman, G. H. Denhardt, and A. Lielausis, *Cold Spring Harbor Symp. Quant. Biol.*, **28**, 375 (1963).

THE MAP

Hershey and Luria reported their first isolation of r and h phage mutants in 1946, at the 11th Cold Spring Harbor Symposium, at which, as was set forth in earlier chapters, the victory of the one-gene-one-enzyme theory was celebrated, and Lederberg and Tatum announced their discovery of bacterial sexuality. But at that same meeting, Delbrück and Hershey presented another finding, which each had made independently, of the utmost importance for the later course of genetic research: Upon infection of an *E. coli* with two or more phages that differ from each other in *two* genetic characters there issue from the infected cell some *recombinant* phages that have obtained one of these two characters from one parent phage and the other of the two characters from the other parent phage. Thus Hershey had carried out *mixed infections* with h and r mutants of T2 and found among the progeny some h^+r^+ wild-type and some hr double-mutant phages, in addition to the two parental types. Here the h^+r^+ wild-type recombinant phage that appeared among the progeny derived its host-range character from the h^+r parent and its rapid-lysis character from the hr^+ parent, whereas the hr double-mutant recombinant phage received its host-range character from the hr^+ and its rapid-lysis character from the h^+r parent. Phages, therefore, can engage in genetic recombination—an activity once considered the prerogative of higher forms that have progressed from vegetative to sexual modes of reproduction.

In the wake of this discovery, Hershey carried out the first detailed study of genetic recombination in phage. For this purpose, he isolated numerous r mutants of phage T2, assigning consecutive numbers, such as $r1$, $r2$, $r3$, to individual r mutants of independent origin. Each of these mutants was then grown in mixed infection with, or *crossed* to, the host-range mutant T2h, and the progeny phages were examined for the presence of the two possible recombinant genotypes hr and h^+r^+. The results of crosses of h to the mutants $r1$, $r7$, and $r13$ are presented in Table 12-2, from which the following conclusions can be drawn:

1. The two recombinant types appear in all three of these crosses, and in each cross the percentage of the total progeny represented by the hr recombinant is equal to that represented by the h^+r^+ recombinant; therefore, complementary recombinant types are formed with equal frequency.

2. The total frequency of recombinant types produced is very different, depending on which r mutant is crossed to the h phage; only 1.7% of all the progeny of the cross $h \times r13$ are recombinants, whereas 12.3% of all the progeny of $h \times r7$ and 24% of all the progeny of $h \times r1$ are recombinants.

These observations were interpreted by Hershey in the terms of the classical notions of gene linkage elaborated 40 years earlier by T. H. Morgan and A. H. Sturtevant. The hereditary material of bacteriophages, Hershey

TABLE 12-2.   Recombinant Frequencies in T2h × T2r Crosses

Cross		Percentage of type in population			
		h^+r^+	hr^+	h^+r	hr
$h \times r1$	input	0	53	47	0
	yield	12	42	34	12
$h \times r7$	input	0	49	51	0
	yield	5.9	56	32	6.4
$h \times r13$	input	0	49	51	0
	yield	0.74	59	39	0.94

From A. D. Hershey and R. Rotman, *Genetics*, **34**, 44 (1949).
Procedure: A growing culture of 2 × 10⁷ *E. coli* per milliliter was infected with an input average of 5 T2h and 5 T2r phages per cell. After allowing 5 minutes for adsorption of phages to their host cells, the infected culture was diluted 10⁴-fold into nutrient broth and incubated for 60 minutes. The progeny phage yield was plated for plaque assay on agar seeded with *E. coli* Tto^s and Tto^r mixed indicators, on which all four phage types can be separately scored.

supposed, consists of linear arrays of genes, each gene carrying the hereditary information for some character of the virus, analogous to the genes in the chromosomes of higher forms. Thus different phage mutants harbor different mutant genes, so that each mutant character is situated in a particular place or *locus* on such a linear structure of linked genes. The genomes of the two mutant phages, h and r, coexisting in the mixedly infected bacterium can therefore be represented as

In one of the parents the host-range gene is of the mutant h type, and the rapid-lysis gene is of the wild r^+ type, whereas the corresponding, homologous genes in the other parent are of the wild h^+ and mutant r types. In accord with classical genetic nomenclature, h and h^+, or r and r^+, are said to be *alleles*, or alternative configurations, of the same gene in different individuals. But in the mixedly infected bacterial cell, the two coexisting phage genomes may proceed to exchange homologous parts with each other, as by breakage of the linear structures at exactly corresponding points between host-range and rapid-lysis genes, followed by crosswise rejoining of the fragments:

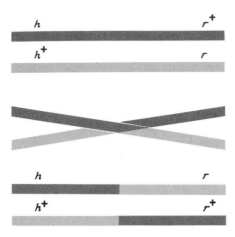

Such crossover of genetic material anywhere between the two mutant genes will then generate the two recombinant hr and h^+r^+ genomes that have derived their host-range gene from one parent and their rapid-lysis gene from the other.

Since the chance event of an exchange can occur at any point along the two parental genetic linkage structures, it follows that the probability that a crossover will actually take place between two given genetic loci depends on the distance between the two loci, or their linkage. The closer the linkage of two genes, the less probable it is that a crossover will occur at an intermediate point, and hence the lower the frequency of phage-progeny recombinant for the characters in question. In contrast, the greater the distance between the two genes, the greater is the chance of a crossover at an intermediate point, and hence the higher the probability of recombinant progeny. That the three r mutants of Table 12-2 produce different yields of recombinants in crosses to the h mutant then implies that in the genome of the T2 phage more than one gene controls the rapid-lysis character and that the mutations carried by T2r1, T2r7, and T2r13 pertain to different genes. Of the three mutations, r13 must pertain to the gene that is most closely linked, and r1 must pertain to the gene that is least closely linked to the host-range gene. These considerations allow one to construct a genetic map of the relative position of the four phage genes on the "chromosome" of the phage. Four conceivable arrangements of the genes of h, r13, r7, and r1 satisfy the condition that r13 is closest to h and that r1 most distant from h:

1. h-r13————————r7————r1
2. r13-h————————r7————r1
3. r1————————————13-h————r7
4. r1————————————h-r13————r7

Further crosses are necessary before it can be decided which of these four alternatives corresponds most closely to the truth. If either map (1) or (4) were correct, the $r13$ gene would be more closely linked to the $r7$ gene than the h gene is linked to $r7$, whereas if map (2) or (3) were correct, $r13$ would be more distant from $r7$ than is h. Thus, in a cross of $r13 \times r7$, the frequency of crossovers between these two genes should be lower under alternatives (1) and (4) and higher under alternatives (2) and (3) than the frequency of crossovers between h and $r7$. If one carries out the cross $r13 \times r7$, then r^+ wild types will appear among the progeny. This confirms, first of all, the earlier inference that the mutations $r13$ and $r7$ are situated at different genes, or are *nonallelic*, since these r^+ individuals can have arisen only by the crossover

that brought the wild-type $r13^+$ and $r7^+$ genes together into the same phage genome.

The proportion of these r^+ wild-type recombinants, furthermore, is less than the proportion of either hr or h^+r^+ recombinants produced in the cross $h \times r7$, showing, secondly, that $r13$ is closer to $r7$ than is h, and thus eliminating alternatives (2) and (3). The exchange event that produces the r^+ wild-type also generates the double mutant $r13r7$, but since the phenotype of the double r mutant is exactly the same as that of single r mutants—that is, it produces an r-type plaque—the double mutant cannot be easily recognized among the progeny. Of course, no r^+ recombinants are produced in the "crosses" $r7 \times r7$, or $r13 \times r13$, since it is impossible to generate a phage genome carrying only r^+ loci by any recombinational event between two genomes carrying an r mutation at exactly corresponding or allelic sites. This allows one to identify the double mutant $r13r7$, inasmuch as this double mutant, in contrast to a single mutant, produces no r^+ recombinants in crosses to either $r7$ or $r13$. Similar mapping studies involving 7 other non-allelic r mutants ($r2$ to $r8$) showed that all seven of these mutations are closely linked to $r7$. This cluster of r mutations constitutes the rII class, which will figure prominently in the discussions of the next chapter.

Hershey was unable to decide between the remaining two alternatives, (1) and (4), for the good reason that, as became clear more than ten years later, *both* alternatives are correct. For as G. Streisinger and Edgar then inferred from a detailed study of recombinant frequencies in crosses involving two parent phages differing from each other in three or more genetic markers, the T-even phage genetic map is *circular*, having neither beginning nor end. This inference, which was at first rather indirect, was soon given direct confirmation

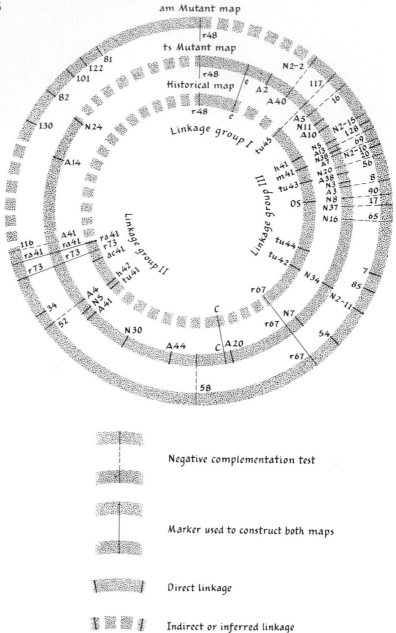

FIGURE 12-4. The circular genetic map of phage T4, based on a comparative linkage analysis of the "historical" map first established by crosses of nonlethal mutant types such as *r* and *h*, and of the *am* and *ts* mutant map established by crosses of conditional lethal mutant types. (The character of the conditional lethal mutants shown on this map and the nature of the genes to which they pertain are listed in Table 12-1.). [From *Molecular Biology of Bacterial Viruses*, by G. S. Stent. W. H. Freeman and Company. Copyright © 1963.]

when Edgar and Epstein's *ts* and *am* mutants became available. These mutants, whose conditionally lethal character pertains to almost all of the genes of the phage genome, were found to occur in sufficient density in all sectors of the map such that the circularity of the phage genome could be established directly by point-to-point linkage measurements around the circumference of the circle. This circular map is shown in Figure 12-4. The distances on this map are expressed in *map units*, one map unit being the distance that separates two genetic loci x and y for which in a standard phage cross between $T2x^+y$ and $T2xy^+$ the recombinants $T2xy$ and $T2x^+y^+$ appear among the progeny at a frequency of 1%. The total extent of the T-even genome is now estimated to be about 1500 map units.

THE MATING POOL

Thus far we have viewed a phage cross as an exchange of genetic material between the genomes of the two infecting parental phages and have pretended that a mixed infection is more or less analogous to the mating of two organisms. Further study of genetic recombination in phage revealed, however, that a phage cross does not represent merely the exchange of hereditary factors between two parent phages but that, instead, recombination involves repeated genetic interactions among the intracellular population of vegetative phage genomes descended from the infecting parents. Taking this fact into consideration, N. Visconti and Delbrück formulated in 1953 a theory of genetic recombination in phages capable of accounting quantitatively for the recombinant frequencies obtained under various experimental conditions. This theory envisages that the vegetative phage genomes exist in an intracellular *mating pool* in which they undergo repeated pairwise matings, each mating leading to an exchange of genetic material by one or more crossovers between the two mated individuals.

From the vantage point of the Visconti-Delbrück theory, the fraction of progeny phages recombinant for two genetic markers introduced into a phage cross depends, therefore, not only on the linkage of the mutant genes in question but also on the number of mating events that have occurred in the mating pool by the time that lysis of the infected bacterium has brought to term the intracellular growth processes. Since linkage of genetic loci cannot, therefore, be simply equated to recombinant frequencies, the "true" linkage, d_{xy}, of two genetic sites x and y is defined as the average number of crossovers that takes place at points between these sites in each mating of two vegetative phages in the mating pool: the greater the distance that separates the sites on the phage "chromosome," the greater the average number of such crossovers per mating. But an observable recombination of genetic characters will result if in any one mating an *odd* number of crossovers occurs at points between the genes determining these characters, for if the number of crossovers happens to be even, then no recombination would be manifested, since the original configuration of the two characters is preserved; for example, the following figure illustrates the nugatory result of two crossovers:

It then follows that the probability p_{xy} of recombination per mating—that is, the probability of an odd number of crossovers between two genes—is related to their "true" linkage, d_{xy}, by the odd terms of the Poisson law (see Chapter 6):

$$p_{xy} = \left(\frac{d_{xy}}{1!} + \frac{d^3_{xy}}{3!} + \frac{d^5_{xy}}{5!} + \dots\right) e^{-d_{xy}} = 0.5\,(1 - e^{-2d_{xy}}) \qquad (12\text{-}1)$$

This expression, which was first derived by J. B. S. Haldane in 1917, implies that for large values of d_{xy} (for very distantly linked genes), p_{xy} approaches the limit 0.5 (Figure 12-5). In other words, if there occur a large number of crossovers per mating between two genes, half of the mating events will end in an odd number and half in an even number of crossovers. For small values of d_{xy}, characteristic of very closely linked loci, equation (12-1) simplifies to $p_{xy} = d_{xy}$; that is, if the average number of crossovers per mating between the two genes is very much less than one, almost all crossovers that occur will be single exchanges. Finally, if the two genes are alleles ($d_{xy} = 0$), then $p_{xy} = 0$; that is, there is no chance whatsoever that a recombinant is produced for two genetic characters that occupy exactly the same site on the phage chromosome.

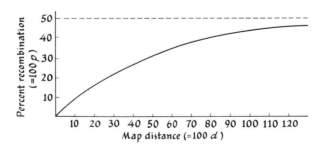

FIGURE 12-5.   Graphical representation of the "mapping function" [equation (12-1)] first derived by J. B. S. Haldane. [From *General Genetics* (2nd ed.), by Srb, Owen, and Edgar. W. H. Freeman and Company. Copyright © 1965.]

Analytical formulation of the postulates of the Visconti-Delbrück theory then allows prediction of the fraction R_{xy} recombinant for the characters x and y among the progeny of a cross involving two parental phage genotypes xy^+ and x^+y. If to that cross one parent contributes a fraction w and the other $1 - w$ of the total phage input and if the number of matings in the line of ancestry of the average phage particle, or *rounds of mating*, is m, then

$$R_{xy} - 2w\,(1 - w)\,(1 - e^{-P_{xy}m})$$
(12-2)

By analyzing the results of numerous crosses of T-even phages, involving a variety of different mutations in the T-even genome, in the light of the relation expressed by equation (12-2) Visconti and Delbrück concluded that $m = 5$ at the end of the normal latent period of T-even phage growth—that is, that there occur about five rounds of mating in the mating pool. Since the map unit is defined as the distance separating two loci x and y for which in a standard phage cross ($m = 5$) recombinants appear in frequency 1% ($R_{xy} = 0.01$), it follows from equations (12-1) and (12-2) that one map unit corresponds to a "true" linkage of 0.004 crossovers per mating in the vegetative phage pool. Or, for the whole T-even genome of 1500 map units, one can reckon that there occur an average of 6 crossovers per mating.

HETEROZYGOTES

Though the Visconti-Delbrück theory successfully accounted for the overall frequencies of recombinants produced in phage crosses, it left unspecified the detailed mechanism of the process that actually results in crossing-over during the postulated elementary mating act. And it was this very process, rather than the population-genetic aspects of phage recombination, that molecular geneticists *really* wanted to understand. One important clue for gaining this understanding was discovered in 1951 by Hershey and Chase: about 2% of the phage progeny issuing from a $T2r \times T2r^+$ cross harbor *both r* mutant and r^+ wild alleles in their genomes and segregate upon further growth to produce the two r and r^+ phage types among their progeny. Such r/r^+ phage particles are called *heterozygotes*, a term that has been borrowed from classical genetics, where, as was set forth in Chapter 1, it denotes organisms that harbor two different alleles of some particular gene among a diploid set of chromosomes. Further studies of these phage heterozygotes showed that their heterozygosity concerns *only about 0.1% of the phage genome*. In other words, the great majority of those few phage particles that are heterozygous for a given gene carry only a single allele of most of their other genes. Furthermore, the state of heterozygosity is not permanent, since heterozygous phage particles do not appear to reproduce themselves as heterozygotes. In the wake of this discovery, C. Levinthal provided strong support for Hershey and Chase's belief that these heterozygotes are not some sort of genetic freaks but must reflect some basic feature of the elementary event of genetic recombination in the vegetative phage pool. When Levinthal examined the phage progeny of the three-factor cross $T2hr2^+r7 \times T2h^+r2r7^+$, he found that nearly all of the detectable $r2/r2^+$ heterozygotes involving the middle $r2$ gene were $hr7^+$ recombinants for the outside genes. The structure of the heterozygote could thus be represented formally as

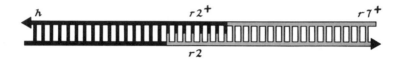

Once it had become clear that the T-even phage genome is a single molecule of DNA, two molecular structures could be considered for this formal structure of the heterozygote. The first of these, the *heteroduplex model*,

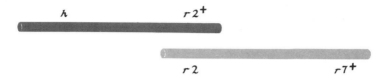

represents a single, covalently bonded DNA molecule containing a limited heterozygous region in which each of the two complementary polynucleotide chains carries one of the two parental alleles. This heterozygote would evidently segregate into *homozygous* $hr2^+r7^+$ and $hr2r7^+$ recombinant DNA molecules upon its semiconservative replication. The second of these structures, the *overlap model*, represents a joint molecule

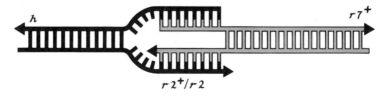

in which two double helices are paired through formation of complementary hydrogen bonds between two pairs of homologous polynucleotide chains in an overlap region bearing the $r2^+$ and $r2$ alleles of the two parental genomes. This heterozygote would also be expected to segregate homozygous $hr2^+r7^+$ and $hr2r7^+$ progeny molecules, though it is not *a priori* obvious just how upon replication of the heterozygote the DNA polymerase would deal with the structural singularity presented to it by the double-helical overlap. One possibility would be that replication of an overlap heterozygote produces a heteroduplex heterozygote that segregates homozygous progeny in the next replication cycle.

When Doermann carried out a detailed study of the segregation pattern of r/r^+ heterozygotes that carry not just one but six closely linked r mutations, $r_a r_b r_c r_d r_e r_f$ in their region of heterozygosity, he found that these six genetic sites do not always segregate together. That is, among the offspring of such sextuple r/r^+ heterozygotes he found homozygous r type progeny that carry some but not all of the six r mutations present in the heterozygote. Furthermore, Doermann noticed that there exists a definite *polarity* in the segregation pattern, in that any segregant carrying the "innermost" r_a mutant locus had a much higher probability of also carrying the proximal r_b locus than the "outermost" r_f locus. Since this segregation pattern was difficult to explain under the covalent continuum of the heteroduplex model and its expected, simple, blockwise segregation by semiconservative DNA replication, Doermann favored the overlap model, which provided for the kind of physical ends necessary for the observed segregational polarity. But in opting for the overlap model, Doermann had to face the following paradox. Since Hershey and Chase had shown, and later work confirmed, that for *any* given 0.1 % stretch of the T-even phage genome some 2 % of the progeny particles are heterozygous, it followed that *every* T-even phage particle must contain $0.02/0.001 = 20$ regions of whatever structural singularity is responsible for heterozygosity. Hence it was to be expected that if the overlap model were correct, the T-even phage DNA molecule should not be a covalent continuum. Instead, it would have to consist of some 20 separate, hydrogen-bonded sectors, which could be expected to fall apart into 40 single polynucleotide chains upon melting of the double helices by the methods described in Chapter 8. But careful molecular-weight determinations had revealed that upon melting, the T-even phage DNA yields only two single polynucleotide chains, each having half the molecular weight of the entire phage genome. Thus the T-even phage DNA does not possess the molecular structure demanded by the overlap model.

This paradox was resolved by George Streisinger in 1963, who then proposed that there exist *two* basically different types of T-even phage heterozygotes: *internal* heterozygotes, which correspond to the heteroduplex model, and *terminal redundancy* heterozygotes, which correspond *formally* to Doermann's favored overlap model, but whose actual molecular structure is to be written as

$r2^+$ h $r7^+$ $r2$

That is, Streisinger proposed that the linear T-even DNA molecule is *terminally redundant*, so that its sequence of genes can be represented as

abcdef . . . wxyzabc

When such a terminally redundant phage DNA molecule replicates in the infected cell, genetic recombination can proceed within the region of terminal redundancy of two daughter molecules, giving rise to molecular concatenates of the type

$$\underline{abcdef \ldots wyzabc}$$
$$\times$$
$$abcdef \ldots wyzabc$$

$$\downarrow$$

$$abcdef \ldots wyzabcdef \ldots wyzabc$$

These concatenates would then have to be cut into phage-genome-sized pieces before incorporation into the heads of infective progeny particles. If this

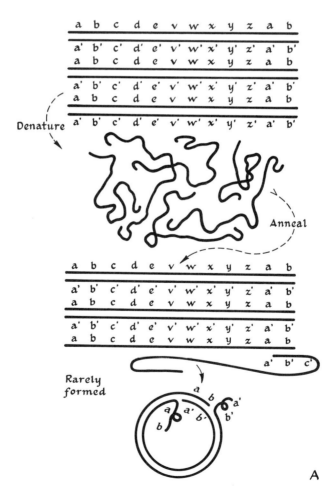

A

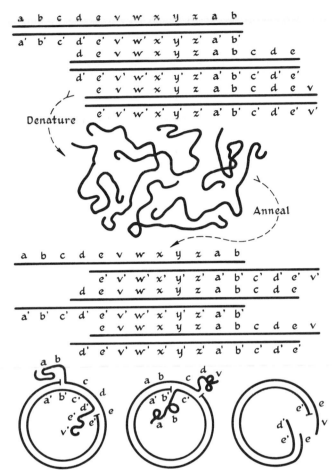

FIGURE 12-6.   Principle of the molecular demonstration of
terminally redundant DNA molecules and of the permuted
nature of the redundancies. **A**. Denaturation and renaturation
of terminally redundant *nonpermuted* linear DNA molecules.
Upon heat-denaturation of a collection of such DNA
molecules, single polynucleotide strands arise which, upon
annealing, would be expected to renature mainly into linear
DNA molecules of the original length. Rarely, however, would
an initial pairing between the short, terminally redundant
sectors at opposite ends of the molecules lead to the
renaturation of a circular molecule. **B**. Denaturation and
renaturation of terminally redundant, permuted linear DNA
molecules. Here the single polynucleotide strands formed by
denaturation can reassociate upon annealing in a variety of
ways, many of which give rise to circular molecules. [After
C. A. Thomas, Jr., *in* G. C. Quarton, T. Melnechuk, and
F. O. Schmitt, (eds.), *The Neurosciences*. Rockefeller
University Press, 1967.]

cutting process were to be of such a nature that it started to take its measure of phage genome length always at the same genetic site, say at *a*, then there would arise only one kind of terminally redundant phage genome—namely, that redundant for the *abc* sector. Hence terminal redundancy heterozygotes could arise only for genes residing in that sector. But if the cutting process were to start taking its measure of genome length at any randomly chosen genetic site, then there would arise a collection of phage genomes whose terminal redundancy would be *circularly permuted*, such as

$$\underline{g \, h \, i \, j \, k \, e \ldots cdefghi}$$

and

$$\underline{mnopqr \ldots ijklmno.}$$

And hence terminal redundancy heterozygotes could arise for *any* phage gene. The genome of a phage particle issuing from an $r \times r^+$ cross will thus be terminally redundant and heterozygous for that r gene if the end of its genome happens to be near the r gene and if a crossover has taken place anywhere on that genome, so that one end carries the r allele and the other end carries the r^+ allele. The circularly permuted nature of the T-even DNA molecule would also explain how that linear molecule could correspond to the circular genetic map shown in Figure 12-4. Streisinger was able to adduce various genetic proofs for the existence of both internal and terminal redundancy heterozygotes, and hence for the notion of circular permutation of the T-even genome. Streisinger could show, furthermore, that the segregation of such terminal redundancy heterozygotes represents a further act of genetic recombination, by means of which one of the two redundant regions becomes replaced by its allele, thus reconstituting a homozygous genome. In contrast, segregation of internal heterozygotes does proceed by means of semi-conservative replication. The fragmentary, polarized segregation observed by Doermann in sextuple heterozygotes evidently pertained to terminal redundancy heterozygotes, which had made a variety of segregational cross-overs *within* one of the two terminally redundant regions.

Soon after Streisinger proposed the circular permutation of the T-even phage genome, C. A. Thomas showed that the nucleotide base sequence of the T-even phage DNA is, as demanded by that model, circularly permuted. For Thomas found that after their separation by melting of the double helices, the two complementary polynucleotide strands of an ensemble of linear T-even phage DNA molecules reassociate to form *circular* double helices, as shown diagrammatically in Figure 12-6.

Thus it became clear that although terminal redundancy heterozygotes *are* produced in every elementary recombination act, they merely reflect the over-all structure of the phage DNA and are not relevant to the molecular basis of the crossover event. The internal heterozygotes, however, bid fair holding the key to understanding that event.

THE MECHANISM OF RECOMBINATION

For some 15 years, one fact was weighing heavily on the minds of all molecular geneticists who tried to understand the mechanism of genetic recombination in phage: *Contrary to the anticipations of exchange by crossing-over, the elementary recombination event generates only one, and not both, of the two complementary recombinants.* This was discovered by Hershey in 1948, when he examined the yield of T2*hr* and T2*h+r+* recombinants liberated by *individual* infected bacteria of T2*h* × T2*r* crosses. These examinations revealed that although these two complementary recombinant types appear in equal frequency among the *total* phage progeny, as can be seen from the data of Table 12-2, their frequencies show very little correlation among the yield of individual bacteria infected with both parental types. This lack of correlation made it appear that any given mating event produces *either* an *hr or* an *h+r+* recombinant, but not, as would be expected from a reciprocal exchange, both types. Following a suggestion by A. H. Sturtevant, Hershey then proposed that genetic recombination in phage may not, after all, derive from breakage and reunion of preformed genetic structures. Instead, he proposed that recombination might occur as an event incidental to the replication of the phage genome, a hypothetical mechanism that later came to be called *recombination by copy choice.* Under the copy-choice doctrine, a mating represents the coming together of two parental genomes at the moment when one of them happens to be serving as a template for the synthesis of a replica. After part of the genetic information of the first parental genome has already been copied into the replica structure, the copy process suddenly switches to the other of the mated parental genomes and begins to incorporate the genetic information of the second parent into the growing *de novo* genome. As illustrated in Figure 12-7, a single recombinant phage thus arises which carries some of the genetic factors of one and some of the genetic factors of the other of the two mated parents. A further, and entirely independent, recombination act is required to produce the complementary recombinant type, which would in no way be correlated with the formation of its complement, though it would occur with equal frequency. If there is a fixed probability per length of phage DNA that a copy-choice switch will occur during a mating, then the chance that a replica genome recombinant for two genes would be produced still depends on the distance d_{xy} between the two loci, just as expressed by equation (12-1).

Though it was not immediately obvious how the copy-choice mechanism

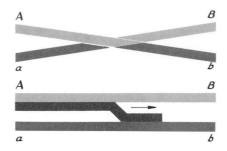

FIGURE 12-7.   Two mechanisms of genetic recombination. *A* and
B represent two genetic loci and *a* and *b* their alleles. Upper
figure: *Breakage and reunion*. The two synapsed parental
chromosomes break between *A* and *B* and between *a* and *b*, and
pieces rejoin crosswise. Lower figure: *Copy choice*. Replication (in
the direction of the arrow) generates a new daughter chromosome,
using first one and then the other of the two synapsed parental
chromosomes as template. A copy-choice switch occurs between
A and *B*. [After M. Delbrück and G. S. Stent, *in* W. D. McElroy
and B. Glass (eds), *The Chemical Basis of Heredity*. The Johns
Hopkins Press, Baltimore, 1957.]

might actually work during the semiconservative replication of the phage
DNA, copy-choice was favored by the majority of molecular geneticists
during the 1950's. An account of the rise and fall of the copy-choice doctrine
would make an interesting chapter in the history of genetics—this account
would begin with the publication in 1931 of the first copy-choice proposal by
J. Belling for chromosomal recombination in higher organisms and Belling's
retraction of that proposal in 1933—but such an account is beyond the scope
of this text. Suffice it to say that hardly anyone believes any longer in copy-
choice.

Besides accounting for the formation of only one of the two complementary
recombinants in a single act of genetic exchange, the copy-choice doctrine
entailed another, rather more basic feature that distinguished it from recombination by breakage and reunion. Under copy-choice, recombinants are
necessarily composed of newly synthesized DNA and should not contain any
of the DNA atoms of the parent phages that entered the cross. Under genetic
recombination by breakage and reunion, however, genetic recombinants containing parental atoms would be expected to arise. Meselson and J. Weigle
provided the first conclusive proof that phage recombinants do contain part
of the DNA of the parental genomes that entered the cross, and hence that
genetic exchange occurs as a result of breakage and reunion rather than copy-
choice. In their experiments they used the *E. coli* phage *lambda* whose
particle weight and DNA complement is only about one quarter of that of the
T-even strains and which will be of great importance in later chapters. The
general features of the growth cycle and genetics of phage *lambda* are rather

similar to those of the T-even phages, including the transfer of about half the atoms of the parental *lambda* phage DNA to the progeny. But one important difference between *lambda* and the T-even phages is that the frequency of genetic exchange in *lambda* is very much less than in the T-even strains. When the results of crosses of genetically marked *lambda* phages are analyzed in terms of the Visconti-Delbrück theory, it becomes evident that by the end of the normal latent period only 0.5 to 1 round of mating has occurred in the line of ancestry of the average *lambda* progeny phage, instead of the 5 rounds of mating experienced by T-even phage. Thus among the progeny of a *lambda* cross there are many individuals without any mating experience whatsoever. In order to examine the connection between the distribution of the transferred parental DNA and genetic recombination in phage *lambda*, Weigle and Meselson infected *E. coli* growing in an ordinary light medium with heavy $^{13}C^{15}N$-labeled *lambda* c^+mi^+ wild-type phages and with light $^{12}C^{14}N$-labeled *lambda* cmi double-mutant phages. (The locations of c and mi, which for the purpose of this discussion can be considered the genes that control the plaque-type are shown on the genetic map of phage *lambda* in Figure 12-8.) The progeny of this cross, which included about 1.5% c^+mi and cmi^+ recombinants, were then subjected to density-gradient-equilibrium sedimentation in concentrated CsCl solution and the titer of the various phage genotypes present at all levels in the density gradient assayed. The result of this experiment is presented in Figure 12-9, from which the following conclusions can be drawn.

1. The distribution of the originally heavy c^+mi^+ parental type exhibits three modes: one main band at the lowest density, consisting of light progeny particles composed of DNA and protein synthesized *de novo* in the light growth medium, and two minor bands at densities corresponding to phage particles possessing one-quarter and one-half of the parental heavy isotopes, respectively. These two minor bands consist of phage particles whose protein is entirely light and whose DNA is either half-parental, half-new (semiconserved), or entirely parental (conserved). The conserved DNA complements have probably not replicated at all, whereas the semiconserved DNA complements were no doubt generated by the semiconservative replication process in the vegetative phage pool.

FIGURE 12-8.   A primitive genetic map of phage *lambda*, showing the percent recombination between four widely separated genes. [From *Molecular Biology of Bacterial Viruses*, by G. S. Stent. W. H. Freeman and Company. Copyright © 1963.]

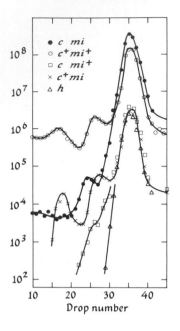

FIGURE 12-9.   Genetic recombination by breakage and reunion in phage *lambda*. Density-gradient centrifugation in concentrated cesium chloride solution of the phage progeny of a cross between "heavy" $^{13}C^{15}N$-labeled λc^+mi^+ wild type and "light" $^{12}C^{14}N$-labeled λcmi, on bacteria growing in $^{12}C^{14}N$-labeled "light" medium. The ordinate shows the titer of progeny phages of parental and recombinant genotypes in drops collected through the bottom of the centrifuge tube after sedimentation. The λh phages are added to the centrifugation mixture as a density reference marker. [After M. Meselson and J. J. Weigle. *Proc. Natl. Acad. Sci. Wash.* **47**, 857 (1961).]

2. The distribution of the originally light *cmi* parental double-mutant type is very different from that of the parental wild type. The double mutant is present mostly in the lowest density band, though its distribution is definitely skewed toward higher densities.

3. The distribution of the c^+mi recombinant resembles that of the c^+mi^+ wild-type parent, in that the recombinant is found in essentially the same three modes as the wild type, whereas the distribution of the reciprocal cmi^+ recombinant, like that of the *cmi* double-mutant parent, shows but a single mode.

The presence of c^+mi recombinants in the two heavy bands of phages containing conserved and semiconserved parental wild-type DNA shows that they contain some of the parental isotope. This demonstrates unequivocally that discrete amounts of original parental DNA appear in recombinant phages, suggesting that recombination occurs by breakage of parental chromosomes followed by the reconstruction of genetically complete chromosomes from the fragments. That the complementary cmi^+ recombinant is practically free of heavy isotope (though containing the mi^+ allele of the c^+mi^+ parent) can be readily explained on the basis of the relative situation of the c and mi genes of the genetic map of *lambda*. Since, as can be seen in Figure 12-8, mi is very nearly at the end of the *lambda* chromosome, the cmi^+ recombinant chromosome would obtain no more than a very small segment of the heavy DNA of the c^+mi^+ wild-type parent in a single exchange. (This explanation presupposes, of course, that the genetic map of *lambda*, unlike that of T-even, has an end and is not circular).

A series of experiments by J. Tomizawa on T4 phage provided further insights into the molecular nature of the genetic exchange process. Tomizawa infected *E. coli* with a mixture of light ^{32}P-labeled and heavy bromouracil-labeled T4 phages and incubated the infected culture for an hour in the presence of the potent metabolic inhibitor potassium cyanide. Under these conditions metabolic reactions that require energy are strongly inhibited; in particular there occurs synthesis of neither protein nor DNA in the infected cells. After this incubation the DNA was extracted from the bacteria and fractionated by centrifugation in a CsCl density gradient, according to the procedure illustrated in Figure 9-10. This fractionation revealed the presence of some ^{32}P-labeled phage DNA molecules of density intermediate between that of light and heavy (BU-labeled) DNA—that is, *joint molecules* containing material derived from both heavy and light parental genomes. Upon being heated to a temperature just sufficient to melt T4 phage DNA, the heavy and light components of these joint molecules separate. The joint molecule thus appears to be composed of one ^{32}P-labeled light DNA segment and one BU-labeled heavy segment joined end-to-end by hydrogen bonds (Figure 12-10). Thus it appeared that a joint DNA molecule that is likely to be an intermediate in the genesis of genetic recombinants can be formed while protein and nucleic acid synthesis are severely reduced. Tomizawa repeated this mixed infection experiment, except that he incubated the infected cells in the presence of the thymidine analog 5-fluorodeoxyuridine (FUDR) rather than cyanide, a condition under which the rate of DNA synthesis is greatly reduced, though protein synthesis can proceed nearly normally. Under this condition, formation of joint ^{32}P-BU-labeled phage DNA molecules was also observed. However, whereas almost all of the joint molecules detected after incubation of the infected cells in FUDR for 20 minutes or less are of the same hydrogen-

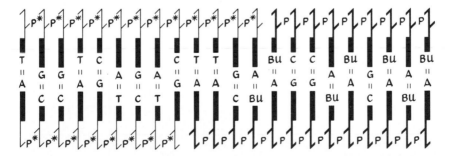

FIGURE 12-10.   Possible structure of the *joint molecule* formed as the first stage in genetic recombination between ^{32}P-labeled light DNA molecules and BU-labeled heavy DNA molecules. (P* = ^{32}P-labeled phosphate diester linkages.)

bonded type as those formed in the presence of cyanide, many joint molecules present after incubation for 45 minutes in FUDR appear to be made up of segments of ^{32}P-labeled light and BU-labeled heavy parental DNA bound in *covalent* linkage, since the heavy and light components can no longer be separated by melting the double helix. Hence the slow progress of DNA synthesis in the FUDR-treated phage-infected cells seems to suffice to convert the hydrogen-bonded joint molecules representing nascent recombinants into the covalently linked DNA continuum in the form of which the phage genome matures in the infective progeny particle. The structure that thus arises at the elementary recombination event is evidently nothing but an internal, hetero-duplex type of heterozygote. Finally, Tomizawa repeated the mixed infection experiment with parental phages carrying various *am* mutations that abolish phage DNA synthesis in the nonpermissive host bacterium used. The result of this experiment was that parental phages carrying *am* mutations in genes 41, 42, 43, 44 and 45 (see Table 12-1) give rise only to hydrogen-bonded joint molecules, but not to covalently linked recombinant molecules. This lends further support to the notion that phage DNA synthesis must be allowed to proceed in order to establish the covalent linkage between polynucleotide chains of mixed parental provenance.

Bibliography

PATOOMB

R. S. Edgar. Conditional lethals.
George Streisinger. Terminal redundancy, or all's well that ends well.
Aaron Novick. Phenotypic mixing.
N. Visconti. Mating theory.

MOBIBAV

Chapters 8 and 9.

HAYES

Chapters 18 and 19.

ORIGINAL RESEARCH PAPERS

Hershey, A. D. Spontaneous mutations in bacterial viruses. *Cold Spring Harbor Symp. Quant. Biol.*, **11**, 67 (1946).

Hershey, A. D., and R. Rotman. Genetic recombination between host range and plaque-type mutants of bacteriophage in single bacterial cells. *Genetics*, **34**, 44 (1949).

Levinthal, C. Recombination in phage T2; its relation to heterozygosis and growth. *Genetics*, **39**, 169 (1954).

Luria, S. E. Mutations of bacterial viruses affecting their host range. *Genetics*, **30**, 84 (1945).

Meselson, M., and J. J. Weigle. Chromosome breakage accompanying genetic recombination in bacteriophage. *Proc. Natl. Acad. Sci., Wash.*, **47**, 857 (1961).

Streisinger, G., R. S. Edgar, and G. H. Denhardt. The chromosome structure in Phage T4. I. The circularity of the linkage map. *Proc. Natl. Acad. Sci., Wash.*, **51**, 775 (1964).

Tomizawa, J., and N. Anraku. Molecular mechanisms of genetic recombination in bacteriophage. *J. Mol. Biol.*, **8**, 516 (1964); **11**, 501 (1965); **12**, 805 (1965).

Visconti, N., and M. Delbrück. The mechanism of genetic recombination in phage. *Genetics*, **38**, 5 (1953).

SPECIALIZED TEXTS, MONOGRAPHS, AND REVIEWS

Stahl, F. W. *The Mechanics of Inheritance* (2nd ed.). Prentice-Hall, Englewood Cliffs, N. J., 1969.

13. Genetic Fine Structure

Since the presentation in Chapter 7 of Avery's discovery of the DNA nature of the bacterial transforming principle, the discussions in this text have been based on the "molecular" view of the gene as a stretch of polynucleotide chain whose nucleotide base sequence specifies the amino acid sequence of a polypeptide chain via a genetic code. But, as a matter of fact, very few of the experiments on the mutation and genetic recombination of bacteria and their viruses that we have considered so far actually depend on that view in any essential way and could have been interpreted almost equally well from the purview of the "classical," indivisible one-gene–one-enzyme gene. We shall now be concerned with work that finally bridged the conceptual gap between the inferences based on purely formal genetic observations pertaining to character differences on the one side, and those based on purely chemical observations pertaining to nucleotides, on the other.

RUNNING THE MAP INTO THE GROUND

The man who, more than any other, brought the new concept of the "molecular" gene to bear directly on genetic experimentation was Seymour Benzer (Figure 13-1). Benzer's point of departure was a finding he made in 1953 that appeared

FIGURE 13-1.   Seymour Benzer
(b. 1921). [Courtesy S. Benzer.]

to be only of mild interest to his fellow phage workers at the time: one class
of closely linked r mutations of T-even phage, which Hershey had previously
termed rII, possess another phenotype in addition to their typical r plaque
morphology on agar plates seeded with the ordinary $E.$ $coli$ strain. Phages
carrying such rII mutations cannot grow at all on special $E.$ $coli$ strains to
which we shall refer to collectively as "K strains." On K strains the r^+ wild-
type phage, as well as other r mutants that do not belong to the rII class,
can grow perfectly well. The rII mutants adsorb to, infect, and even initiate
the synthesis of phage DNA and other products of phage metabolism in K
strain host cells, but such rII-infected K strain bacteria do not produce
infective progeny phages, nor do they ever lyse. Benzer realized that this
conditionally lethal growth defect of rII mutants can serve as a powerful
selective agent for detecting the presence of a very small proportion of rII$^+$
phages within a large population of rII mutants, since on agar plates seeded
with a K strain all of the rII$^+$ wild type can form plaques, but none of the
rII mutant individuals can. Though the detailed biochemical reasons for the
inability of rII mutants to grow on K strains remain unexplained to this
day, it is known that K strains owe their inability to propagate rII phage
mutants to the incorporation of the genome of another phage into their
chromosome. That other phage is $lambda$, which was mentioned briefly in the
preceding chapter and which will figure prominently in the following chapters.
In the decade from 1953 to 1963, Benzer accumulated a collection of rII

mutants of phage T4, and these mutants were to play a key role in the development of molecular genetics. This collection first comprised dozens, then hundreds, and finally thousands of mutant individuals, all of whom had arisen independently from one stock of T4rII$^+$ wild type phage.

While the efforts recounted in Chapter 12 were under way—that is, the efforts to chart the total extent of the T-even genome by establishing point-by-point linkage of an ever greater number of different phage genes—Benzer set out in the opposite direction, intending to run one particular sector of the genetic map, namely the domain of the rII mutations, into the ground. That is to say, Benzer proceeded with the construction of a *fine-structure genetic map* of the T4 genome. For this purpose Benzer crossed members of his rII mutant collection two-by-two and selectively scored for the frequency of rII$^+$ wild-type recombinants produced by plating the phage progeny of each cross on plates seeded with an *E. coli* K strain. In this way he hoped to detect very rare recombinants between adjacent genetic sites, since the limit of resolution of this method would have allowed him to find rII$^+$ recombinants in frequency as low as 0.0001%—a frequency so low that if the nonselective method of simply hunting for r^+ plaque types on ordinary, rII-permissive *E. coli* were employed, 10^6 progeny r plaques would have to be inspected for every r^+ recombinant found. The frequencies of r^+ recombinants produced in these very numerous crosses allowed Benzer to arrange all his mutants in a linear order and thus to construct the map shown in Figure 13-2. Among this first set of 60 mutants, Benzer found some pairs that produced no r^+ recombinants at all when crossed. Such mutant pairs must carry *recurrences* of a mutation at precisely the same site of the phage genome; they must be *exact alleles*. More important, the *lowest frequency* with which r^+ recombinants were formed in any cross between two different rII mutants that produced *any* r^+ recombinants at all was 0.01%. No rII mutations were found that had occurred at two nonidentical genetic sites so closely linked that in their cross r^+ recombinants arose in the hundredfold frequency range from 0.01% to 0.0001%, within which r^+ recombinants would have still been readily detectable. Thus Benzer seemed to have discovered the minimum nonzero distance between two genetic sites still separable by recombination. (Later work by others showed that rII mutant pairs *can* be found that yield r^+ recombinants in Benzer's empty 0.01% to 0.0001% frequency range. This finding does not mean, however, that there exist genetic sites that are separated from each other on the map by less than Benzer's "minimum nonzero distance." Instead, the low probability of recombination between such exceptional mutant pairs appears to be attributable to a secondary effect that the precise molecular configuration of the contiguous mutant sites has on the recombination process—an effect that reduces the chance of generating a functional r^+ recombinant gene.)

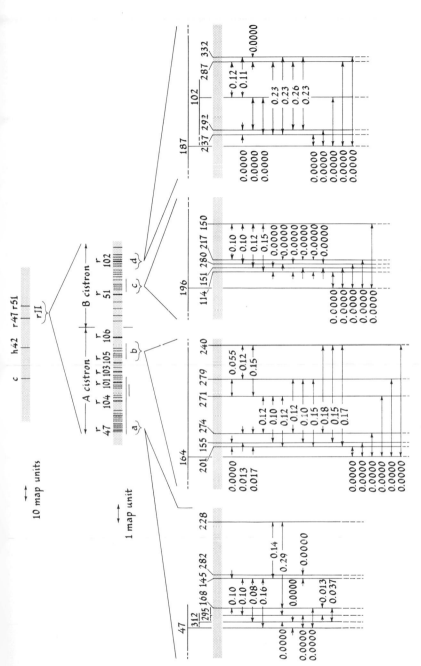

FIGURE 13-2. The first fine-structure genetic map of the rII region of T4, based on the frequency of r⁻ recombinants produced in pairwise crosses between a set of 60 rII mutants of independent origin. Successive levels in this figure correspond to progressively greater magnifications of the viral genome. In the lowest level, numbered vertical lines represent individual rII mutants, and the decimals indicate the percentage of r⁺ recombinants found in crosses between two mutants connected by an arrow. The horizontal bars shown in the middle and lowest level represent the genetic extent of long-span mutations, or deletions. [After S. Benzer, *Proc. Natl. Acad. Sci. Wash.* **41**, 344 (1955).]

It is possible to form a rough idea of the fraction of the whole T4 phage genome represented by the minimum nonzero distance found by Benzer. As was seen in Chapter 12, the circular genetic map of T4 extends over about 1500 map units, or a distance 1500 times that separating two genetic sites for which complementary recombinants appear in frequency 1 % in a "standard" phage cross. Hence two rII mutations that produce 0.01 % r^+ wild-type recombinants are separated by 0.02 map units, or by about $0.02/1500 = 1.3 \times 10^{-5}$ of the total phage genome. The meaning of this minimum, nonzero recombination distance can now be translated into chemical terms, if it is assumed that genetic recombination is equally probable at all points of the phage DNA that represents that genome. Then, since the DNA complement of a single T4 particle contains about 2×10^5 nucleotide base pairs, the minimum recombinational distance can be reckoned as $1.3 \times 10^{-5} \times 2 \times 10^5$, or about three nucleotide base pairs. That is to say, two phage mutants whose mutated sites are separated by no more than three nucleotide pairs on the phage "chromosome" are able to produce recombinant phage progeny when crossed with each other. This estimate of three nucleotide pairs may still be too high, since if the assumption of equal probability of crossover per length of DNA were incorrect, the number of nucleotide base pairs per map unit would be even less. Benzer's map showed, therefore, that genetic recombination is a process that can separate genetic sites represented by virtually contiguous nucleotides on the phage DNA macromolecule.

A mutant that owes its phenotype to the alteration of a single nucleotide base pair should correspond to a *point mutant*. That is to say, a point mutant in the rII region of the T4 DNA ought to yield r^+ recombinants in crosses to at least one of two other nonallelic rII test mutants previously shown to yield r^+ recombinants in crosses to each other. Most of the rII mutants isolated by Benzer, and all those represented by vertical lines on the map in Figure 13-2, do indeed satisfy this criterion for a point mutant. Furthermore, most of these rII mutants sport r^+ wild-type reverse mutants during their

FIGURE 13-3.   Segmental subdivision of the rII region by means of deletions. First level: The whole T4 genetic map. Second level: Seven sections define seven segments of the rII region. Third level: Three deletions define four subsegments of the A5 segment. Fourth level: Three deletions define four subsegments of the A5c subsegment. Fifth level: Tentative order and spacing established by pairwise crosses of seven point mutants in the A5c2a2 subsegment. The site of each of these point mutants probably corresponds to a single base pair in the viral DNA, drawn in the bottom level approximately to the same scale as the genetic map. [After "The Fine Structure of the Gene," by S. Benzer, *Scientific American*, January 1962. Copyright © 1962 by Scientific American, Inc. All rights reserved.]

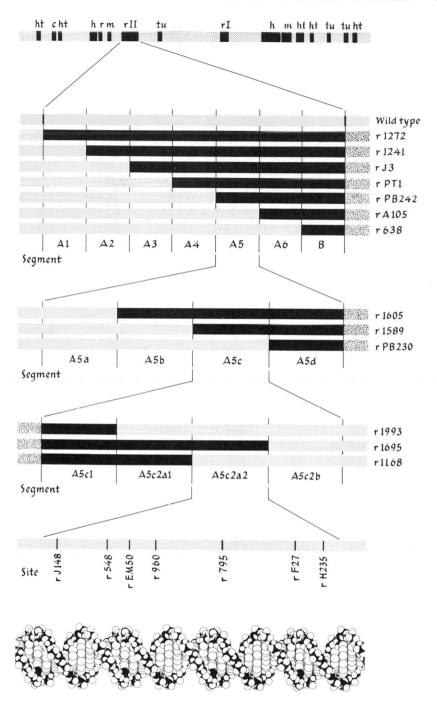

ht c ht   h r m rII   tu   rI   h m ht ht tu tu ht

Wild type
r 1272
r 1241
r J3
r PT1
r PB242
r A105
r 638

A1   A2   A3   A4   A5   A6   B
Segment

r 1605
r 1589
r PB230

A5a   A5b   A5c   A5d
Segment

r 1993
r 1695
r 1168

A5c1   A5c2a1   A5c2a2   A5c2b
Segment

Site   r J148   r 548   r EM50   r 960   r 795   r F27   r H235 growth, although the frequency with which they do so varies within very wide limits. Some of the *r*II mutants isolated by Benzer do not, however, behave as point mutants: they fail to yield r^+ recombinants in crosses with each of two or more *r*II mutants previously identified as nonallelic point mutants; moreover, they sport no r^+ revertants. This second class of mutants, therefore, appears to have arisen by the simultaneous mutational alteration of several genetic sites. The extent of the mutation of such a long-span mutant can be ascertained by crossing it to the set of previously mapped point mutants in the *r*II region and observing which point mutants do not yield r^+ recombinants in the cross—that is, are "covered" by the mutation. The genetic sites occupied by some of Benzer's long-span mutants are shown in Figure 13-2. It can be seen that these mutations vary considerably in length, some cover only a small sector of the map, whereas others extend over the entire *r*II region and are incapable of yielding r^+ recombinants in crosses to any other *r*II mutant. M. Nomura and Benzer were eventually able to show that these long-span mutations are, in fact, *deletions* of parts of the phage genome. Since some of these deletions are seen to cover more that 1 map unit of the *r*II region, they can be estimated to entail the loss of more than $2 \times 10^5/1500$, or about 100 nucleotide base pairs.

Deletion mutants were put to good use by Benzer for establishing the detailed map of the *r*II region in a way that made it unnecessary to cross every mutant of the collection to very other mutant; without this trick, the millions of phage crosses otherwise required would have rendered the construction of this map a prohibitively laborious effort. Benzer divided the *r*II region of the T4 phage genome into segments (in the manner shown in Figure 13-3), each segment being defined by the length of the map covered by one particular deletion but not by another. It was then a simple task to make a preliminary placement of each *r*II mutant to be mapped into its appropriate segment by establishing the deletions with which the mutant does and does not produce wild-type recombinants in crosses. By use of deletion mutants having suitable starting and ending points, Benzer proceeded to assign each of his *r*II mutants to one of 47 segments of the *r*II region, as shown in Figure 13-4. For this purpose, an unknown *r*II mutant was first crossed to seven standard deletion mutants that defined the seven main segments; once its main segment was known, the mutant was crossed to an appropriate secondary set of reference deletions. Thus, in only two steps, any point mutation could be placed into one of the 47 segments. It is fair to say that without this astute exploitation of deletion mutants for rapid mapping, our knowledge of the genetic fine structure of the phage genome would still be very rudimentary; progress would have been hamstrung by the geometric increase in the number of crosses required for the mapping of an arithmetically increasing number of mutants available for study.

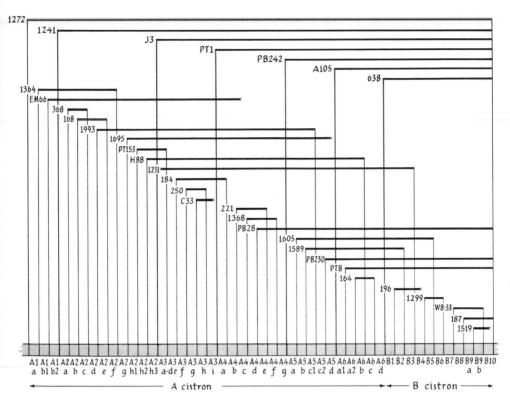

FIGURE 13-4. Map of *r*II deletions that serve to divide the *r*II region into 47 ordered segments. (Some ends are not used to define a segment; they are shown as fluted.) [After S. Benzer, *Proc. Natl. Acad. Sci. Wash.* **47**, 403 (1961).]

THE CISTRON

Benzer's second, and probably most important, use of the *r*II mutants provided an experimental definition of the gene. During the growth of classical genetics, the word "gene" had gradually come to represent the unit of genetic material that is passed on from parent to offspring and can be recognized operationally through its ability to *mutate* to alternative states, to *recombine* with other, similar units, and to *function* in the endowment of the organism with some particular phenotype. From the "classical" point of view these three aspects—mutation, recombination and function—were thus the attribute of one and the same hereditary unit: the gene. But as the preceding discussions must have made clear already, there can be no such thing as the "classical gene," since the unit of mutation and recombination is evidently the individual nucleotide base pair of the DNA, whereas the unit of function is that sequence of hundreds or thousands of nucleotides that specify the sequence of the twenty standard amino acids that make up the primary structure of the protein coded by the gene.

Benzer next addressed himself to the question whether the phenotype of the rII mutants of his collection is attributable to genetic lesions in more than a single functional unit. The mere fact that two r mutants manifest the same phenotype under some set of experimental conditions does not automatically guarantee that their mutational alterations pertain to the same functional unit. For instance, we have already noted that the r plaque type on the ordinary E. coli strains is produced by mutations at such widely separated sites of the T4 genome that they could hardly belong to one and the same functional unit. And thus the failure of different rII mutants to grow on the nonpermissive K strain also need not necessarily reflect one and the same functional defect in their hereditary material. In order to examine whether two different rII mutants belong to the same functional unit, Benzer adapted to phages the "cis-trans" or complementation test (Figure 13-5), which had been developed previously with higher organisms for the very purpose of probing the nature of the functional unit of the gene. Benzer's complementation test is based upon the finding that in a strain K bacterium infected jointly with rII mutant and r^+ wild-type phages, both types can grow normally. The normal rII^+ gene of the wild-type parent is, therefore, able to supply the function necessary for growth in the strain K bacterium, not only for itself but also for the defective rII mutant. Or, in genetic parlance, in mixed infection the r^+ wild-type gene is dominant over its rII mutant allele. In the complementation test a strain K bacterium is infected with a pair of rII mutants in order to examine whether the two mutants, each unable to grow alone in a strain K bacterium, are able to cooperate in the mixedly infected cell to produce a litter of infective progeny. If the mutant pair is able to cooperate in this way, then it may be concluded that the two different mutations have not affected the same functional unit of the phage genome. The failure of one mutant to grow in a strain K bacterium (its rII phenotype) must mean that it cannot carry out one particular function, or elicit the synthesis of one particular protein necessary for growth in the infected strain K cell, whereas the rII phenotype of the other mutant must mean that it cannot carry out some other necessary function pertaining to a different functional unit of the genome. The two mutants are able to reproduce in joint growth because each supplies that necessary function in which the other is deficient. (It goes without saying that either one of these putative functions of the rII region can be necessary only for growth in strain K cells, and not in cells of the ordinary E. coli strain, in which rII mutants can grow perfectly well.) But if the two mutants are unable to cooperate, then it can be concluded that both mutations have occurred at genetic sites belonging to the same functional unit of the phage genome. In that case the rII phenotype of both

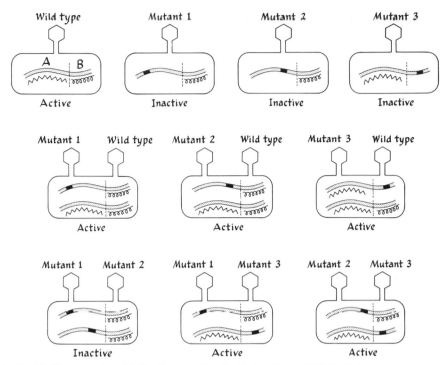

FIGURE 13-5.   The cis-trans or complementation test of rII mutants in *E. coli*, strain K.
Top: A and B genes are intact in the wild-type r^+ phage genome, so that both A-gene
and B-gene polypeptides (shown here as sawtooth and spiral lines, respectively) are
formed. Mutants 1 and 2 bear defects at two different sites of their A gene and can
form only the B-gene polypeptide; mutant 3 bears a defect in its B gene and can form
only the A-gene polypeptide. Since the presence of both polypeptides is required for
growth in bacteria of strain K, all three mutants are inactive on this strain. Middle: In
mixed infection with any of the three defective mutants, the wild type can supply both
A-gene and B-gene polypeptides, and hence growth proceeds. Bottom: Since neither
mutant 1 nor mutant 2 can form the A-gene polypeptide, no growth proceeds in their
mixed infection. In mixed infection of mutant 1 or mutant 2 with mutant 3, however,
the former supply the B-gene polypeptide, and the latter supplies the A-gene polypeptide;
hence growth proceeds. [After "The Fine Structure of the Gene," by S. Benzer, *Scientific
American*, January 1962. Copyright © 1962 by Scientific American, Inc. All rights
reserved.]

mutants is caused by a deficiency in exactly the same function, which neither
mutant can supply to the other in joint growth.

On the basis of this test, Benzer found that rII point mutants fall into
two functional groups, A and B; all mutants belonging to one group complement any member of the other group in the production of infective progeny
in joint growth in strain K, but do not complement in this way any member
of their own group. It turned out, furthermore, that the two groups A and

B could be assigned definite positions on the genetic fine-structure map of the T4 phage: as Figure 13-2 shows, all rII mutants whose mutational sites are located to the left of the vertical line belong to **group A**, and all those located to the right belong to group B (middle level). Similar tests carried out with deletion mutants show that these obey the same rules. Mutants carrying deletions that map entirely within one group do not complement any mutants of that group but do complement all mutants of the other group; and deletion mutants whose deletion extends across the boundary separating the two groups do not, save for an interesting exception to be considered in Chapter 18, complement any mutants of either group. The two mutant groups thus signal the existence of two functional units within the rII region of the phage genome; each functional unit presumably governs the synthesis of a specific polypeptide necessary for growth in strain K bacteria. Benzer referred to the genetic unit of function revealed by grace of the cis-trans test as the *cistron*—a term whose invention in 1955 marked an important advance in the dialectics of heredity. In the first few years after its invention, the term cistron seemed to be well on its way toward replacing the old-fashioned and, by then obviously ambiguous term "gene" in molecular genetic parlance. But the use of "gene" seems to have prevailed, and it is "cistron" that is slowly falling into disuse. In any case, the terms gene and cistron are now fully equivalent in referring exclusively to the genetic unit of *function*. They are employed in this sense in the remainder of this text.

The approximate length of the rIIA and rIIB cistrons, or genes, can be inferred from the recombination frequencies of mutant sites identified at their extreme ends; these frequencies indicate that the distance between the two most distant loci of gene A is about 6 map units, and of gene B about 4 map units. Using the assumptions concerning the total extent of the T4 genetic map and the DNA content of the phage previously employed for the estimate of the minimum unit of genetic recombination, one reckons that gene A corresponds to a sequence of $(6/1500) \times 2 \times 10^5$, or about 800, and gene B to $(4/1500) \times 2 \times 10^5$, or about 500, nucleotide base pairs of the phage DNA double helix, which is in good accord with the rough *a priori* estimate of the length of the genetic unit of function inferred from first principles in Chapter 8.

Use of the cis-trans complementation test made it possible for Edgar and Epstein to assign their large ensemble of conditionally lethal *ts* and *am* mutants, which had filled in the T-even phage genetic map, to individual genes. In order to test whether two different *ts* mutants pertain to the same unit of function, *E. coli* were infected jointly with the two mutants and maintained at the restrictive temperature of 42°C. If the two mutants complemented each other to produce infective progeny phages, they were assigned to different genes, and if they did not complement they were assigned to the same gene. Similarly, nonpermissive *E. coli* strains were infected jointly with two different *am* mutants, and the mutants assigned to different genes or to the same gene, according to whether they do or do not complement each other in the production of infective progeny. Finally, the outcome of mixed infection of nonpermissive *E. coli* with a *ts* and an *am* mutant at the restrictive temperature of 42°C allowed determination of the mutual complementation capacity, and hence gene assignment, of these two mutant types. In this way, some 60 different T even phage genes were identified by 1965; their disposition over the circular genetic map is shown in Figure 13-6. One striking feature of this functional genetic map is that genes which pertain to the related phage functions appear to lie in the same general area of the map. Furthermore, the disposition of these genes appears to be correlated with the *time* of intracellular phage growth at which the gene product makes its appearance or starts to function. Thus the genes that govern early functions, such as most (though not all) of the genes which control enzymes that pertain to the replication of the phage DNA, map in the sector of the circle between 8 o'clock and noon, whereas all the other genes, which govern such late functions as cell lysis or structural proteins of the phage particles, map in the remainder of the circle.

This extension of Benzer's complementation analysis to other parts of the T-even phage genome showed, however, that the results are not always as clear as they had been in the original definition of the A and B genes of the *r*II region. Although the gene assignments of all *am* mutants were unambiguous, Edgar and Epstein encountered some *ts* mutants that, on the basis of their position on the genetic map and the results of functional cis-trans tests with closely linked *am* mutants, clearly belonged to the *same* gene yet gave good complementation in the cis-trans test with each other. Similar examples of *intragenic complementation* were soon encountered in attempts to define units of genetic function in the genome of bacteria and other organisms.

The existence of intragenic complementation does not really detract from the basic validity of Benzer's definition of the gene and can be readily accounted for in terms of quaternary protein structure. As was set forth in Chapter 4, many proteins carry out their biological function, not as a single polypeptide chain, but as part of a quaternary aggregate of two or more polypeptide chains. For example, it was mentioned that the *E. coli* β-galactosidase is an aggregate of four identical polypeptide chains. Let us now consider a *ts* mutation in a gene specifying a protein that owes its catalytic activity to a quaternary complex of four identical polypeptide chains. Here the *ts* mutant phenotype evidently devolves from the introduction of a noxious amino acid into some site of the mutant polypeptide chain. Consequently, the temperature range in which the four-chain quaternary aggregate can take on its physiologically active quaternary structure has been narrowed. That is to say, although the quaternary aggregate of mutant polypeptide chains retains activity at the permissive temperature of 25°C, it takes on a denatured form at 42°C. Suppose now that *two* copies of the gene specifying that protein, each carrying a different *ts* mutation, are present within the same cell, as they might be in a cis-trans complementation test. Here there would arise a *hybrid* quaternary aggregate of the mutant protein in which

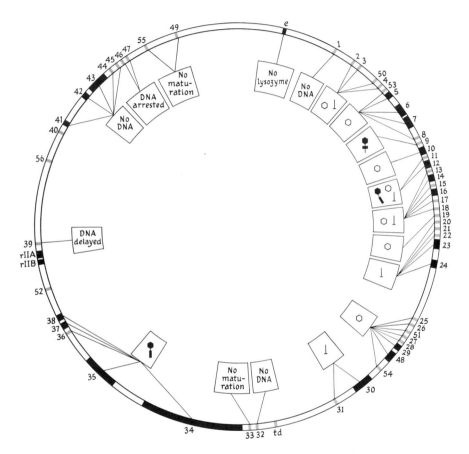

FIGURE 13-6. The gene map of T4 phage. The relative positions of the 60 genes identified by 1965 and the major physiological properties of mutants defective in various genes are shown. Each gene was defined by the existence of at least one *am* mutant capable of giving complementation in the cis-trans test with *am* mutants belonging to every other previously identified gene. The minimum length is shown for some genes (black segments), based on recombination frequencies of mutant sites belonging to the same gene. The long length shown for gene 34 is now known to derive from some as yet not fully understood high probability of genetic exchange in that region, and therefore represents an overestimate of the size of that gene. The boxes indicate the deficiencies in synthesis associated with mutations in some genes, or, in other genes, the phage components that are present in defective lysates of corresponding mutants. [From "The Genetics of a Bacterial Virus," by R. S. Edgar and R. H. Epstein, *Scientific American*, February 1965. Copyright © 1965 by Scientific American, Inc. All rights reserved.]

some of the four polypeptide chains are contributed by one and some by the other *ts* mutant gene. And now there exists the possibility that the temperature range over which the hybrid quaternary mutant protein aggregate takes on a physiologically active structure is wider than the active temperature range of either of the two pure quaternary aggregates made up exclusively of one or the other of the *ts* mutant polypeptide chains. That is to say, the two different noxious amino acid replacements in primary protein structure brought about by the two *ts* mutations may exert a reciprocally compensatory effect, so that the hybrid mutant quaternary aggregate has the broad temperature-stability profile characteristic of the wild-type protein.

Intragenic complementation is encountered also in ordinary *loss-of-function* mutations (see Chapter 6). In such mutations two mutant versions of the same gene, neither capable of forming a polypeptide chain that is functional at *any* temperature, complement to yield a functional hybrid quaternary protein structure. The fact that Benzer did not encounter any intragenic complementation among his large ensemble of *r*II phage mutants suggests (though does not rigorously prove, of course) that the protein products of both *r*IIA and *r*IIB genes play their physiological roles in quaternary structures that contain only a *single* polypeptide chain of either kind. But the failure of *am* mutants to manifest any intragenic complementation, and hence the possibility of unambiguously assigning such mutations to particular genes on the basis of the cis-trans test, has an entirely different explanation, whose consideration must await our later discussions of Chapter 18.

MOLECULAR MECHANISM OF MUTATION

We are now ready to take up a question that was raised in Chapter 7 but deferred until now. That question concerned the mechanism by which those changes that represent the molecular basis of mutation arise in the DNA nucleotide sequence. For fine-structure studies on the manner of appearance of T-even phage mutants were to provide some of the crucial insights into the mutation process. Phages offer one important advantage over bacteria for this purpose: it is possible to study mutation of the phage DNA both in its resting, or extracellular, state in the infective phage particle, as well as in its replicating, or intracellular, vegetative state. The earliest studies of Hershey and Luria had already shown that the rate of spontaneous mutation of the resting phage DNA is very low, so low that for many years it was believed (incorrectly as it turned out) that the extracellular phage particles do not mutate at all over periods of months or years. Hence it is mainly during the vegetative multiplication of the phage in the host cell that new mutants appear on the scene. If, for instance, a culture of *E. coli* is infected with an inoculum of 10^6 T2r^+ particles per milliliter and phage multiplication is allowed to proceed through several growth cycles, until all the bacteria in the culture have been lysed and the phage titer has risen to a final level of 10^{10} particles per milliliter, then it is found that the proportion of r mutants among the whole phage population rises with each growth cycle from an initial level of about 10^{-4} r mutants per r^+ wild type to a final level of about 10^{-3} r mutants per r^+ wild type. It can be inferred, therefore, that phage mutants arise as *copy errors* during the intracellular replication of the genome. That is, replication of the DNA of the parent phage is necessarily of very high fidelity; occasionally an error does occur in the replication process, however, which engenders a change in nucleotide sequence, or mutation, in one of the vegetative replicas. The mutant replica of the parental DNA later matures into an intact, infective progeny particle, which, in its turn, infects another bacterial cell. In the course of this next growth cycle, it is the mutant information that is now replicated faithfully to produce a litter of mutant phage progeny. Since the vegetative replication of phage DNA proceeds by the semiconservative template mechanism of Watson and Crick, it is possible to consider the multiplication of the phage genome as a binary fission process, and hence for purposes of statistical analysis entirely analogous to the multiplication of the bacterial genome. Hence the expression previously derived in Chapter 6 for bacterial mutation, relating the number of mutant individuals n per total number of descendants N from a single parent after growth for g generations to the mutation rate a,

$$n/N = ga \qquad (6\text{-}2)$$

is also applicable to mutation of the vegetative phage. Thus in the $r^+ \rightarrow r$ mutation considered here, an initial number N_0 of T2r^+ phages gave rise to a final yield of N progeny after each had undergone $g = \ln (N/N_0)/\ln 2$ fissions. Hence according to equation (6-2) there should have arisen a total of $Na \ln (N/N_0)/\ln 2$ new mutant progeny, while multiplication of the initially present n_0 mutant parents produced $n_0 N/N_0$ mutant progeny. The final number of phage mutants n, is therefore

$$n = Na \ln (N/N_0)/\ln 2 + n_0 N/N_0$$

or

$$a = (n/N - n_0/N_0) \, [\ln 2/\ln(N/N_0)] \qquad (13\text{-}1)$$

Thus we may reckon the rate of the mutation as

$$a = (0.001 - 0.0001) \times \frac{\ln 2}{\ln(10^{10}/10^6)} = 7 \times 10^{-5}$$

per phage per duplication. Similar measurements of the rate of the T2$h^+ \rightarrow$ T2h

host-range mutation revealed that this mutation occurs at the rate of only 10^{-8} mutations per phage per duplication. Thus the $r^+ \to r$ mutation rate of the phage turns out to be more than 1000 times higher than that of the $h^+ \to h$ mutation. The reason for the relatively high $r^+ \to r$ mutation rate appears to be that the rapid lysis mutation represents a *loss* of function, which can occur in any one of several phage genes. Hence change of any one of very many nucleotide bases of the phage DNA can lead to the r mutant type. The low $h^+ \to h$ mutation rate, on the other hand, is attributable to the fact that the host-range mutation is a *change* of function. Hence change of only a very limited number of nucleotide bases of the phage DNA can be expected to preserve the indispensable function of the phage attachment organs while changing their structure to match the changed T2 receptor spots of the Tto^r *E. coli* mutant cell.

How such rare copy errors might arise during the replication of the DNA was given a molecular explanation by Watson and Crick in their 1953 paper. It will be recalled that under the Watson and Crick replication scheme the two complementary polynucleotide strands of the parental DNA molecules separate. Each parental strand then serves as the template for the ordered synthesis of a new complementary polynucleotide chain through the formation of specific hydrogen bonds between purine and pyrimidine bases of the parental polynucleotide and the mononucleotide building blocks available for polymerization in the intracellular metabolic pools. It is to be noted that the formation of the specific Watson-Crick hydrogen bonds between adenine and thymine and between guanine and cytosine (or HMC) requires that these bases be in the particular *tautomeric* forms shown in Figure 8-6. But, in fact, in an aqueous medium each base exists in a state of equilibrium between various tautomeric forms, of which those shown in Figure 8-6, are merely the most probable. In some other, less probable form of each base, the shift of a crucial hydrogen atom makes possible the formation of specific hydrogen bonds to a base different from that of the normal complementary base-pairing scheme. For instance, the rare *imino* form of adenine and the rare *enol* form of thymine shown in Figure 13-7 allow "illegitimate" complementary bonding of these bases to cytosine and guanine, instead of the "legitimate" base-pairing to thymine and adenine, respectively. Watson and Crick thought it possible, therefore, that mutations derive from the rare circumstance that a purine or pyrimidine base attached either to the parental polynucleotide chain or to the free nucleotide building block to be incorporated into the replica chain happens to be in its rare tautomeric form at the very moment that it participates in the replication act. In this way, an "illegitimate" base-pairing would result in the introduction of an incorrect nucleotide into the growing replica polynucleotide; there would thus be a permanent change in nucleotide sequence, or a mutation. This mutation could be reversed only by another infrequent accident in which, just when at some future time the mutated site is about to act as a template, one of the two bases involved in the replication act happens to be in that rare tautomeric form which restores the original nucleotide to its rightful position in the polynucleotide chain by another "illegitimate" base-pairing. Although, as we shall soon see, this proposal of Watson and Crick describes the manner in which only some of the spontaneous mutants arise, it nevertheless provided the first molecular basis for understanding the mutation process.

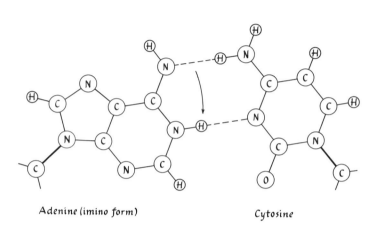

Adenine (imino form)   Cytosine

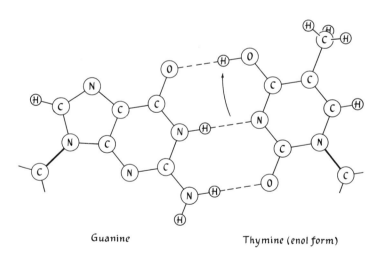

Guanine   Thymine (enol form)

FIGURE 13-7.   "Illegitimate" base pairing of the rare *imino* form of adenine with cytosine, and of the rare *enol* form of thymine with guanine. [From *Molecular Biology of Bacterial Viruses*, by G. S. Stent. W. H. Freeman and Company. Copyright © 1963.]

MUTAGENS

Since the discovery in 1927 by H. J. Muller and by L. J. Stadler that X-rays and ultraviolet light can induce mutations, or raise the frequency of mutants above the spontaneous background, many physical agents and chemical substances have been found to act as mutagens on the genetic material of living cells. The study of mutagenesis in phages however, was rather slow in getting under way, and it was not until 1955—when Rose Litman and A. B. Pardee discovered that addition of the heavy thymine analogue 5-bromouracil (Figure 9-9) to the culture medium of T2-infected bacteria results in the appearance of a very greatly increased proportion of mutants among the phage progeny—that experimental studies began on the molecular basis of the mutation process through the use of mutagens.

Under the conditions in which it acts as a mutagen, bromouracil is incorporated into the phage DNA in place of thymine. The introduction of bromouracil residues into the DNA polynucleotides in place of thymine cannot, however, in itself constitute a mutation, since upon removal of the mutagen from the growth medium (as upon assaying the bromouracil-labeled phage stock on ordinary nutrient agar plates), bromouracil nucleotides in the parental phage DNA are in turn replaced by thymine nucleotides in the replica chains. Hence the replacement of thymine by bromouracil entails no permanent change in nucleotide base sequence, or informational content, of the phage DNA. Rather, the mutagenic action of bromouracil must derive from its propensity to raise the probability of copy errors that result in permanent replacement of one purine and pyrimidine base pair by another. How does bromouracil raise this probability? The most frequent tautomer of bromouracil, like that of thymine, is its *keto* form, so that bromouracil generally forms its specific base-pairing hydrogen bonds with adenine. But the presence of the bromine atom at the 5-position of the pyrimidine ring causes bromouracil to be in the rarer *enol* form, which is suitable for an "illegitimate" base-pairing with guanine, a much greater fraction of the time than is thymine (Figure 13-7). Thus, bromouracil should be able to increase the frequency of copy errors in two different ways; (a) by promoting *template transitions*, in which residues of bromouracil already incorporated into DNA (in place of thymine) make more probable incorrect insertions of guanine (in place of adenine) into the replica chain growing on the bromouracil-labeled parental template strand, and (b) by promoting *substrate transitions*, in which bromouracil residues attached to free nucleotide building blocks make more probable erroneous insertions of bromouracil (and, hence, ultimately of thymine) in place of cytosine into *de novo* polynucleotides. The mutational event corresponds in the first case to the replacement of an adenine-thymine base pair by guanine-cytosine, and in the second case to the

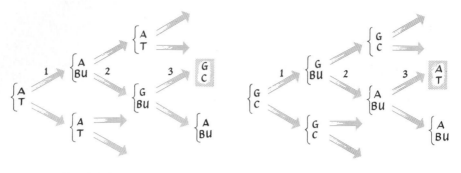

Template transition   Substrate transition

FIGURE 13-8.   Two mechanisms of bromouracil-induced mutagenesis in replicating DNA. *Template transition*. (1) Bromouracil (BU) in its *keto* form is incorporated into a growing replica strand in place of T by pairing with A of the template strand. (2) At some subsequent replication cycle, the BU happens to be in its rare *enol* form and directs the illegitimate incorporation of G into the growing replica strand. (3) At the next replication cycle, the falsely incorporated G directs the incorporation of C into the growing replica strand, leading to the permanent replacement of the original A-T pair by a G-C pair at the mutant site. Here, the actual mutagenic step 2 *can* proceed in the absence of BU from the growth medium. *Substrate transition*. (1) In its rare *enol* form, BU is incorporated into a growing replica strand in place of C, by illegitimate pairing with G in the template strand. (2) At the next replication cycle, the falsely incorporated BU, now once more in its *keto* form, directs the incorporation of A into the growing replica strand. (3) Sometime later, after removal of BU from the growth medium, BU is replaced by its natural analog T, leading to the replacement of the original G-C base pair by an A-T pair at the mutant site. Here, the actual mutagenic step 1 *cannot* proceed in the absence of BU from the growth medium. [From *Molecular Biology of Bacterial Viruses*, by G. S. Stent. W. H. Freeman and Company. Copyright © 1963.]

replacement of a guanine-cytosine base pair by adenine-thymine (Figure 13-8) It is possible to show experimentally that bromouracil does exert its mutagenic action on the T-even phage genome in these two different ways; one class of mutations, the template transitions, is found to occur upon replication of bromouracil-substituted phage DNA in the absence of any bromouracil in the culture medium, whereas another class of mutations, the substrate transitions, occurs only while bromouracil is present in the culture medium.

A *purine base analogue* that came into use in T-even phage mutagenesis soon after bromouracil is 2-aminopurine, which, as is shown in Figure 13-9, can form a pair of hydrogen bonds with thymine and a single hydrogen bond with cytosine. The mutagenic action of 2-aminopurine appears to derive from its propensity to be incorporated into the phage DNA in place of adenine, and more rarely in place of guanine, and thus for promoting both template and substrate transitions.

Some years before the discovery of the mutagenic action of these nucleotide base analogues, it had been found that temporary exposure of T-even

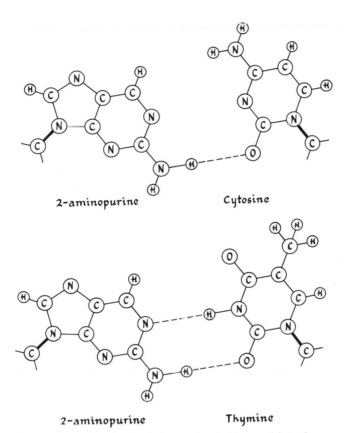

2-aminopurine   Cytosine

2-aminopurine   Thymine

FIGURE 13-9.   The double-hydrogen-bonded base pairing of
2-aminopurine with thymine and its single-hydrogen-bonded base
pairing with cytosine.

infected *E. coli* to acridine dyes such as proflavin and acridine orange (Figure
13-10) causes the appearance of mutants in greatly increased frequency
among the phage progeny. For many years, the nature of the mutagenic
action of acridines remained rather obscure, except that proflavin and acridine orange were found to *intercalate* between the purine and pyrimidine
base-pairs of the DNA double helix (Figure 13-10). How this distortion of the
DNA molecule results in changes in nucleotide base sequence will be considered later in this chapter.

The first chemicals found to be highly mutagenic for extracellular rather
than vegetative T-even phages were the *alkylating agents* ethyl methane
sulfonate, or EMS, and ethyl ethane sulfonate, or EES.

$$CH_3SO_3CH_2CH_3$$

Ethyl methane sulfonate
(EMS)

$$CH_3CH_2SO_3CH_2CH_3$$

Ethyl ethane sulfonate
(EES)

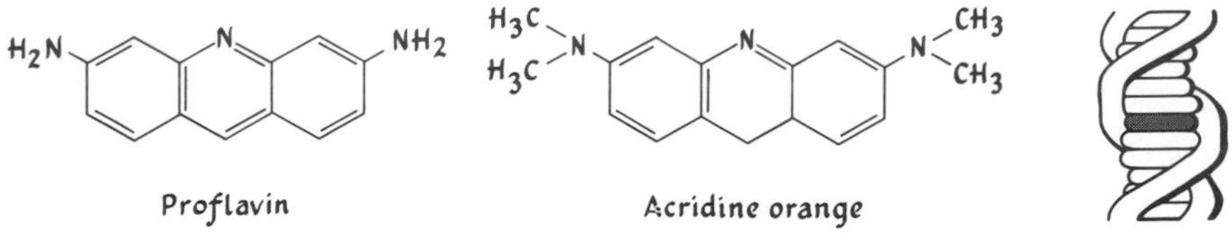

Proflavin   Acridine orange

FIGURE 13-10. The chemical structure of two acridine dye mutagens and the manner in which they intercalate in the purine pyrimidine base stack and thus distort the structure of the DNA double helix. [Intercalation diagram after L. S. Lerman, *Proc. Natl. Acad. Sci. Wash.* **49**, 94 (1963).]

These agents appear to react with the phage DNA by ethylating the 7-position of the purine ring of guanine or adenine. This reaction is then followed by hydrolysis of the purine-deoxyribose bond and thus by an eventual loss of the whole purine base from the polynucleotide chain. Upon replication of the viral DNA containing such an ethylation-induced gap in one of its strands at a site at which there had previously resided one of the purines, either the correct complementary pyrimidine or an incorrect purine or pyrimidine may be inserted at the homologous site in the growing complementary *de novo* strand. If the correct pyrimidine happens to be inserted, the original genetic information has been restored; but if an incorrect base happens to be inserted, a permanent change in base sequence results—that is, a mutation occurs.

Another highly effective *in-vitro* mutagen for bacteriophages is nitrous acid, HNO_2. Incubation of suspensions of T-even phages at 25°C with 0.05 M nitrous acid at pH 4.0 inactivates the phage population with a half-life of 10 minutes; among the survivors, an ever-increasing proportion of diverse mutants is found. The kinetics of appearance of these mutants suggest that they result from the interaction of a single molecule of nitrous acid with the phage DNA. Direct chemical study of the reaction of nitrous acid with model nucleotide substances and intact nucleic acid molecules shows that the most probable reactions are the conversions by oxidative deamination of cytosine to uracil (or, in T-even phage DNA, of HMC to 5-hydroxymethyluracil), of adenine to hypoxanthine, and of guanine to xanthine (Figure 13-11). It is obvious that the conversion of HMC to hydroxymethyluracil at any site of the parental phage DNA would almost certainly entrain the incorporation of adenine, instead of the rightful guanine, into the first replica polynucleotide chain at the site complementary to that which had experienced the *in vitro* conversion, since the hydrogen-bonding facilities of hydroxymethyluracil are identical to those of thymine. In the next replication cycle of the phage DNA, the adenine at the mutated site would attract a thymine into its complementary polynucleotide, and the replacement of the HMC guanine base pair by a thymine-adenine base pair would be the ultimate issue of this *in vitro* deamination of HMC in the viral DNA. It is more difficult to predict the behavior in further replication cycles of the hypoxanthine residues produced

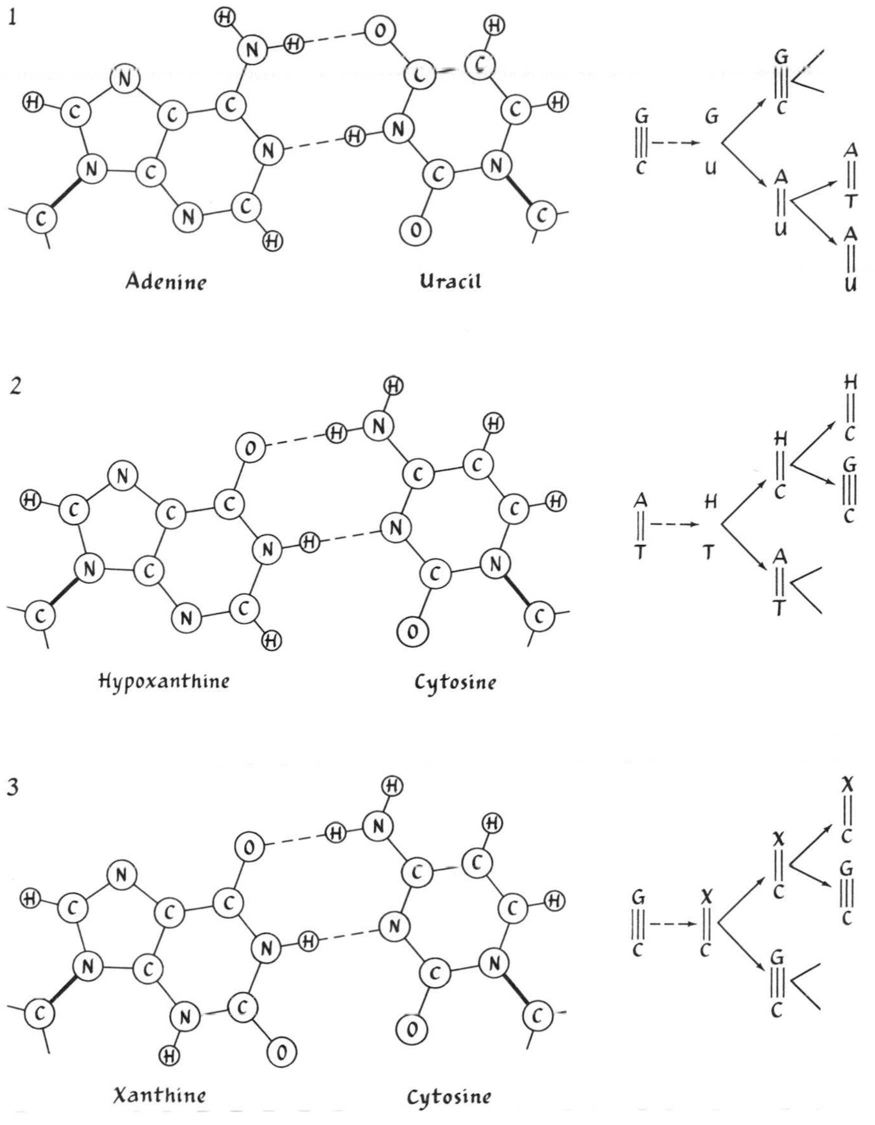

FIGURE 13-11. The chemical basis of the mutagenicity of nitrous acid. (1) The 6-amino group of cytosine in a G-C base pair is deaminated to yield uracil; upon replication of the HNO₂-treated DNA, uracil pairs with adenine, ultimately leading to the permanent replacement of the G-C pair by an A-T pair. (2) The 6-amino group of adenine in an A-T base pair is deaminated to yield hypoxanthine; upon replication of the HNO₂-treated DNA, hypoxanthine pairs with cytosine, ultimately leading to the permanent replacement of the A-T pair by a G-C pair. (3) The 2-amino group of guanine in a G-C base pair is deaminated to yield xanthine; upon replication of the HNO₂-treated DNA, xanthine still pairs with cytosine, thus entailing no permanent change in base pair configuration. [From E. Freese, *In Structure and Function of Genetic Elements*, Brookhaven Symposia in Biology, no. 12, p. 63 (1959).]

by the reaction of nitrous acid with adenine and guanine, since the hydrogen bonding facilities of hypoxanthine and xanthine are unlike those of either adenine or guanine. But if hypoxanthine and xanthine pair with cytosine (or HMC) at the time of DNA replication, as they well might, then the deamination of adenine, in analogy with the deamination of cytosine, would ultimately result in the replacement of the adenine-thymine base-pair by a guanine-HMC base-pair at the mutated genetic site, whereas the deamination of guanine would not be mutagenic at all.

From the chemical point of view one of the most specific *in vitro* mutagens available appears to be hydroxylamine, NH_2OH. Hydroxylamine reacts preferentially with cytosine (or HMC), and causes its ultimate replacement by thymine. Hence T-even phage mutants induced by hydroxylamine treatment can be supposed to derive from the replacement of an HMC-guanine base pair by a thymine-adenine pair.

REVERSE MUTATION

It was for the experimental clarification of the molecular mutation process that Benzer made a third important use of his rII mutants. For Benzer realized that the study of the rII → r^+ *reverse mutation* of an ensemble of rII mutants of diverse origins ought to provide a deep insight into the nature of the events that produce the r^+ → rII *forward mutations*. Benzer and his colleague E. Freese had isolated hundreds of rII mutants of T4 phage, some of which had arisen spontaneously from the T4rII$^+$ wild type and others of which had been induced by one or the other of the mutagens described in the preceding pages. Each rII mutant was then tested for the rate at which it mutates back, or reverts, to the rII$^+$ wild type, either spontaneously or during growth in the presence of a nucleotide base analogue or acridine dye mutagens. For this purpose, a lysate of an appropriately treated bacterial culture infected with the rII phage mutant was plated on agar seeded with strain K indicator bacteria on which only the rII$^+$ revertants can grow. These studies resulted in the following findings. First, the ensemble of *spontaneous* rII mutants showed an enormous range with respect to the rate at which individual mutants back mutate spontaneously to the rII$^+$ state, some reverting as frequently as 10^{-3} per phage per duplication and others reverting as rarely as 10^{-9} per phage per duplication, intermediate spontaneous reversion frequencies being also represented as a nearly continuous spectrum. About 10% of the spontaneous mutants, furthermore, did not revert *at all* at any detectable frequency. This finding implied that different spontaneous rII mutants must owe their mutant character to basically different alterations at the nucleotide level, whose restoration to the original wild type configuration likewise requires very different molecular events. Second, the rII mutants *induced* by phage growth in the presence of the nucleotide base analogues or by treatment of the extracellular phages with the alkylating agents, nitrous acid or hydroxylamine, were found to be much more homogeneous than the spontaneous mutants with respect to their rate of spontaneous reverse mutation, most of them reverting at rates of about 10^{-9} per duplication. Third, and most important, it was found that there is a strong relation between the origin of a particular *r*II mutant and the ability of any mutagen to induce its reversion. This is evident from the data presented in Table 13-1. Each of

TABLE 13-1.   Induced Reverse Mutation of a Set
of Spontaneously Reverting T4*r*II Mutants

Mutagen used for induction of $r^+ \to r$II (forward mutation)	No. of *r*II mutants tested	Percentage of *r*II mutants found inducible to revert to r^+ by base analogue mutagens
Aminopurine	98	98
Bromouracil	64	95
Hydroxylamine	36	94
Nitrous acid	47	87
Ethyl ethane sulfonate	47	70
Proflavine	55	2
Spontaneous	110	14

From E. Freese, 5th International Congress of Biochemistry, Moscow 1961.

the mutants studied there represents an alteration of a different site of the phage DNA (since it can be located at a different site on the fine structure map in Figure 13-2) and reverts spontaneously to the *r*II$^+$ state. For some of these mutants the reversion rate during growth in the presence of the nucleotide base analogue mutagens was essentially the same as the spontaneous reversion rate: the reverse mutation of such mutants evidently cannot be induced by these mutagens. For other mutants, the reversion rate was tens, hundreds, or thousands of times greater in the presence of the mutagen than in its absence; such mutants are considered to be "inducible to revert." The data of Table 13-1 show that practically all the *r*II mutants originally induced from the T4*r*$^+$ wild type by bromouracil, 2-aminopurine, hydroxylamine, and nitrous acid, and most of the EES-induced mutants, can be induced to revert by base analogue mutagens; but practically none of the proflavine-induced mutants and only about 10% of the spontaneous mutants respond to the base analogue mutagens in regard to their reversion rate to the r^+ state.

This dichotomy of mutant types suggested to Freese that there must exist two basically different sorts of chemical alterations of the viral DNA responsible for revertible rII mutations: mutations of the first kind that arise by the action of base analogue mutagens (and of EES, nitrous acid, and hydroxylamine) and that are revertible by base analogue mutagens, and mutations of the second kind that arise by the action of proflavin (and rarely, of EES), and that are not revertible by base analogue mutagens. Only about 10% of spontaneous rII mutants carry mutations of the first kind, whereas the remainder of spontaneous rII mutants carry mutations of the second kind.

From what has already been said about the way in which bromouracil, 2-aminopurine, nitrous acid, and hydroxylamine are likely to exert their mutagenic effect, it seems most plausible that mutations of the first kind correspond to an alteration that results in the replacement of one pyrimidine by another pyrimidine, or of one purine by another purine, at some site of the DNA polynucleotide chain. Such changes are now called *transitions*. They can be represented by the schema

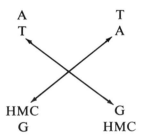

It is easy to understand why base analogue mutagens induce both forward and reverse mutations of the first kind. Bromouracil, for instance, once incorporated into the viral polynucleotide chains in place of thymine, greatly raises the probability of the template transition that incorrectly attracts guanine instead of adenine into the replica chain by illegitimate base-pairing at the time of replication. But bromouracil can also induce the reverse mutation that once more replaces the mutant guanine-HMC pair by the original adenine-thymine pair, by greatly raising the probability of the substrate transition of incorrectly incorporating bromouracil (and hence, ultimately thymine) into the growing replica chain by illegitimate base-pairing with the guanine at the mutant site. The original Watson-Crick proposal for the occurrence of spontaneous mutations by tautomerism of purine and pyrimidine bases thus probably describes correctly the origin of that 10%

of spontaneous mutants that carry mutations of the first kind, and also accounts for the spontaneous reversion to the r^+ wild type of rII mutations of the first kind (albeit at a rate considerably lower than in the presence of base analogue mutagens).

The molecular explanation of mutations of the second kind is not as obvious as that of mutations of the first kind. Freese proposed that mutations of the second kind represent what he called *transversions*, in which a pyrimidine has been replaced by a purine, or a purine by a pyrimidine, in the DNA polynucleotide chain, according to the schema

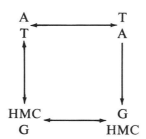

Later, more detailed studies of acridine-dye-induced mutations showed however, that mutations of the second kind correspond to *insertions* or *deletions* of one or a few nucleotide residues into or from the DNA polynucleotide chain, rather than to the transversions proposed by Freese. That is not to say, however, that transversions do not occur; they do occur, in fact, but are to be included among mutations of the first kind, and are thus inducible as well as revertible by base analogue mutagens. Most of the rII mutants that do not revert at all, either spontaneously or in the presence of any mutagen, are long-span mutations, which, as was explained earlier in this chapter, derive from deletion of hundreds, or even thousands, of nucleotides from the DNA polynucleotide chain. Deletion mutants do not revert to the rII$^+$ state because, once lost, whence could their missing DNA segment be restored?

MUTATIONAL HETEROZYGOTES

If the elementary mutational event corresponds to the introduction of an incorrect nucleotide at some site of the growing replica polynucleotide chain, and if the DNA of the vegetative phage replicates through the semiconservative mechanism of Watson and Crick, then it is possible to predict a feature of the nascent mutant genomes that could hardly have been anticipated before the molecular basis of these events was understood. Let us suppose that during the growth of one of the replica chains a copy error occurs—for example,

FIGURE 13-12.   Genesis of mutational heterozygotes in semi-conservative replication of the phage DNA. A copy error causes the introduction of an incorrect G (instead of the correct A shown in heavy type) into the nascent polynucleotide chain of one of the two DNA molecules at a site where the parental DNA molecule possessed an A-T base pair (shown in heavy type). This error produces a "heterozygous" DNA double helix that carries the original, nonmutated base sequence in one of its strands and the mutant base sequence in the other of its strands. At the next replication cycle all strands of the first-generation daughter DNA molecules separate, and each strand acts as template for the synthesis of its complement. This produces one fully mutant and three nonmutant structures among the second-generation DNA molecules.

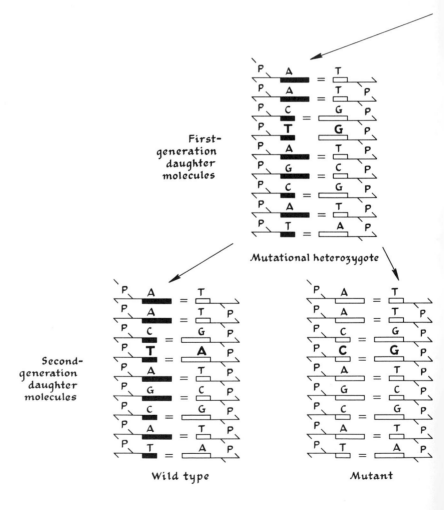

Wild type   Wild type

that a thymine residue in the parental strand makes one of the rare illegitimate pairings with guanine, instead of the legitimate pairing with adenine. The double-helical DNA molecule resulting from this mutagenic replication act would then carry the mutant information in its *de novo* strand, whereas, the old, unaltered parental strand would still carry the original, nonmutant information (Figure 13-12). At the next replication cycle the complementary strands of this nascent mutant molecule would separate again, with each strand acting as the template for the synthesis of its complementary polynucleotide chain. This would result in the appearance of one DNA double helix that carries the mutant information in both of its strands and one DNA double helix that is entirely nonmutant. The nascent mutant DNA molecule would thus represent a *heteroduplex heterozygote* that carries two alleles, one mutant and the other nonmutant, at one of its genetic sites, and that segregates into mutant and nonmutant genomes in a later replication cycle. One could expect, therefore, that during intracellular phage growth some phage DNA molecules that carry a copy-error mutation of the last replication cycle are withdrawn from the vegetative pool for encapsulation into structurally intact, infective phage particles. Such particles would be mutational heterozygotes.

In 1959 D. Pratt was able to show that most, if not all, bromouracil-induced r^+ reverse mutants of a nucleotide-base-analogue-induced rII mutant do initially arise as genetically heterozygous rII/r^+ genomes that only later segregate into homozygous r^+ revertants. For the purpose of this demonstration bromouracil mutagen was added to bacteria infected with the T4rII mutant just before the end of the eclipse period of intracellular phage growth, and the first infective progeny particles that appeared intracellularly just after the end of the eclipse period were released by artificial lysis of the cells. This program of mutagenesis insured that any r^+ reverse mutants induced and withdrawn from the phage DNA precursor pool during the short exposure of the culture to the mutagen must have arisen only during the very last replication cycle. The copy error responsible for restoring to the phage DNA the genetic information of the r^+ wild state occurred, therefore, so late that no further replication (and hence segregation into homozygous mutant structures) could have occurred. The result of this experiment was that more than 80% of all r^+ revertants induced by a brief exposure of the rII-infected bacterial culture to bromouracil are indeed mutational heterozygotes, carrying both the original mutant rII as well as the reverted r^+ genetic site. Hence, just as the mechanism of Watson and Crick demanded, and contrary to the expectation of some other conceivable DNA replication mechanisms that predict a conservative distribution of the substance of the parental DNA molecule, copy errors in the synthesis of replica polynucleotides do appear to generate a heteroduplex that, in one of its complementary polynucleotide strands, still preserves intact the hereditary information of its progenitor.

GENERAL NATURE OF THE GENETIC CODE

It was taken for granted in our discussion of reverse mutation that the r^+ revertants sported by various rII mutants all owe their wild phenotype (their ability to grow on strain K) to a genuine act of reverse mutation in which the "incorrect" purine-pyrimidine base pair extant at the mutant site of the viral DNA has been replaced by the "correct" base pair of the wild-type genome. Such "true" reversions, however, are not the only way in which revertant phage genomes that exhibit the wild phenotype can arise. Instead of a restoration of the mutant site to its original, wild-type state, it is also possible that there occurs a *suppressor mutation* at another genetic site which "suppresses" the mutant phenotype. The interaction of such suppressor mutations with the original mutant locus to restore the wild phenotype can have a variety of entirely different physiological explanations, as was previously discussed in connection with bacterial mutation in Chapter 6. In any case, Benzer and Freese were not unaware of the possibility that suppressor mutations might obscure the conclusions they hoped to draw from such data as those presented in Table 13-1. Hence they attempted to establish in various ways that the r^+ revertants found in their studies on mutagenesis actually represent "true" reversions and not suppressor mutations. It now seems likely that most r^+ revertants sported by rII mutants of the *first kind* (i.e., transitions and transversions) do derive from reverse mutation at the mutant site. But in 1961 it was discovered by Crick that many r^+ revertants sported by rII mutants of the second kind (i.e., short deletions and insertions) are not true revertants but *double mutants* that owe their ability to grow on strain K to a juxtaposition of two mutations, the original and its suppressor, in the rII region of the T4 genome. The recognition of this fact was then exploited in a most ingenious way by Crick and Sydney Brenner to secure the first experimental evidence for the by then eight-year-old doctrine of the nucleotide triplet genetic code. They showed, furthermore, that the correct reading of the long nucleotide sequence of a gene is assured by beginning the translation into amino acid sequence from a fixed starting point and proceeding triplet by triplet to the end. Hence if the triplet *reading frame* happens to be shifted by one nucleotide at any intermediate point, then the reading of all subsequent triplets is shifted in "phase" and a totally incorrect translation produced from that point onward. The results of these experiments suggested also that the genetic code is rich in synonyms, in that most of the 20 standard amino acids must be represented by more than one of the 64 possible nucleotide triplets.

Crick and Brenner's genetic code experiments began with the observation that most of the spontaneous r^+ wild-type revertants sported by the T4rII mutant designated as FC0 do not carry true reversions of the mutated genetic site (located in the rIIB gene), but instead are double mutants that carry a suppressor mutation in the vicinity of the original rII mutation. That is, these revertants are not really wild r^+ genotypes at all, but only "pseudo-wild" phenotypes in which the juxtaposition of two particular rII mutant sites, FC0 and its *direct intragenic* suppressor, renders the phage capable of growth on the restrictive strain K host. The presence of these suppressor mutations can be demonstrated by crossing such a pseudowild revertant with the authentic T4r^+ wild-type phage.

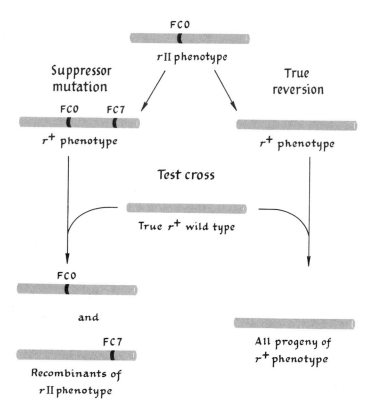

From this cross of two parent phages of r^+ phenotype there issue rII recombinants that form typical r plaques on the permissive *E. coli* strain and cannot grow at all on the restrictive strain K. These rII recombinants are of two types: one type carries the original FC0 mutation and the other type carries a new rII mutation, the suppressor to FC0. (No such r recombinants would, of course, be produced by a cross of a true r^+ revertant with the T4r^+ wild type.) It is possible to ascertain the location of these rII suppressor mutations to FC0 by means of the fine-structure mapping methods; the sites of 7 different such suppressors to FC0 have been indicated on line B of Figure 13-13. It can be seen that these suppressors are all close, but not very close, to FC0. They cluster at two separate sites on either side of FC0, the

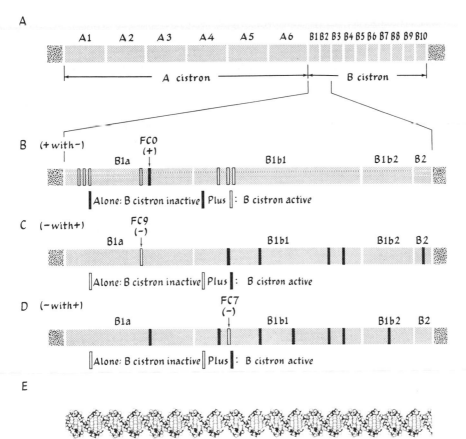

FIGURE 13-13.   The ensemble of mutually suppressive mutations in the *r*II region of the T4 genome. **A.** The two genes, A and B, of the *r*II region, shown subdivided into their segments. **B.** Sites of the mutation FC0 in the B1a subsegment of the B1 segment of the B cistron, and of seven mutations isolated as suppressors to FC0. **C,D.** Sites of twelve further mutations in B1 and B2 segments isolated as suppressors to the suppressors. **E.** The viral DNA drawn to approximately the same scale as the genetic map. The two segments contain about 100 nucleotide base pairs. [After "The Genetic Code," by H. C. Crick, *Scientific American*, October 1962. Copyright © 1962 by Scientific American, Inc. All rights reserved.]

distance between the two clusters being about one-tenth the whole length of the *r*IIB gene, in whose B1 segment FC0 is situated.

Since each of these suppressors is an *r*II mutant in its own right, it is possible to study also its reverse mutation to the r^+ state, in the same manner as that already used for FC0. It was found in this way that these suppressor mutants, like FC0, do not generally revert to the true r^+ wild type either, but instead again sport suppressed double mutants capable of growing on strain K. Lines C and D of Figure 13-13 show the location of a number of *r*II mutants that were isolated as suppressors to the two suppressors FC9 and FC7. It is evident that these secondary suppressor mutations also occur in the general neighborhood of the original FC0 mutant site in the B gene. Finally, suppressors to the suppressors to the suppressors can be isolated in the same way. In this manner a set of some 80 independent rII mutants was isolated, including FC0, all of which are suppressors of some mutants in the set and carry their mutation within a limited region of the rIIB gene. The double mutants that carry a mutation and its suppressor (and can thus form plaques on strain K) form, however, a variety of plaque types on the ordinary *E. coli* strain. Some of these plaque types are indistinguishable, or hardly distinguishable, from true r^+, whereas the mutant character of others is easily recognized and rather resembles the r plaque type.

The key to understanding the genetic interaction of this set of suppressor mutations was the fact that FC0 is an rII mutant induced by treatment of a T4r^+-infected bacterium with an acridine dye. Since acridine dyes derive their mutagenic action from promoting either the insertion or the deletion of nucleotides from replicating DNA chains, it can be supposed for the sake of discussion that the mutation of FC0 represents the insertion at the FC0 site of an additional nucleotide into the wild-type purine-pyrimidine base sequence. What effect would such an insertion have on the primary structure of the polypeptide chain coded by the wild-type rIIB gene if, as set forth in the preceding paragraphs, the nucleotide sequence is read triplet by triplet, from one end (say the "left" end) of the B gene to the other? As is evident from Figure 13-14 the insertion of an extra nucleotide at the FC0 site will obviously cause the reading of all the triplets to the "right" of FC0 to be shifted by one nucleotide, and hence be incorrect. In other words, from that point onwards, the amino acid sequence coded by the rIIB gene polypeptide will be completely altered, and if any protein is formed at all, it will bear little resemblance to that produced by the normal, wild-type gene. Crick and Brenner next postulated that a suppressor mutation to FC0—for example, FC7—corresponds to the deletion of a nucleotide. When the FC7 deletion mutation is present by itself, all triplets to the "right" of FC7 will be read incorrectly, and thus the normal rII$^+$ function is absent: FC7 is an rII mutant. But when both FC0 and FC7 mutations are present within the same genome, as in the pseudowild double mutant, then, although the reading of triplet words between FC0 and FC7 will be still altered, the correct reading will be restored to the part of the gene to the right of FC7. And if the rII enzyme protein can tolerate some alterations in amino acid sequence in the short region of the B-gene polypeptide chain coded in the vicinity of FC0 without losing all of its catalytic function, the protein synthesized by the suppressed double mutant may possess sufficient enzymatic activity to allow successful phage growth in strain K bacteria. The function of this modified enzyme, however, might not be wholly "normal," in which case the double mutant,

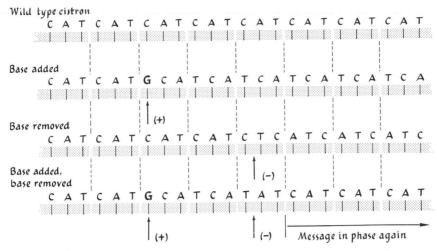

FIGURE 13-14.   The effect of mutational insertions or deletions of single nucleotides on the genetic message, if transcription of the message proceeds triplet by triplet from a fixed point to the right. The hypothetical "correct" message of the wild-type poly-nucleotide segment shown here is CAT, CAT, CAT, ... Insertion of the nucleotide G distorts the remainder of the message to TCA, TCA, TCA, ... Deletion of the nucleotide A distorts the remainder of the message to ATC, ATC, ATC, ... Insertion of G at one site and deletion of A at another restores the correct message CAT, CAT, CAT, ... [After "The Genetic Code," by F. H. C. Crick, *Scientific American*, October 1962. Copyright © 1962 by Scientific American, Inc. All rights reserved.]

though capable of growth on strain K, would not necessarily manifest the true r^+-plaque phenotype when plated on the ordinary rII-permissive *E. coli* strain.

The following convention can now be established. The mutant FC0, which for the sake of discussion was assumed to carry an insertion of a single nucleotide, is assigned the symbol +, and its suppressors, among them FC7 and FC9, are assigned the symbol −. It does not matter here whether FC0 is really an insertion and FC7 and FC9 are really deletions: it is important only that a mutation and its suppressor be of opposite sign. Accordingly, the suppressors to the suppressors of type − are again assigned the symbol + (see Figure 13-13). It follows from these considerations that all combinations of two mutants of the same sign, such as + + or − −, brought into the same genome should still be of mutant phenotype—a prediction that was verified.

Although in the foregoing considerations it was assumed that each amino acid is coded by a nucleotide triplet, the conclusions do not really depend on this assumption and would be equally applicable if each amino acid were coded by four, five, or any other larger number of nucleotides. It was only essential that the reading of the nucleotide sequence proceed from one end of the gene to the other, code word by code word, so that juxtaposition of insertions and deletions of single nucleotides restores the correct reading frame to regions of the gene beyond the two mutant sites. But by constructing triple mutants of the type $+ + +$ or $- - -$ it was possible to show that the coding is done by nucleotide triplets. Such triple mutants manifest the wild or pseudowild phenotype, provided that the frame-shifted nucleotide sequence between the two outermost mutant sites does not encompass any "unacceptable triplets," which would lead to a nonfunctional protein product. For instance, the triple $+ + +$ mutant carrying FC0 and two suppressors of suppressors to FC0 forms plaques on strain K, even though either singly or in pairs each of these mutant sites produces an rII mutant phage that is quite unable to grow on strain K. It is thus the combination of all three mutations of the same sign within the same DNA polynucleotide chain that restores function, or partial function, to the mutant rIIB gene. This result could be expected only if the code words are really triplets, since only if they are could the net result of deletion or insertion of three nucleotides be the restoration of the correct reading frame beyond the three mutant sites. (Figure 13-15). Strictly speaking, this result proved only that the number of nucleotides per code word is a multiple of three, since the whole analysis is based upon the assumption that the original FC0 mutant had undergone the insertion or deletion of a single nucleotide. Thus, as Crick and Brenner pointed out, the more conservative inference to be drawn from these observations was that the number of nucleotides per code word is $3n$, where n is the number of nucleotides inserted or deleted at a time by the mutagenic action of acridines. In the following discussions we will continue to pretend, however, that the code words are simply nucleotide triplets, or that $n = 1$. Definitive justification of this pretension will be offered in Chapter 18.

These experiments suggested, finally, that it is unlikely that only 20 of the 64 possible nucleotide triplets represent the standard set of 20 amino acids

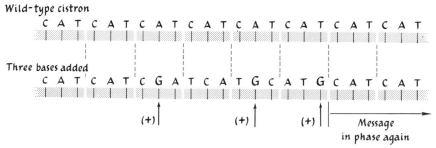

FIGURE 13-15.   Insertion of three nucleotides at closely situated sites distorts the genetic message over a short stretch of the genetic segment but leaves the remainder of the message unaffected. Deletion of three neighboring nucleotides produces the same result. [After "The Genetic Code," by F. H. C. Crick, *Scientific American*, October 1962. Copyright © 1962 by Scientific American, Inc. All rights reserved.]

and that the remaining 44 triplets are "nonsense" and correspond to no amino acid at all. If this were so then the genetic region in which suppressor mutations of the FC0 family could occur should be very much smaller than that actually found. For if most of the 64 possible nucleotide triplets were nonsense, then the shift in reading frame that obtains between any + and − mutant sites separated by more than a very few nucleotides would invariably lead to at least one nonsense codon that would interrupt the continuity of assembly of the polypeptide chain and hence would prevent the assembly process from ever reaching those regions of the gene to which the correct reading frame had been restored. It became therefore quite likely that the code is rich in synonyms, in that many of the 20 standard amino acids are represented by more than one kind of nucleotide triplet. Since only a minor fraction of the 64 possible nucleotide triplets would then be nonsense, rather lengthy shifts in reading frame could thus be tolerated without the generation of nonsense codons that interrupt the continuity of the amino acid assembly process.

Bibliography

PATOOMB

Seymour Benzer. Adventures in the rII region.

HAYES

Chapters 7, 8, and 13.

MOBIBAV

Chapter 10.

ORIGINAL RESEARCH PAPERS

Benzer, S. Fine structure of a genetic region in bacteriophage. *Proc. Nat. Acad. Sci., Wash.,* **41**, 344 (1955).

Freese, E. The specific mutagenic effect of base analogues on phage T4. *J. Mol. Biol.,* **1**, 87 (1959).

Lerman, L. Structural considerations in the interaction of DNA and acridines. *J. Mol. Biol.*, **3**, 18 (1961).

Litman, R. M., and A. B. Pardee. Production of bacteriophage mutants by a disturbance of deoxyribosenucleic acid metabolism. *Nature,* **178**, 529 (1956).

Nomura, M., and S. Benzer. The nature of the "deletion" mutants in the *r*II region of phage T4. *J. Mol. Biol.*, **3**, 684 (1961).

Pratt, D., and G. S. Stent. Mutational heterozygotes in bacteriophages. *Proc. Nat. Acad. Sci. Wash.*, **45**, 1507 (1959).

Schuster, H. The reaction of nitrous acid with deoxyribonucleic acid. *Biochem. Biophys. Research Comm.*, **2**, 320 (1960).

SPECIALIZED TEXTS, MONOGRAPHS, AND REVIEWS

Benzer, S. Genetic fine structure. In *Harvey Lectures*, Series 56. Academic Press, New York, 1961.

Drake, J. W. *The Molecular Basis of Mutation*. Holden-Day, San Francisco, 1970.

Fincham, J. R. S. *Genetic Complementation*. Benjamin, New York, 1966.

Freese, E. The molecular mechanism of mutations. In *Proceedings of the 5th International Congress of Biochemistry,* Moscow, 1961.

Orgel, L. E. The chemical basis of mutation. *Advan. Enzymol.*, **27**, 289 (1965).

14. Lysogeny and Transduction

So far, phages have been regarded as bacteriocidal agents that invade bacteria, take over and monopolize the synthetic capacities of the host cell for their own ends, and multiply to yield several hundred progeny phages within each infected cell; intracellular growth of the phage parasite kills the infected bacterium, since the host cell is lysed to allow escape of the nascent phage progeny into the world outside. This view demands that survival of phages is wholly dependent on the concomitant death of sensitive host bacteria. But such a morbid bite-the-hand-that-feeds existence of phages is by no means their only way of life, inasmuch as phage genomes can also be perpetuated as part of the hereditary apparatus of bacteria during normal growth of the host cell. It is this "symbiotic" relation between bacteria and their phages with which we will now be concerned.

LYSOGENIC BACTERIA

Since the early 1920's strains of *E. coli* were known that permanently "carry" phages, and hence are *lysogenic*, or capable of causing lysis of other bacterial strains sensitive to the action of the carried phage. Since the concept of *maintenance* of any phage by bacteria did not seem to fit with d'Hérelle's view of the phage as a virus inexorably lethal for the host cell on which it grows, he declared that the so-called lysogenic bacteria are merely contaminated with phage and that they can readily be rid of their viruses by appropriate purification procedures. This view of lysogeny as a non-phenomenon was shared by Delbrück's school of phage workers during the 1940's, mainly because none of the T-phage strains on which their efforts were then concentrated, happened to manifest the lysogenic state. Nevertheless, there *had* been a spate of publications in the 1930's which showed that lysogeny cannot possibly be dismissed all that easily. In particular, Eugène Wollman and F. M. Burnet had already adduced some proof that lysogenic bacteria perpetuate the phages they carry in a *noninfectious* state.

Lysogeny was rescued from oblivion by André Lwoff (Figure 14-1). Soon after the end of World War II, Lwoff took up work on a lysogenic strain of *Bacillus megaterium.* At the outset of his work Lwoff posed to himself these questions: (1) Can the capacity of lysogenic bacteria to produce phages be perpetuated without intervention of exogenous phages? (2) How do lysogenic bacteria liberate the phages they produce? (3) What factors induce the production of phages in a population of lysogenic bacteria?

To Lwoff it seemed evident that the study of mass cultures of lysogenic bacteria, as it had been practiced by his predecessors, could furnish only partial and not definitive answers to these questions, and that only the observation of individual bacteria, or of microcultures containing a small number of individuals, could lead to unambiguous conclusions. Lwoff, therefore, proceeded to cultivate an individual cell of *B. megaterium* in a microdrop and to watch its division under the microscope. Immediately after the division, one of the two daughter cells was withdrawn from the microdrop by means of a micropipette held in a micromanipulator and plated on agar to determine whether it would give rise to a colony of lysogenic descendant bacteria. As soon as the cell that remained in the culture fluid had once more divided, one of the two new daughter cells was withdrawn and plated on agar, and the process of withdrawing and plating one of the cells generated by each division was repeated for a total of 19 divisions. In the same experiment, samples of the culture medium were also withdrawn from the microdrop at various times and assayed on a nonlysogenic, phage-sensitive strain of *B. megaterium* for the presence of any free phage. The result of this experiment was that every colony derived from the cells in the microdrop was lysogenic, and that none of the samples of the culture fluid contained any infective phage. Lysogeny was thus demonstrated to persist for at least 19 successive divisions in the absence of any exogenous phage.

In the course of several such microscopic studies of the division of individual bacteria, spontaneous lysis of the cell in the microdrop was occasionally observed. Whenever the culture fluid was assayed for the presence of free phage after such lysis, some hundred phages were found to be present in the drop. It could thus be inferred with some confidence that lysogenic bacteria

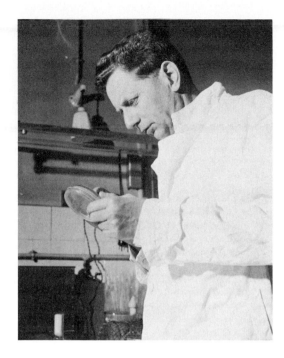

FIGURE 14-1.   André Lwoff
(b. 1902).

liberate their phage by lysis, each lysing cell yielding a burst of many phage particles. These findings allowed Lwoff to describe lysogeny in the terms in which the phenomenon is now understood: each bacterium of a lysogenic strain harbors and maintains a noninfective structure, the *prophage*, which endows the cell with the ability to give rise to infective phage without the intervention of exogenous phage particles. In a small fraction of a population of growing lysogenic bacteria the prophage becomes *induced* to produce a litter of infective phage, and hence prophage induction leads to death and ultimate lysis of the cell.

INDUCTION

Having thus disposed of the first two of the three questions that Lwoff had set out to answer, he and two of his pupils, L. Siminovitch and N. Kjeldgaard, turned their attention to the third. They had reason to believe that induction of the prophage is under the control of factors external to the lysogenic cell and began, therefore, a search for agents or conditions that would raise the small fraction of lysogenic bacteria that produce phage spontaneously. After trying without avail a great variety of chemical and physical treatments, they finally found that irradiation of a growing culture of *B. megaterium* with small doses of ultraviolet light (UV) induces phage production in nearly the whole

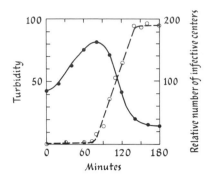

FIGURE 14-2.   Induction of the prophage of lysogenic *E. coli* strain
K12(λ) by UV light. A culture of exponentially growing bacteria is
exposed to an optimal dose of UV that induces vegetative phage
development in more than 90% of the bacteria. The irradiated culture
is then incubated. The turbidity in arbitrary units (plotted here as filled
circles) and the titer of infective centers, or ratio of the number of
plaque forming units to the number of irradiated bacteria (plotted here
as open circles), of the culture are determined after the number of
minutes shown on the abscissa have elapsed since its irradiation. [After
F. Jacob and C. R. Fuerst, *J. Gen. Microbiol.* **18**, 518 (1958).]

population of lysogenic cells. The result of an analogous induction experiment done with a lysogenic strain of *E. coli* is presented in Figure 14-2. It can
be seen that for the first 70 minutes after UV-irradiation, the lysogenic
bacteria grow in size, as indicated by the continuous increase in turbidity of
the culture, and that lysis of nearly all the bacteria suddenly ensues after that
time, as indicated by a rapid loss of turbidity, or clearing, of the culture. It
can be seen, furthermore, that plaque assays made throughout this period
show that there are very few free, infective phage particles in the culture until
the onset of bacterial lysis. Thereafter the number of free phages increases
rapidly, until an infectivity plateau is finally attained that represents an
average yield of nearly 200 phages per lysogenic cell in the irradiated culture.
There is little doubt, therefore, that the phages are liberated by lysis of induced lysogenic cells. It thus appeared not only that the lysogenic bacterium
passes on the faculty for producing phage to all of its descendants, a few of
which might at some future time lyse and liberate infective phage particles, but
also that every bacterium of the culture can be induced at will to begin producing infective phage without the intervention of any exogenous, free phage
particles.

The events that ensue in the lysogenic cell after induction of its prophage
are entirely analogous to those of the vegetative phase of phage development
following infection of a sensitive baterium with an exogenous free phage

particle. There first occurs an eclipse during which no infective phage can be detected in the cell but during which the substance of the progeny, both their genetic DNA as well as their proteinaceous structural components, is synthesized; later, about halfway through the latent period, these various phage precursors begin to combine, forming mature infective progeny phages. These progeny, which can also be released prematurely by artificial lysis of the cells, appear in the culture medium after spontaneous lysis at the end of a latent period.

TEMPERANCE AND VIRULENCE

Lwoff's predecessors in the study of lysogeny had already found that a fraction of nonlysogenic bacteria infected with phage carried by a lysogenic strain do not lyse, but instead survive the infection to give rise to clones of lysogenic bacteria that henceforth carry the infecting phage as prophage. "Artificial" lysogenic strains were thus created, which, according to Lwoff represented "the first example of a specific hereditary property being conferred to an organism by a specific extrinsic particle." Infection of a nonlysogenic bacterium with a phage may, therefore, result in two very different responses.

1. The *lytic response*, in the course of which the infecting phage enters the vegetative phase, multiplies, forms mature progeny phage particles, and ultimately lyses the host cell at the end of the latent period. This is, of course, the sequence of events which was the subject of Chapter 11.

2. The *lysogenic response*, in the course of which the infecting phage does not multiply and is, instead, "reduced" to the prophage state, allowing the host cell to survive as a lysogenic bacterium.

All phages worthy of their name are capable of giving the first of these responses, since a phage that *never* multiplies to form infective progeny particles could not be said to be a "virus." Only *temperate* phages, however, are capable of giving the lysogenic response, in contradistinction to the so-called *virulent* phages, which do not give the lysogenic response and are never found in lysogenic bacteria in prophage guise.

Lysogeny is usually a very stable character of a bacterial strain, nearly as stable as any other of its hereditary characters. Thus in most populations of lysogenic cells only a very small fraction of individuals appear to have lost their prophage—that is, have been *cured* of their lysogeny. Nevertheless, nonlysogenic derivatives of erstwhile lysogenic strains *can* be isolated. It is difficult to make a precise determination of the frequency of curing, but it has been estimated for at least one lysogenic strain of *E. coli* that the spontaneous rate of prophage loss is about 10^{-5} cured cells per cell per generation.

IMMUNITY

Ever since their discovery, it has been known that lysogenic bacteria are *immune* to infection by phage particles of the same type as their prophage. Such immunity is almost a logical necessity for the very existence of lysogeny, since any culture of lysogenic bacteria sensitive to infection and lysis by the temperate phage it carries would soon be destroyed by a chain reaction of successive phage growth cycles initiated by the first infective phage particle that appeared in the culture upon spontaneous prophage induction of one of the lysogenic cells. Lysogenic immunity does not prevent the attachment of the temperate phage particles to the bacterial surface. Immunity, therefore, is to be distinguished from phage *resistance*, which, as was discussed in connection with the life cycle of virulent phages in earlier chapters, bacteria can acquire through hereditary modifications of their cell envelope that block the stereospecific reaction of cell wall receptor spots with the phage tail-attachment organs.

What happens when a temperate phage infects an immune bacterial cell? The bacterium continues to grow and divide as though no infection had occurred, while the genetic material of the infecting phage, though injected

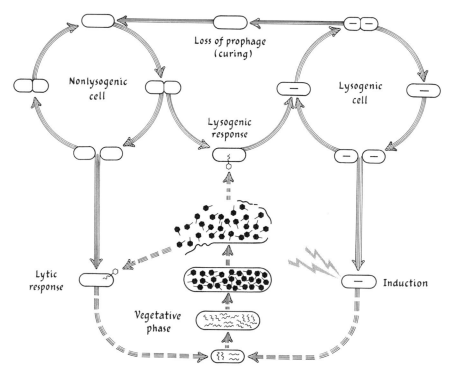

FIGURE 14-3.   General view of lysogeny. [After A. Lwoff, *Bacteriol. Rev.* **17**, 269 (1953).]

into the cell, does not multiply; the phage genome remains singular and is passed on to one of the two daughter cells in successive bacterial divisions. Immunity thus prevents the infecting temperate phage from entering the vegetative state. Immunity, furthermore, is highly specific and extends only to infecting phages of a type identical to or very closely related to that carried as prophage; a cell lysogenic for one phage strain is usually sensitive to infection by all other phage strains to which the corresponding nonlysogenic bacterium is sensitive. It is the prophage itself that must be responsible for the specific immunity of the lysogenic cell, because after loss of the prophage the non-lysogenic, cured derivative is fully sensitive to the temperate phage type formerly carried.

The attempt to gain a deeper understanding of these facts will be deferred until the latter part of this chapter; they are summarized, however, in Figure 14-3, which shows Lwoff's diagrammatic representation, the Device of Lysogeny.

NATURE OF THE PROPHAGE

What kind of organelle is the prophage that perpetuates in lysogenic bacteria the faculty to produce phage in the absense of exogenous phage? Not only does it not possess the infectivity of the intact, free phage, but it does not even carry any of the proteins of the mature virus particle, since noninduced lysogenic bacteria do not contain any materials that can react specifically with the antibodies of an antiserum directed against the homologous infective temperate phages carried by the strain. But inasmuch as the prophage must contain the phage genome it seemed most probable that the prophage, like the noninfective vegetative phage of the lytic response, comprises the phage DNA.

In considering the relation of the prophage to the bacterial cell, two general alternative hypotheses were entertained (Figure 14-4).

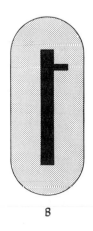

FIGURE 14-4. Schematic representation of two alternative hypotheses on the relation of the prophage to the lysogenic bacterial cell. **A.** The prophage consists of many "cytoplasmic" particles. **B.** The prophage is incorporated into the bacterial chromosome. [After F. Jacob, *Les bactéries lysogènes et la notion de provirus*. Monographies de l'Institut Pasteur. Masson, Paris, 1954.]

A   B

1. The prophage is a "cytoplasmic" bacterial particle whose rate of reproduction is adjusted to that of the host cell. It is distributed at random to the two daughter cells at each bacterial division.

2. The prophage is directly associated with and replicates in synchrony with the bacterial chromosome.

It became possible to study the relation of prophage and bacterial chromosome after Esther Lederberg found in 1951 that *E. coli* K12, the very strain in which J. Lederberg and E. L. Tatum had previously discovered bacterial conjugation, is, in fact, lysogenic and carries a temperate phage, which she called *lambda*. The lysogeny of K12 became manifest after nonlysogenic, or "cured," derivatives of it had been accidentally isolated. On these nonlysogenic derivatives, samples of the culture fluid of the parent K12 strain produce plaques of phage *lambda*. According to the convention now in use under which lysogenic bacteria are designated by the name of the bacterial strain followed in parentheses by the symbols representing the phage strain carried as prophage, the Lederbergs' lysogenic strain of *E. coli* was henceforth referred to as K12(λ). This is the strain with which we have already become familiar under the name K in preceding chapters, in connection with the genetics of the *r*II locus of the T-even phages.

A decade of study of the structure, physiology, and genetics of phage *lambda* revealed that, except for its ability to exist in both infective as well as prophage states, *lambda* has many general traits in common with the T-even phages. The *lambda* phage particle has a DNA-filled head and a slender tail (Figure 14-5). The *lambda* head, however, contains a DNA molecule only 50,000 nucleotide base pairs in length, and hence carries about one-fourth as much genetic information as is carried by the T-even phage chromosome. The chemical composition of *lambda* phage DNA resembles that of the DNA of its *E. coli* host, being free of the idiosyncratic glucosylated HMC residues of the T-even phage DNA. Like the T-even DNA, the *lambda* DNA is terminally redundant, but the terminal redundancy of *lambda* DNA consists of only a short single-stranded segment of 20 nucleotides that are complementary in base sequence (Figure 14-6). The "cohesive ends" represented by this terminal redundancy allow conversion of the linear DNA molecule of the infecting phage particle into the circular DNA molecule of the intracellular vegetative phage. An enzyme that "nicks" the circular vegetative DNA molecules at two specific internucleotide bonds, one on each polynucleotide chain, regenerates the linear DNA molecules with their cohesive ends to provide the chromosomes of the progeny phages.

The earliest mutants of *lambda* to have been isolated concerned host range and plaque type, just as did the first T-even phage mutants. By means of genetic crosses among these mutants, a primitive genetic map of the *lambda* chromosome (Figure 12-8) was constructed. But by 1960, Alan Campbell had

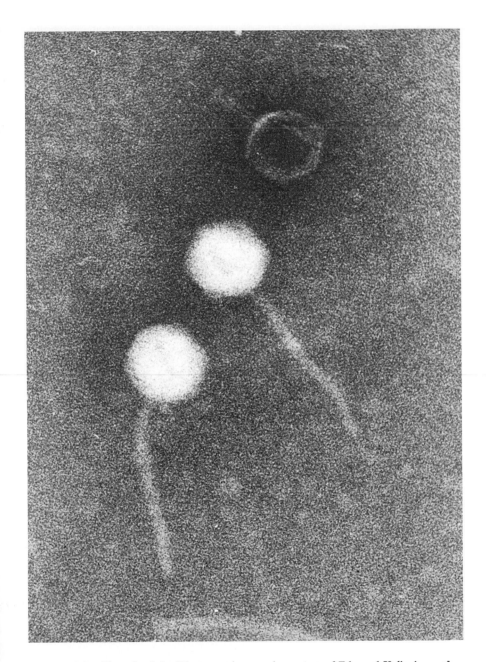

FIGURE 14-5.   Phage lambda. [Electron micrograph courtesy of Edouard Kellenberger.]

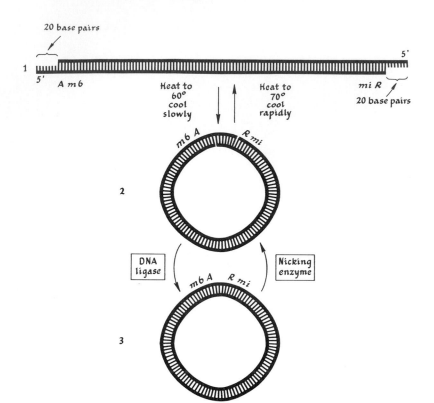

FIGURE 14-6.   Circularization and cohesive ends of the *lambda* phage DNA. (1) The chromosome of the infective *lambda* phage is a linear DNA molecule 50,000 nucleotide base pairs in length, carrying gene *A* at one end (the " left " end) and gene *R* at the other (the " right " end). At both ends, the polynucleotide chain carrying the 5'-phosphate terminus is 20 base pairs longer than that carrying the 3'-hydroxyl terminus. (2) The base sequence of the two terminal producing ends is complementary, thus providing the entire DNA molecule with "cohesive ends." Upon heating a solution of *lambda* DNA molecules to 60°C and then cooling it slowly, the two cohesive ends join by complementary base pairing and convert the linear DNA molecule into a circle. Heating the resulting circles to 70° and cooling the solution rapidly results in the melting of the joint formed by the cohesive ends and in the restoration of the linear molecule. (3) In the cohesive end joint, the 5'-phosphate termini of both polynucleotide chains are adjacent to their own 3'-hydroxyl termini, thus presenting the aspect of single-strand interruptions of a DNA double helix. Action of the DNA ligase enzyme (see Figure 9-19) seals these interruptions by catalyzing the formation of the internucleotide phosphate diester bonds, giving rise to a fully covalently bonded, circular double helix without any interruptions. Within five minutes of the injection of *lambda* DNA into an *E. coli* host cell, the linear DNA molecule of the mature phage particle has been converted into a covalently linked circular molecule, in which form it replicates for the generation of progeny DNA molecules. Upon the encapsulation of these progeny DNA circles into the heads of progeny phage particles, a base-sequence-specific *nicking enzyme* recognizes part of the base sequence between genes *A* and *R* and hydrolyses a unique phosphate diester bond among the 50,000 such bonds of each polynucleotide chain to regenerate right and left cohesive ends of the linear chromosomes. The nicking process can be inferred to concern a unique phosphate diester bond because the 20 nucleotides of the cohesive ends of every *lambda* DNA molecule carry exactly the same base sequence. [After A. D. Kaiser, *in* F. O. Schmitt, ed., *The Neurosciences: Second Study Program*, Rockefeller Univ. Press, New York, 1971.]

isolated many conditional lethal mutants of *lambda* whose fundamental genetic character was soon found to correspond to that of the then recently discovered T-even *amber*, or *am*, mutants. Temperature-sensitive, or *ts*, mutants of *lambda* were also found, and the availability of an ensemble of conditional lethal *am* and *ts* mutants soon made possible the identification and mapping of most of the genes of the *lambda* phage genome (Figure 14-7). This *lambda* map shows even more strikingly than the T-even map that, in general, genes of related function cluster, and, in particular, that genes which function early in the period of intracellular phage growth occupy a single continuous sector of the chromosome. Among these early-function genes there are three, *N*, *O*, and *P*, whose products evidently play a paramount role in the control of phage development, in that under restrictive conditions *none* of the genome of a *lambda* phage carrying a conditional lethal mutation in genes *N* or *O* or *P* can be expressed (except that small portion of the genome which lies between *N* and *O*). The cyclical closing and opening of the circular *lambda* DNA molecule outlined schematically in Figure 14-6 explains why the *lambda* genetic map is *linear*. For unlike the circularly permuted T-even phage DNA, the *lambda* DNA circle is opened only between the gene pair *A* and *R*, resulting in a unique linear structure.

The discovery of lysogenic and nonlysogenic strains of the sexually fertile *E. coli* K12 opened the way for bacterial crosses in which the distribution of the *lambda* prophage among recombinant bacteria could be followed. The first such crosses were carried out at a time when the nature of the bacterial conjugation process was still incompletely understood and when the frequency of bacterial recombinants was only of the order 10^{-7} per cell. Nevertheless, these first experiments showed that, in at least one respect, the *lambda* prophage does appear to behave as a chromosomal hereditary determinant, in that some of the recombinant bacteria were lysogenic and others were not. The pattern of prophage distribution among the recombinant bacteria indicated, moreover, that *lambda* lysogeny is linked to the *gal* genes responsible for the fermentation of galactose: most of the recombinant bacteria had derived their *gal* and *lambda* lysogeny characters from the same parental strain. In contrast to the segregation pattern to be expected of a chromosomal gene, however, the *lambda* prophage showed an *asymmetrical* distribution in reciprocal crosses between lysogenic and nonlysogenic bacteria: whereas the nonlysogenic character of an F^+ donor parent appeared in the recombinants of crosses with a lysogenic $F^-(\lambda)$ recipient, the lysogenic character of an $F^+(\lambda)$ donor parent was almost never found among recombinants of crosses with nonlysogenic F^- recipients.

Once "high-frequency" (Hfr) fertility mutants of F^+ donor lines of K12 had been isolated (and it had been recognized by Wollman and Jacob that an Hfr bacterium transfers its chromosome in an oriented manner to the conjugal F^- cell, beginning with a definite point of origin), genetic loci of the Hfr chromosome could be mapped by measuring their times of entry into the zygote. In this way it was shown by crossing a nonlysogenic Hfr donor to a

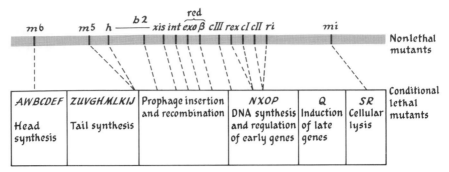

FIGURE 14-7. The definitive genetic map of phage lambda. Each of the capital letters represents a gene pertaining to the function indicated on the map and defined by cis-trans tests of conditional lethal mutants; *m6*, *m5*, and *mi* = abnormally small plaque type; *b2* = attachment region to bacterial chromosome; *xis* = excision protein; *int* = integration protein; *red* (*exo* and *β*) = genetic recombination proteins; *cI*, *cII*, and *cIII* = immunity (and clear plaques); *rex* = gene controlling exclusion of T-even *r*II mutant phage growth. [After A. Campbell, *Episomes*, Harper & Row, New York, 1969.]

lysogenic F⁻(λ) recipient that the nonlysogeny character enters the zygote after the *gal* genes and before the *trp* genes and segregates among the recombinants like any other gene. Furthermore, crosses in which both Hfr and F⁻ were lysogenic but carried different genetically marked *lambda* prophages showed that the Hfr prophage enters the zygote and segregates among the recombinants in exactly the same way as the nonlysogeny character of the preceding cross. It could be concluded from these results that the prophage occupies a site on the *E. coli* chromosome between *gal* and *trp* genes and that the characters lysogeny, or (λ), and nonlysogeny are to be considered as alternative states of the same chromosomal locus.

ZYGOTIC INDUCTION AND THE REPRESSOR

The distribution of the *lambda* prophage among the recombinants is very anomalous, however, when the Hfr parent is lysogenic and the F⁻ parent is nonlysogenic. In such crosses the prophage is almost never inherited by the recombinants, whose frequency is very much lower than, and whose genetic character is very different from, that found in crosses of the types Hfr × F⁻(λ) or Hfr(λ) × F⁻ (λ). Jacob and Wollman could show that these abnormalities are the consequence of a phenomenon they called *zygotic induction*: whenever a chromosomal segment of the lysogenic Hfr carrying the *lambda* prophage is introduced into the nonlysogenic F⁻ recipient, the prophage becomes induced, enters the vegetative state, and causes the growth of one to two hundred infective *lambda* phage particles, which are finally liberated upon lysis of the zygote. In such crosses, most zygotes into which the Hfr had introduced the chromosomal segment carrying the *lambda* prophage are lost, and only those genes of the donor that are transferred earlier than the prophage and are not followed by it, and hence are proximal to the origin, can be expected to appear among the recombinants. No zygotic induction occurs, however, when the F⁻ is lysogenic, whether the Hfr parent is lysogenic or not, and no anomalous segregation of parental markers is observed among the issue of such crosses. Hence the presence of the *lambda* prophage renders the F⁻ recipient immune not only to superinfection by the homologous free *lambda* phage but also to vegetative development of a homologous chromosomal *lambda* prophage transferred to it from an Hfr donor parent.

Since it is evidently the "cytoplasm" of the F⁻ cell that determines whether induction of the zygote occurs, it may be concluded that the immunity of lysogenic bacteria is attributable to a cytoplasmic factor. This factor, whose existence had been first postulated by G. Bertani (Figure 10-5), not only prevents vegetative multiplication of an exogenous phage genome introduced into the lysogenic cell but, even more important, also holds in check the endogenous prophage and thus makes possible the very maintenance of the lysogenic condition. This cytoplasmic factor is itself the product of a phage gene carried by the prophage, and has been called *immunity repressor*. As will be considered in more detail in Chapter 20, the immunity repressor inhibits the expression of genes *N*, *O*, and *P* whose products are essential for the initiation of vegetative phage growth. Induction of prophage development is then to be understood as a consequence of the reduction in the intracellular concentration of the immunity repressor to a level sufficiently low for expression of genes *N*, *O*, and *P* to occur and for vegetative phage multiplication to be irreversibly initiated. It is by destroying or inhibiting the function of the immunity repressor that agents such as UV light probably exert their inducing action.

According to Jacob, it could be concluded, therefore, that every introduction of a temperate phage genome into a nonlysogenic bacterium, whether by infection or by conjugation, "results in a 'race' between the synthesis of the immunity repressor and that of the 'early' [gene] proteins required for vegetative phage multiplication. The fate of the host cell, survival with lysogenization or lysis as a result of phage multiplication, depends upon whether the synthesis of the immunity repressor or that of the protein is favored. Changes in the cultural conditions favoring the synthesis of the immunity repressor . . . would favor lysogenization, and vice versa." The synthesis of its own self-repressor is thus the very essence of the temperance of temperate phages.

GENETIC CONTROL OF LYSOGENIZATION

The plaques of phage *lambda*, like those of other temperate phages, have turbid centers that are caused by the secondary growth of lysogenized bacteria of the *lambda*-sensitive indicator strain. Mutants of *lambda* can readily be found, however, that make plaques with clear rather than turbid centers. These so-called *c* mutants owe their variant plaque morphology to a heredi-tary defect in the reactions leading to the reduction of the phage to prophage; the frequency with which *c* mutants give the lysogenic response is very much less than the corresponding frequency for the c^+ wild type, and hence much less, or no, secondary growth of lysogenized bacteria takes place in the center of the *c* mutant plaques (Figure 14-8). (The λc mutant and λc^+ wild type figured in Meselson and Weigle's experiment on the molecular basis of genetic recombination described in Chapter 12.) The *c* mutants were studied by A. D. Kaiser, who found that they pertain to three genes, *c*I, *c*II, and *c*III, in the early-function region of the genetic map of *lambda* (Figure 14-7). Of these genes, it is the *c*I gene on which the ability of *lambda* to lysogenize the sensitive host cell primarily depends, whereas *c*II and *c*III genes play a more complicated, and as yet not fully fathomed role in the control of the lyso-genization process. Thus mutational alteration of .the *c*I gene leads to a blockade of the train of reactions which leads to the reduction of the infecting

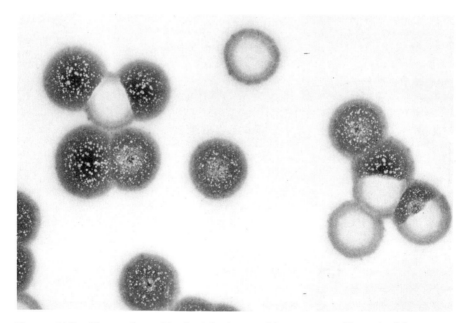

FIGURE 14-8.   Plaques formed by *lambda* phage and its *c* mutants. The λc^+ wild type produces turbid plaques, attributable to heavy secondary growth of lysogenized bacteria of the nonlysogenic indicator strain. The *c* mutants have a greatly reduced capacity to establish the lysogenic response and hence form clear plaques with little or no secondary bacterial growth. (The tiny isolated colonies visible in the clear plaques represent clones of *lambda*-resistant *E. coli* mutants). [After J. J. Weigle, *Proc. Natl. Acad. Sci. Wash.*, **39**, 628 (1953).]

phage genome to the prophage state. Since from mixed infection of sensitive bacteria with λc^+ and λcI phages *doubly* lysogenic bacteria carrying both c^+ and cI prophages can be recovered, it follows that the c^+ state is dominant over the cI state. This suggests that the cI gene controls the synthesis of the specific immunity repressor and that cI mutants carry a hereditary defect for synthesis of an effective immunity repressor. In spite of their inability to synthesize the immunity repressor, cI mutants are, however, still sensitive to the inhibitory action of the normal repressor elaborated by the c^+ wild-type phage.

CHROMOSOMAL INTEGRATION OF THE PROPHAGE

How, in the course of the lysogenic response, does the phage genome manage to become integrated into the bacterial chromosome? Almost as soon as this problem was posed, following the demonstration of the specific chromosomal location of the prophage, it seemed very likely that prophage integration is the consequence of an act of genetic exchange between homologous sectors of phage and host chromosomes. This view soon received strong support from the finding by G. Kellenberger, M. Zichichi, and Weigle that *lambda* mutants carrying a deletion of the b2 sector of the phage genome (Figure 14-7) cannot be integrated into the bacterial chromosome, though these mutants are perfectly capable of lytic growth. Thus it appeared to be the b2 sector that is involved in the genetic exchange between the *lambda* chromosome and the site of the *E. coli* genome situated between the *gal* and *trp* genes, which is referred to as the *lambda attachment*, or *attλ*, locus. Phage mutants lacking the b2 region cannot, therefore, find the *attλ* locus on the host chromosome at which they are to be integrated. Despite these insights the topological nature of the product of genetic exchange between phage and host chromosomes remained unclear for nearly another ten years. Finally, in 1962, Campbell proposed that crossing-over between phage and host chromosome results in *insertion* of the entire phage chromosome into the continuity of the bacterial chromosome, in a manner exactly analogous to the chromosomal insertion of the F, or sex factor discussed in Chapter 10. Campbell proposed (at a time when the facts presented in Figure 14-6 were not yet known) that the vegetative *lambda* chromosome is circular and that at the moment of lysogenization the b2 region of the phage and the *attλ* locus of the bacterial chromosome synapse. The phage is then inserted as a linear structure into the continuity of the chromosome by a reciprocal crossover within the synapsed regions (Figure 14-9). But since it was then already known that the linear genetic map of *lambda*, established by crosses in the lytic response, has one end near m6 and the other near *mi*, Campbell's insertion hypothesis led to an important prediction: since in lysogenization the circular vegetative *lambda* chromosome is broken by a crossover in the b2 region and not between m6 and *mi*, one would expect the gene order of the *lambda*

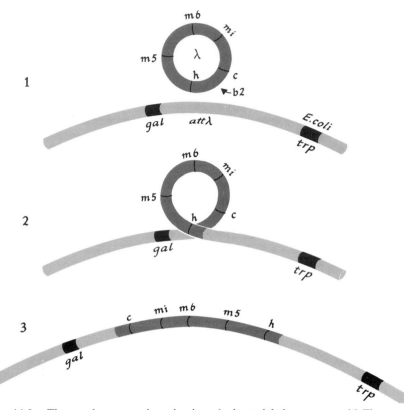

FIGURE 14-9.   The prophage as an insertion into the bacterial chromosome. (1) The circular vegetative *lambda* phage chromosome synapses with the *lambda* attachment site of the bacterial chromosome. (2) The phage chromosome breaks between *h* and *c* genes in the *b2* region, the bacterial chromosome breaks between *gal* and *trp* genes, and heterologous pieces rejoin. (3) The crossover generates one continuous genetic structure containing the *lambda* genome interposed between the bacterial *gal* and *trp* genes. (The distance between the genetic loci in Figures 14-9 through 14-13 is not drawn to scale.) [After A. Campbell, *Adv. Genet.* **11**, 101 (1962).]

prophage map to be different from that of the *lambda* phage map. And this is exactly what turned out to be the case: charting the genes of the *lambda* prophage by bacterial conjugation experiments involving genetically marked lambda prophages in donor and recipient cells yields a map whose ends are near *c* and *h* rather than near *m6* and *mi*, and which thus represents a circular permutation of the vegetative phage map of Figure 14-7. Later studies revealed that the crossover between chromosomes of phage and host cell is achieved by means of a special *integration protein* encoded in the *lambda int* gene. This protein is capable of specific recognition of the DNA base sequence of some part of the *b2* sector of the *lambda* genome.

Induction of the prophage and onset of vegetative phage growth are the reverse of the integration process. Inactivation of the immunity repressor triggers events that lead to the escape of the phage genome by reversing the steps shown in Figure 14-9 and thus regenerates the circular genome of the vegetative *lambda* phage. The *int* gene protein also participates in this reversal of the integration process, but a second, or *excision protein*, encoded in the *xis* gene is now known to be required in addition to the *int* gene protein for release of the prophage from the chromosome of the lysogenic bacterium.

The foregoing considerations have shown that temperate phages such as *lambda* are elements that are endowed with genetic continuity and which, upon being introduced into a nonlysogenic bacterium, can exist in both an *autonomous*, or vegetative, state, in which the phage genome reproduces independently of the host genome, and in an *integrated*, or prophage, state, in which the phage genome forms part of the host genome and is reproduced in synchrony with it. According to Wollman and Jacob's definition, given in Chapter 10, temperate phages, like the F sex factors, thus belong to the class of *bacterial episomes*. From an evolutionary point of view, therefore, phages may be considered as that class of episomes which has developed the hereditary information necessary for synthesis of a proteinaceous soma into which the genome can be encapsulated. In this way the phage DNA has evolved into an *infectious* episome that, in the guise of a phage particle, is capable of extra-protoplasmic existence for the transit from host to host.

RESTRICTED TRANSDUCTION

As will now be seen, the biological role of infectious episomes is much more significant than the mere broadcast and propagation of phage genomes in bacteriadom. For phages turn out to be *vectors of intercellular transfer of bacterial genes*.

This aspect of phages was discovered by Joshua Lederberg and his students, who referred to the process by which phages transfer bacterial genes as *transduction*. Their work was to show, furthermore, that there exist two fundamentally different transduction mechanisms. One of these, *restricted* transduction, was discovered in 1956, in the course of an experiment designed to test whether phage *lambda* can transfer *E. coli* genes from donor to recipient cells. This experiment consisted of inducing the *lambda* prophage of a culture of prototrophic, wild-type K12(λ) bacteria with UV light in order to generate a lysate of *lambda* phages. A variety of nonlysogenic mutant cultures of K12 bacteria was then infected with these *lambda* phages and plated for colony count on various selective agars in order to test whether any of the wild-type alleles of the genes of the K12(λ) donor cells had been transferred to the lysogenized mutant recipients. The results of this survey were mostly negative: none of the donor characters was found to have been transferred to the lysogenized recipients by *lambda* phage. There was one important exception, however: in the experiment in which the nonlysogenic recipient bacteria had been Gal⁻, about 10^{-6} of the *lambda*-infected Gal⁻ bacteria had actually acquired the Gal⁺ character of the wild-type donor strain. It was thus found that the *lambda* phage *is* capable of gene transfer, or transduction, but only of a restricted sort that pertains exclusively to the *gal* genes in the vicinity of the *lambda* attachment locus, or *attλ*, on the *E. coli* chromosome (Figure 10-21). Phage *lambda* does not seem capable of transducing genes from any other part of the *E. coli* genome.

Upon closer examination of the clones of transduced bacteria or *transductants*, which had acquired the Gal⁺ character of the prototrophic donor cell, it was found, first of all, that all the Gal⁺ transductants were either actively lysogenic (liberated infective *lambda* phages on induction), or were at least immune to infection by exogenous *lambda*, showing that the nonlysogenic K12 Gal⁻ recipient cell had acquired the *lambda* prophage along with the donor *gal* genes. Second, it was found that most of these Gal⁺ transductants were *genetically unstable*; each Gal⁺ colony contained about 1–10 % of individuals that had lost the Gal⁺ character and were once more of the same Gal⁻ phenotype as their recipient ancestor. This character of the Gal⁺ transductants is immediately obvious when they are plated on EMB-galactose agar; nearly every red transductant colony contains white sectors representing clones of Gal⁻ segregants (thus presenting an aspect quite similar to that of colonies of the F–Lac⁺/Lac⁻ diploids on EMB-lactose agar shown in Plate III). The frequent segregation of recipient Gal⁻ types by the Gal⁺ transductants (estimated to be about 2×10^{-3} per bacterial division) led to the inference that these transductants are actually Gal⁺/Gal⁻ partial heterozygotes. In these heterozygotes the *gal* donor fragment brought in by the transducing *lambda* phage has been *added* to the recipient genome, rather than exchanged for its alleles, and the Gal⁻ segregants represent descendants of such heterozygotes from which the *gal* donor fragment has been lost.

But a most remarkable finding was made when the lambda prophage of a culture of such lysogenic Gal⁺/Gal⁻ heterozygotes was induced by UV light: the lysate of infective *lambda* phages resulting from this induction possessed an extraordinarily high transducing power, since nearly half of the phage particles appeared to be capable of transducing the Gal⁺ character into new Gal⁻ recipient cells. Such lysates were called HFT, for *high-frequency transduction*, in contrast to the original LFT, or low-frequency transduction, lysates from which the heterozygote transductants had first been obtained at a frequency of only about 10^{-6} transductants per *lambda* phage. Like the LFT lysates, HFT lysates can transduce only the *gal* character, and the transductants produced are mostly unstable lysogenic Gal⁺/Gal⁻ heterozygotes, which yield HFT lysates upon UV induction.

Further study of the phages present in an HFT lysate revealed that its *infective lambda* phages are devoid of any transducing power and that the *gal* genes of the donor bacteria are carried by *defective lambda* phages. For it was found that if nonlysogenic Gal⁻ recipient bacteria are infected at a low multiplicity of infection with the *lambda* phages of an HFT lysate, so that each bacterium that is infected at all is infected with only one phage particle, then practically all heterozygote Gal⁺/Gal⁻ transductants produced appear to carry a defective prophage; though they liberate no infective *lambda* phages, they are immune to superinfection by exogenous *lambda* particles. Such transductants carry a defective prophage, since just those phage particles in the HFT lysate which carry the *gal* genes of the donor cell are themselves defective. This defectivity renders the *gal*-carrying transducing phages incapable of vegetative reproduction, and hence unable to form plaques.

These defective, *gal*-transducing *lambda* phage particles are designated by the symbol λ*dg*. The defective λ*dg* phages of the HFT lysate can multiply vegetatively and give rise to structurally intact progeny phages, as long as they are together in the same cell with nondefective, normal *lambda helper* phages that can supply the missing synthetic or morphogenetic functions to the defective phage genome. The λ*dg* phages can, therefore, be crossed to nondefective, genetically marked λ phages, and the site of the defectivity in the transducing genome can be located on the *lambda* genetic map. Such crosses showed that the defectivity of the λ*dg* particles, as well as their transducing power, arises from an exchange of the *gal* region of the bacterium for about one-third to one-fourth of the phage genome, so that the defective λ*dg* phages are missing some of their own essential genes. The missing phage genes belong to the *h* region of the *lambda* phage genome, as shown in Figure 14-10. Nondefective, or actively lysogenic, Gal⁺ transductants are therefore obtained only if the nonlysogenic Gal⁻ bacteria are infected at a *high multiplicity* of infection with the phages of the HFT lysate, so that every cell that has been infected with a λ*dg* transducing phage has, at the same time, also been infected with a normal, nontransducing λ particle.

The mechanism of integration of the *lambda* prophage into the bacterial chromosome shown diagrammatically in Figure 14-9 readily explains the origin and genetic structure of the λ*dg* phage. For λ*dg* appears to be created by precisely the same process by which bacterial genes are incorporated into the F sex factors to generate such F' episomes as F-*lac* (see Chapter 10). Evidently the rare creation of λ*dg* particles upon UV induction of an ordinary K12(λ)Gal⁺ bacterium results from an improper looping-out during recircularization

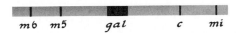

m6   m5   gal   c   mi

FIGURE 14-10.   Map of the chromosome of a defective λ*dg* phage, showing the replacement of the phage *h* gene region by the bacterial *gal* genes.

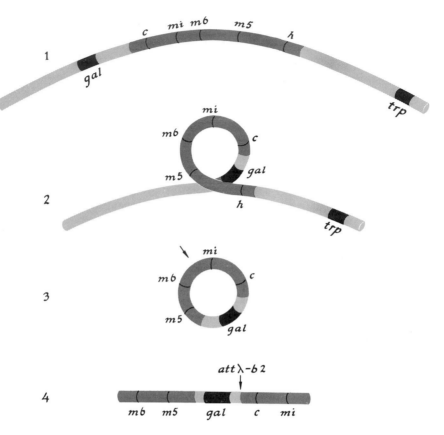

FIGURE 14-11.   Creation of a Gal⁺ defective λdg transducing phage. (1) The chromosome
of the lysogenic K12(λ)Gal⁺ bacterium. (2) Improper looping out of the prophage sector
following UV-induction of the lysogenic cell. (3) Excision of the improper loop gives
rise to a circular phage chromosome that includes the bacterial *gal* genes but leaves
behind in the bacterial chromosome the *h* gene region of the phage. Betwen *gal* and *c*,
at its boundary of bacterial and phage genome, the excised chromosomal loop carries a
nucleotide sequence derived partly from the *E. coli attλ* and partly from the *lambda b*2
region. (4) Normal opening of the excised circle between *m*6 and *mi* by the "nicking
enzyme" gives rise to the linear λdg chromosome of Figure 14-10.

attending the excision of the *lambda* prophage from the *E. coli* chromosome, as shown in Figure 14-11. This improper looping-out causes the
inclusion of the bacterial *gal* genes within the excised circle, while excluding
from the circle and leaving behind in the bacterial chromosome those sectors
of the *lambda* genome that are most distal to *gal* in the prophage structure—
that is, the *h* region. The phage genome thus created is evidently defective, in
that it lacks a block of essential phage genes. Upon infection of the Gal⁻
recipient, the *b*2 region of the circularized λdg genome synapses with *attλ* locus

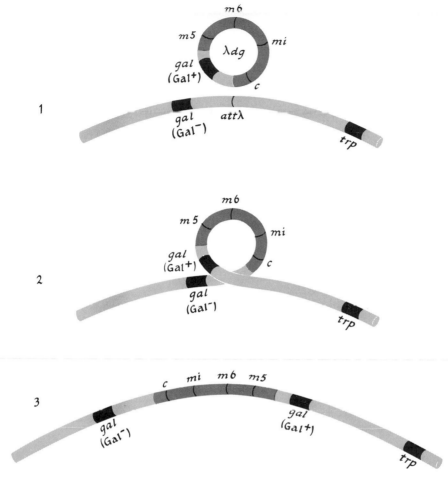

FIGURE 14-12. Transduction of a nonlysogenic K12 Gal⁻ recipient cell by a Gal⁺λ*dg*
phage into a K12(λ*dg*) Gal⁺/Gal⁻ partially heterozygous transductant. (1) Recipient
chromosome and λ*dg* phage chromosome synapse in the homologous *att*λ and *b*2
regions. (2) Phage and bacterial chromosome crossover in their synapsed regions.
(3) The crossover generates a continuous genetic structure containing a defective *lambda*
prophage interposed between two bacterial *gal* genes, one of Gal⁺ donor type and the
other of Gal⁻ recipient type.

of the recipient chromosome and by means of the crossover event shown in
Figure 14-12 integrates the λ*dg* genome into the recipient genome to give rise
to a Gal⁺/Gal⁻ partial heterozygote carrying the defective *lambda* prophage.
Occasional reversal of this integration process would lead to escape of the
defective prophage and hence to the appearance of a Gal⁻ segregant. In case
the Gal⁻ recipient cell had been infected with *both* infective λ and defective
λ*dg* phages, production of a λ/λ*dg* lysogenic Gal⁺/Gal⁻ heterozygote would

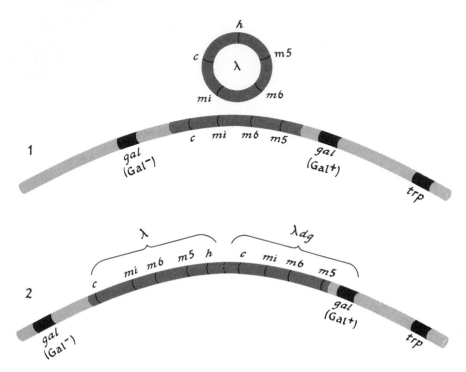

FIGURE 14-13.   Production of a doubly lysogenic K12(λ/λdg)Gal$^+$/Gal$^-$ transductant upon mixed infection of a K12 Gal$^-$ recipient cell by a λdg Gal$^+$ and a normal λ phage. (1) The circular chromosome of the normal λ phage may synapse in any region of its genome (except in the h gene region) with the homologous region of the λdg prophage already inserted into the recipient chromosome by the process outlined in Figure 14-12. (2) Normal λ phage and defective λdg prophage chromosomes crossover in their synapsed regions, giving rise to a continuous genetic structure containing a normal and a defective *lambda* prophage inserted between two *gal* genes. UV-induction of a bacterium carrying such a chromosome gives rise to an HFT lysate containing equal proportions of normal λ and transducing λdg phages, by reversal of the reaction sequences of Figures 14-13 and 14-12.

result if the crossover of Figure 14-12 were followed by the event shown in Figure 14-13. Upon UV-induction of such a doubly lysogenic cell both λ and λdg prophages can escape from the host chromosome upon recircularization and reversal of the processes of Figures 14-12 and 14-13. The presence of the intact *lambda* genome provides all the enzymes and structural proteins required for vegetative *lambda* phage growth, so that at the end of the normal latent period the induced bacteria lyse, yielding a mixed progeny phage population carrying in part normal λ and in part defective λdg genomes—that is to say, an HFT lysate.

The incorporation of fragments of the host genome into the phage genome can also be demonstrated by physicochemical techniques, since the density of λdg phages differs from that of normal λ phage particles. That is, the improper looping-out process that exchanges the *gal* region of the bacterial chromosome for the *h* gene region of the phage chromosome does not appear to conserve the normal circumference of the excised circle, and hence may result in defective transducing phages that comprise either more or less DNA than the normal *lambda* genome. Since the phage protein, whose amount per *lambda* phage particle appears to be constant, is less dense than the phage DNA, λdg phages carrying either more or less DNA than normal λ phages have a correspondingly greater or lesser weight per unit volume than normal infective phage particles (Figure 14-14). This difference in density permits the physical separation of mixtures of defective and normal phage particles by CsCl density-gradient sedimentation, and hence the preparation of pure stocks of λdg—that is, stocks free of the infective λ helper phages, which if absent from the host cell render the defective genome incapable of maturing as an intact phage.

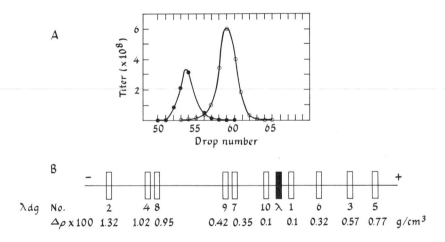

FIGURE 14-14.   Density differences between normal λ and λdg transducing phages.
A. Density-gradient centrifugation in a concentrated CsCl solution of an HFT lysate obtained by UV-induction of a Gal$^+$/Gal$^-$ transductant culture. After sedimentation equilibrium has been reached and the phages have moved into bands corresponding to their density, the centrifuge is stopped and the contents of the centrifugate tube collected drop by drop through a hole in the bottom of the tube. The drops are assayed for their titers of normal plaque-forming λ phages (open circles) and λdg transducing phages (solid circles). An increasing drop number on the abscissa reflects decreasing density. It is apparent that the λdg phages of the HFT lysate are more dense than normal λ phages.
B. Densities of some λdg stocks of independent origin. Each vertical bar represents the band position of one of ten different λdg transducing strains, as well as that of the λ wild type, in a CsCl density gradient. The difference between the densities of each strain and of the λ wild type, $\Delta\rho$, is shown here multiplied by 100. [After J. J. Weigle, M. Meselson, and K. Paigen. *J. Mol. Biol.* **1**, 379 (1959).]

GENERAL TRANSDUCTION

The first phage-mediated transfer of bacterial genes to be discovered was not the restricted transduction by λdg, but a phenomenon now called *general transduction*. In 1951, Lederberg and his student N. Zinder were looking for genetic recombination in *Salmonella typhimurium*, hoping to find a process analogous to the bacterial conjugation that Lederberg and Tatum had found in *E. coli* some five years earlier. They employed the same experimental procedure that had been so successful in those earlier experiments: about 10^9 cells each of two auxotrophic mutant strains of *S. typhimurium*—one unable to synthesize phenylalanine, tryptophan, and tyrosine (Phe$^-$ Trp$^-$ Tyr$^-$), and the other unable to synthesize methionine and histidine (Met$^-$ His$^-$)—were plated together on an amino-acid-free minimal medium like the one specified in Table 2-1, containing only inorganic salts and a carbohydrate as a source of carbon and energy. The result of this experiment was that about 10^{-5} of the bacteria plated grew into prototrophic colonies that no longer required any added amino acids for growth. Since none of the bacteria of either auxotrophic strain plated alone was able to grow into colonies on this minimal medium in the absence of their required amino acids, the amino-acid-independent prototrophs that appear in the mixed plating could not have derived from reverse mutants in which the auxotrophic Phe$^-$ Trp$^-$ Tyr$^-$ or Met$^-$ His$^-$ mutant genes had spontaneously reverted to their prototrophic state. Instead, these prototrophs must have been genetic recombinants that united the wild-type Phe$^+$ Trp$^+$ Tyr$^+$ alleles of one and the wild-type Met$^+$ His$^+$ alleles of the other parent strain within the same cell. In further experiments, Zinder and Lederberg found that other characters can also be exchanged between the two *Salmonella* strains: genetic recombinants also arose during mixed growth of mutants that were unable to ferment galactose (Gal$^-$) or xylose (Xyl$^-$), or mannitol (Mtl$^-$), or maltose (Mal$^-$), or that were resistant to streptomycin (Str^r). *Salmonella*, like *Escherichia*, thus appeared to be capable of engaging in a process of genetic exchange that results in the appearance of recombinants endowed with genetic characters from bacteria of different lines of descent.

   Growth of the two parent cultures in the Davis **U**-tube (Figure 10-3) revealed, however, that in contrast to the requirements for *E. coli* conjugation, cell contact between the parent strains is *not* necessary for recombination to occur. Genetic exchange in *Salmonella* thus appeared to be mediated by a filtrable agent (FA), small enough to pass through the sintered glass pores, that brings hereditary factors from donor to recipient cells. Closer study of the FA revealed that it is, in fact, the temperate *Salmonella* phage P22, which one of the parent strains happened to carry as a prophage. It was possible to reach this conclusion from the following correlations between FA and P22:

1. The infectivity of P22 and the genetic competence of FA are protected from inactivation by hydrolytic enzymes, such as deoxyribonuclease and trypsin.

2. The dimension and mass of P22 and FA are the same.

3. Treatment with anti-P22 serum or with heat inactivates the infectivity of P22 and the genetic competence of FA at the same rate.

4. Phage-resistant strains of *Salmonella* that no longer adsorb phage P22 cannot interact genetically with FA; P22 and FA have the same host range.

Zinder and Lederberg thus discovered transduction as a third mechanism of genetic exchange in bacteria, and one that differs radically from either transformation or conjugation. Since this transduction mediated by phage P22 can pertain to *any* sector of the *Salmonella* genome, it is called *general transduction*, in contradistinction to the restricted transduction mediated by phage *lambda*, which, as was seen in the preceding discussions, pertains only to sectors of the *E. coli* genome adjacent to the *lambda* attachment locus.

Once the role of the temperate phage P22 in transduction had been established, the following basic experimental procedure in the study of transduction became established. The nonlysogenic donor strain is infected with phage P22, lytic multiplication of the phage on the sensitive bacteria is allowed, and the progeny phages produced are harvested and freed of any remaining unlysed donor bacteria. A culture of the lysogenic or nonlysogenic recipient strain is then infected with this P22-phage lysate, and the bacteria are plated on a selective agar that allows growth of only those types into which some particular character of the donor strain has been transduced. In case the recipient is lysogenic and carries the P22 prophage, it is immune to P22, and hence all of the recipient cells will survive infection by the phages of the transducing lysate. In case the recipient strain is nonlysogenic, and hence not immune to P22, then transduced bacterial clones can descend only from those bacteria in which infection by phages of the transducing lysate has not resulted in the lytic response. Sensitive bacteria giving the lytic response are of course lost and cannot give rise to transductants. This procedure makes possible a quantitative estimate of the frequency with which transduction occurs. The relative efficiency of transduction of any P22-phage lysate is evidently the ratio of the number of transductions produced to the number of P22 phage particles with which the recipient bacteria had been infected. Such estimates show that this efficiency is very low: the ratio of P22 phages that can transduce into the recipient bacteria a given character of the donor strain—for example, capacity for amino acid synthesis or sugar fermentation—is only about 10^{-5}.

Zinder and Lederberg addressed themselves next to the question of how much of the genome of the donor cell is carried by one of these evidently very rare transducing phage particles. For this purpose, a stock of P22 phage was grown on a prototrophic, streptomycin-sensitive (Trp$^+$ Gal$^+$ Xyl$^+$ Str^s) donor strain and a multiple mutant Trp$^-$ Gal$^-$ Xyl$^-$ Str^r strain was infected with this phage lysate. The appearance of Trp$^+$ or Gal$^+$ or Xyl$^+$ transductants was then scored on three different media: (1) minimal glucose agar without tryptophan, (2) EMB-galactose agar, and (3) EMB-xylose agar. Each of these media selects for transduction of only one of the donor characters, the other characters remaining unselected, inasmuch as the bacterial clone can grow or give a positive color on EMB agar whether the other characters have been transduced or not. The result of this experiment was that roughly equal numbers of transductants appeared on all three types of agar, showing that the chances were about equal that a P22 phage particle carried either the Trp$^+$, or the Gal$^+$, or the Xyl$^+$, character of the donor bacterium. When these transductants were then tested for their unselected characters, however, it was found that practically none of the bacteria transduced for a selected character had also at the same time acquired one of the unselected characters of the donor; for example, all the Trp$^+$ transductants that grew on the minimal glucose agar still had the Gal$^-$, Xyl$^-$, Str^r characters of the recipient strain. Since similar results were also found when other fermentation or auxotrophic mutations were employed as selected or unselected characters, it could be inferred that each transducing phage particle carries and brings into the recipient cell only a small part of the genome of the donor bacterium on which it had grown. This inference is of course eminently plausible, since the total amount of DNA making up the genome of the *Salmonella* bacterium is about a hundred times greater than the amount of DNA in the head of the P22 phage. It would thus be difficult to imagine that any one phage particle could contain very much more than a few percent of the total DNA, and hence of the genome, of the donor cell.

The following notion of the transduction phenomenon was thus developed. As the temperate P22 phage grows in its lytic cycle on the sensitive donor bacterium, a small fragment of the bacterial DNA manages to insinuate itself into the head of a progeny phage at the time that the phage DNA molecules are withdrawn from the vegetative pool for maturation into infective phage particles. When such an unusual phage particle infects a recipient bacterium after its release from the donor cell, the stowaway bacterial DNA is injected into the cell as if it were a phage DNA molecule. If the bacterium survives this infection then the transduced bacterial DNA has an opportunity to undergo genetic exchange with its homologous alleles on the recipient host chromosome and thus to produce the rare recombinant cell that carries a small part of the genome of the donor cell.

It was obvious that the behavior of linked genetic loci in transduction should be of great significance for understanding how the phage particles actually acquire parts of the donor genome and how these parts are later integrated into the recipient genome. Hence it seemed a pity that transduction had been encountered only in *Salmonella*, which in the 1950's was still largely a genetic *terra incognita*, and not in *E. coli*, for which a variety of mutant characters were available whose linkage relations on the bacterial chromosome had already been established by conjugational analysis. It was good news, therefore, when E. Lennox found in 1955 that the temperate phage P1 is capable of transducing genetic markers from donor to recipient cells in *E. coli*. This discovery made it possible to show that none of the genetic characters previously known to be situated at distant loci of the *E. coli* chromosome are ever cotransduced. But it was also possible to show that the two linked genes *thr* and *leu* (see Figure 10-21), which govern the synthesis of the amino acids threonine and leucine, *are* cotransducible with a frequency of about 1%; that is to say, about 1% of Thr⁻ recipient bacteria into which the *thr* allele of the donor bacterium is transduced also obtains the (unselected) *leu* allele of the donor strain. The distance between *thr* and *leu* is approximately 2% of the total length of the *E. coli* genome; although they act as very closely linked genetic sites in the recombinational events ensuing upon bacterial conjugation, the linkage of these genes is thus barely demonstrable in transduction.

ORIGIN OF THE GENERAL TRANSDUCING FRAGMENT

The discovery that the restricted transducing *lambda* phages are defective stimulated analogous experiments designed to test the possible defectivity of the general transducing P1 phage. In earlier work on general transduction, the transductants recovered after infection of nonlysogenic recipients were usually lysogenic (i.e., they carried the prophage of the transducing phage), because proper precautions had not been taken to avoid multiple infection of the nonlysogenic recipients by the transducing phage stock. When conditions of true single infection were at last realized, it was found that transducing P1 phages likewise lack a functionally or structurally normal genome. But in contrast to the defective lysogenic transductants recovered after single infection of nonlysogenic Gal⁻ recipients with *λdg*, nearly all transductants that were recovered after single infection of nonlysogenic recipients with P1 transducing phages turned out to be nonlysogenic, still sensitive to the P1 phage type that transduced them. Lysogenic transductants can, of course, be obtained if the recipient bacteria are infected with several P1 phages per cell. Phage populations liberated by induction of such lysogenic P1 transductants, furthermore, do not possess the high transducing power of HFT lysates generated by the induction of the Gal⁺/Gal⁻ heterozygotes produced by restricted transduction with *λdg*.

In 1965 Ikeda and Tomizawa carried out an experiment designed to probe the basis of this difference between the properties of transductants generated by restricted and those generated by general transduction. In particular, this experiment was directed toward inquiring into the origin of the bacterial genetic markers carried by P1-transducing phage particles. For this purpose, one culture, *a*, of a thymine-requiring (Thy⁻) donor strain of *E. coli* was grown in a light medium in the presence of thymine (T), and two other cultures *b*, and *c*, of the same strain were grown in a heavy medium containing 5-bromouracil (BU) instead of the required T. All three cultures were then infected with P1 phages, the bacteria of culture *b*, however, having been transferred from a heavy BU-supplemented medium to a light T-supplemented medium at the moment of their infection. The three resulting P1 phage lysates were subjected to CsCl density-gradient sedimentation, the phages present at various levels of the gradient at sedimentation equilibrium were collected by the method described in Figure 9-10, and each fraction was assayed for its content of both *infective* (i.e., plaque-forming) and *transducing* (i.e., capable of transducing Leu⁻ auxotrophic recipients to Leu⁺ prototrophy) P1 particles. The result of this experiment is presented in Figure 14-15, which shows that both infective and transducing P1 particles produced in the light culture *a* have a similar density (about 1.47 g/cm³), which is true also for both infective and transducing P1 particles (both having a density of about 1.49 g/cm³) produced in the heavy culture *c*. Thus replacement of T by BU in the DNA of the P1 phage causes the density of the whole phage particle to increase by about 0.02 g/cm³. Among the P1 phage produced in culture *b*, however, in which BU was present only *before* infection, the transducing particles are evidently of the higher density characteristic of the BU-culture *c*, whereas the infective particles are of the lower density characteristic of the T-culture *a*. The high density of the transducing phage particles produced in culture *b* indicates that their DNA contains fragments of the BU-labeled bacterial DNA that had undergone no replication after P1 phage infection of the donor cell and were incorporated directly into the head of the transducing phage particle. It can be inferred, furthermore, that the transducing phage particles carry very little, if any, *phage* DNA, since, as is indicated by the low density of the infective P1 particles, all the phage DNA produced in culture *b* was of the light or T-labeled kind.

The absence of phage DNA from the P1-transducing phages now readily accounts for two facts: that transductants produced by single infection of nonlysogenic recipient cells are nonlysogenic, and that induction of lysogenic transductants that *do* carry the P1 prophage does not generate HFT lysates. Evidently the P1-transducing particles carry pristine fragments of the bacterial genome that, in contrast to the bacterial genes carried by *λdg*, are *not recombined* into any defective phage genomes, and hence neither endow the recipient cell with immunity against P1 nor undergo vegetative multiplication in the presence of an infective P1 helper genome.

Figure 14-15 makes it clear that transducing and infective phage particles produced in the uniformly T-labeled light culture *a* do not have *exactly* the same density, and hence can be resolved in the CsCl gradient on the basis of their density. Thus it is possible to estimate the fraction of all P1 phage particles present in the lysate that are of the transducing type. For that purpose, the DNA of the donor bacteria was labeled with ^3H by growth of the culture in a medium supplemented with ^3H-labeled thymidine. The

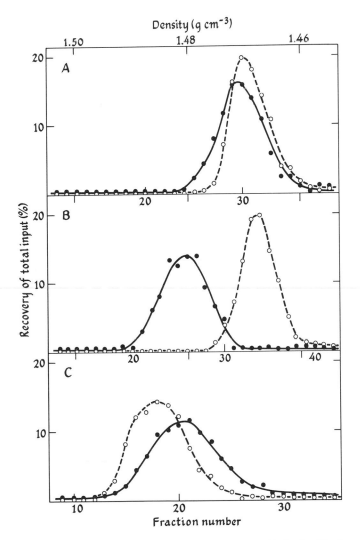

FIGURE 14-15. CsCl density-gradient-sedimentation profiles of infective P1 phage particles (open circles) and Leu$^+$ transducing P1 phage particles (filled circles) present in P1 phage lysates of Leu$^+$ Thy$^-$ *E. coli* donor cells grown under three different bromouracil (BU) versus thymine (T) labeling programs. A. T present both during growth of *E. coli* and after their P1 infection. B. BU present during growth of *E. coli*; T present after their P1 infection. C. BU present both during growth of *E. coli* and after their P1 infection. [Redrawn from H. Ikeda and J. Tomizawa, *J. Mol. Biol.* **14**, 85 (1965).]

culture was then transferred to a medium containing nonradioactive thymidine and $^{32}PO_4^{-3}$, so that any new DNA synthesized after this time would be ^{32}P-labeled. The culture was then infected with P1 phages and the resulting P1 phage lysate was fractionated by CsCl density-gradient sedimentation. The result of this experiment was that the amount of ^3H-labeled DNA found to have the density of the P1 particles capable of transducing the Leu$^+$ marker of the donor bacteria represented 0.3 % of the total ^{32}P-labeled DNA present in the phage particles analyzed in the gradient. Hence it could be concluded that about 0.3 % of the total phage particles present in P1 lysate are of the transducing kind. This number now makes it possible to make a rough estimate of the frequency of P1 particles carrying a transducing segment of the bacterial DNA for any *given* genetic marker of the donor genome. The P1 phage DNA complement (of both infective and transducing particles) comprises about 10^5 nucleotide base-pairs, and the total *E. coli* DNA genome was seen to comprise about 3×10^6 nucleotide base-pairs, so that the fraction of the *E. coli* genome carried by one transducing phage particle is $10^5/3 \times 10^6$ or 3 %. This estimate is in good agreement with Lennox's earlier finding that cotransduction of *thr* and *leu* genes of *E. coli*, whose distance of separation represents about 2 % of the bacterial genome, is just barely demonstrable. Genes separated by *more* than 3 % of the total genome thus could never be cotransduced, since the DNA molecule representing that linkage group could not be accommodated within a single P1 phage particle. Now if 0.3 % of all phages in a P1 lysate are of the transducing kind, then the frequency of phage particles carrying any *given* donor gene is $0.03 \times 0.003 = 9 \times 10^{-5}$. This estimate, in turn, is in reasonable agreement with the finding that the efficiency of obtaining Trp$^+$ transductants upon infecting a Trp$^-$ auxotroph with P1 phages grown on a Trp$^+$ wild-type donor strain is about 3×10^{-5} transductants per P1 phage adsorbed.

FINE-STRUCTURE MAPPING

Extensive P1-transduction experiments involving many genes previously mapped on the *E. coli* chromosome on the basis of conjugational recombination showed that the linkage of two genes on the bacterial chromosome can be inferred from the relative frequency with which they are cotransduced: the greater the frequency of cotransduction, the closer the linkage. This is eminently plausible, for the closer the linkage of two genes, the greater the chance that they should happen to reside on the same 3 % fragment cut from the bacterial genome and thus be incorporated into one and the same transducing phage particle. But if one examines the cotransduction of genetic markers so closely linked that they should nearly always be carried in the same phage particle (if they are transduced at all), one finds that such markers do not invariably show up in the same transductant bacterium. This transductional segregation of very closely linked markers is, no doubt, a reflection of the process of genetic recombination, by which the transduced genetic donor sites become integrated into the genome of the recipient cell. As is shown in Figure 14-16, a double crossover is required for every integration act. It follows, therefore, that two closely linked genetic markers of the donor genome brought into the same recipient cell are integrated into one and the same recombinant genome only if neither of the two required crossovers occurs anywhere between them. Since the chance that one of the two required exchanges will not occur between the two markers increases with their linkage, it is possible to infer

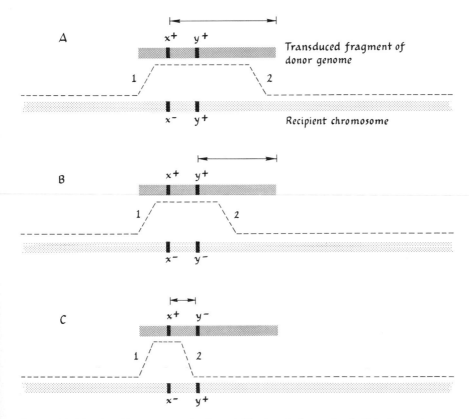

FIGURE 14-16. Double crossovers in the neighborhood of two closely linked genetic sites x and y required to generate x^+y^+ transductants. A. If the x^+y^+ wild type acts as donor and the x^-y^+ single mutant acts as recipient. B. If the x^+y^+ wild type acts as donor and the x^-y^- double mutant acts as recipient. C. If the x^+y^- single mutant acts as donor and the x^-y^+ single mutant acts as recipient. The arrow in each panel indicates the length of the genetic segment within which the second crossover must occur. [From *Molecular Biology of Bacterial Viruses*, by G. S. Stent. W. H. Freeman and Company. Copyright © 1963.]

the map distance of very closely linked genetic sites from the frequency of their cotransduction, or to obtain in this way a fine-structure map of a small segment of the bacterial chromosome.

On the basis of this principle Charles Yanofsky and Lennox constructed a fine-structure genetic map of the *trp* region of the *E. coli* chromosome in 1959. This region, as can be seen from the overall map of Figure 10-21, is located near minute 24 of the circular chromosome, close to the *ton* and *cys*B genes which control, respectively, the T1 phage receptors, and synthesis of the amino acid cysteine. To construct this map, they focused their attention on Yanofsky's set of Trp$^-$ auxotrophs, whose growth requirements and metabolite accumulations divided these mutants into five distinct classes (Table 5-1) according to the exact step in the terminal part of the pathway of tryptophan biosynthesis (Figure 3-10) at which each mutant is blocked. Now every one of these Trp$^-$ auxotrophs can be converted into a Trp$^+$ prototroph by infection with phages of a P1-transducing lysate grown on a Trp$^+$ *E. coli* wild-type donor strain. Upon selecting such prototroph transductants by plating the P1-infected Trp$^-$ recipients on minimal tryptophan-free agar it is found that about 3×10^{-5} Trp$^+$ transductants appear per P1 phage infecting a Trp$^-$ recipient cell, regardless of which mutant figured as the recipient. For the purpose of fine-structure mapping, however, transductions were performed in which Trp$^-$ auxotrophs of the five different classes figured both as donors and as recipients. For the first set of such transductions P1 phage was grown on a donor Trp$^-$ auxotroph belonging to one of the five classes—for instance, TrpE$^-$, whose members carry a mutation in the anthranilate synthetase gene and hence can grow on anthranilic acid, or on indole, or on tryptophan. A recipient Trp$^-$ auxotroph belonging to another class that responds to fewer growth supplements than the donor—for instance, TrpB$^-$, which carries a mutation in the tryptophan synthetase B protein gene and hence can grow *only* on tryptophan—was then infected with that transducing phage. Transductants were selected by plating the P1-infected recipient bacteria on minimal agar containing a supplement on which the donor, but not the recipient, can grow, such as anthranilic acid in the example chosen here. Next the transductant colonies were picked and retested for their capacity to grow on completely unsupplemented minimal glucose agar—that is, to see whether any were transduced to Trp$^+$ prototrophy. As can be seen in line C of Figure 14-16 all prototrophic transductants—those that can dispense with the growth requirement of the donor—must have arisen by a crossover between the TrpE$^-$ (or y^-) mutant site of the donor genome and the TrpB$^-$ (or x^-) mutant site of the recipient genome. But all those transductants that on retesting show the growth requirement of the donor (i.e., anthranilic acid, or indole, or tryptophan) must have resulted from transductional events in which the TrpE$^-$ (y^-) mutant site and TrpB$^+$ wild-type allele (x^+) of the donor were cotransduced. Since the probability that the two donor loci *are* cotransduced rises with their linkage, the percentage of transductants that have acquired the donor growth requirements is evidently a measure of the relative distance between the TrpE⁻ and TrpB⁻ mutations on the *E. coli* chromosome. Part I of Table 14-1 presents the result of that first set of P1

TABLE 14-1.   Transductional Fine-structure Mapping of the *trp* Gene Cluster of *E. coli*

Part	Donor class	Recipient class	Supplement in selective medium	Average (in percent) of Trp⁺ among transductants
	TrpE⁻	TrpC⁻	anthranilic acid	6.7
	TrpE⁻	TrpA⁻	anthranilic acid	14.3
I	TrpE⁻	TrpB⁻	anthranilic acid	16.6
	TrpC⁻	TrpB⁻	indole	6.2
	TrpA⁻	TrpB⁻	indole	3.0
II	TrpE⁻,B⁻	TrpA⁻	anthranilic acid	88
	TrpE⁻	CysB⁻	tryptophan	37
III	TrpC⁻	CysB⁻	tryptophan	47
	TrpA⁻	CysB⁻	tryptophan	54
	TrpB⁻	CysB⁻	tryptophan	53

Data from C. Yanofsky and E. S. Lennox, *Virology*, **8**, 725 (1959).

transductions (which did not include the TrpD⁻ class). The data show, first of all, that the genes that govern enzymes of the tryptophan pathway are very closely linked, since in all cases more than 80% of the transductants had received both mutant (y^-) and wild-type (x^+) alleles of the donor genome. Second, it is apparent that the *trp*E gene is at one end of this group of linked genes and that *trp*A and *trp*B are at the other end, with the *trp*C gene situated in between.

These data do not, however, make it possible to decide which of the two evidently very closely linked *trp*A and *trp*B genes is actually at that end of the linkage group. In order to settle this point, a *double-mutant* transduction experiment (or, in classical genetic parlance, a *three-factor cross*) was performed, in which a TrpE⁻ TrpB⁻ donor strain carried *two* Trp⁻ mutations, one in the *trp*E gene and the other in the *trp*B gene, and a TrpA⁻ recipient carried a mutation in the *trp*A gene. Transductants were then selected on minimal agar supplemented with anthranilic acid: all such transductants *must* be of TrpA$^+$ TrpB$^+$ type, and hence must have obtained the *trp*A wild-type allele (x^+) of the donor and the *trp*B wild-type allele (y^+) of the recipient in order to be able to grow on anthranilic acid. That is, a cross-over between TrpA$^-$ and TrpB$^-$ mutant sites *must* have occurred. These transductants were then tested for their capacity to grow on completely unsupplemented agar in order to determine whether in acquiring the wild-type *trp*A allele of the TrpA$^+$ donor they were cotransduced also for the donor *trp*E mutant gene. The result of that double-mutant transduction is shown in part II of Table 14-1. It can be seen there that 88 % of the TrpA$^+$ TrpB$^+$ transductants were prototrophs—that is, had *retained* the *trp*E gene of the TrpE$^+$ recipient.

We can now consider the meaning of this result. If the order of the three genes under study here were *trp*E-*trp*B-*trp*A, then according to part A of Figure 14-17 crossovers 1 and 2 are in any case required to generate the selected TrpA$^+$ TrpB$^+$ type. These crossovers would retain the TrpE$^+$ allele of the recipient, and two additional crossovers, 3 and 4, would be necessary before the mutant TrpE$^-$ alleles of the donor could appear in the transductant. If the gene order were *trp*E-*trp*A-*trp*B, however, then according to part B of Figure 14-17 required crossover 1 would bring the mutant TrpE$^-$ allele of the donor into the transductant genome, unless required crossover 2 happens to occur between the linked *trp*E and *trp*A genes. Since it would be much more likely that crossover 2 occurs somewhere to the left of the *trp*E gene, one would not expect frequent retention of the recipient TrpE$^+$ allele. That the great majority of TrpA$^+$ TrpB$^+$ transductants of this experiment are of TrpE$^+$ recipient type thus indicates that *trp*B is closer to *trp*E than is *trp*A. The results of later P1-transduction experiments in which TrpD$^-$ mutants were also used indicated that the *trp*D gene is between *trp*E and *trp*C. Hence the map order of the five tryptophan genes is

$$trpE - trpD - trpC - trpB - trpA$$

In order to determine how this linked group of tryptophan genes is oriented in the *E. coli* chromosome, a further set of transductions was undertaken. For this purpose, a cysteine-requiring CysB$^-$ auxotroph was infected as recipient with P1-transducing phages grown on various Trp$^-$ donor auxotrophs, and Cys$^+$ transductants were selected on minimal agar supplemented with tryptophan. The transductants were then picked and retested for their capacity to grow in the absence of tryptophan, in order to determine whether in acquiring the wild-type *cys* donor gene they were also cotransduced for the mutant *trp* donor gene. The result of this experiment is presented as part III of Table 14-1. It can be seen from these data that the percentage of prototrophs resulting from this transduction is much greater than in any transduction recorded in part I of the table. Thus the *cys*B gene is separated by a greater distance from any *trp* gene than the *trp* genes are separated from each

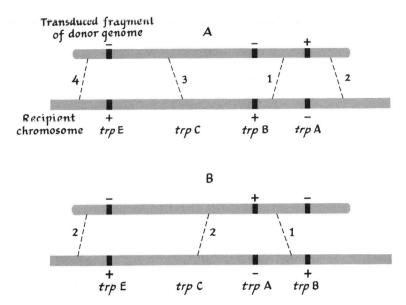

FIGURE 14-17.   Principle of the double mutant transduction experiment, or three-factor cross, that establishes the relative order of three genetic sites on the genetic map. In the experiment considered here the TrpA⁺ allele of the *trp*A gene of a TrpE⁻ TrpB⁻ *E. coli* donor strain was transduced into TrpA⁻ recipients and selection was made for TrpA⁺ TrpB⁺ transductants. **A.** If the order of the genes were *trp*E-*trp*B-*trp*A then the two required crossovers 1 and 2 would leave the recipient TrpE⁺ allele of the trpE gene in the transductant chromosome, whereas two additional crossovers, 3 and 4, would be required for incorporation of the donor TrpE⁻ allele of the *trp*E gene into the transductant chromosome. **B.** If the order of the genes were *trp*E-*trp*A-*trp*B then the two required crossovers 1 and 2 would leave the recipient TrpE⁺ allele of the trpE gene in the transductant chromosome only if crossover 2 happened to occur between the linked *trp*E and *trp*A genes, whereas they would incorporate the donor-TrpE⁻ allele of the *trp*E into the transductant chromosome if crossover 2 happened to occur at any site beyond the *trp*E gene.

other. Nevertheless, it appears that such Cys⁺ Trp⁺ prototrophic transductants are least common when TrpE⁻ mutants act as donor strains and most common when TrpA⁻ and TrpB⁻ mutants act as donor strains. Hence it can be concluded that the relative position of these genes is

$$\text{cysB} - - trp\text{E} - trp\text{D} - trp\text{C} - trp\text{B} - trp\text{A}$$

Finally, Yanofsky and Lennox, in a manner analogous to that which Benzer had previously used for his fine-structure mapping of the two *r*II genes of the T4 phage genome, initiated what was to become a detailed map of mutant sites within a single gene of this cluster of linked *trp* genes. For this purpose P1-transducing phage was grown on four different donor strains, the Trp⁺ prototroph and three different auxotrophs, *trp*B1, *trp*B2, and *trp*B3, all carrying their mutation in the *trp*B gene. Two Trp⁻ His⁻ *double auxotrophs*, one carrying the *trp*B1 mutation, the other carrying the *trp*B3 mutation, and both carrying a second mutation in the very distant *his* gene [located at minute 39 of the *E. coli* chromosome (Figure 10-21)], were then infected with P1 phages grown on the four different donor strains. Selections were made for His$^+$ transductants by plating on minimal agar supplemented with tryptophan, and for Trp$^+$ transductants by plating on minimal agar supplemented with histidine.

As is to be expected from lines A and C of Figure 14-16, use of the Trp$^+$ wild type as donor in this experiment should yield many more Trp$^+$ transductants than use of any of the Trp$^-$ auxotrophs as donors, and if the latter auxotrophs are used the closer the linkage between donor and recipient mutant loci, the less the frequency of Trp$^+$ transductants. But since the general transducing potency of different P1 phage stocks is subject to great variation, it is difficult to infer accurate linkage relations from absolute frequencies. For that reason, transduction for the very distant *his* gene, which all four donor strains ought to be able to provide with equal facility, was introduced into this experiment to allow normalization with respect to the general transducing potency of the various P1 phage stocks. The result of this experiment is presented in Table 14-2, where the frequency of Trp$^+$

TABLE 14-2.   Fine-structure Mapping of the *trp*B Gene

	His$^-$ Trp$^-$ recipient					
	*trp*B1			*trp*B3		
	No. of transductants			No. of transductants		
His$^+$ donor	Trp$^+$	His$^+$	Trp$^+$/His$^+$ (in percent)	Trp$^+$	His$^+$	Trp$^+$/His$^+$ (in percent)
*trp*B1	0	5000	<.02	251	14,959	1.7
*trp*B2	403	12,358	3.3	996	18,239	5.5
*trp*B3	55	2978	1.8	0	∼5000	<0.02
Trp$^+$	1917	1026	187	—	—	—

From C. Yanofsky and E. S. Lennox, *Virology*, **8**, 425 (1959),

transduction is expressed as the percent ratio of Trp$^+$ to His$^+$ transductants resulting from any given infection with a particular P1 phage stock. It can be seen, first of all, that the transduction efficiency for the Trp$^+$ His$^+$ wild-type donor is 187%. (For reasons that still remain unclear P1 can transduce the *trp* genes slightly more efficiently than *his* genes.) Second, it can be seen that, as expected, the normalized Trp$^+$ transducing frequency for the Trp$^-$

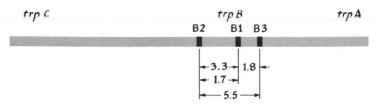

FIGURE 14-18.   Fine structure map of three mutant sites B1, B2, and B3
in the *trp*B gene of *E. coli*. Numbers represent the percentage probability
of a crossover between any mutant site pair spanned by the arrow, as
inferred from the transduction data of Table 14-2.

auxotrophic donors is dramatically less than for the wild type donor, which
attests to the very close genetic linkage of these nonidentical mutant sites
within the *trp*B gene. Third, it can be seen that, as expected, no Trp$^+$ transductants at all appear when donor and recipient bacterium carry the very
same mutation. Fourth, and last, it follows from these transduction frequencies that the order of the three mutant sites within the *trp*B gene is
evidently B2-B1-B3—that is, the genetic map shown in Figure 14-18 can be
drawn.

In later years, Yanofsky was to isolate very many more Trp$^-$ mutants,
particularly mutants that carried a mutation in the *trp*A gene. A much more
extensive fine-structure map was then constructed of the *trp*A gene using
essentially the same method as that employed for establishing the order of
the first mutant sites within the *trp*B gene. This fine-structure map of the *trp*A
gene is shown in Figure 14-19.

COLINEARITY OF GENE AND PROTEIN

In previous discussions of the general nature of the genetic code (Chapters
8 and 13) it was taken for granted that the DNA nucleotide base sequence
is colinear with the protein amino acid sequence that it specifies. That is,
it was assumed implicitly that the order of the bases specifying an amino acid
sequence is the same as the order of the amino acids being specified. But no
evidence has been adduced so far in these pages which would prove that this
most fundamental of assumptions about the informational relation between
gene and protein is actually true. The delay in providing this evidence here
mirrors an analogous delay that was encountered in molecular genetic research, for more than ten years were to elapse between the first clear formulation of this assumption and its verification by Yanofsky in 1966. His proof
of gene-protein colinearity was based on his transductional fine-structure
genetic map of the *E. coli trp*A gene. While carrying out these formal genetic
mapping experiments, Yanofsky addressed himself also to the extraction,

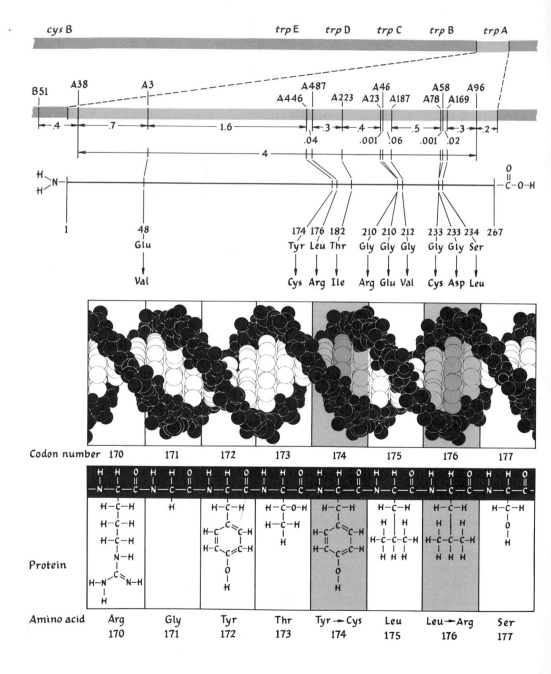

FIGURE 14-19. Facing page: Fine-structure map of a number of mutant sites in the *trp*A gene of *E. coli*. Numbers represent the percentage probability of a crossover between any mutant site pair spanned by the arrow. The line drawn below the genetic map represents the tryptophan synthetase A protein polypeptide chain, and the numbered amino acid interchanges indicate the position and nature of the normal amino acid residue in that chain which have been replaced by an abnormal amino acid residue as a consequence of each TrpA⁻ mutation. Below: Scale drawings of the double-helical DNA molecule of that part of the *trp*A gene in which amino acid residues 170 to 185 of the A protein are encoded and of the corresponding polypeptide chain. [After "Gene Structure and Protein Structure," by C. Yanofsky, *Scientific American*, May 1967. Copyright © 1967 by Scientific American, Inc. All rights reserved.]

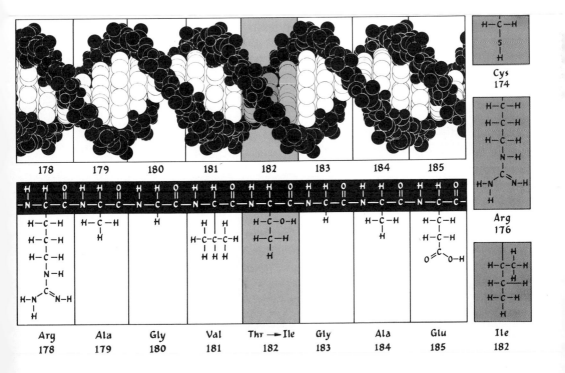

purification, and structural analysis of the tryptophan synthetase A protein coded by the *trp*A gene. This work showed that, as was brought out in Chapter 4, the A protein consists of a polypeptide chain of 267 amino acids arranged in the exact sequence shown in Figure 4-5. Thus the length of the *trp*A gene should correspond to a stretch of 267 × 3, or about 800, nucleotides. Yanofsky then focused his attention on the "right-hand" third of the *trp*A gene (as drawn in Figure 14-19), where the sites of eight different Trp⁻ mutations had been located by transductional mapping; two of these mutations, namely *trp*A46 and *trp*A23, can be seen to have occurred at very close, though not identical, genetic sites.

Yanofsky next determined the primary structure of the functionally defective tryptophan synthetase A proteins of these eight Trp⁻ mutant strains of *E. coli* in order to ascertain the nature of the chemical alteration responsible for the loss of catalytic function. The result of this analysis is also presented in Figure 14-19, where it can be seen that in every mutant the nonfunctional A protein carries an abnormal amino acid at some definite polypeptide site. For instance, the mutation *trp*A446 has resulted in a change of the 174th amino acid of the A protein polypeptide chain: whereas tyrosine is normally present at that site, mutant *trp*A446 carries cysteine there. Furthermore, in confirmation of the genetic inference that mutations *trp*A46 and *trp*A23 occurred at very close genetic sites, the protein analysis revealed that both mutations have resulted in a replacement of the glycine residue normally present as the 210th amino acid in the A protein. As can be seen, *trp*A46 carries glutamic acid and *trp*A23 carries arginine as the 210th amino acid of the mutant A protein polypeptide. Most importantly, however, it is evident from the results shown in Figure 14-19 that *the relative order of the mutant sites on the fine structure map of the trp*A *gene* (and hence of the nucleotides representing these mutant sites) *is exactly the same as the relative order of the amino acid residues in the polypeptide chain which these mutations have affected.* Thus was proven true the molecular genetic Article of Faith that the nucleotide sequence of the gene is colinear with the polypeptide amino acid sequence that it specifies.

A further important observation was made when structural analyses were undertaken of the tryptophan synthetase A protein of Trp⁺ *reverse* mutants of the Trp⁻ mutant *trp*A23. In agreement with the considerations of Chapter 13, where it was envisaged that a reverse mutation restores the original nucleotide sequence to the mutant gene, and hence the normal amino acid sequence to the corresponding protein, the A protein of some of these Trp⁺ reverse mutants was in fact found to contain the normal glycine instead of the noxious arginine of mutant *trp*A23 as its 210th amino acid residue. The A protein of some other reverse mutants, however, turned out to contain not the normal glycine but serine as its 210th amino acid. This finding offers direct proof of the existence of "silent mutations," in which, as was proposed in Chapter 6, the mutational replacement of one amino acid residue by another escapes detection. For, as can be seen from the present example, although some amino acid substitutions in the primary polypeptide structure (such as the replacement of glycine by arginine at site 210) entirely abolished the catalytic function of tryptophan synthetase A protein, other substitutions at the same site (such as replacement of glycine by serine) still permit catalytic function of the resultant mutant enzyme.

Bibliography

PATOOMB

André Lwoff. The prophage and I.
A. D. Kaiser. On the physical basis of genetic structure in bacteriophage.
J. Weigle. Story and structure of the λ transducing phage.

MOBIBAV

Chapters 12 and 13.

HAYES

Chapters 17 and 21.

ORIGINAL RESEARCH PAPERS

Jacob, F., and E. L. Wollman. Induction of phage development in lysogenic bacteria. *Cold Spring Harbor Symp. Quant. Biol.*, **18**, 101 (1953).

Kaiser, A. D. Mutations in a temperate bacteriophage affecting its ability to lysogenize *E. coli. Virology*, **3**, 42 (1957).

Kaiser, A. D., and F. Jacob. Recombination between related temperate bacteriophages and the genetic control of immunity and prophage localization. *Virology*, **4**, 509 (1957).

Kellenberger, G., M. L. Zichichi, and J. J. Weigle. A mutation affecting the DNA content of bacteriophage lambda and its lysogenizing properties. *J. Mol. Biol.*, **3**, 399 (1961).

Lederberg, E. M. Lysogenicity in *E. coli* K-12. *Genetics*, **36**, 560 (1951).

Lwoff, A., and A. Gutmann. Recherches sur un Bacillus megathérium lysogène. *Ann. Inst. Pasteur*, **78**, 711 (1950). In English translation *in* G. S. Stent (ed.), *Papers on Bacterial Viruses*. Little, Brown, Boston, 1960.

Morse, M. L., E. M. Lederberg, and J. Lederberg. Transduction in *Escherichia coli* K12. *Genetics*, **41**, 142 (1956).

Yanofsky, C., B. C. Carlton, J. R. Guest, D. R. Helinski, and V. Henning. On the, colinearity of gene structure and protein structure. *Proc. Natl. Acad. Sci. Wash.* **51**, 266 (1964).

Zinder, N. D. Infective heredity in bacteria. *Cold Spring Harbor Symp. Quant. Biol.*, **18**, 261 (1953).

Zinder, N. D., and J. Lederberg. Genetic exchange in *Salmonella*. *J. Bacteriol.*, **64**, 679 (1952).

SPECIALIZED TESTS, MONOGRAPHS, AND REVIEWS

Campbell, A. M. *Episomes*. Harper & Row, New York, 1969.

Jacob, F., and E. L. Wollman. The relationship between the prophage and the bacterial chromosome in lysogenic bacteria. *In* G. Tuneval (ed.), *Recent Progress in Microbiology*. Almqvist and Wiksell, Stockholm, 1959, p. 15.

Lwoff, A. Lysogeny. *Bacteriol. Rev.*, **17**, 269 (1953).

15. DNA Transactions

Once the view of DNA as the genetic material had become fixed in the minds of molecular geneticists, the DNA double helix tended to assume the image of a sacrosanct information store whose purine-pyrimidine base sequence ought to be immune from any metabolic transactions other than those required for its auto- and heterocatalytic roles. But the discovery of genetic recombination by breakage and reunion of DNA molecules had already shown that the integrity of the double helix is *not*, in fact, inviolate within the cell, as was recounted in Chapter 12. In this chapter we shall consider some further examples of the variety of enzymatic reactions in which DNA molecules participate, but that do not directly concern the replication or expression of genetic information. Some of these reactions were to provide insights into the molecular mechanisms by which genetic recombination by breakage and reunion is achieved.

HOST-CONTROLLED RESTRICTION AND MODIFICATION

Like other viruses, phages can "adapt" themselves to better proliferation on different kinds of host cells. It was noticed by many of the early phage workers that phage types which at first grow only very poorly on some bacterial strains will grow much better after one or more "passages" on the new hosts. This adaptive capacity of phages was considered by d'Hérelle to be one of the strongest arguments in favor of their "living" nature. It was commonly thought in those days that viruses, and indeed microbes in general, are endowed with a plastic sort of heredity that can respond directly, and hence adapt itself, to the environment—a notion which, as was mentioned in Chapter 6, caused Luria to regard microbiology as "the last stronghold of Lamarckism." With the rise of phage genetics, however, the idea gained currency that the adaptation of phages consists of the selection of spontaneous "fitter" phage mutants during repeated passage on the new host and that their hereditary plasticity derives only from the variety of mutant types always present in large phage populations. But just about the time that these more sophisticated genetic notions finally replaced old-time Lamarckian naïvetés, it was found that there does exist a type of adaptive variation in phage that cannot be accounted for by mutation and selection, and which does appear to take place under direct influence of the host cell. This kind of variation, which is now called host-controlled restriction and modification, was discovered more or less independently and almost simultaneously by several different workers in 1952.

A typical example of host-controlled restriction and modification might be described as follows: all but a small minority of the individuals in a stock of phage P grown on a bacterial strain A are unable to grow on a second bacterial strain B. Hence P phages grown on strain A, designated as P·A, are *restricted* in their host range. Those rare P·A phages that *are* able to grow on strain B give rise in their first cycle of growth on strain B to a population of *unrestricted* phage progeny of type P·B that are capable of growing with full efficiency on both strains A and B. The few restricted P·A phages that do manage to grow on bacteria of the refractory strain B are not exceptional (for example, mutant) particles. Instead, it is the few productive B bacteria that are exceptional, since their frequency varies with the physiological conditions of growth. After one cycle of growth of the unrestricted P·B phage on strain A, however, all but a small minority of the phage progeny are once more of the *restricted* P·A type that can grow only on strain A but not on strain B. The dependence of the host range on the bacterial strain in which the phage has undergone its *last reproductive cycle* is thus due to an adaptive change that is phenotypic rather than genetic, and is entirely distinct from the mutational extension of phage host range discussed in Chapter 12.

Figure 15-1 presents a schematic summary of the example of host-controlled restriction and modification of phage *lambda* discovered by J. Bertani and J. J. Weigle and later worked out in detail by W. Arber. As is shown in Figure 15-1, most particles in a stock of *lambda* phages grown on *E. coli* K12 (designated hereafter as λ·K) are unable to multiply on lysogenic *E. coli* K12(P1) carrying the P1 prophage (which had previously figured in our considerations of generalized transduction). All of the progeny of those

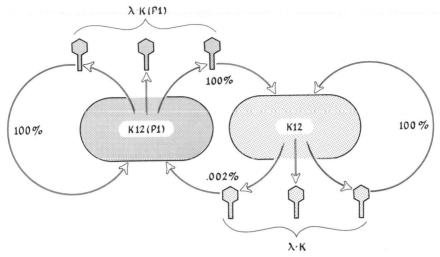

FIGURE 15-1. Host-controlled restriction and modification of bacteriophage *lambda* in K12(P1). Stipple pattern in the phages indicates phage structures whose specificity is determined by the host cell, similarly stippled, in which the phage was grown. The percentages indicate the efficiency of plaque formation on the type of host indicated by the arrow. [After G. Bertani and J. J. Weigle, *J. Bacteriol.* **65**, 113 (1953).]

rare *lambda* particles that do succeed in growing on K12(P1) [designated hereafter as λ·K(P1)] are able to propagate themselves with full efficiency on both lysogenic K12(P1) as well as nonlysogenic K12. It is thus the presence of the P1 prophage that, at one and the same time, causes restriction of the growth of most of the λ·K phages in the *E. coli* bacterium and modification of the progeny of those rare *lambda* particles that do succeed in growing into the unrestricted λ·K(P1) form. When a single unrestricted λ·K(P1) phage infects a nonlysogenic K12 bacterium, there then appears about one λ·K(P1) phage per cell still capable of growing on K12(P1) among the 100 to 250 progeny particles liberated by each infective center; all the rest of the progeny phages have been modified to the restricted λ·K type unable to grow on K12(P1) bacteria. This finding suggested that, the rare unrestricted λ·K(P1) progeny phages produced by the nonlysogenic K12 bacteria owe their character to some transfer of material from their λ·K(P1) parent. The nature of this transferred material could be identified by infecting nonlysogenic K12 bacteria with heavy D$_2$O-labeled unrestricted λ·K(P1) phages and measuring the density of the progeny phage populations. The result of this experiment is presented in Figure 15-2. It may be seen there that whereas the restricted λ·K progeny population was completely light, the few unrestricted λ·K(P1) individuals among the progeny were heavy; their position in the cesium chloride density gradient corresponds to a density of about one-fourth the

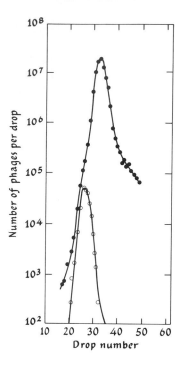

FIGURE 15-2.   Transfer of the host specificity of unrestricted λ·K(P1) phages through one cycle of growth on K12. A culture of nonlysogenic K12 bacteria growing in ordinary H₂O medium is infected with fully D-labeled λ·K(P1) phages [grown on K12(P1) bacteria in a D₂O medium] at a multiplicity of about 0.01 phages per cell. Intracellular phage growth is allowed to proceed for about one hour, at which time the infected bacteria yield an average burst of 80 λ·K and 0.5 λ·K(P1) phages per cell. The progeny phage population is subjected to CsCl density-gradient centrifugation, as described in the legend of Figure 9-10. The drops collected through the bottom of the centrifuge tube are assayed for their content of unrestricted *lambda* phages on K12(P1) bacteria (open circles) and for their content of total *lambda* phages on K12 bacteria (solid circles). [Redrawn from W. Arber and D. Dussoix, *J. Mol. Biol.* **5**, 18 (1962).]

deuterium content of the fully deuterated λ·K(P1) parent phages. It may be concluded, therefore, that the transferred material that preserves the K(P1) host specificity during growth of phage *lambda* in nonlysogenic K12 bacteria is none other than the parental phage DNA. Half of that DNA—probably one of the two strands of the semiconservatively replicated parental *lambda* phage DNA moiety—is carried by each of the rare unrestricted λ·K(P1) progeny phages. [Though the λ·K(P1) progeny carry half of the parental DNA, they contain only one-fourth of the parental deuterium because the parental phage protein, which makes up about half of the mass of the phage, is not transferred at all.]

What is the nature of the block that prevents the restricted λ·K phages from growing on K12(P1) bacteria? Arber was able to show that λ·K phages not only adsorb to K12(P1) bacteria but even inject their DNA into the refractory host cell. Shortly after this injection, however, the λ·K DNA is broken down into smaller molecular fragments. Subsequent studies by Arber, by Meselson, and by others have shown that the restriction process is the work of a nuclease enzyme that is encoded in the P1 prophage genome and produces a limited number of *double-strand* breaks in the λ·K DNA. These breaks are generated at specified sites of the lambda genome and thus reflect a specific recognition of certain nucleotide base sequences by P1 restriction nuclease. Mutants of P1 phage have been isolated that have lost the capacity to form the nuclease, which means that λ·K can grow freely on lysogenic bacteria carrying such a mutant P1 as their prophage.

The phenomenon of host-controlled restriction and modification can thus be explained in terms of two apparently separate and distinct functions of the P1 phage genome: P1 causes a modification of the DNA of *lambda* phages growing in its presence; this modification protects the *lambda* DNA against a breakdown by the P1 restriction nuclease that breaks ordinary, or non-modified *lambda* DNA. When *lambda* phages carrying P1-modified DNA molecules infect nonlysogenic K12, all the replicas of the parental *lambda* DNA are necessarily of the nonmodified, or breakdown-sensitive type, and hence give rise to restricted *lambda* progeny viruses. The modified parental *lambda* DNA chains always retain their breakdown-resistance however, and reappear in unrestricted descendant phages on transfer.

DNA TRANSMETHYLATION

In order to fathom the nature of the modification process that is evidently responsible for protecting *lambda* DNA against nuclease action, an amendment must first be made to an oversimplification embodied in the discussions of DNA chemistry in preceding chapters. For we have pretended until now that DNA polynucleotide chains contain only the four bases adenine, guanine, thymine, and cytosine (except for mention of the special case of the T-even phage DNA, which contains 5-hydroxymethylcytosine instead of cytosine). But as the analytical work of Chargaff and of Hotchkiss showed in the late 1940's, DNA from various biological sources also contains a minor proportion of other nucleotide bases. Two such bases are 5-methylcytosine and 6-methylaminopurine (Figure 15-3), which constitute about 0.1% of all the bases in the DNA of some *E. coli* strains. For some years, the presence of these minor bases in DNA seemed difficult to reconcile with the Watson-Crick mechanism of DNA replication, since it was not apparent how such bases could be consigned to a definite site in the polynucleotide chain by the complementary base-pairing process. Finally, in the early 1960's, it was

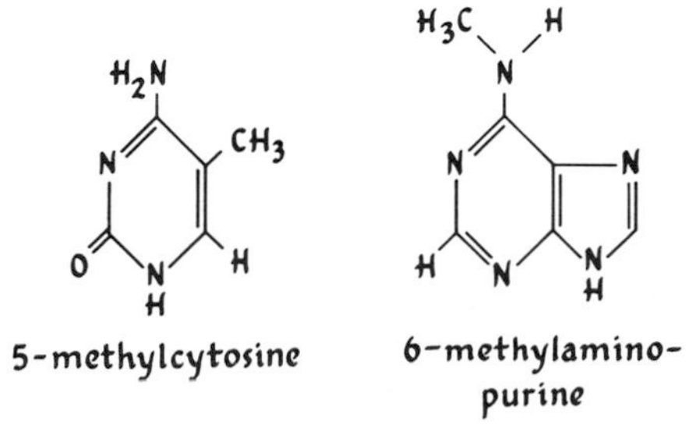

5-methylcytosine   6-methylamino-
                        purine

FIGURE 15-3. Two methylated, "minor" nucleotide bases found in DNA.

demonstrated that 5-methylcytosine and 6-methylaminopurine arise by the *transmethylation in situ* of cytosine and adenine bases of intact macro-molecular DNA polynucleotide chains. This reaction, summarized in Figure 15-4, is catalyzed by bacterial *transmethylase* enzymes, which transfer the methyl group of S-adenosylmethione to either of the two common nucleotide bases, thus converting the methyl donor to S-adenosylhomocysteine. Further studies soon showed that these DNA base transmethylation reactions are endowed with a high degree of specificity: the transmethylase enzymes convert only a definite and limited set of adenine and cytosine bases of any particular intact DNA molecule into their methylated derivatives, indicating that these enzymes, like nuclease restriction enzymes, are able to recognize the specific nucleotide-base-sequence surround of the adenine or cytosine base on which they act. Furthermore, the transmethylase enzymes carried by different bacterial species, or even by different strains of the same species, often differ in the exact adenine-cytosine methylation pattern they achieve upon acting on methylation-pristine DNA molecules of identical nucleotide base sequence. For instance, it can be easily shown that after the transmethylases of *E. coli* strain B have been allowed to methylate all the bases of a sample of DNA on which they are able to act according to the dictates of their nucleotide-base-sequence-recognition specificity, the transmethylase enzymes of *E. coli* strain K12 'can still methylate an additional set of bases on those same DNA methyl-acceptor molecules.

Just as the origin of the methylated bases in DNA remained unknown for many years, so did the physiological functions that these bases might play. Indeed, these functions are *still* largely unknown, although one of them at least has now been elucidated: transmethylation, we now know, is the chemical basis of the host-controlled modification process. The first indication of this function was obtained by Arber in 1965, when he showed that nascent *lambda* phage DNA is not modified (and hence is subject to restriction breakdown) during methionine-starvation of its *E. coli* host cell (a condition under which the methyl donor S-adenosylmethionine is in short supply). It was not possible, however, to substantiate by means of direct chemical analyses the inferred connection between transmethylation and host modification of *lambda* phage DNA, because only a small fraction of the hundred or so methylated bases of its complement of 10^5 nucleotide bases actually figures in the modification and restriction processes that occur in the K12(P1) host. Arber therefore addressed himself to another host-restricted and modified phage—namely the minute, single-stranded DNA phage fd (which is a relative of the phage f1 to be discussed in Chapter 19). For the purpose of this discussion it will suffice to state that in both the size of its phage DNA complement and the manner of DNA replication, phage fd is similar to phage ϕX174 (as was set forth in Chapter 11): upon infection, the single plus strand of fd phage DNA, some 6000 nucleotides in length, enters the host bacterium, where it

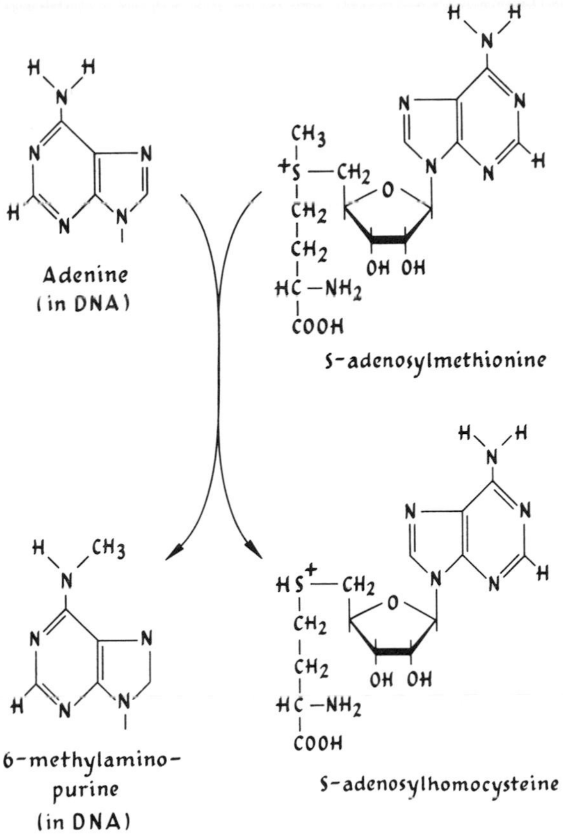

FIGURE 15-4.   The conversion of adenine into 6-methyl-aminopurine catalyzed by transmethylase enzymes.

serves as template for the synthesis of a complementary minus strand. The resulting double-strand replicative form then serves as template for the synthesis of replica plus strands. Phage fd·K (grown in strain K12) is restricted on the *E. coli* strain B; but about 0.07% of such fd·K phages do manage to grow on strain B cells and give rise to unrestricted fd·B progeny. These fd·B phages can grow with full efficiency on both strain K12 and strain B bacteria. This finding indicates that the genome of the strain B bacterium includes some genes that are not present in the genome of strain K12 and which specify both restriction and modification enzymes directed against one or more nucleotide sequences of the fd phage genome. Chemical analysis of the DNA plus strand extracted from restricted fd · K phages revealed that it contains two residues of 6-methylaminopurine among its 6000 nucleotide bases and no detectable 5-methylcytosine. The plus strand extracted from unrestricted fd·B phages is likewise free of 5-methylcytosine but contains *four*, rather than only two, residues of 6-methylaminopurine. This finding suggested that the restriction-modification system present in strain B and absent from strain K12 addresses itself to two specific adenine nucleotides of the fd phage DNA, whose methylation is the modification that provides immunity against action of the homologous restriction nuclease.

The suggestion that two genetic sites of phage fd take part in the restriction-modification phenomenon was given strong support by Arber's discovery that some *mutants* of fd, even though grown on strain K12, are *not* restricted on strain B. Such mutants arise by two successive, independent mutational events, the first of which allows 3%, rather than 0.07%, of fd·K phages to grow successfully on strain B and the second of which allows 100% of fd·K phages to grow successfully on strain B. Thus it could be concluded that the mutational alteration of the nucleotide base sequence of two critical sites in the fd DNA base sequence renders these sites immune from attack by the strain-B-specific restriction nuclease.

The complementary processes of host-specific restriction and modification apply to a range of phenomena much wider than mere phage growth, in that they provide the cell with a general means for recognizing and resisting the introduction of foreign DNA molecules. Thus restriction and modification actually play an important role in all of the mechanisms of interbacterial gene transfer considered in the preceding chapters—in transformation, in conjugation, and in transduction. For it is apparent that if the recipient bacterium contains restriction nucleases directed against nucleotide base sequences of the donor DNA that were not methylated by modification enzymes of the donor bacterium, then there is little chance of integrating the donor genes into the recipient genome in any of these genetic recombination processes. Or, from an even more general point of view, the acquisition by an organism of an idiosyncratic DNA restriction-modification system can be the first step toward the formation of a new species, in that the resultant protection against crossbreeding with otherwise quite similar organisms provides the reproductive isolation necessary for speciation. In any case, the discovery of site-specific DNA methylation and breakage added a new dimension to the concept of specificity of the hereditary substance, which had until then been assumed to derive exclusively from the permutation of only four purine and pyrimidine bases in the DNA polynucleotide chain.

DNA REPAIR

It was pointed out in Chapter 8 that the double-helical, self-complementary nature of DNA implies that every DNA molecule carries *two* complete sets of its genetic information, albeit written in complementary notation. As will now be seen, nature has taken advantage of this aspect of DNA structure and evolved processes that use the redundancy of the two polynucleotide chains to raise enormously their stability as information carriers. The principle of error-detection and error-correction based on redundant components is a familiar one to engineers concerned with the design of very complex engines such as electronic computers or space vehicles, for which reliability of operation rather than economy of construction is of paramount importance. It is also appreciated by anyone who sends a telegram in duplicate, a procedure that at only twice the cost makes certain that the recipient will be able to detect any chance errors that might have occurred in the telegraphic transmission. Thus the possibility exists that if a sector of one polynucleotide chain of a DNA molecule is damaged, the other chain can serve not only for retrieval of the genetic information but also for repair of the damage.

The existence of such DNA repair processes first came to the notice of students of the lethal effects of ultraviolet light on bacteria. It has been known since 1877 that UV-irradiation kills bacteria, and ghostly blue germicidal UV lamps are by now a familiar fixture wherever sterile air or surfaces are wanted for hygienic reasons. How does the absorption of UV light kill bacteria? As early as 1928, experiments in which bacteria were irradiated with *monochromatic* UV (i.e., light resolved into quanta of a narrow range of wavelength) showed that the most lethal UV quanta have the wavelength 260 mμ, which are also the quanta which happen to be most readily absorbed by the purine and pyrimidine bases of DNA. This finding led to the inference that bacterial DNA, rather than protein molecules, are the actual "targets" of the UV and that some chemical alteration of the DNA engendered by the absorption of UV quanta is responsible for the lethal UV effect. When, nearly 20 years later, it was at last discovered that DNA is the genetic material, it could be readily envisaged that a cell would not survive the production of photochemical lesions in its hereditary polynucleotide. Chemical analyses of UV-irradiated DNA soon revealed a variety of light-induced modifications of purine and pyrimidine residues. But the photochemical lesion now known to be mainly responsible for UV death was discovered only in 1960; it is the *dimerization* of two adjacent thymine residues belonging to nucleotides on the same DNA polynucleotide chain, as shown in Figure 15-5. This linkage of adjacent thymine residues, which occurs when one residue absorbs a quantum of UV, engenders a local distortion of the secondary structure of the DNA double helix and aborts function of the gene of whose nucleotide sequence the two adjacent thymines form part.

Detailed studies of the survival of bacteria after their exposure to UV eventually showed that the chance that a given dose of UV is lethal to any cell is strongly dependent on the precise physiological conditions to which that cell has been subject after its irradiation. This finding, in turn, led to the conclusion that bacteria can *repair* some of the photochemical lesions in their

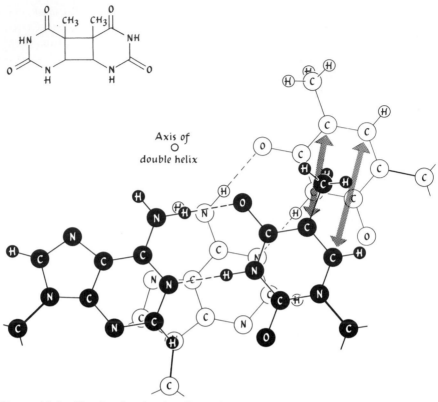

FIGURE 15-5.   The thymine-thymine dimer photoproduct of UV-irradiation of DNA. Above: Chemical structure of the dimer. Below: A dimer formed between contiguous thymine residues of two successive adenine-thymine base pairs in the DNA. The UV-induced bonds that weld successive levels of the double helix are shown as heavy arrows. [From *Molecular Biology of Bacterial Viruses*, by G. S. Stent. W. H. Freeman and Company. Copyright © 1963.]

DNA and that the lethal lesions are those which do not happen to be repaired. This view was given strong support when it was discovered that mutation of several *uvr* genes of *E. coli* (located in three widely separated regions of the genetic map) greatly raises the UV-sensitivity of the mutant cell, since it could now be inferred that the normal allele of the mutated gene presides over the formation of an enzyme that takes part in the repair process. The nature of one of these repair processes was discovered by R. Setlow in 1964. Setlow had labeled the DNA of both normal Uvr+ and UV-sensitive Uvr− mutant bacteria with ³H-thymine and irradiated both types of cells with equal doses of UV. He incubated these cells for about half an hour and then examined their extracts for the presence of ³H-labeled thymine dimers. These analyses showed that both extracts contained the same total number of thymine dimers per cell, which meant that despite their difference in UV-sensitivity, both normal Uvr⁺ and UV-sensitive Uvr⁻ mutant bacteria sustain the same number of primary photochemical lesions per incident UV quantum. However, whereas in the UV-sensitive mutant strain all of the thymine dimers were found to be part of structurally intact DNA molecules, in the normal strain most of the thymine dimers were found in short nucleotide chain fragments no more than half a dozen nucleotides in length. This finding led Setlow to conclude that normal *E. coli* bacteria posses an enzymatic system that can repair most of the primary UV lesions by *excising* the noxious thymine dimers from the irradiated polynucleotide chains and replacing them with wholesome thymine nucleotides. The UV-sensitive mutant evidently owes its reduced capacity to survive a given UV dose to the failure of its mutant *uvr* gene to produce an active enzyme essential for excising the thymine dimer.

The replacement of the excised nucleotides enveloping the thymine dimer occurs by *repair replication*. This process can be followed by carrying out the DNA replication order experiment described earlier in Figure 9-11 with UV-irradiated Thy⁻ *E. coli*. For this purpose, the culture of UV irradiated bacteria is exposed to radioactively labeled bromouracil (BU) for various time periods in order to allow incorporation of the denser BU into the replicating DNA in place of its lighter thymine analog. The DNA is then extracted from the bacteria, and the extract is subjected to CsCl density-gradient-equilibrium sedimentation in order to ascertain the density of the DNA molecules into which the radioactively labeled BU was incorporated. The result of this experiment is that the BU is found only in DNA molecules whose density is indistinguishable from that of ordinary light DNA. This means that, in contrast to normal replication, which, as was shown in Chapter 9, proceeds at only one single replication Y-fork, repair replication proceeds at many separate points of the genome. For here BU has evidently been incorporated into many short polynucleotide segments embedded in long stretches of unreplicated polynucleotide chains, so that the few isolated BU residues do not have any appreciable effect on the density of the DNA molecules of which they form part.

The overall "cut-and-patch" repair process of thymine dimer lesions, as proposed by Setlow and by P. Howard-Flanders, is presented schematically in Figure 15-6. Under this scheme, one or more enzyme molecules constantly runs the entire circular bacterial genome to monitor the DNA double helix for structural abnormalities, much as a railroad test trolley runs over the track to monitor the rails for possible structural defects. When such an enzyme monitor encounters a distortion of the double helix caused by a thymine-dimer lesion, it effects two cuts of the polynucleotide chain which excise the thymine dimer and a few of its adjacent nucleotides. The resulting gap is then refilled by action of a repair DNA polymerase, which could, in fact, be *the* DNA polymerase isolated by Kornberg, as described in Chapter 9. This polymerase adds the nucleotide patch to the 3'-OH of the end nucleotide of the old polynucleotide chain and uses as template the intact, complementary DNA

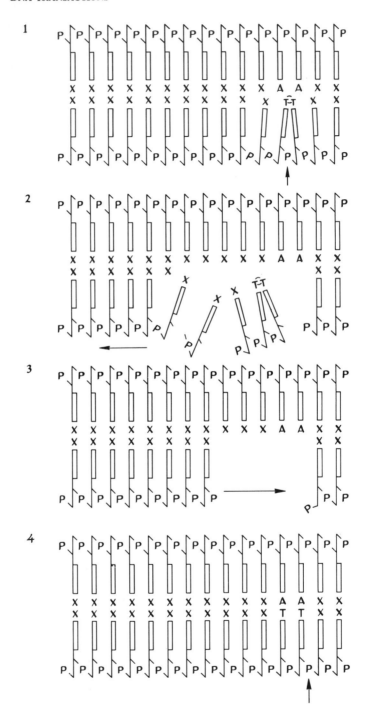

strand that did not happen to sustain a UV lesion in the same sector. Finally, the reconstruction of the DNA double helix is completed by forming a phosphate diester bond between the 3'-OH of the last nucleotide to have been added by repair replication and the 5'-OH of the nucleotide at the other end of the old polynucleotide chain. This act is performed by the DNA ligase enzyme, whose action was described in Figure 9-19.

The therapeutic function of this DNA repair process is not confined to the removal of UV-induced thymine dimer lesions, but extends to the correction of a wide variety of other potentially lethal disturbances of the cell genome. For instance, some of the noxious alterations produced by exposure of the bacterial DNA to X-rays (which break polynucleotide chains) or to mustard gas (which causes chemical cross-links between neighboring purine bases) can be detected and given remedy by the "cut-and-patch" repair system. Indeed, it has been shown that the structural disturbance caused by the noncomplementary base mismatch extant in a recombinational or mutational *heteroduplex heterozygote*, as discussed in Chapters 12 and 13 (see Figure 13-12), is subject to repair correction. Thus by excising one of the two bases forming an "illegitimate" pair (in the Watson-Crick base-pairing sense) from one of the two polynucleotide strands and replacing it through repair replication by the "legitimate" nucleotide called for by the nonexcised repair template strand, the "cut-and-patch" process can convert a heterozygote into a homozygote. It is to be noted, however, that the chance of actually rectifying any such structural disturbance is subject to wide variations. First, the efficiency of operation of the enzymatic DNA repair system depends on the genetic constitution of the organism: some organisms possess very efficient repair systems and hence are very resistant to injury of their DNA, whereas others are inefficient in repair, or lack the repair system altogether, and hence are prone to die after the slightest trauma. Second, in even those organisms that do have an efficient repair system, the physiological conditions that obtain during the time that repair is to be carried out, such as temperature or nutritional state, exert a great influence on the probability that the damaged DNA will be actually restored to a normal condition.

FIGURE 15-6.   The "cut-and-patch" repair process of thymine dimer lesions proposed by R. Setlow and by P. Howard-Flanders. (1) Segment of a UV-irradiated double-stranded DNA molecule containing a thymine dimer lesion. (2) A single-stranded fragment containing the dimer lesion is excised by an enzyme. (3) A repair DNA polymerase inserts nucleotides complementary to those of the opposite, intact strand into the gap produced by excision. (4) The reconstruction of the DNA double helix is completed by formation of the last phosphate diester bond catalyzed by DNA ligase enzyme. DNA molecules containing lesions other than UV-induced thymine dimers, such as those produced by various chemical mutagens mentioned in Chapter 13, can also be repaired by this mechanism. [After P. Howard-Flanders and R. P. Boyce, *Radiation Research*, Supplement 6, 156, (1966).]

REPAIR AND RECOMBINATION

The mechanism of genetic recombination by breakage and reunion of DNA molecules (discussed in Chapter 12) evidently includes some reaction steps that bear a strong formal resemblance to those that have just been considered in connection with DNA repair. For in order that two homologous DNA molecules can form the "joint molecule" found by Tomizawa as an intermediate in the genesis of a recombinant structure, internucleotide bonds must first be broken within sectors of corresponding nucleotide sequence and on complementary single strands of both molecules. The liberation of these complementary strands from the double-helical embrace of their old partner strands and the reformation of a hybrid double helix with their new partner strands can then give rise to the overlapping "joint molecule." The recombination act must now be completed by conversion of the "joint molecule" into a covalently linked hybrid DNA double helix. In this conversion the three repair steps—excision, repair replication, and ligase action—bid fair coming into play. Figure 15-7 presents a schematic summary of this process, as envisaged by Howard-Flanders.

The idea that the same bacterial enzymes are actually involved in DNA repair and recombination first found direct experimental support in 1965 with A. J. Clark's discovery of Rec⁻ mutants of *E. coli* that are unable to undergo genetic recombination in either conjugational or transductional crosses. This defect can be traced back to mutations in several *rec* genes of which one, *rec*A is located between minutes 50 and 55 of the *E. coli* genetic map (Figure 10-21). The Rec⁻ mutants conjugate (or absorb the transducing phage) and accept the donor DNA normally, but they are unable to integrate it into the recipient genome unless the Rec⁺ allele of the donor *rec* gene has also entered the recipient cell. Thus the *rec* genes appear to control formation of enzymes necessary for the recombination process. In addition to their inability to undergo genetic recombination, Rec⁻ mutants manifest a second striking property: they show an abnormally high sensitivity to UV-irradiation, and thus resemble *uvr* gene mutants. Study of the DNA metabolism of *rec* gene mutants after exposure to UV shows, however, that unlike *uvr* gene mutants,

FIGURE 15-7. The role of excision, repair replication and DNA ligase action in genetic recombination. (1) An alternative, less-precisely arranged version of the "joint molecule" shown in Figure 12-10 is formed as the first stage in genetic recombination. The cut ends of a pair of homologous DNA molecules overlap, and two single strands of opposite polarity, each terminating in a 5'-end, pair and form complementary hydrogen bonds. (2) The redundant, nonpaired 3'-ends of the two DNA molecules are excised by a nuclease enzyme. (3) A repair DNA polymerase inserts nucleotides complementary to those of the opposite strand into the two gaps produced by excision. (4) Formation of phosphate diester bonds between the repair patch and old strands by DNA ligase enzyme joins the two recombinant DNA molecules in covalent linkage. [After P. Howard-Flanders and R. P. Boyce, *Radiation Research*, Supplement 6, 156 (1966).]

1

2

3

4

they are capable of excising and repairing the UV-induced thymine dimers.

Thus the high UV-sensitivity of Rec⁻ mutants that carry their mutation in the *rec*A gene cannot be attributed to an incapacity to carry out the normal thymine-dimer repair process. Instead, the explanation of the phenotype of this Rec⁻ mutant class is found in another connection between DNA repair and genetic recombination. Since it can be shown that UV-irradiated *E. coli* cells can survive to produce viable progeny cells even though they still harbor some *unrepaired* thymine dimers in their DNA, it can be concluded that bacteria must also possess some means for their genetic rescue other than the "cut-and-patch" process. As Howard-Flanders was able to show, these means consist of a capacity for repair, not of the lesions present in the irradiated DNA, but of the *defective DNA daughter polynucleotides* that are produced upon replication of the unrepaired, irradiated parent DNA. By examining the physichochemical characteristics of the radioactively labeled DNA synthesized by UV-irradiated *E. coli* carrying about 100 unrepaired thymine dimers per genome, Howard-Flanders found that the first post-irradiation replica DNA is synthesized in the form of interrupted poly-nucleotide chains. The spacing of these interruptions indicated that each replica strand has a gap opposite each thymine dimer on the parental template strand. This suggests that the presence of a thymine dimer on the DNA template chain provides a block for the further progress of the DNA poly-merase. Within an hour of their synthesis, however, the interrupted replica chains are converted into continuous, full-length DNA strands. The gaps in the replica chain are thus closed by a *post-replication repair process*, shown schematically in Figure 15-8. Howard-Flanders has proposed that this post-replication repair process occurs by genetic recombination of interrupted complementary sister strands, as shown schematically in Figure 15-9. According to this scheme, the interrupted daughter strand pairs with its uninterrupted sister strand (which may, however, itself contain a gap else-where), after the latter strand breaks at an appropriate place to allow forma-tion of an incestuous "joint molecule." The gap can now be filled by repair replication through template service of the uninterrupted strand. Upon rewinding of parent and daughter strands and ligase sealing, two gapless double helices are generated. A series of such recombinational events, one for each unrepaired thymine dimer in the irradiated DNA, would ultimately reconstitute a lesion-free genome capable of normal function.

It would appear, therefore, that the inability to undergo genetic recombina-tion and the high UV-sensitivity of the *rec*A gene mutants both derive from an inability to carry out some step other than repair replication in the post-replication repair by recombination process. The schema outlined in Figure 15-9 also explains the discovery first made by Wollman and Jacob in 1955 that UV-irradiation of a phage greatly decreases the linkage that its genes manifest in genetic crosses. For it is evident that the placement of an unrepaired

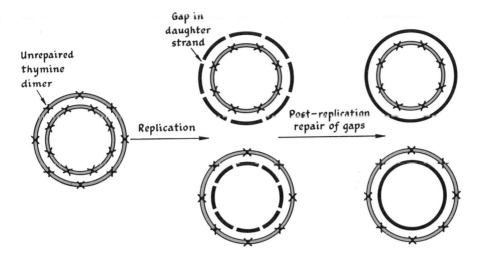

FIGURE 15-8.   " Post-replication " repair. Replication of the circular bacterial DNA molecule bearing unrepaired thymine dimer lesions (represented here as crosses) gives rise to daughter molecules whose polynucleotide strands are interrupted by gaps at sites opposite each thymine dimer on the parental template strand. The gaps are then closed by "post-replication" repair. [After E. Witkin, Proc. XII Intern. Congr. Genetics, Tokyo, 1969. vol. 3, p. 225.]

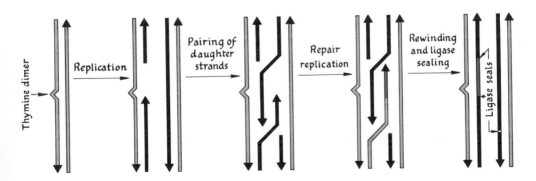

FIGURE 15-9.   "Post-replication" repair by genetic recombination. The interrupted daughter strand of a parental DNA molecule bearing a thymine dimer lesion pairs with its uninterrupted sister replica strand of opposite polarity, after the latter breaks at an appropriate site, giving rise to an incestuous recombinant "joint molecule" like that shown in Figure 15-7. The gap in the interrupted daughter strand is then filled by repair replication. Parent and daughter strands rewind, and ligase action seals the daughter strands.

thymine dimer between two genetic sites x and y will greatly increase the chance d_{xy} (as defined in equation 12-1) that a crossover will take place between them.

UV-MUTAGENESIS

In 1914 V. Henri found among the survivors of UV-irradiated bacteria a high proportion of what appeared to him to be hereditary variants, differing from the normal type in such aspects as colony shape and pathogenicity. Henri concluded from this observation—13 years before Muller's demonstration of the mutagenic effect of X-rays on Drosophila—that UV is mutagenic for bacteria. The proof of this proposition had to await the rise of bacterial genetics in the 1940's, however, when Demerec showed that among the 10^{-4} survivors of a Ton^s (T1 phage-sensitive) strain of $E.$ $coli$ exposed to a given dose of UV the proportion of Ton^r mutants is more than a thousandfold higher than the spontaneous level among unirradiated bacteria. UV-irradiation soon became one of the most commonly used mutagens for the isolation of bacterial mutants, and many of the $E.$ $coli$ mutant strains mentioned in the preceding chapters—for example, Wollman and Jacob's multiple mutant Hfr and F$^-$ strains used in the conjugation experiments of Chapter 10, or Yanofsky's Trp$^-$ mutants used in the fine structure genetic studies of the trp genes of Chapter 14—were recovered from among the survivors of UV-irradiated nonmutant parent strains. But whereas the molecular basis of spontaneous mutation and of base analog and acridine dye mutagenesis was fairly well understood by 1960 (as recounted in Chapter 13), clarification of the mechanism underlying mutagenesis by UV—historically the first bacterial mutagen to be recognized and long the most widely used—had to wait until the DNA repair mechanisms had been elucidated.

One of the persons mainly responsible for working out the mechanism of UV-mutagenesis was Evelyn Witkin, who began studying the genetic effects of UV on $E.$ $coli$ in 1946. Her work soon showed that the mutagenic effect of most UV lesions suffered by the bacterial DNA is only $potential$, in that the chance that induced mutants will appear is, like the chance of cell survival, strongly influenced by the postirradiation treatment of the bacteria. This led her to posit the existence of physiology-dependent repair processes long before their molecular basis (described in the preceding section) had been discovered. Once the roles of thymine dimers, of their "cut-and-patch" repair, and of uvr mutants in cell survival became known, Witkin carried out the experiment reported in Figure 15-10. In this experiment two Str^s strains of $E.$ $coli$, one a normal Uvr$^+$ strain and the other a UV-sensitive Uvr$^-$ mutant, were irradiated with various doses of UV. Samples of the irradiated cultures were then assayed both for cell survival by colony formation on ordinary nutrient agar and for the proportion of UV-induced Str^r mutants among the surviving cells by colony formation on streptomycin-supplemented agar. As can be seen, the

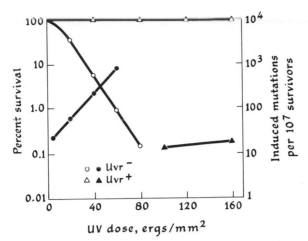

FIGURE 15-10.   Effect of UV-irradiation on two Str^s *E. coli* strains, one a wild type Uvr$^+$ of normal UV-sensitivity and the other a Uvr$^-$ mutant of abnormally high UV-sensitivity. The left-hand ordinate shows the fraction of the bacteria still capable of forming a colony upon plating (open symbols) after sustaining the UV dose shown on the abscissa, and the right-hand ordinate shows the fraction of UV-induced Str^r mutants among the survivors (filled symbols). No induced Str^r mutants were detected in the Uvr$^+$ strain at doses below 90 ergs per mm^2. [From E. Witkin, Proc. XII Intern. Congr. Genetics, Tokyo, 1969, vol. 3, p. 225.]

two strains differ so greatly in their UV-sensitivity that a UV dose which kills all but 0.1 % of the cells of the Uvr$^-$ mutant strain kills none of the normal Uvr$^+$ cells. But whereas that dose raises nearly 100-fold the proportion of Str^r mutants among the few survivors of the *uvr* gene-mutant strain, even twice that dose does not raise above the spontaneous background the proportion of Str^r mutants among the hale population of irradiated normal cells. This finding allows two inferences to be made about the origin of the UV-induced mutants: (1) the thymine dimer must be the lesion primarily responsible for the potential mutagenic event, and (2) the "cut-and-patch" repair sequence cannot be the process that converts the potential into the *actual* mutagenic event (as would have been the case, for instance, if the repair replication step of Figure 15-6 were one of low fidelity and hence prone to copy-errors). It could be concluded, therefore, that the mutation is engendered by *unrepaired* thymine dimers in cells that survive despite the presence of that unrepaired lesion. But since the survival of such cells appears to depend on the post-replication repair reaction outlined in Figure 15-9, Witkin reasoned that the UV-induced mutations must be attributed to strand-mismatching or to copy-errors that occur in the molecular recombination sequence that underlies this process.

Figure 15-11 shows the result of an experiment that enabled Witkin to demonstrate the verity of this inference. This experiment is carried out in the same manner as that presented in Figure 15-10, except that here both of the

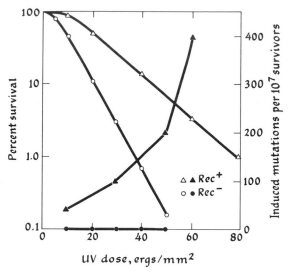

FIGURE 15-11.   Effect of UV-irradiation on two Uvr⁻mutant strains of Strˢ *E. coli*, one of normal Rec⁺ phenotype and the other of Rec⁻ phenotype. The left-hand ordinate shows the fraction of bacteria still capable of forming a colony upon plating (open symbols) after sustaining the UV dose shown on the abscissa, and the right-hand ordinate shows the fraction of UV-induced Strʳ mutants among the survivors (filled symbols). [From E. Witkin, Proc. XII Intern. Congr. Genetics, Tokyo, 1969, vol. 3, p. 225.

two UV-irradiated Strˢ strains used are Uvr⁻ mutants and that one of them carries a second mutation affecting the DNA recombination process. That is to say, although neither strain used in this experiment can carry out the "cut-and-patch" repair process, the double-mutant strain has an even higher UV-sensitivity than the single mutant because, owing to its Rec⁻ phenotype, it cannot even effect the postreplication repair by recombination. As can be seen, the UV dose dependence of the frequency of UV-induced Strʳ mutants among the surviving cells of the Uvr⁻, Rec⁺ mutant is similar to that found in the experiment of Figure 15-10. But among the survivors of the most highly UV-sensitive Uvr⁻, Rec⁻ double mutant, the frequency of Str⁺ mutants *does not rise above the spontaneous background*. Hence it can be concluded that, in full accord with Witkin's proposal, failure to carry out the normal post-replication repair process prevents the eventual realization of the mutagenic potential of unrepaired thymine dimers.

*   *   *

As the examples set forth in this chapter have shown, DNA is not to be thought of as a mere inert information store whose polynucleotide chains are called on only for template service in auto- and heterocatalytic function.

Instead, DNA is also the substrate of a multitude of enzymes that modify its chemical structure in highly specific ways so as to allow a distinction between self and nonself, that monitor the integrity of the information store and, if found wanting, repair it, and that break and rejoin homologous nucleotide sequences to produce the recombinant genomes, and hence the genetic diversity, on which evolution feeds.

Bibliography

HAYES

Chapter 15.

MOBIBAV

Chapters 11 and 14.

ORIGINAL RESEARCH PAPERS

Bertani, G., and J. J. Weigle. Host controlled variations in bacterial viruses. *J. Bacteriol*, **65**, 113 (1953).

Boyce, R. P., and P. Howard-Flanders. The release of UV-induced thymine dimers from DNA in *E. coli* K12. *Proc. Natl. Acad. Sci. Wash.*, **51**, 293 (1964).

Clark, A. J., and A. D. Margulies, Isolation and characterization of recombination-deficient mutants of *E. coli* K12. *Proc. Natl. Acad. Sci. Wash.*, **53**, 451 (1965).

Luria, S. E. Host-induced modifications of bacterial viruses. *Cold Spring Harbor Symp. Quant. Biol.*, **18**, 237 (1953).

Setlow, R. B., and W. L. Carrier. The disappearance of thymine dimers from DNA. An error correcting mechanism. *Proc. Natl. Acad. Sci. Wash.*, **51**, 226 (1964).

Witkin, E. M. Radiation induced mutations and their repair. *Science*, **152**, 1345 (1966).

SPECIALIZED TEXTS, MONOGRAPHS, AND REVIEWS

Arber, W. Host-controlled modifications of bacteriophage. *Ann. Rev. Microbiol.*, **19**, 365 (1965).

Arber, W., and S. Linn. DNA modification and restriction. *Ann. Rev. Biochem.*, **38**, 467 (1969).

Witkin, E. M. The role of DNA repair in recombination and mutagenesis. *Proc. XII Intern. Congr. Genetics*, **3**, 225 (1969).

16. DNA Transcription

We now proceed to an examination of the processes that actually effect the realization of the genetic information carried by the nucleotide sequence of DNA as the amino acid sequence of the corresponding polypeptide chain. That is, we shall now lift the lid off of what has so far remained the black box within which there occurs the heterocatalytic function of DNA.

RIBONUCLEIC ACID, OR RNA

Here we encounter the second kind of nucleic acid present in all living cells —*ribonucleic acid*, or RNA (Figure 2-7). As was set forth in Chapter 2, RNA is composed of polynucleotide chains closely resembling those of DNA, the main chemical differences between DNA and RNA being that RNA contains *ribose* (Figure 2-5) instead of deoxyribose as the sugar backbone of its nucleotides, and that RNA contains the pyrimidine *uracil* (Figure 2-5) instead of thymine as one of its four nucleotide bases. These two rather slight chemical differences (an additional hydroxyl group on the sugar of RNA and an additional methyl group on one of the pyrimidines of DNA) turn out to have momentous consequences for the biological roles played by the two types of polynucleotides.

The studies that were to elucidate the biological role of RNA began in the late 1930's, in the laboratories of T. Caspersson and Jean Brachet. At that time, Caspersson had perfected microscopic methods for localizing the two types of nucleic acids in the nucleus and cytoplasm of intact, single eukaryotic cells. This method took advantage of two facts known about RNA and DNA: (1) both absorb very strongly ultraviolet light of wavelength 260 mμ and (2) of the two nucleic acids only DNA reacts with the *Feulgen reagent* to produce an intensely colored stain at its site of localization. Thus microscopic topographic surveys of the UV-absorption and Feulgen reagent color of various regions of eukaryotic cells showed that almost all of the cellular DNA (i.e., Feulgen-positive nucleic acid) is in the nucleus and almost all of the cellular RNA (i.e., Feulgen-negative nucleic acid) is in the cytoplasm. These microscopic surveys indicated, furthermore, that most of the cytoplasmic RNA is concentrated in small particles. At about the same time, Brachet had developed methods for extracting and fractionating the organelles of eukaryotic cells for the purpose of making a direct chemical analysis of their components. This method made it possible to separate the nucleus from the cytoplasm and to show directly that the former contains most of the DNA and the latter most of the RNA. Further fractionation of the cytoplasm resulted in the isolation of the RNA-containing particles. These particles were found to contain a considerable amount of protein, in addition to their RNA. Both Caspersson and Brachet had noted upon examination of cells from different animal tissues that there appears to be a strong positive correlation between the content of the cytoplasmic RNA-protein particles and the rate at which the cell is known to synthesize proteins. For instance, the cytoplasm of immature reticulocytes—the rapidly dividing, red blood cells of the bone marrow, which are very active in hemoglobin protein synthesis—was seen to be rich in RNA-protein particles, whereas that of the mature, synthetically quiescent red blood cells of the circulating blood was seen to be practically devoid of them. The idea was put forward, therefore, that there is some necessary connection between protein synthesis and RNA, although the nature of that link was not very clearly spelled out. Brachet went so far as to suggest, however, that the RNA-protein particles *are the cellular site of protein synthesis*. This suggestion was proven to be correct in the early 1950's, when radioactively labeled amino acids had become generally available for biochemical experimentation. By injecting rats with a labeled amino acid and then extracting the RNA-protein particles from samples of tissue taken a short time after injection, it was found that most of the few radioactive atoms that managed to show up in protein in that short time are contained in the RNA-protein particles. That is to say, the nascent polypeptide chains into which the labeled amino acid is incorporated are under construction in or on these particles.

RIBOSOMES

Nucleic acid analyses of prokaryotic cells, such as *E. coli*, soon showed that they too contain both DNA and RNA, the weight of RNA per cell being from three to ten times that of DNA. And in 1952 it was discovered by both electron microscopic and physicochemical studies of bacterial lysates that most of the prokaryotic RNA is, like the eukaryotic RNA, associated with protein to form small particles (Figure 16-1). These particles, as well as the corresponding particles present in the cytoplasm of eukaryotic cells, came to be called *ribosomes*. Prokaryotic ribosomes have a diameter of about 20 mμ and contain about twice as much RNA by weight as protein. Eukaryotic ribosomes are slightly larger and contain a somewhat lower proportion of RNA than do prokaryotic ribosomes. In an *E. coli* cell there may be present anywhere from 1500 to 15,000 such ribosomes (the actual number per cell depends on the rate of cell growth, as will be discussed in Chapter 20), which occupy a major fraction of the total intracellular space. Structural studies carried out on *E. coli* ribosomes by James Watson and his collaborators eventually revealed that such ribosomes are composed of two *subunits*, both of which contain RNA and protein (Figure 16-2): a larger one called "50S" and a smaller one called "30S." Together, they constitute the "70S" ribosome. (The designations 30S, 50S, and 70S refer to the velocity at which each of these bodies sediments in a centrifuge under a set of standard conditions. It is to be noted that sedimentation velocities, unlike masses, are not simply additive.) The 30S subunit contains one RNA molecule, which is about 1500 nucleotides in length and sediments at a velocity of 16S, whereas the 50S subunit contains two RNA molecules, one about 100 and the other

FIGURE 16-1. The electron micrograph on which *E. coli* ribosomes were first identified. The spheres in this picture are the ribosomes, and the rod is a tobacco mosaic virus particle (see Chapter 20) that serves as a standard for comparison. Magnification 140,000 diameters. [From H. K. Schachman, A. B. Pardee, and R. Y. Stanier, *Arch. Biochem. Biophys.* **38**, 245 (1952). Photograph courtesy of Robley C. Williams.]

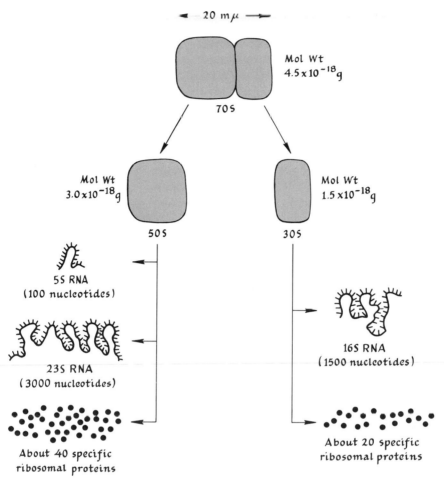

FIGURE 16-2.   Schematic representation of the structure of the *E. coli* ribosome. At Mg^{++} concentrations below 0.01 M the 70S ribosome breaks down into its two unequal 30S and 50S subunits. The subunits can, in turn, be broken down into their RNA and protein components.

about 3000 nucleotides in length, which sediment at velocities of 5S and 23S, respectively. Table 16-1 presents the purine and pyrimidine base composition of the RNA of *E. coli* ribosomes. As can be seen, the compositional equivalence rule valid for DNA [A] = [T] and [G] = [C] does not hold here, since neither is [G] equal to [C] nor is [A] equal to [U] (U being the structural analog of T). Consequently, the ribosomal RNA molecules cannot (and do not) exist in the complementary, base-paired Watson-Crick double helix. Instead, they are present as single polynucleotide chains, of which neither the exact conformation nor the nature of their function in the intact ribosomes has as yet been elucidated.

TABLE 16-1.   Nucleotide Base Composition of *E. coli* Ribosomal RNA, Post-infection RNA, and T4 Phage DNA

	Base composition (mole percent)			
	[A]	[U] or [T]	[G]	[C]
E. coli ribosomal RNA	25	21	32	22
Post-infection RNA	29	29	25	17
T4 phage DNA*	32	32	18	18

* The DNA of phage T4 does not actually contain cytosine but instead the cytosine analog 5-hydroxymethyl cytosine.

Figure 16-3 presents the result of an experiment which demonstrates that the *E. coli* ribosomes are the sites of bacterial protein synthesis. In this experiment, radioactive $^{35}SO_4^{-2}$ was added to a growing *E. coli* culture in order to allow assimilation of ^{35}S isotope into the amino acids methionine and cysteine, and hence into bacterial protein. Fifteen seconds later, further assimilation of ^{35}S isotope into bacterial protein was arrested by adding an excess of nonradioactive $^{32}SO_4^{-2}$ to the culture medium. Either immediately or 120 seconds after addition of the nonradioactive $^{32}SO_4^{-2}$, the bacteria of samples of this pulse-labeled culture were broken open. The sedimentation characteristics of the ^{35}S-labeled protein present in the cells were then ascertained by means of *sucrose density-gradient centrifugation*, a technique whose invention in the mid-1950's greatly facilitated the study of the structure and function of ribosomes in particular, and of subcellular structures in general. This technique, illustrated schematically in panel A

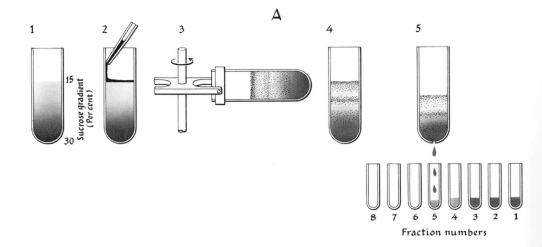

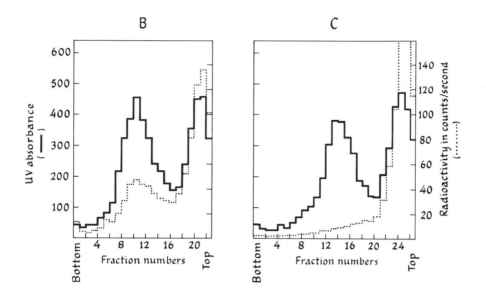

FIGURE 16-3.   Demonstration that *E. coli* ribosomes are the sites of cellular protein synthesis. Radioactive $^{35}SO_4^{-2}$ was added to an *E. coli* culture growing in a synthetic medium. Fifteen seconds later, an excess of nonradioactive $^{32}SO_4^{-2}$ was added to the culture medium. The bacteria were then broken open either immediately or 120 seconds after addition of the $^{32}SO_4^{-2}$. The cell extracts were then subjected to sucrose density-gradient centrifugation. A. Principle of the sucrose density-gradient sedimentation technique. (1) A centrifuge tube is filled slowly with a sucrose solution whose sucrose concentration varies continuously from an initial value of 30% to a final value of 15%. The resulting density gradient is very stable against disturbances by convection or mechanical shock. (2) The ^{35}S-labeled cell extract is carefully layered on top of the density-gradient solution. (3) The tube is spun in a centrifuge for 75 minutes at 37,000 rpm. (4) Centrifugation has caused the ribosomes of the extract to move toward the bottom of the tube as a single band characteristic of their 70S sedimentation velocity. Meanwhile materials sedimenting faster than ribosomes have formed bands further toward the bottom, and materials sedimenting more slowly than ribosomes have formed bands further toward the top of the tube. (5) The band-sedimentation profile is ascertained by punching a hole in the bottom of the centrifuge tube, collecting fractions dropwise, and analyzing their content of UV-absorbing material (i.e., nucleic acids) and ^{35}S radioactivity. [After " Polyribosomes," by A. Rich, *Scientific American*, December 1963. Copyright © 1963 by Scientific American, Inc. All rights reserved.] Sedimentation profiles of the extract of cells broken open immediately after the 15-second labeling period, or 120 seconds after the addition of $^{32}SO_4^{-2}$, are shown in parts **B** and **C**. In both profiles, the left-hand (or faster-moving) peak of UV-absorbance represents the 70S ribosomes and the right-hand (or slower-moving) peak, free nucleic acid molecules, which sediment with velocities of less than 30S. The ^{35}S radioactivity peaks indicate the sedimentation bands of nascent polypeptide chains. [Data from K. McQuillen, R. B. Roberts, R. J. Britten, *Proc. Natl. Acad. Sci. Wash.* **45**, 1437 (1959).]

of Figure 16-3, makes it possible to resolve into bands mixtures of materials which sediment at different velocities. The sucrose density-gradient sedimentation profiles of the two cell extracts are shown in panels B and C of Figure 16-3. As can be seen in panel B, at the end of the 15-second labeling period, more than half of the ^{35}S tracer atoms are present in nascent protein molecules that sediment with the 70S ribosomes, the remainder being present in already finished, or free, protein molecules, which sediment much more slowly. The results of panel C show that within 120 seconds after the addition of $^{32}SO_4^{-2}$ and the arrest of assimilation of ^{35}S atoms, most of the ^{35}S label formerly associated with nascent protein carried by the 70S ribosomes has now passed into the slowly-sedimenting fraction, which consists of finished protein molecules. It could be concluded, therefore, that ^{35}S tracer atoms first appear in nascent polypeptide chains bound to ribosomes and later pass out of the ribosomes into the "soluble" (i.e., nonribosomal or general) protein of the bacterium. Since the rate at which the ^{35}S tracer atoms are seen to flow in and out of the ribosome-bound nascent polypeptides in the experiment of Figure 16-3 can be shown to suffice for the entire synthesis of bacterial protein, it may be inferred that the ribosome provides the scene for the assembly of amino acids into polypeptides of predetermined sequence.

MESSENGER RNA

We have yet to ask how the information encoded in DNA for the assembly of specific amino acid sequences is made available to the ribosomes during protein synthesis. Since the ribosomes in the cytoplasm of eukaryotic cells are certainly not in physical contact with DNA molecules in the nucleus, the simplest answer to this question—namely, that the DNA polynucleotide chains themselves govern the polypeptide assembly process—cannot be generally correct. The idea was proposed, therefore, that the heterocatalytic function of DNA is a *two-stage process*. In the first stage of this process, each DNA gene would serve as the template for the synthesis of RNA molecules onto which the precise nucleotide sequence making up that gene, and hence its encoded amino acid sequence information, is *transcribed*. After their transcription on the DNA template, these RNA molecules were then imagined to migrate to the cytoplasm, where in the second stage of the heterocatalytic function their nucleotide sequences would be *translated* into polypeptide chains of predetermined primary structure. It is no longer clear today who was actually responsible for formulating this general conception of the heterocatalytic function—Alexander Dounce published one, albeit imperfect, form in 1952—but these notions must certainly have figured in the thoughts of Watson and Crick soon after they discovered the structure of DNA and solved the problem of its autocatalytic function in 1953. In any case, under this early version of the informational essence of protein synthesis, the RNA transcript was thought to provide the RNA moieties for newly formed ribosomes. Hence each gene was imagined to give rise to the formation of one *specialized* kind of ribosome, which in turn would direct the synthesis of one and only one kind of protein—a scheme that Brenner, Jacob, and Meselson epitomized as the "one gene-one ribosome-one protein" hypothesis.

Earlier studies of the metabolism of phage growth carried out by S. S. Cohen in 1948 seemed difficult to reconcile with this hypothesis from the very outset of its formulation. As was seen in Chapter 11, the kinds of proteins made after T-even infection of *E. coli* differ radically from those made previously by the uninfected cell. Hence under the "one gene-one ribosome-one protein" hypothesis one would have expected a burst of ribosomal RNA synthesis following phage infection, while the phage-infected cell is renovated for future production of the polypeptide chains encoded in the phage DNA. But contrary to that expectation Cohen had found that upon infection of *E. coli* with T2 phage, net synthesis of RNA, and hence of ribosomes, not only does not accelerate but *comes to a stop*, indicating that synthesis of new kinds of ribosomes is not a precondition for synthesis of new kinds of proteins. Later experiments, however, showed that although there is no *net* synthesis of RNA in the phage-infected cell, *some* post-infection synthesis of RNA does take place during phage growth. In these experiments, transient exposure of T2-infected *E. coli* to radioactive $^{32}PO_4^{-3}$ resulted in the finding that some of the ^{32}P tracer atoms first enter and later leave the RNA of the infected cell. Hence the post-infection RNA formed is evidently *unstable*, being constantly made and later broken down, so that its synthesis does not lead to any net increase in the amount of RNA in the infected culture. Closer examination of this unstable, post-infection RNA revealed that its purine-pyrimidine base composition is significantly different from that of *E. coli* ribosomal RNA and, instead resembles more closely that of the T2 phage *DNA*, provided that for the purpose of this comparison the uracil (U) and cytosine (C) in RNA are considered the equivalents of their structural analogs thymine (T) and hydroxymethyl cytosine (HMC) in DNA (Table 16-1). Hence it appeared likely that the post-infection RNA is synthesized under the direction of the phage DNA, whose nucleotide sequences it probably harbors. Careful fractionation of extracts of phage-infected *E. coli* revealed that most of the post-infection RNA is attached to ribosomes. The post-infection RNA, however, is clearly not a structural member of the ribosomes, since, unlike ribosomal RNA, it can easily be released from ribosomes by treatments that leave the ribosomes intact, and it manifests a broad spectrum of molecular sizes rather than the three distinct size classes characteristic of ribosomal RNA.

These facts (as well as considerations of the genetic regulation of enzyme synthesis to be discussed in Chapter 20) led François Jacob and Jacques Monod to propose in 1961 that contrary to the "one-gene–one-ribosome–one-protein" hypothesis, ribosomes are *not* congenitally specialized in their capacity to synthesize specific polypeptide chains. Instead of regarding the ribosomal RNA as the direct template for the orderly assembly of amino acids, they proposed that the nucleotide sequence of each DNA gene is transcribed onto *messenger* RNA molecules. Jacob and Monod further envisaged that these messenger RNA molecules then enter into temporary combination with unspecialized ribosomes already endowed with their own ribosomal RNA, and that it is the messenger RNA-ribosome complex that is competent to synthesize the protein encoded into the DNA gene represented by its messenger RNA transcript. From this viewpoint, the ribosome is the "workshop" for protein synthesis and the messenger RNA the "blueprint." Jacob and Monod proposed, finally, that the messenger RNA has only a limited functional lifetime, and hence can serve for the construction of only a few protein molecules; thus during active protein synthesis the messenger RNA molecules are in a state of rapid turnover. One and the same ribosome may, therefore, synthesize one kind of protein molecule at one moment and another kind at the next, depending on the kind of messenger the ribosome happens to have engaged.

Before these *a priori* conceptions of Jacob and Monod had even appeared in print, Brenner, Jacob, and Meselson were able to demonstrate their applicability to the events attending protein synthesis in T4-infected bacteria. They carried out an experiment in which *E. coli* were grown in a medium containing ^{15}N and ^{13}C as the only sources of nitrogen and carbon, so that all the cell constituents were labeled with these heavy isotopes. After growth in the heavy medium, the bacteria were infected with T4 phage and immediately transferred to a medium containing the ordinary isotopes ^{14}N and ^{12}C, so that all cell constituents synthesized after infection were labeled with light nitrogen and carbon atoms. Radioactive $^{32}PO_4^{-3}$ was also added to the infected cultures, so that the density distribution of newly synthesized, radioactively labeled post-infection RNA among the cell components liberated by artificial lysis of the infected bacteria could be followed by means of equilibrium sedimentation in CsCl density gradients.

The results of this experiment confirmed the previous observation that most of the post-infection RNA is associated with ribosomes. As can be seen in Figure 16-4, however, density analysis of the ribosomes carrying the post-infection ^{32}P-labeled RNA reveals that these ribosomes contain the heavy, pre-infection ^{15}N and ^{13}C isotopes rather than the light, post-infection ^{14}N and ^{12}C isotopes. In other words, this experiment showed that *the phage-induced, post-infection RNA enters old ribosomes that are already present in the cell before its infection with T4 phage*; indeed, no light ribosomes at all are seen to appear in the infected cell, showing that phage infection really brings ribosome synthesis to term. This result eliminates the possibility that the amino acid sequence information is encoded in the ribosomal RNA, since the ribosomes of the uninfected bacterium could hardly possess the plans for constructing phage proteins *before* the phage DNA, primary repository of that knowledge, has even entered the cell. Instead, it can be concluded that the phage-induced, post-infection RNA plays the role of the

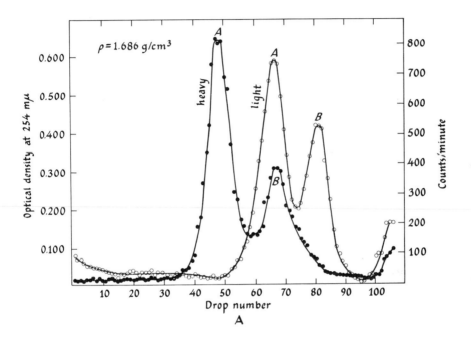

FIGURE 16-4.   Demonstration of the existence of messenger RNA. Cesium chloride density-gradient centrifugation of ribosomes of normal and T4 phage-infected bacteria, and of the phage-induced, post-infection RNA. Open circles represent the optical density at a   wavelength of 254 m, and hence total nucleic acid content, and solid circles the radioactivity, and hence content of labeled RNA, of drops collected through the bottom of the centrifuge tube. A. Density of heavy and light bacterial ribosomes. *E. coli* cells, grown in an ^{15}N-^{13}C-^{32}P-labeled medium, are mixed with a 50-fold excess of cells grown in ordinary nutrient broth. The ribosomes of this mixed bacterial population are extracted from the cells by alumina grinding, isolated from the extract by differential centrifugation, and centrifuged into bands in a concentrated CsCl solution. Each of the two species of ribosomes can be seen to form two bands, *A* and *B*. The ^{32}P-activity peak of band *B* of the labeled heavy ribosomes falls just under the optimal density peak of band *A* of the unlabeled light ribosomes. (Figure continues on next two pages.)

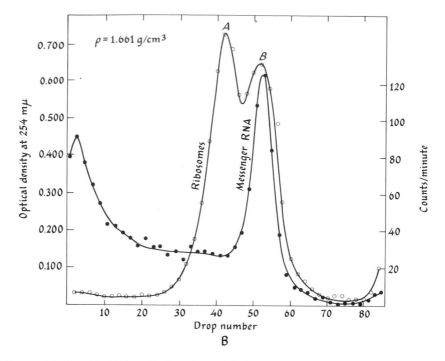

FIGURE 16-4 (cont.)   **B**. Association of post-infection RNA with ribosomes. A culture of *E. coli* is infected with an average of 30 T4 particles per cell and exposed to ^{14}C-labeled uracil in its medium from the third to the fifth minute of intracellular phage growth. The ribosomes are then extracted, purified, and centrifuged in concentrated CsCl solution, as above. It can be seen that the ^{14}C-activity peak of the post-infection RNA falls just under the optical-density peak of the band. Hence the phage induced RNA appears to be associated only with the ribosomes of band *B*, which, as independent experiments show, is the band of ribosomes actively engaged in protein synthesis.

postulated messenger RNA on which the phage DNA devolves the task of bringing the information on amino acid sequence encoded in its genes to the host-cell ribosomes, the sites at which the phage messenger RNA molecules preside over the assembly of phage polypeptides. In the following discussions we will refer to ribosomal RNA and messenger RNA by their commonly used abbreviations rRNA and mRNA.

Jacob and Monod's proposal led to the search of an analogous mRNA in uninfected *E. coli* cells, since it seemed plausible that the bacterial DNA would carry out its heterocatalytic function in the same manner as the phage DNA. These searches soon revealed that about half the weight of RNA synthesized by normal *E. coli* at any moment is, in fact, of the unstable

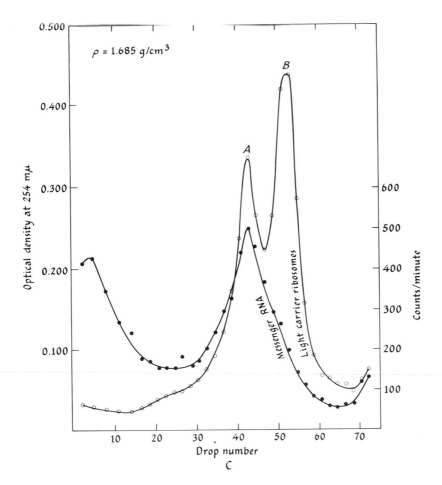

FIGURE 16-4 (cont.) **C.** Density of post-infection RNA formed in bacteria switched from heavy to light medium immediately after infection. *E. coli*, grown in an ^{15}N-^{12}C-labeled heavy medium, are infected with an excess of T4 phages and immediately brought into ordinary light broth. The infected bacteria are exposed to ^{32}P in their growth medium from the second to the seventh minute of intracellular phage growth and then mixed with a 50-fold excess of nonlabeled light T4-infected bacteria. The ribosomes are then extracted from this mixed culture, purified, and centrifuged in concentrated cesium chloride solution, as above. It can be seen that the ^{32}P-activity peak of the labeled post-infection RNA falls just under the optical-density peak of band *A* of the light carrier ribosomes. That position, as was seen in **A**, corresponds to band *B* of the heavy ribosomes. Since, as was evident in **B**, the postinfection RNA is associated with band *B*, it can be concluded that here the ^{32}P-labeled post-infection RNA enters the heavy ^{15}N-^{13}C-labeled ribosomes constructed before infection during growth of the culture in the heavy medium. [Parts **A**, **B**, and **C** after S. Brenner, F. Jacob, and M. Meselson, *Nature* **190**, 576 (1961).]

mRNA type, most of the remainder being represented by the stable rRNA. The average lifetime of the unstable mRNA of *E. coli* can be estimated to be about one-tenth of the generation time of the cell. Hence the mRNA present in the cell at any time was formed mainly just during the preceding 0.1 of the generation time and thus amounts to only about $0.5 \times 0.1 = 0.05$ of the total RNA content of the cell.

RNA-DNA HYBRIDIZATION

That the post-infection RNA formed in T4-infected bacteria is, in fact, a transcript of nucleotide sequences of the phage DNA was proved soon after the promulgation of the mRNA concept, when Sol Spiegelman developed methods for producing specific molecular hybrids between complementary RNA and DNA polynucleotide chains. The principle of the experimental formation of such molecular hybrids is illustrated in Figure 16-5. It should be recalled, first of all, that heating double-helical DNA molecules to temperatures above their "melting point" causes the rupture of the complementary purine-pyrimidine hydrogen bonds (see Chapter 8). This rupture, in turn, leads to separation of the two complementary polynucleotide strands. If the hot solution containing the dissociated DNA strands is gradually cooled to room temperature, complementary polynucleotide chains will reassociate by matching homologous base sequences and eventually reconstitute the original double-helical molecule. If, however, single-stranded mRNA molecules bearing purine and pyrimidine base sequences complementary to some of the DNA strands are added to the hot solution during the gradual cooling, or annealing process, then RNA-DNA hybrid double helices will form in addition to the original DNA double helices. The formation of these hybrid double helices can be detected in a variety of ways, one

FIGURE 16-5.   Schematic outline of the formation and detection of specific molecular hybrids between DNA and RNA polynucleotide chains. Two species of mRNA molecules are added to a solution of DNA molecules that have served as transcripton templates for mRNA1 but not for mRNA2. Upon heating the mixture to 100°C, the complementary DNA template strands separate. Upon slow cooling mRNA1 matches its nucleotide base sequence with the complementary base sequence of the DNA template strand from which it had been transcribed, and both strands associate to form an RNA-DNA hybrid double helix. Since mRNA2 finds no matching nucleotide base sequence in this solution, it remains in the single-stranded form. Upon filtration of the annealed mixture through filters with very fine pores, the single-stranded, randomly coiled polynucleotide chains, in particular mRNA2, pass through the pores into the filtrate, whereas the stiff double helices, both renatured DNA and the mRNA1-DNA hybrid, are retained on the filter surface. The formation of the specific hybrid double helices can be detected by radioactive labeling of mRNA and counting the radioactivity retained on the filter.

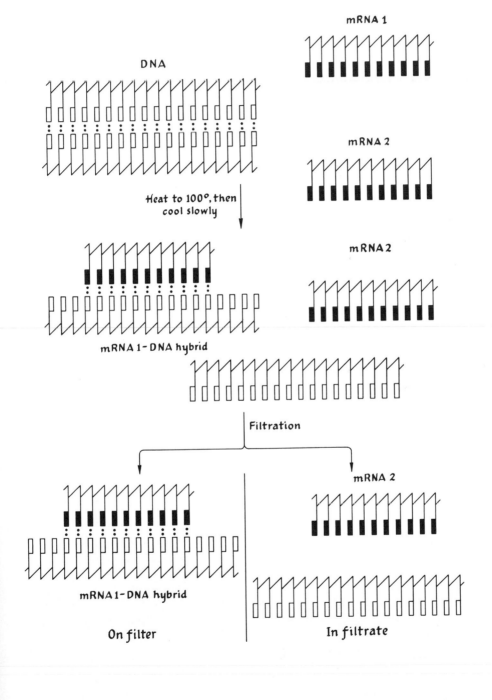

mRNA 1

DNA

Heat to 100°, then
cool slowly

mRNA 2

mRNA 1 – DNA hybrid

mRNA 2

Filtration

mRNA 2

mRNA 1 – DNA hybrid

On filter

In filtrate

of the simplest being their retention on special filters that permit the free passage of single-stranded RNA molecules.

When this method became available, the ^{32}P-labeled post-infection RNA extracted from T4-infected *E. coli* growing in a medium to which $^{32}PO_4^{-3}$ was added after infection could be removed from its complex with the ribosomes and tested in an RNA-DNA molecular hybridization reaction. For this purpose samples of the radioactive RNA extract were added to a hot solution of either phage DNA extracted from infective T4 phage particles or *E. coli* DNA extracted from uninfected bacteria. The solutions were then annealed, in order to allow formation of hybrid double helices between any DNA and RNA strands of complementary base sequence. After cooling, the solutions were filtered, and the amount of ^{32}P radioactivity retained on the filters was measured. The result of this experiment was that the filters retained most of the ^{32}P-labeled post-infection RNA added to the solution containing T4 DNA and very little of the same RNA added to the solution containing *E. coli* DNA. That is to say, the nucleotide base sequences of the post-infection RNA are complementary to those of the phage DNA and not to those of the host-cell DNA, in complete accord with the earlier inference that the phage DNA serves as template for synthesis of an mRNA transcript of the phage genes.

We can now inquire whether *both* of the complementary polynucleotide strands making up a given gene in the phage DNA figure as templates for mRNA formation or whether *only one* of them is active in the heterocatalytic function. The latter possibility seemed more likely from first principles. For it would be most improbable that both of two mRNA molecules of complementary nucleotide sequence could be translated into functional polypeptide chains, unless there existed some special feature of the decoding machinery that generates the same amino acid sequence upon translation of mRNA molecules of complementary nucleotide sequence. One of the first replies to this inquiry was obtained by Spiegelman and his collaborators in studies of the minute phage ϕX174. As was set forth in Chapter 11, the infective ϕX174 phage particle carries only a single-stranded, circular DNA molecule, the "plus" strand. And upon its entry into the *E. coli* host cell, the plus DNA strand serves as template for a complementary "minus" DNA strand to generate the circular, double-stranded RF DNA molecule. The RF DNA then directs synthesis not only of replica plus strand DNA molecules to provide the genetic material for the progeny phages but also of phage-specific mRNA molecules to provide for the synthesis of phage proteins in the infected cell. This phage mRNA transcript can be labeled with ^{32}P by transient exposure of the ϕX174-infected *E. coli* to $^{32}PO_4^{-3}$, and then extracted from the cells. Hybridization of the extracted RNA to either single-stranded plus DNA isolated from intact ϕX174 phage particles or to double-stranded RF DNA extracted from ϕX174-infected *E. coli* then reveals that the ^{32}P-labeled phage mRNA is hybridizable only to the RF DNA but not to the plus strand. That is to say, the nucleotide base sequence of the phage mRNA is complementary only to the minus strand, and not to the plus strand, of the double-stranded RF DNA. It can be concluded therefore, that in agreement with expectation, only one of the two complementary DNA strands—namely, the minus strand—is active in the heterocatalytic function. Transcription of the minus strand evidently gives rise to mRNA molecules that are *identical* in nucleotide sequence to the plus strand carried by the infective ϕX174 phage particle.

Later, more complicated hybridization experiments carried out with T4 mRNA (as well as with the mRNA that phage *lambda* produces in its host cell) showed that here too only one of the two complementary DNA strands is transcribed *in the domain of any one gene*. As it turned out, however, these larger phages differ from ϕX174, in that in some of their genes one of the two DNA strands, and in other of their genes the other of the two DNA strands, serves as the transcription template.

It is also possible to demonstrate the complementary relation between the E. coli DNA and the mRNA present in the uninfected cell by means of the molecular hybridization test. The results of hybridization tests between bacterial RNA and DNA, however, are much more difficult to interpret than of those between phage-induced RNA and phage DNA because of the much greater number of genes, and hence greater diversity of nucleotide sequences, carried by the bacterial genome. That is to say, since (as will be seen in Chapter 20) different bacterial genes are transcribed at widely differing rates, there exists in the cell a wide range in the number of mRNA molecules whose nucleotide sequence is complementary to any one single bacterial gene. But since the DNA added to the hybridization reaction mixture contains equal numbers of all bacterial genes, there exists the possibility that the most abundant bacterial mRNA species in the cell extract fail to hybridize completely because they have saturated the complementary DNA genes available to them in the hybridization test. With this cautionary exhortation in mind, we may now enquire into the origin of the *stable* RNA species that make up the bulk of the bacterial RNA. In particular, we may ask whether the 16S and 23S molecules of rRNA carried by the 30S and 50S ribosomal subunits, respectively, are also transcribed from the bacterial DNA, or, more specifically, whether there exist bacterial genes complementary in nucleotide sequence to these two species of rRNA. We omit from this enquiry the 5S rRNA because its presence in the 50S ribosomal subunit was not yet known at the time the experiment now to be described was carried out. And since there can be more than 10,000 ribosomes (and hence more than that number of 16S and 23S molecules of rRNA) per cell, it is obvious that if these rRNA species are to be hybridized efficiently, an enormous excess of bacterial DNA over RNA must be added to the reaction mixture.

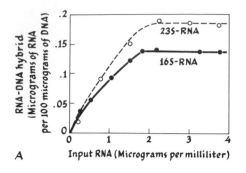

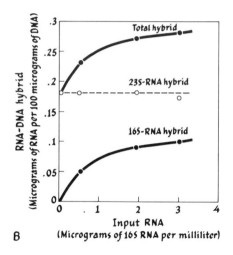

FIGURE 16-6 Hybridization of 16S and 23S rRNA extracted from bacterial ribosomes with the homologous DNA. A. Saturation test. The amounts of ^{32}P-labeled 16S or 23S rRNA shown on the abscissa were added to a fixed amount of bacterial DNA and the hybridization-filtration sequence was allowed to proceed. The abscissa shows the amount of ^{32}P-labeled RNA retained on the filter, and hence of the amount of rRNA-DNA hybrid formed. B. Competition test. The amounts of ^{3}H-uracil-labeled 16S rRNA shown on the abscissa were added to hybridization mixtures containing a fixed amount of bacterial DNA and a saturating amount of ^{32}P-labeled 23S rRNA. After hybridization-filtration, the amounts of ^{32}P and ^{3}H radioactivity retained on the filter were assayed separately in a detector capable of resolving the radiation emitted by the two isotopes and converted into the amounts of 16SrRNA-DNA, 23SrRNA-DNA, and total rRNA-DNA hybrid recorded on the ordinate. [After "Hybrid Nucleic Acids," by S. Spiegelman, *Scientific American*, May 1964. Copyright © 1964 by Scientific American Inc. All rights reserved.]

Figure 16-6,A presents the result of an experiment in which increasing amounts of purified 16S or 23S rRNA molecules were added to various samples of a hybridization mixture containing a fixed amount of homologous bacterial DNA, and the amount of RNA hybridized was measured in each sample. As can be seen, both 16S and 23S species of rRNA are definitely hybridized to the bacterial DNA, and the amount of RNA entering the hybrid increases with the amount of RNA added to the reaction mixture until a plateau is reached. This plateau evidently signals the saturation of the DNA genes complementary to either of the two rRNA species. Hybridization of the 16S rRNA attains its plateau when 0.14 μg RNA is hybridized per 100 μg DNA present in the reaction mixture. This indicates that 0.14 percent of the bacterial DNA contains nucleotide sequences capable of serving as templates for the transcription of 16S rRNA. Similar inferences drawn from the hybridization plateau of the 23S rRNA lead to the conclusion that 0.18 percent of the bacterial DNA is complementary to 23S rRNA.

Figure 16-6,B presents a modification of the preceding experiment designed to test whether the 16S and 23S rRNA molecules arise from the same or from different template regions of the bacterial DNA. That is to say, a test was made to see whether 16S and 23S rRNA molecules share a common nucleotide sequence, or whether they are entirely different. For this purpose, samples of a hybridization mixture were set up containing a fixed amount of bacterial DNA and a *saturating* amount of ^{32}P-labeled 23S rRNA. To these samples were then added increasing amounts of ^{3}H-uracil-labeled 16S rRNA, and the amount of both ^{32}P and ^{3}H radioactivity present in the resulting RNA-DNA hybrids was measured. As can be seen, the amount of the ^{3}H-labeled 16S rRNA hybridized reached the previous plateau despite the simultaneous presence here of 23S rRNA. And the amount of the ^{32}P-labeled 23S rRNA hybridized remained at its saturation level, despite the presence of increasing amounts of 16S rRNA molecules. Hence it can be concluded that the two species of rRNA molecules do not compete for the same DNA nucleotide sequences in the hybrid test and hence that they arise from distinctly separate genes of the bacterial DNA.

We may now make a rough estimate of the number of genes encoding the rRNA nucleotide sequences. Since the *E. coli* genome consists of about 3×10^6 DNA nucleotide base pairs, of which 0.14 percent were seen to form part of double helices complementary on one strand to the 1500 nucleotide base sequence of 16S rRNA, it follows that there are about $0.0014 \times 3 \times 10^6/1500$, or 3 genes coding for 16S rRNA per genome. Similar calculations lead to an estimate of $0.0018 \times 3 \times 10^6/3000$, or about 2 genes coding for 23S rRNA per genome. Since, as was set forth earlier in this chapter, about half the weight of RNA synthesized by *E. coli* at any moment is rRNA, it follows that the few rRNA genes are $1/(0.0014 + 0.0018) = 300$ times more active as transcription templates than the average gene of the remainder of the bacterial genome that gives rise to mRNA.

RNA POLYMERASE

Just as the details of the mechanism of DNA replication came to be worked out after the discovery of the DNA polymerase enzyme (as set forth in Chapter 9), so was the mechanism of DNA transcription into RNA elucidated upon the nearly simultaneous discovery in 1960 by S. Weiss, by J. Hurwitz, and by Audrey Stevens of the *RNA polymerase* enzyme in extracts of various eukaryotic and prokaryotic cells, including *E. coli*. In the presence of DNA template molecules, RNA polymerase catalyzes the conversion of ribonucleoside triphosphates into ribopolynucleotide chains according to the following reaction:

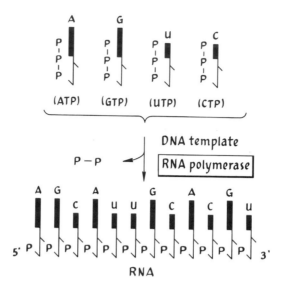

That is to say, ATP, GTP, CTP, and UTP are polymerized into RNA by formation of an ester link between the innermost of the three phosphates attached to the 5'-hydroxyl of one ribonucleotide and the 3'-hydroxyl group of another, with the attendant liberation of the outermost two phosphates as inorganic pyrophosphate. The chemistry of RNA synthesis, therefore, closely resembles the chemistry of DNA synthesis, as set forth in Chapter 9.

The overall chemical reaction of RNA synthesis could actually occur in two different ways. As is outlined in Figure 16-7, growth of the nascent RNA molecule might proceed by incorporating the next nucleoside triphosphate either through linking its 3'-hydroxyl group to the 5'-triphosphate of the last nucleotide incorporated into the growing chain (in which case the chain would be said to grow *from* the 3'-end) or linking its 5'-triphosphate to the 3'-hydroxyl group of the last nucleotide incorporated into the growing chain (in which case the chain would be said to grow *from* the 5'-end). In order to decide between these two alternatives, an *in vitro* RNA-synthesis reaction catalyzed by RNA polymerase was allowed to proceed for 8 minutes in the presence of nonradioactive ribonucleoside triphosphates. By that time, growth of all RNA chains that were ever going to be synthesized in that reaction mixture had been initiated. Next, ^{3}H-labeled ribonucleoside triphosphates were added to the reaction mixture, and chain growth was allowed to continue for another 2 minutes. The RNA molecules formed were then isolated from the reaction mixture and analyzed for the presence of ^{3}H-label in both their 3'- and 5'- ends. The result of this experiment was that only the 3'-ends contained any ^{3}H label, whereas the 5'-ends were entirely nonradioactive. Thus it could be concluded that *growth of the RNA*

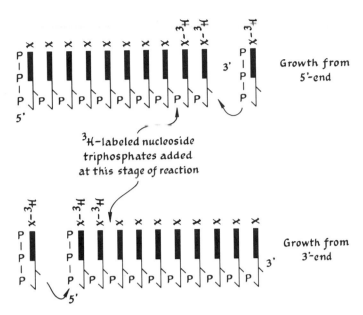

FIGURE 16-7.   Two possible directions of RNA chain growth.

chains begins at the 5'-end, because once a molecule has laid down a non-labeled nucleotide at its starting end at the earliest stages of the reaction, no radioactivity can appear at that starting end when ^{3}H-labeled ribonucleoside triphosphates are added to the reaction mixture at later stages. Similarly, it followed that *chain growth proceeds at the 3'-end,* since radioactivity will appear at that growing end as soon as ^{3}H-labeled ribonucleoside triphosphates are added to the reaction mixture. In other words, *RNA chains grow from their 5'-end.*

That this *in vitro* RNA synthesis catalyzed by RNA polymerase mimics the actual *in vivo* DNA transcription was first indicated by the observations that the reaction will not proceed in the absence of a DNA template and, moreover, that the purine and pyrimidine base composition of the RNA produced depends on the nature of the DNA template supplied. Later, a more convincing demonstration of the biological relevance of this reaction came from experiments in which RNA polymerase was presented with circular, double-stranded RF DNA extracted from *E. coli* infected with φX174 phage. For this purpose radioactively labeled ribonucleoside triphosphate substrates of the enzyme were introduced into the reaction mixture, and the molecular hybridization character of the radioactive RNA produced was then tested. These tests showed that such synthetic RNA molecules can form hybrid double helices upon annealing their hot mixture with RF DNA. They fail to hybridize, however, with the single plus strand DNA molecules extracted from intact φX174 virus particles. Thus it followed that in the *in vitro* reaction mixture, RNA polymerase not only gives rise to an RNA chain of purine-pyrimidine base sequence complementary to the template DNA with which it has been presented, but even selects as template that same minus strand of the two complementary RF DNA strands, which also happens to be the unique strand active in the *in vivo* transcription reaction.

Thus DNA replication and DNA transcription came to be viewed as basically similar processes, in that in both reactions the double-stranded template DNA was thought to unwind. In transcription, however, only one of the two DNA strands would serve for the assembly of ribonucleotide building blocks into RNA polynucleotides; that is, each purine and pyrimidine base of the DNA template chain would attract and hold in place, through the specific hydrogen bonds shown in Figure 8-5, a free nucleotide bearing the complementary purine or pyrimidine base. (It is to be remembered, of course, that uracil replaces thymine in RNA, so that in the transcription process the free uracil ribonucleotide is attracted and held in place by the adenine.) One important difference between DNA replication and DNA transcription, of course, is that in DNA replication the two parental polynucleotide strands remain permanently separated after they unwind and serve as templates for the synthesis of replica strands (Figure 9-1), whereas in DNA transcription the two unwound DNA strands must ultimately rewind after the transcribed RNA molecule has left its DNA template to serve in protein synthesis. A schematic representation of the transcription process is presented in Figure 16-8, where the components have been drawn approximately to scale. Here it can be seen that the DNA double helix moves across the RNA polymerase, unwinding its two strands as it encounters the enzyme. Within the domain of the polymerase, one of the two unwound DNA strands serves as the template for the ordered assembly of the ribonucleotide building blocks by formation of Watson-Crick

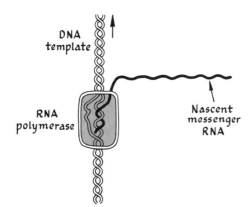

FIGURE 16-8. Schematic representation of the DNA transcription process by RNA polymerase. The arrow indicates the direction in which the DNA template moves through the polymerase enzyme. [From G. S. Stent, *Proc. Roy. Soc.* (London) Ser. B. **164**, 181 (1966).]

complementary hydrogen bonds, giving rise to a short segment of *hybrid RNA-DNA helix*. But the nascent RNA molecule eventually unwinds from the hybrid double helix and leaves the domain of the polymerase, thus permitting rewinding of that sector of DNA double helix which had served as the transcription template a few moments earlier.

RNA CHAIN GROWTH RATE

We may now examine the *rate* at which RNA chain growth actually proceeds within the cell, or the speed with which the transcription process outlined in Figure 16-8 advances along the DNA template. For this purpose, we will consider a modification of the experiment described in connection with Figure 16-7, which had shown that growth of RNA chains proceeds from the 5'-end. In this experiment, the RNA of growing *E. coli* is labeled by adding ^{3}H-labeled purine and pyrimidine precursors of nucleoside triphosphates to the culture medium. (It is not useful in this experiment to add the ^{3}H-labeled ribonucleoside triphosphates themselves, because *E. coli* cannot take them up from the medium.) At various times, t seconds shortly after adding the label, growth of samples of the culture is stopped, the ^{3}H-labeled RNA is extracted from the cells, and the amounts of both total radioactivity, T millicuries, and of radioactivity in nucleotides at the 3'-ends, E millicuries, of the RNA molecules is determined. Now if the extract contains a total of N RNA molecules that were growing at a rate of r nucleotides/second, and if the specific radioactivity of the ^{3}H label is S millicuries/nucleotide, then we may write

$$T = rtNS \qquad (16\text{-}1)$$

But since each growing RNA molecule carries one radioactive nucleotide at its 3'- (or growing) end, we may also write

$$E = NS \qquad (16\text{-}2)$$

From equations (16-1) and (16-2) it follows that

$$r = \frac{T}{tE} \qquad (16\text{-}3)$$

That is to say, the rate of RNA chain growth can be reckoned directly from the ratio of the total and 3'-terminal radioactivity found in RNA molecules that had been allowed to incorporate labeled nucleotides for a known short time. The results of such experiments showed that in *E. coli* RNA chain growth proceeds at the rate of 43 nucleotides/second at 37°C.

Comparison of this estimate of the speed of transcription with the speed of DNA replication inferred in the discussions of Chapter 9 shows that these two processes, both of which occur on the DNA template, are dynamically incompatible. For as we had seen, the growing fork at which DNA replication occurs must advance around the circular *E. coli* genome at a speed of about 1400 nucleotide base pairs/second. Hence rotation of the as yet nonreplicated DNA double helix in advance of the growing fork must proceed at the rate of 140 revolutions/second to provide for the *pari passu* separation of ten nucleotide base pairs per turn of double helix. But the 30-fold slower speed of RNA transcription from the same DNA double helix would demand the correspondingly lower rate of rotation of only about 4 revolutions/second. Hence, since it is improbable in the extreme that the entire apparatus of transcription (and, as we shall see presently, of translation) of nascent RNA chains rotates *with* the DNA double helix, it would follow that the sector of the unreplicated DNA just ahead of the growing fork must rotate independently of the rest of the circular bacterial genome. That is, there must exist a *swivel* (probably a single-strand interruption) between that sector of the parental DNA which is about to be replicated and the other sectors that are meanwhile being transcribed into RNA.

INITIATION OF TRANSCRIPTION

As was already stated, in some genes of phages T4 and *lambda*, one, and in other genes the other, of the two complementary DNA strands serves as the transcription template. Furthermore, it has been mentioned in anticipation of the discussions of Chapter 20 that different genes of the bacterial genome are transcribed at quite different rates. These facts make it seem very unlikely that RNA polymerase molecules initiate their transcription of phage or bacterial DNA either uniquely at one single point or haphazardly at many random points of the genome. Instead it can be inferred that the DNA structure must embody a set of well-defined transcription starting points at which polymerase enzymes not only initiate growth of RNA chains but also select as template the "correct" one of the two complementary DNA strands. Studies on the initiation of RNA synthesis by *E. coli* RNA polymerase on ϕX174 RF, T4 and *lambda* phage DNA templates in *in vitro* reaction mixtures have led to the conclusion that the double-helical structure of the DNA molecules is essential for the recognition of starting points by the enzyme. These studies showed that only a limited number of RNA polymerase molecules per phage genome can start their transcription simultaneously—about one per thousand nucleotide base pairs of double-helical DNA template supplied. This supports the view that RNA chain initiation does not occur at random points. But if the complementary strands of the DNA template are separated by "melting" the double helix (by the procedure described in

Chapter 8) before it is used in the reaction mixture, then a much greater number of polymerase molecules can simultaneously begin transcription. Most important, the transcription proceeding on the melted templates is no longer confined to the "correct" one of the two complementary DNA strands but instead gives rise to RNA molecules embodying the nucleotide sequences of *both* DNA strands.

The simplest explanation of these observations is that, although RNA polymerase can bind to and begin transcription at any ordinary sector of a *single* DNA strand, it has virtually no affinity for the ordinary sectors of the intact DNA double helix. The limited number of starting points on the intact DNA double helix for which RNA polymerase does have affinity could then be thought to constitute special sectors in which particular purine and pyrimidine base sequences allow the polymerase to open up the DNA double helix. Upon opening of the double helix, the polymerase would attach itself to one of the two complementary single strands and begin using that strand as a transcription template. These particular starting-point base sequences, however, would have to be of such a character that the sequence carried by the "incorrect" strand within the domain of the starting point would deny the polymerase any possibility of starting growth of an RNA chain upon it.

Analyses of the nucleotide sequences of various phage DNA's used in these transcription experiments have brought to light a highly suggestive fact concerning the possible nature of the starting points: at an average interval of about a thousand nucleotides, there occur *clusters of pyrimidine bases* in that DNA strand which is known to act as the transcription template. The average frequency of such pyrimidine clusters in the template strand is thus similar to the average frequency of transcription starting points. Furthermore, artificial DNA double helices that carry only pyrimidines in one strand and only purines in the other strand are known to "melt" more readily than double helices in which purine and pyrimidine bases are evenly distributed over both strands. Hence it does not seem very far-fetched to imagine that the RNA polymerase starting points are coded into the DNA as short sectors in which the heterocatalytically active strand carries clusters of pyrimidines and the complementary, heterocatalytically inactive strand carries clusters of purines.

It was eventually found that the RNA polymerase molecule recognizes the starting point by means of a special protein component. For some years after its discovery pure preparations of active RNA polymerase proved difficult to obtain. Finally, efforts to isolate RNA polymerase in pure form from *E. coli* succeeded in 1968, and the structure of the enzyme could be elucidated. These studies showed that the enzyme consists of an aggregate of three different polypeptide subunits, α, β and σ. The partial aggregate of α and β polypeptides appears to be responsible for the catalysis of RNA chain growth, whereas the presence of the σ polypeptide in the aggregate is required only for the initiation of transcription on intact double helical DNA templates. Once the polymerase has initiated chain growth, the σ subunit is released from the enzyme aggregate and RNA chain growth proceeds in its absence. It seems likely at present that there is only one kind each of α and β polypeptide subunits present in *E. coli* RNA polymerase, and that they jointly serve not only for the transcription of all genes of the bacterial DNA but also for the transcription of any phage DNA that happens to infect that cell. It now seems equally likely, however, that there exist several different types of σ polypeptide subunits in *E. coli*, each capable of recognizing the singular structure of different kinds of starting points and thus conferring upon individual RNA polymerase molecules a certain degree of differentiation in their initiation specificities. It seems even more likely at present that T4 phage induces the synthesis of its own σ polypeptide in the infected cell, so as to render the α and β subunits of the *E. coli* RNA polymerase capable of using a particular class of starting points on the T4 DNA.

Bibliography

HAYES

Chapter 12.

ORIGINAL RESEARCH PAPERS

Bremer, H., and M. W. Konrad. A complex of enzymatically synthesized RNA and template DNA. *Proc. Natl. Acad. Sci. Wash.*, **51**, 801 (1964).

Brenner, S., F. Jacob, and M. Meselson. An unstable intermediate carrying information from genes to ribosomes for protein synthesis. *Nature*, **190**, 576 (1961).

Hurwitz, J., J. J. Furth, M. Anders, P. J. Ortiz, and J. T. August. The enzymatic incorporation of ribonucleotides into RNA and the role of DNA. *Cold Spring Harbor Symp. Quant. Biol.*, **26**, 91 (1961).

McQuillen, K., R. B. Roberts, and R. J. Britten. Synthesis of nascent protein by ribosomes in *Escherichia coli. Proc. Natl. Acad. Sci. Wash.*, **45**, 1437 (1959).

Schachman, H. K., A. B. Pardee, and R. Y. Stanier. Investigations on the macromolecular organization of microbial cells. *Arch. Biochem. Biophys.*, **38**, 245 (1952).

Weiss, S. B. Enzymatic incorporation of ribonucleoside triphosphates into the interpolynucleotide linkages of ribonucleic acid. *Proc. Natl. Acad. Sci. Wash.*, **46**, 1020 (1960).

SPECIALIZED TEXTS, MONOGRAPHS, AND REVIEWS

Peterman, M. L. *The Physical and Chemical Properties of Ribosomes.* Elsevier, Amsterdam, 1964.

Stent, G. S. Genetic transcription. *Proc. Roy. Soc.* (*London*) Ser. B., **164**, 181 (1964).

Watson, J. D. The involvement of RNA in the synthesis of proteins. *Science*, **140**, 17 (1963).

Synthesis and Structure of Macromolecules. XXVIII Cold Spring Harbor Symposium on Quantitative Biology, New York, 1963.

17. RNA Translation

It was first suggested on the basis of general considerations (Chapter 8) and later demonstrated by genetic analysis of certain T4 phage mutants (Chapter 13) that the primary structure of proteins is encoded in DNA in the form of sequences of nucleotide triplets, or codons, each of which represents one of the 20 standard amino acids. After the preceding chapter has set forth how that sequence of nucleotide triplet codons is transcribed from its permanent repository in DNA onto ephemeral mRNA molecules as the first stage of the heterocatalytic function, we are now ready to consider how in the second stage of the heterocatalytic function amino acids are actually assembled into polypeptide chains of predetermined order on the ribosomal scene. That is to say, we will now examine the actual *translation* of the information embodied in the mRNA sequence.

DIRECTION OF POLYPEPTIDE CHAIN GROWTH

The discussion of the structure of polypeptides in Chapter 4 showed that at one end of the chain there is present a free α-carboxyl group and at the other end a free α-amino group. For unlike all the amino acids that make up the body of the chain and are held to their neighbors by a pair of peptide bonds, each of the two chain-terminal amino acids, the *carboxy-* and the *amino-*

terminal amino acid, is held by only one peptide bond to its sole neighbor. The simplest way in which the assembly of amino acids into a polypeptide chain can be imagined to occur is that the nascent chain grows from one of its ends by gradual addition of one amino acid after another. Finally, when the requisite number of amino acids has been joined, growth of the nascent chain terminates and the completed polypeptide chain is released from the ribosome to allow it to assume whatever functional role it is to play in the life of the cell. Let us suppose now that ^{3}H-labeled amino acids are added to the medium of a cell culture at time t_1 and that at various times shortly thereafter, t_2, t_3, etc., the appearance of these ^{3}H-amino acids in *completed* polypeptide chains no longer attached to ribosomes is ascertained. Then under the postulated mechanism of chain growth, one would expect to find the results shown schematically in Figure 17-1. At t_1 the ribosomes would contain nonlabeled nascent peptide chains at various stages of completion. By t_2, a certain number of ^{3}H-labeled amino acids would have been added to the growing end of these nascent chains. Furthermore, a few of the oldest nascent chains would have been completed and released from the ribosomes, and growth of an equal number of new peptide chains would have been initiated with ^{3}H-labeled amino acids between t_1 and t_2. By t_3, more ^{3}H-labeled amino acid would have been added to the nascent chains, and growth of additional chains would have been both initiated and completed between t_2 and t_3. Thus the specific content of ^{3}H-labeled amino acids in completed polypeptide chains present in the cells at t_2 or t_3 would be lowest at that end at which chain growth is initiated and would increase continuously toward the opposite end at which chain growth is completed.

In order to test whether polypeptide chain growth does proceed in this manner, and if it does, to ascertain whether the carboxy- or the amino-terminal amino acid is the point at which chain growth is initiated, H. M. Dintzis actually carried out the experiment schematized in Figure 17-1. For this purpose a culture of reticulocytes (or immature red cells) from rabbit blood was maintained at 15°C, and ^{3}H-leucine was added to the culture for various periods lasting from 4 to 60 minutes. Hemoglobin, the principal protein synthesized by these reticulocytes, was then extracted from the cells and separated into its two constituent α and β polypeptide chains. Since much of the sequence of the hundred and fifty amino acids making up each of these chains had been previously ascertained by the methods of primary structure determination described in Chapter 3, it was possible to break up the ^{3}H-labeled hemoglobin polypeptides into characteristic peptide fragments of known order and thus to measure the specific ^{3}H-content of the leucine residues at various known sites of the chain. The result of Dintzis' experiment is presented in Figure 17-2, where it can be seen that the specific ^{3}H-content of the leucine residues of both hemoglobin chains increases continuously with the distance between the leucine site and the amino-terminal chain end.

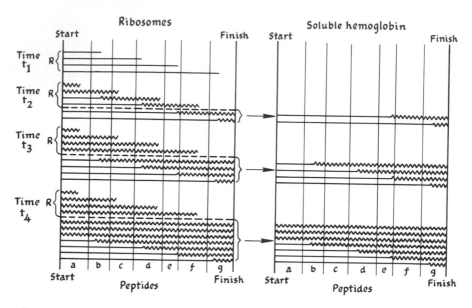

FIGURE 17-1. Stepwise nature and direction of hemoglobin polypeptide chain growth in blood cells to which radioactively labeled amino acids were added at time t_1. The left-hand part of this diagram shows the pattern of unlabeled (straight lines) and labeled (zigzag lines) nascent polypeptide chains present on the ribosomes at t_1 and at progressively later times t_2, t_3, and t_4. The groups of chains designated as R represent *incomplete* nascent polypeptides. The other chains, having reached the finish line, are assumed to be released at once from their ribosomes of origin and to appear as soluble, complete hemoglobin protein in the cell. The contemporaneous labeling pattern of the soluble hemoglobin is shown in the right-hand part of this diagram. For instance, in the ribosome compartment at time t_2, the two completely zigzag lines represent chains formed completely from amino acids between t_1 and t_2. The middle two lines represent chains in which amino acids have been added to pre-existing chains during that time interval. These chains have not yet reached the finish line and are therefore still attached to ribosomes. The bottom two lines represent chains that have crossed the finish line, left the ribosomes, and become mixed with the soluble, complete hemoglobin. The lower-case letters symbolize the sequence of peptide fragments of the polypeptide chains produced upon digestion of the protein with the enzyme trypsin. [After H. M. Dintzis and P. M. Knopf, *in* H. Vogel, V. Bryson, and J. Lampen (eds.), *Informational Macromolecules*. Academic Press, New York, 1963.]

This effect, furthermore, is the more pronounced the shorter the labeling period. Thus comparison of these results with the schema of Figure 17-1 shows that the polypeptide chains do grow in a stepwise manner and that the *amino-terminal amino acid is the point of chain-growth initiation*. These findings were soon generalized for the synthesis of other proteins in both eukaryotic and prokaryotic cells. In particular, it could be shown that the assembly of the *E. coli* polypeptide chains is initiated with the amino-terminal amino acid and terminated with the carboxy-terminal amino acid.

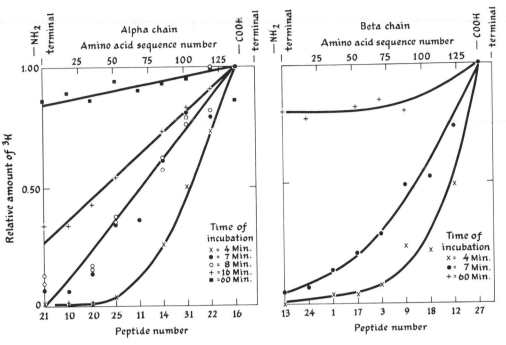

FIGURE 17-2.   Distribution of ^{3}H-leucine residues among the peptide fragments generated by trypsin digestion of separated α and β chains, beginning with the amino-terminal and ending with the carboxy-terminal residue. The numbers on the bottom abscissa represent individual tryptic peptide fragments positioned to match the amino acid sequence scale on the top. The ordinate represents the specific content of labeled leucine (^{3}H-radioactivity per total leucine) found in each of the individual peptide fragments listed on the bottom abscissa. [After H. M. Dintzis and P. M. Knopf, *in* H. Vogel, V. Bryson, and J. Lampen (eds.), *Informational Marcomolecules*. Academic Press, New York, 1963.]

POLYRIBOSOMES

It was set forth in Chapter 16 that of the total bacterial RNA, about 80% is made up of rRNA and only about 5% of mRNA. That is to say, there are present only $0.05 \times 4600/0.8 = 290$ nucleotides of mRNA for every unit of 4600 nucleotides of rRNA that resides in a single ribosome. But molecular size determinations had shown that mRNA molecules are much longer than 290 nucleotides; consequently, this calculation leads to the paradoxical inference that the bacterium does not contain enough mRNA molecules to provide polypeptide assembly information to all of its ribosomes. This apparent paradox was soon resolved, however, by the discovery of the stepwise growth of polypeptide chains. For since the amino acids are added one at a time to the nascent chain, there is no need for the *entire* mRNA molecule to be in contact at all times with the ribosome that it services. The possibility then came to mind that the mRNA might run through the ribosome in much the same way that a magnetic tape runs through the head of a tape recorder. In this way, only a short segment of the mRNA (about a dozen or so nucleotides) would be in contact with the ribosome at any time. And among these few nucleotides would be that nucleotide triplet codon which specifies the next amino acid to be added to the nascent polypeptide chain under construction on that ribosome. As soon as the next amino acid has been added, the mRNA tape would advance through its client ribosome by three nucleotides in order to bring the second-next triplet codon into position. In this way, a single mRNA molecule might run through several ribosomes simultaneously, in the manner of a magnetic tape running through the tandem heads of a tape recorder designed to generate an echo chamber effect. Thus the apparent deficiency in the number of mRNA molecules compared to the number of ribosomes would be rectified by having several ribosomes share the services of a single messenger.

The verity of these conceptions was soon proven with the finding that clusters of ribosomes are, in fact, attached to one and the same molecule of mRNA. The discovery of such clusters, or *polyribosomes*, came with the development of more gentle methods of extracting ribosomes from cells than had previously been available. Figure 17-3 shows the result of centrifugation in a sucrose density gradient of such a ribosome extract from rabbit reticulocytes engaged in hemoglobin synthesis. It is immediately apparent that this ribosome sedimentation profile differs significantly from that shown in Figure 16-2, in that here there is manifest not only the sedimentation band of the single ribosomes but also a variety of bands formed by ribosomes that sediment at much higher velocities—200S and higher. Each of these faster sedimenting bands corresponds to a polyribosomal cluster of particular size—that is, clusters composed of 2, 3, 4, 5, or more ribosomes, the rule being that the faster a band sediments the greater the number of ribosomes per cluster it contains. Figure 17-4 shows three electron micrographs of single ribosomes and polyribosomal clusters composed of 3 and 5 ribosomes per cluster, recovered in the first, third, and fifth bands, respectively, of the sucrose density-gradient-sedimentation profile of Figure 17-3. That the ribosomes of the polyribosomal cluster are actually connected by a communal molecule of mRNA is demonstrated by the sedimentation profile in Figure 17-5. Here the same ribosome extract analyzed in the experiment of Figure 17-3 was first exposed to the enzyme ribonuclease, which catalyzes the hydrolysis of the mRNA polynucleotide chain, before sucrose density-gradient sedimentation. As can be seen, such hydrolysis of the mRNA chain breaks up the polyribosomal cluster, as reflected by the conversion of the faster-moving bands into the slowest moving band made up of single ribosomes.

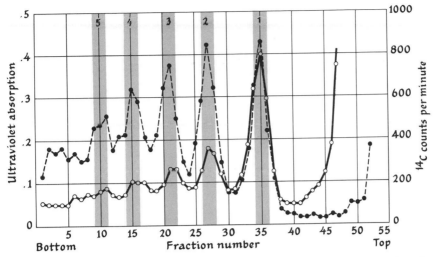

FIGURE 17-3. Sucrose density-gradient sedimentation profile of the ribosomes released by lysis and gentle grinding of rabbit reticulocyte cells exposed to ^{14}C-labeled amino acids for the last 45 seconds (at 37°C) before lysis. The left-hand ordinate shows the total UV-absorption at 260 mμ (solid line), and hence nucleic acid content of the fraction collected from the centrifuge tube registered on the abscissa, and the right-hand ordinate shows the ^{14}C radioactivity (broken line) found to be incorporated into nascent hemoglobin polypeptide chains in the same fraction. The individual peaks correspond to bands containing (from right to left) clusters of one, two, three, four, and five ribosomes. The higher ratio of ^{14}C-radioactivity to total nucleic acid content found in the faster sedimenting polyribosomal clusters indicates that they are relatively more active in protein synthesis than are the single ribosomes. [After "Polyribosomes," by A. Rich, *Scientific American*, December 1963. Copyright © 1963 by Scientific American, Inc. All rights reserved.]

Two possibilities may now be considered concerning the dynamics of formation of bacterial polyribosomes. One is that a molecule of mRNA is first transcribed in its entirety before any ribosomes attach to it and start translating its nucleotide sequence into polypeptide chains. The other possibility is that ribosomes attach themselves one by one to the head, or 5'-end, of the nascent mRNA chain while that chain is still being transcribed from its own DNA template; here formation of polyribosomes upon mRNA would proceed concomitantly with its synthesis, and the head of the nascent messenger would have served as template for polypeptide assembly long before its tail had been transcribed. These two possibilities can be distinguished experimentally by adding a radioactively labeled RNA precursor, such as ^{14}C-uracil, to a growing bacterial culture and measuring the length of time required for achieving the maximal amount of labeled mRNA per ribosome in very small polyribosomes consisting of only one or two ribosomes each, on the one hand, and in very large polyribosomes consisting of very many ribosomes, on the other hand. Under the first possibility of

FIGURE 17-4.   Electron micrographs of the ribosomes recovered from bands 1 (right), 3 (center), and 5 (left) of the sucrose density-gradient-sedimentation experiment of Figure 17-3. These bands can be seen to contain single ribosomes, and clusters of three and five ribosomes, respectively. [Courtesy of Alexander Rich.]

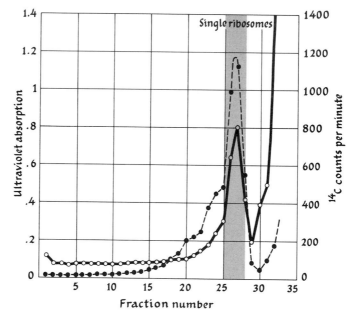

FIGURE 17-5.   Sucrose density-gradient-sedimentation profile of ribosomes released from rabbit reticulocytes processed in the manner described in the legend of Figure 17-3, but subjected to brief treatment with ribonuclease before sedimentation analysis. Enzyme-catalyzed hydrolysis of the mRNA chains connecting the ribosomes of polyribosome clusters can be seen to have converted the clusters into single ribosomes that sediment as a single ribosomal band. [After "Polyribosomes," by A. Rich, *Scientific American*, December 1963. Copyright © 1963 by Scientific American, Inc. All rights reserved.]

ribosome attachment exclusively to completed mRNA molecules, the maximum amount of labeled mRNA per ribosome would be expected to be attained in both large and small polyribosomes only after all the mRNA present in the cell had reached its final steady-state labeling. Under the second possibility of stepwise ribosome attachment to nascent mRNA, however, the maximum mRNA labeling of small polyribosomes would be expected to be reached long before steady-state labeling of the entire mRNA had been attained, since here it would be true that the fewer the number of ribosomes in the cluster, the more recent the synthesis of the segment of mRNA with which they are actually in contact. But the maximum mRNA labeling of large polyribosomes would, as under the first possibility, only be reached upon steady-state labeling of the entire mRNA. Experiments designed with these alternatives in mind demonstrated conclusively that the small polyribosomes attain maximal labeling of their mRNA long before the large polyribosomes do, which means that, in accord with the second possibility, transcription and translation of mRNA are dynamically coupled processes.

The general schema of the heterocatalytic function of DNA that evolved from the preceding insights, has been summarized graphically in panels A, B, and C of Figure 17-6. The conception of this schema permitted a clear formulation of the factors determining the functional life time of a nascent mRNA molecule (which in the example of Figure 17-6 was arbitrarily chosen to last for template service in the assembly of only four protein molecules). For it could be envisaged that the life expectancy of an mRNA molecule is governed by the relative probability of two competing events at the 5'-end of a nascent mRNA chain: either an unemployed ribosome may attach there and thus initiate the assembly of another polypeptide chain, or a special degradative enzyme may attack that 5'-end and start a stepwise breakdown of the mRNA chain into its constituent nucleotides, thus rendering impossible its template service for assembly of any more polypeptide chains. The details of the simple example chosen for illustration in Figure 17-6 are not to be accepted as having universal validity, or even as being typical. First of all, as was stated in Chapter 16, many mRNA molecules contain tandem transcripts of two or more contiguous genes, in which case the ribosomes of a single polyribosomal cluster would initiate sequentially the assembly of two or more different polypeptide chains. Second, the average functional life span of many mRNA species is ten or twenty times longer than that shown here, thus allowing the genesis of polyribosomes made up of dozens of ribosomes working in tandem.

Some years after the general schema shown in Figure 17-6 (panels A, B, and C) had been worked out on the basis of biochemical analyses of growing bacterial cells, a direct visualization of the overall heterocatalytic function of the *E. coli* DNA was finally achieved. Figure 17-7 presents an electron micrograph of the transcription-translation process in which the main features

A

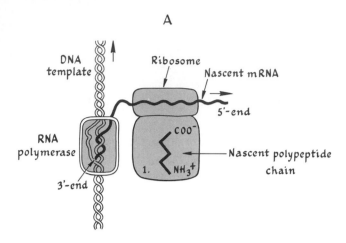

FIGURE 17-6.   Coupled transcription and translation in the heterocatalytic function of bacterial and phage DNA. Schematic summary of the process, as inferred from biochemical studies. **A.** The head, or 5′-end, of an mRNA molecule has just been transcribed and nucleotides are being added to the nascent polynucleotide chain at its tail, or 3′-end, at the rate of about 45 nucleotides per second, as one of the two complementary DNA strands serves as template for the chain growth catalyzed by RNA polymerase. The first ribosome has just engaged the chain-initial 5′-end of the nascent mRNA chain and already carries a short polypeptide representing the amino-terminal (and hence chain-initial) end of an encoded polypeptide chain about 150 amino acids in length. The arrows indicate the directions in which the DNA template moves through the RNA polymerase and the mRNA transcript moves through the ribosome. DNA double helix, RNA polymerase and ribosome are drawn approximately to scale. Polypeptide and mRNA chains are not drawn to scale. **B.** Five seconds later the nascent mRNA chain has grown by about $5 \times 45 = 225$ nucleotides. That length of mRNA chain has moved through the first ribosome and directed, nucleotide triplet by nucleotide triplet, the addition of about $225/3 = 75$ more amino acids to the nascent polypeptide chain being assembled on that ribosome. Meanwhile, a second, third, and fourth ribosome have successively engaged the 5′-end of the nascent mRNA chain, and proceeded to assemble younger versions of the same polypeptide chain. Thus a polyribosomal cluster has been formed. **C.** Another five seconds later, transcription of the mRNA molecule has been completed, and it is no longer in contact with the DNA template whence it arose. The first ribosome has by now completed assembly of its nascent polypeptide chain, has released that chain to the cell, and has itself been released from the mRNA chain and from the polyribosomal cluster of which it formed part. Being presently idle, that ribosome is ready—or as will be seen shortly, its 30S and 50S subunits are ready—to engage the 5′-end of another nascent mRNA molecule. The second, third, and fourth ribosomes are nearing completion of their own younger polypeptide chains. The completed polypeptide chain and the amino termini of the nearly complete nascent polypeptide chains on the second and third ribosomes fold to assume their eventual tertiary protein structure. In this latter 5-second period no further ribosomes have attached to the 5′-end of the mRNA molecule. That is to say, that messenger is now nearing the end of its functional life, during which in the example shown here, it managed to spawn four new polypeptide chains.

B

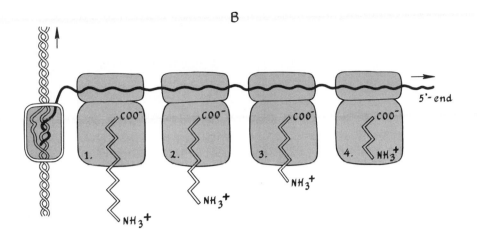

C

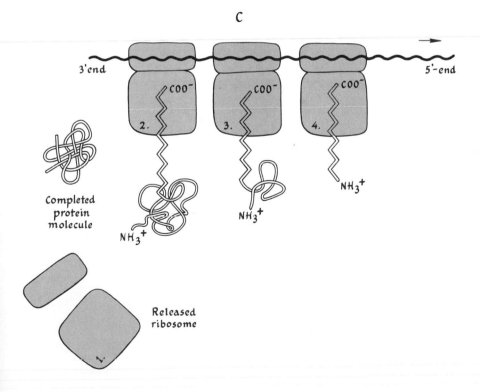

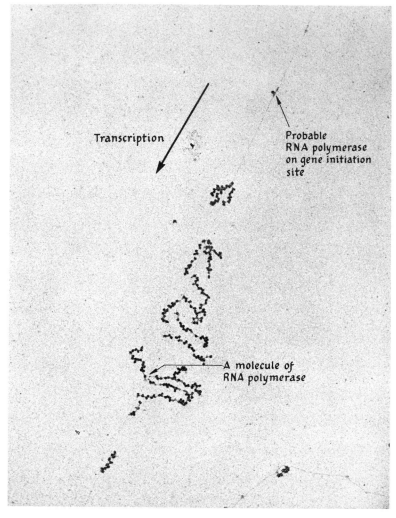

Transcription

Probable
RNA polymerase
on gene initiation
site

A molecule of
RNA polymerase

FIGURE 17-7.   Direct electron-microscopic visualization of the coupled transcription-translation process in an *E. coli* cell extract. The straight line running diagonally across the picture is the bacterial DNA. The prominent black dots are ribosomes on the lateral nascent mRNA chains, seen as thin threads connecting the polyribosomal ribosomes. Transcription of the DNA fiber is evidently proceeding from top right to bottom left, since the nascent mRNA molecules emerging from the DNA fiber at the lower sectors are longer (and carry more ribosomes) than those emerging at the upper sectors. One arrow points to the location of the probable RNA polymerase initiation site, or to the *promoter* (see Chapter 20), for the transcription of the gene (or linked genes) visualized here. The other arrow points to a structure visible at the junction of DNA and mRNA, which, in all probability, is a molecule of RNA polymerase. [Photograph courtesy of O. O. Miller, Jr., Oak Ridge National Laboratory, and Barbara A. Hamkalo and Charles A. Thomas, Jr., Department of Biological Chemistry, Harvard Medical School.]

of the schematic drawings of panels A, B, and C of Figure 17-6 can be seen to have their counterpart in the real world.

The great difference in lifetimes between the *unstable* mRNA, which is in constant turnover, and the *stable* rRNA, which, once formed, is a more or less permanent fixture of the cell, can now be fathomed in terms of the difference in intracellular condition of the 5'-ends of these two classes of ribopolynucleotide chains. For whereas the 5'-end of mRNA is constantly being re-exposed to the degradative enzyme by running through its client ribosomes, the 5'-end of rRNA is afforded permanent protection from the degradative enzyme once it combines with the affined species of ribosomal proteins with which it constitutes the intact ribosomal subunit.

THE ADAPTOR

Granted that mRNA molecules provide the information for protein synthesis in the ribosomes, how are the amino acids actually assembled by the messenger into a sequence of the correct predetermined order? In 1958, Crick analyzed this problem in these terms: "One's first naive idea is that the RNA will take up a configuration capable of forming twenty different 'cavities,' one for the side-chain of each of the twenty amino acids . . . [but] on physical-chemical grounds, the idea does not seem in the least plausible. Apart from the phosphate-sugar backbone, which we have assumed to be regular and perhaps linked to the structural protein of the [ribosome] particles, RNA presents mainly a sequence of sites where hydrogen bonding could occur. One would expect, therefore, that whatever went on to the template in a specific way did so by forming hydrogen bonds. It is therefore a natural hypothesis that the amino acid is carried to the template by an adaptor molecule, and that the adaptor is the part which actually fits on to the RNA. In its simplest form [this hypothesis] would require twenty adaptors, one for each amino acid."

"What sort of molecules such adaptors might be is anybody's guess," wrote Crick, though he went on to say that "there is one possibility which seems more likely than any other—that they might contain nucleotides. This would enable them to join on to the RNA template by the same 'pairing' of bases as is found in DNA, or in polynucleotides." Crick imagined further that "a separate enzyme would be required to join each adaptor to its own amino acid and that the specificity required to distinguish between, say, leucine, isoleucine and valine would be provided by these enzyme molecules, instead of by cavities in the RNA. Enzymes, being made of protein, can probably make such distinctions more easily than can nucleic acid."

The adaptor hypothesis thus leads to the following image of the amino acid assembly process: before its incorporation into the nascent polypeptide chain, each amino acid molecule is fitted out with a nucleotide adaptor that contains a nucleotide triplet *or anticodon*, complementary in its nucleotide sequence to the nucleotide triplet, or *codon*, that codes for the same amino acid in the messenger RNA. The amino-acid-nucleotide complexes then diffuse into the ribosomes, where they are held in their proper places on the messenger template by pairs of hydrogen bonds between complementary purines and pyrimidines of adaptor and messenger RNA molecules. Once lined up in this way along the messenger RNA in the "correct" order, the individual amino acid residues are joined to each other through peptide bonds, by means of a chemical rearrangement that simultaneously liberates the amino acid from its bond to the nucleotide adaptor and joins it to the nascent polypeptide chain.

While these purely speculative notions were under discussion, biochemical investigations into the enzymology of protein biosynthesis began to turn up an ensemble of specific reactions and cell components that gradually came to resemble more and more the postulated adaptor system. Thus, in 1957 a hitherto unknown type of RNA was discovered, first in eukaryotic cells and soon thereafter in bacteria, which, like rRNA (and unlike the then as yet unknown mRNA) is a *stable* cell component, but which unlike rRNA (and like mRNA) is not a structural member of ribosomes. The polynucleotide chains of this new type of RNA were found to be only about 80 nucleotides in length. These short RNA molecules were eventually given the name *transfer RNA*, or tRNA, because, as was then found in the laboratory of Paul Zamecnik and as will be set forth now, they act as a transfer agent of amino acids in protein synthesis. In *E. coli*, tRNA accounts for about 15% of the total bacterial RNA, and thus makes up the remainder not accounted for by the 80% of rRNA and 5% of mRNA. Since one tRNA chain contains only 80/4600 as many nucleotides as the rRNA complement of one ribosome, there are present in the bacterium $0.15 \times 4600/(0.80 \times 80)$, or about ten tRNA molecules for every ribosome. The involvement of tRNA in protein synthesis came to light when M. Hoagland, Zamecnik, and their colleagues found that the very first step in the utilization of an amino acid for protein biosynthesis is its "activation" by means of an *amino-acid-activating enzyme* that catalyzes the reaction between the amino acid and ATP shown as step 1 in Figure 17-8. The product of this reaction is an *amino acid adenylate* in which the amino acid is linked to adenosine through a phosphate ester bond. As the next step that very same activating enzyme molecule, with which the amino acid adenylate is still combined, was found to catalyze the attachment of the amino acid to a tRNA molecule. More precisely, the amino acid adenylate was found to react with the free 3'-OH group of the ribose of *adenylic acid* that tRNA molecules carry at their 3' end, so as to link the α-carboxyl group of the amino acid by an acyl bond to the tRNA molecule, according to the reaction shown as step 2 in Figure 17-8. As we shall see shortly,

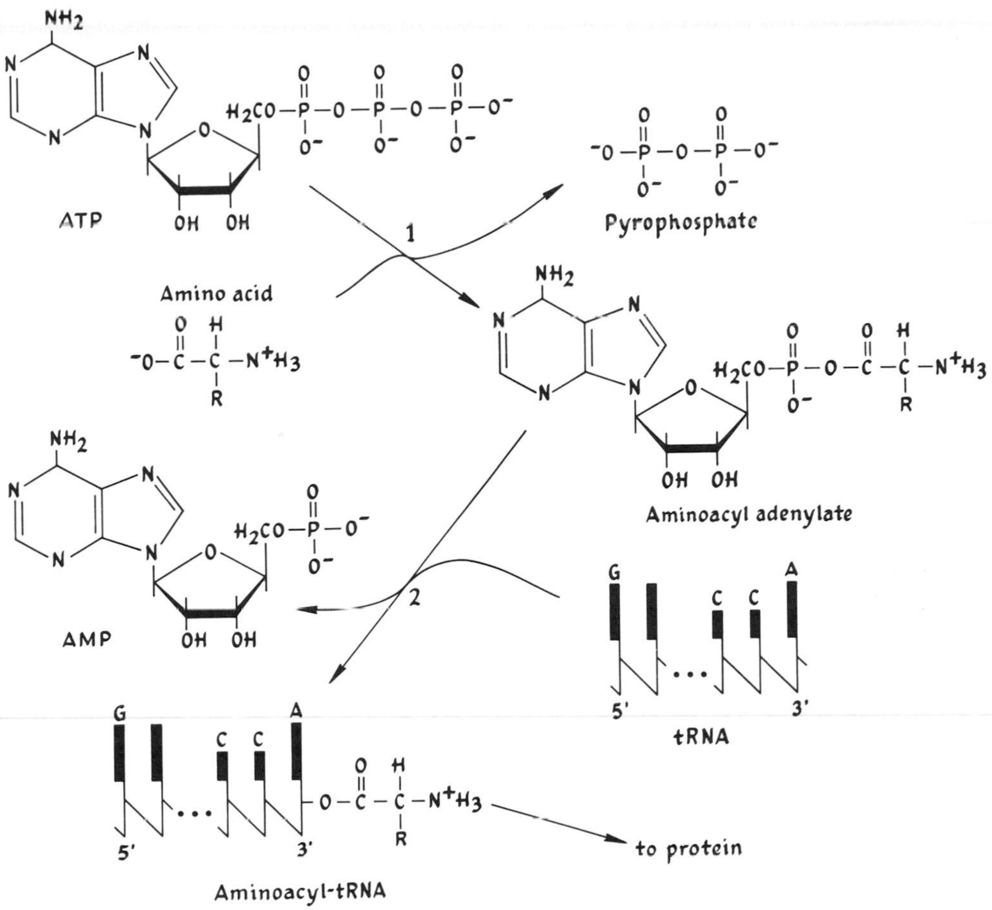

FIGURE 17-8.   The first two reaction steps in the utilization of an amino acid for protein biosynthesis, both catalyzed by an amino acid activating enzyme (or aminoacyl-tRNA synthetase). (Step 1) "Activation" of the amino acid. (Step 2) Transfer of the amino acid from the aminoacyl adenylate to the 3′-OH terminus of its affined tRNA molecule (the approximately 75 internal nucleotide residues of the tRNA polynucleotide chain have been omitted from this drawing).

it is through this *aminoacyl-tRNA* complex that the amino acid reaches its ultimate cellular destination—namely, the nascent polypeptide chain.

Quantitative studies soon showed that the attachment reaction of amino acids to tRNA is of a highly specific nature, in that to each and every one of the twenty standard amino acids there corresponds at least one affined species of tRNA molecule that will accept that and only that amino acid. This conclusion followed from the observation that the maximum amount of any given amino acid that can be attached to the tRNA of a cell extract is independent of the presence of the other nineteen standard amino acids. That maximum amount represents the *acceptor capacity* for the amino acid under consideration and is a measure of the relative abundance of its specifically affined tRNA molecules in the extract. Table 17-1 shows the results

TABLE 17-1.   Amino-acid-acceptor Capacity of *E. coli* tRNA

Amino acid	mμ moles accepted per mg tRNA	Amino acid	mμ moles accepted per mg tRNA
Alanine	2.97	Methionine	0.95
Arginine	2.97	Phenylalanine	0.88
Aspartate	0.76	Proline	1.50
Glutamate	2.33	Serine	1.49
Glycine	3.28	Threonine	1.86
Histidine	1.06	Tyrosine	0.94
Leucine	3.58	Valine	2.79
Lysine	2.59		

From C. D. Yegian, G. S. Stent, and E. M. Martin. *Proc. Natl. Acad. Sci. Wash. U.S.* **55**, 839 (1966).
Procedure: To a salt solution buffered at *p*H 7 were added ATP, purified extracts of *E. coli* tRNA and aminoacyl-tRNA synthetases, and one of the amino acids labeled with ^{14}C at a known specific radioactivity. The reaction mixture was incubated at 30°C for 20 minutes, after which time the tRNA and its attached amino acid was precipitated from the solution by addition of ethanol to a final concentration of 67%. The content of ^{14}C-amino acid of the precipitate was then assayed in a radiation counter.

of a measurement of the acceptor capacity of *E. coli* tRNA for 16 of the 20 standard amino acids. As can be seen there, the acceptor capacities vary over a fivefold range. They indicate, therefore, that *E. coli* contains about five times as many tRNA molecules capable of accepting leucine (the highest capacity) as molecules capable of accepting aspartate (the lowest capacity). In 1958, Robert W. Holley, one of the original discoverers of tRNA, began to develop methods for the isolation of individual tRNA species from their mixture in a cell extract. These methods depended on the finding that the rather slight chemical differences distinguishing one tRNA species from another engender slight differences in relative distribution of their water solutions over the two immiscible solvents formamide and isopropyl alcohol. The plot shown in Figure 17-9 gives the result of one of Holley's fractionations of tRNA extracted from yeast on the basis of this principle. After a very large number of sequential extractions and re-extractions of one solvent phase by another, the original tRNA mixture became distributed among a

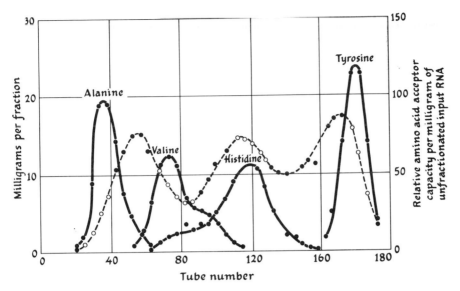

FIGURE 17-9. The resolution of a mixture of tRNA molecules extracted from yeast by means of the *countercurrent distribution* method, in which the two immiscible solvents formamide and isopropyl alcohol are used. The countercurrent apparatus consists of a series of 200 tubes, each containing equal volumes of formamide and isopropyl alcohol. One of the solvents forms a lower phase and the other an upper phase. A small volume of a water solution of the tRNA extract is introduced into a tube in the middle of the series, and the apparatus is set in motion. The apparatus automatically shakes the tubes to allow equilibration of the tRNA solution over the two phases. After equilibration the phases are separated, and the upper phase of each tube is transferred to the tube on the left and the lower phase to the one on the right. A new equilibrium distribution between the two phases is allowed to occur after the attainment of which upper and lower phases are once more transported to neighboring tubes on opposite sides. This procedure is repeated 200 times, by which time the tRNA solution is spread over the entire tube series, while upper and lower phases have flown in opposite directions through the apparatus. The tRNA solution present in each tube (represented on the abscissa) is then analyzed for its UV-absorption at 260 mμ, or total nucleic acid content (left-hand ordinate, broken line) and for its capacity to accept radioactively labeled alanine, valine, histidine, or tyrosine (right-hand ordinate, solid lines). [After "The Nucleotide Sequence of a Nucleic Acid," by R. W. Holley, *Scientific American*, February 1966. Copyright © 1966 by Scientific American, Inc. All rights reserved.]

series of 200 tubes, tube 1 containing the RNA molecules most soluble, and tube 200 the RNA molecules least soluble, in one of the two solvents. The solutions in each of the 200 tubes were then analyzed both for their total RNA content and for their tRNA acceptor capacity for each of the four amino acids alanine, valine, histidine, and tyrosine. As can be seen, the total RNA became distributed rather broadly over the set of tubes, although peaks of concentration were manifest in tubes 60, 110, and 170. The individual amino-acid-acceptor capacities, however, are seen to have become much more narrowly distributed. In particular, the two species of tRNA molecules affined to alanine and to tyrosine have been completely separated from those affined to valine and histidine. The rather broader distribution of the valine and histidine tRNA species signaled a further important fact—namely, that to some of the 20 standard amino acids there corresponds more than a single species of affined tRNA molecule.

While these efforts to fractionate tRNA into its component species were under way, Paul Berg and his collaborators succeeded in resolving the amino-acid-activating enzyme proteins present in an *E. coli* extract into a number of fractions, each fraction being capable of preferentially catalyzing the attachment of one particular amino acid to its affined species of tRNA. As is shown in Table 17-2, they found that the "valine" enzyme fraction cata-

TABLE 17-2.   Relative Efficiency of Catalysis of Amino Acid Attachment to *E. coli* tRNA by Individual *E. coli* Aminoacyl-tRNA Synthetases

Amino acid tested	μmoles amino acid attached per mg tRNA per hr			
	Valine enzyme	Isoleucine enzyme	Leucine enzyme	Methionine enzyme
Valine	25	< 0.03	0.18	< 0.01
Isoleucine	< 0.01	3.3	< 0.01	< 0.01
Leucine	< 0.01	< 0.07	3.2	< 0.01
Methionine	< 0.01	0.07	< 0.01	3.5

From P. Berg, F. H. Bergmann, E. J. Ofengand, and M. Dieckmann, *J. Biol. Chem.* **236**, 1726 (1961).

Procedure: By means of chromatography on ion exchange columns, a mixture of *E. coli* aminoacyl-tRNA synthetases was resolved into a number of fractions. Each of four of these enzyme fractions (referred to here as "valine enzyme," "isoleucine enzyme," "leucine enzyme" and "methionine enzyme") was added to a buffered salt solution containing ATP and purified *E. coli* tRNA. Each of these four reaction mixtures was then subdivided into four aliquots, to each of which one of our ^{14}C-labeled amino acids of known specific content of ^{14}C label was added. The rate of amino acid attachment was then followed in all samples by measuring the amount of ^{14}C-label co-precipitable with tRNA as a function of time.

lyzes the attachment of valine to tRNA more than 2500-fold faster than the attachment of isoleucine, leucine, or methionine. Furthermore, each of the "isoleucine," "leucine," and "methionine" enzyme fractions show a similarly high degree of discrimination in favor of the attachment of their cognate amino acids. Thus it could be concluded that the specificity of attachment of an amino acid to its affined species of tRNA depends on the work of a specific activating enzyme that consummates this union. As soon as it was recognized that this is actually their main physiological function, the amino-acid-activating enzymes came to be known as *aminoacyl-tRNA synthetases,* the name under which they will be referred to in the remainder of this text.

With the availability of these insights it required little further imagination to infer that the tRNA molecules do in fact figure as Crick's postulated adaptors, and that three of the 80 or so nucleotides making up the tRNA chain represent the anticodon by means of which the amino acid is recognized by the complementary mRNA codon in the polypeptide assembly process at the ribosomal scene. Since, under this view, the aminoacyl-tRNA synthetases succeed in matching each amino acid with its correct adaptor, they represent the intracellular agency that "knows" the genetic code. But since the amino acid sequence of the aminoacyl tRNA synthetases, and hence the specificity of *their* structure, is in turn inscribed into the DNA, the DNA is, in the last analysis, also privy to the code.

After abundant indirect support for the verity of these conceptions had already been accumulated, G. von Ehrenstein, B. Weisblum, and Benzer were finally able to deliver a direct experimental proof of the adaptor role of tRNA in 1963. In this experiment ^{14}C-labeled cysteine was attached to its affined tRNA in the usual reaction mixture containing ATP and aminoacyl-tRNA synthetases. The ^{14}C-labeled cysteine-tRNA complex was then treated with hydrogenated nickel at room temperature. This procedure releases the —SH group from cysteine and thus converts it into alanine, according to the reaction

$$\begin{array}{cc} \overset{\displaystyle H}{\underset{\displaystyle \underset{\displaystyle SH}{\overset{\displaystyle |}{CH_2}}}{\overset{\displaystyle |}{H_2N-C-COOH}}} + Ni\,(H) \longrightarrow & \overset{\displaystyle H}{\underset{\displaystyle \underset{\displaystyle}{\overset{\displaystyle |}{CH_3}}}{\overset{\displaystyle |}{H_2N-C-COOH}}} + H_2S + Ni \\[2em] \text{Cysteine} & \text{Alanine} \end{array}$$

Thus in this way was produced the highly unnatural complex between ^{14}C-labeled alanine and cysteine-affined tRNA molecules. This complex was then added to another reaction mixture containing polyribosomes extracted from rabbit reticulocytes, ATP, the set of standard amino acids in non-radioactive form, and more tRNA and aminoacyl-tRNA synthetases. In such a reaction mixture, the rabbit polyribosomes carrying the hemoglobin mRNA continue their assembly of those nascent hemoglobin polypeptide chains in whose synthesis they were engaged at the moment of extraction from the reticulocyte blood cells. Von Ehrenstein, Weisblum, and Benzer found that the ^{14}C-labeled alanine they had added to the reaction mixture as a complex with cysteine specific tRNA was actually incorporated into hemoglobin α-polypeptide chains completed *in vitro.* They then proceeded to ascertain the exact positions in the α-chain at which that ^{14}C-alanine was located and found that it had been introduced into positions at which *cysteine* (and not alanine) is normally present. Hence it could be concluded that, in full accord with the adaptor hypothesis, once an amino acid is attached to the 3'-OH end of a tRNA molecule it is recognized in the polypeptide assembly process for what it is only by the structural specificity of that tRNA adaptor and no longer by the nature of its side chain.

TRANSFER RNA STRUCTURE

How do the tRNA adaptor molecules arise in the cell? Although the details of the mechanism of their synthesis still remain to be worked out, there now is left little doubt that the short tRNA polynucleotide chains are transcribed from DNA templates by RNA polymerase, just as are transcribed the much longer chains of mRNA and rRNA. First of all, molecular RNA-DNA hybridization experiments between tRNA and DNA of *E. coli* (analogous to those described in Figure 16-8,A for rRNA and DNA) have shown that 0.046 percent of the bacterial DNA contains nucleotide sequences complementary to those of tRNA. Hence, by the same reasoning as that employed in Chapter 16 for estimating the number of genes in the *E. coli* genome coding for 16S and 23S rRNA molecules, it can now be reckoned that among the 3×10^6 nucleotide base pairs of the *E. coli* genome, there are about $3.6 \times 10^6 \times 0.00046/80 = 20$ segments 80 nucleotides in length that are complementary to the nucleotide sequences of tRNA molecules. Furthermore, competitive RNA-DNA hybridization experiments (analogous to those described in Figure 16-8,B for 16S and 23S rRNA) carried out with individual species of *E. coli* tRNA fractionated according to the method described in Figure 17-9 have shown that tRNA molecules affined to different amino acids do not compete for the same DNA nucleotide sequences. Hence each individual tRNA species is a polynucleotide chain of unique primary structure, whose provenance is a gene into which that sequence is encoded. In confirmation of that conclusion, *E. coli* mutants were later isolated whose phenotypes (which will soon figure in our discussion) could be traced back to mutations of just such genes specifying the structure of tRNA molecules.

Chemical analysis of the nucleotides recovered upon hydrolysis of tRNA molecules was to provide a surprise, however. And that surprise was that in addition to the four ordinary nucleotides, adenylic (A), guanylic (G), cytidylic (C), and uridylic (U) acids characteristic of RNA, tRNA was found to contain also a number of extraordinary nucleotides not previously found in natural nucleic acids (Figure 17-10). One of these extraordinary nucleotides,

FIGURE 17-10.   Structure of seven extraordinary ribonucleotides found in tRNA polynucleotide chains and the pathway of their *in situ* formation from the ordinary nucleotides A, G, and U. Only the bases have been drawn in their full chemical structure. The ribose-phosphate moiety of each nucleotide is shown in the conventional short-hand symbolism.

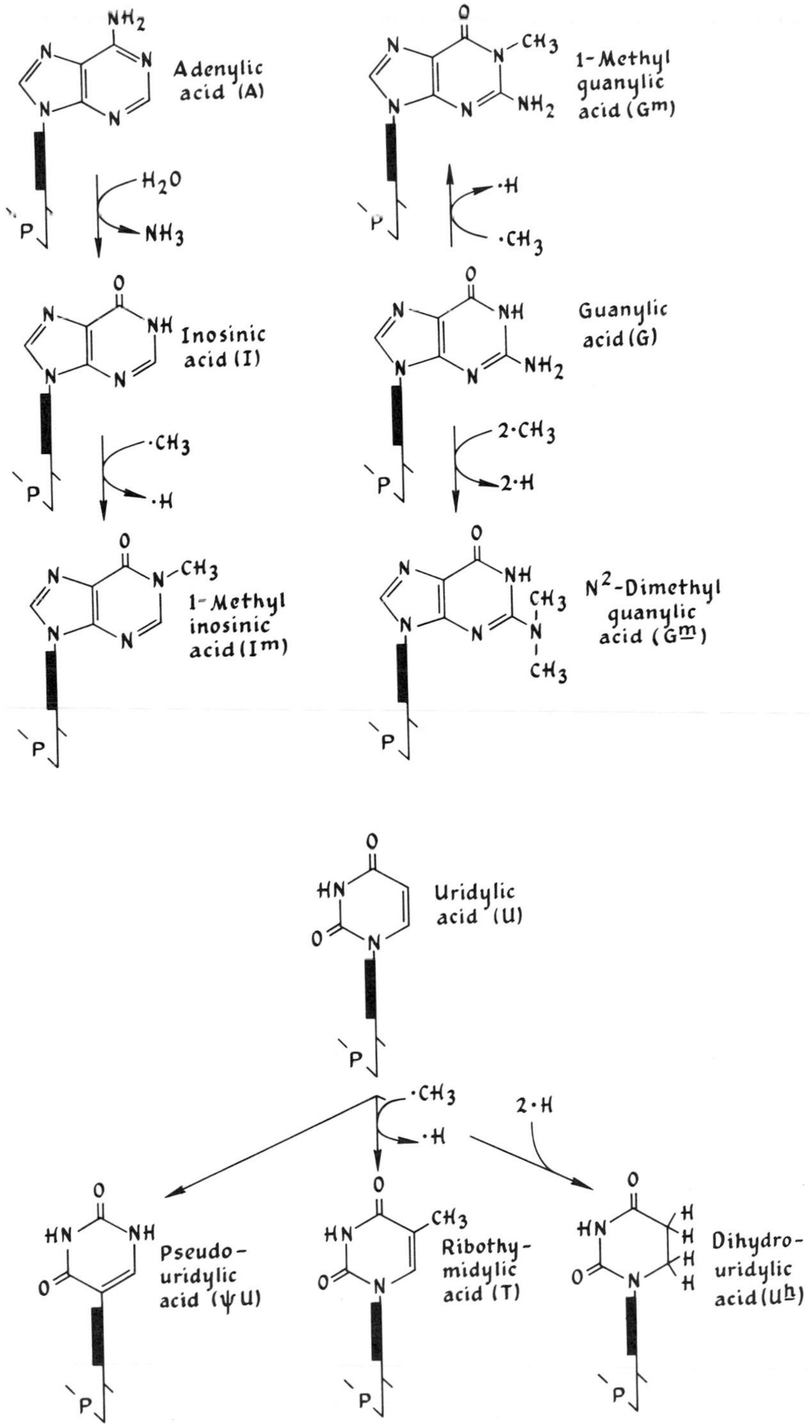

Adenylic acid (A)

H_2O

NH_3

Inosinic acid (I)

·CH_3

·H

1-Methyl inosinic acid (I^m)

1-Methyl guanylic acid (G^m)

·H

·CH_3

Guanylic acid (G)

2·CH_3

2·H

N^2-Dimethyl guanylic acid ($G^{\underline{m}}$)

CH_3

Uridylic acid (U)

·CH_3

·H

2·H

Pseudo-uridylic acid (ψU)

Ribothy-midylic acid (T)

CH_3

Dihydro-uridylic acid ($U^{\underline{h}}$)

inosinic acid (I), bears the purine hypoxanthine, whose hydrogen-bonding capacity for forming complementary base pairs is that of neither adenine nor guanine. In three others of these extraordinary nucleotides, 1-methyl-inosinic (I^m), 1-methylguanylic (G^m), and N^2-dimethylguanylic ($G^{\underline{m}}$) acids, presence of methyl groups interferes with the formation of *any* complementary base pairs. And in yet another one, pseudouridylic (Ψ) acid, the uracil pyrimidine ring is attached to ribose, not by its 1-nitrogen, as in normal uridylic acid, but by its 5-carbon. Another extraordinary nucleotide related to uracil is ribothymidilic (T) acid, in which the 5-carbon atom of uracil bears a methyl group—a substitution that converts uracil into thymine, the very pyrimidine base whose natural occurrence had long been thought to be restricted to *DNA*. Finally, the extraordinary nucleotide dihydrouridylic acid ($U^{\underline{h}}$) carries an extra hydrogen on both the 5-carbon and the 6-carbon of its heterocyclic ring. This ring is not really a pyrimidine, since the link between its 5- and 6-carbon is a single rather than a double bond.

How can the presence of these extraordinary nucleotides in tRNA be reconciled with the inference just drawn that these short polynucleotide chains arise by transcription of DNA templates? Surely, these nucleotides cannot be introduced into the growing polynucleotide chain by any complementary base-pairing mechanism. The answer to that question was provided by the discovery of enzymes in both bacteria and eukaryotic cells which catalyze the conversion of ordinary nucleotides of nascent tRNA molecules into extraordinary nucleotides, as shown in Figure 17-10. Thus one enzyme was found that converts the —NH_2 group on the 2-carbon of tRNA adenylic acid into an —OH group, thus creating an I residue at some site of the already formed polynucleotide chain. Other enzymes were found to transfer the methionine methyl group of S-adenosyl methionine to the relevant ring positions of tRNA hypoxanthine, G, or U to create I^m, G^m, $G^{\underline{m}}$, and T at the corresponding sites of the intact polynucleotide. Finally, still other enzymes could be shown to act on tRNA to rearrange the bond between the pyrimidine ring and ribose to convert U into Ψ or to hydrogenate that ring to convert U into $U^{\underline{h}}$.

Holley's purpose in devising the tRNA fractionation method described in Figure 17-9 had been to purify one single species of tRNA molecule affined to a given amino acid in order to establish its primary structure, or precise nucleotide base sequence. He focused his attention on the alanine-specific molecules of yeast tRNA, and in 1965, seven years after he set out on this project, he had reached his objective. The general principles by which Holley established the exact sequence of the 77 nucleotides of the purified alanine-specific tRNA polynucleotide resembled those that F. Sanger had first employed a decade earlier for ascertaining the sequence of the 51 amino acids of the beef insulin polypeptide (as was set forth in Chapter 3). That is to say, Holley first broke down the polynucleotide chain into a number of shorter fragments by allowing two different ribonuclease enzymes to act on separate samples of the tRNA. One of these enzymes, pancreatic ribonuclease, splits the polynucleotide chain at the 3'-side of all pyrimidine nucleotides, whereas the other enzyme, takadiastase ribonuclease II, splits the chain at the 3'-side of all guanine and hypoxanthine purine nucleotides. Digestion of the alanine-specific tRNA by either of these ribonuclease enzymes yielded two separate sets of either 19 or 17 different kinds of mono- and polynucleotides, ranging in length from one to eight nucleotides. These nucleotide fragments were then separated by chromatography and the nucleotide sequence of each was established by stepwise degradation with still another nuclease enzyme that splits the internucleotide bonds one at a time. On the basis of the overlaps observed in the primary structure of the two sets of different kinds of nucleotide fragments obtained in this way, Holley was able to reconstruct the whole sequence of the entire alanine-specific tRNA molecule, as shown in Figure 17-11. The presence of the few widely dispersed extraordinary nucleotides (one each of I, I^m, G^m, G^m, and T; two Ψ and three U^h) provided very convenient landmarks for these reconstruction efforts. In the representation of the nucleotide sequences shown in Figure 17-11, a further (conventional) reduction in symbolism has been employed. Here a sequence of letters such as GpApUpCp is equivalent to the schematic representation first introduced in Figure 8-5, namely

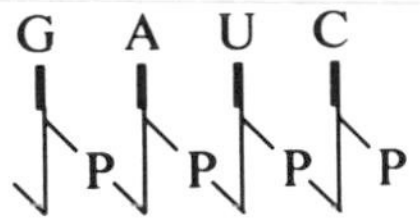

$$
\begin{array}{cccc}
G & A & U & C \\
| & | & | & | \\
\end{array}
$$

$$\sqrt{}\, P \sqrt{}\, P \sqrt{}\, P \sqrt{}\, P$$

That is to say, that particular sequence of letters represents a polynucleotide chain whose nucleotide base sequence beginning at the 5'-end and proceeding toward the 3'-end of the chain is guanine (G), adenine (A), uracil (U), and cytosine (C). Each lower case p represents a phosphate diester group linking the 3'-OH of the nucleotide to its left and the 5'-OH of the nucleotide on its right.

Physicochemical studies of the kind that had previously established the hydrogen bonding of purine and pyrimidine bases in the DNA double helix (see Chapter 8) suggested that a considerable fraction of the bases of a tRNA molecule are actually hydrogen-bonded to each other. Holley therefore contemplated the tRNA nucleotide base sequence he had worked out and tried to see how this sequence could give rise to a partially base-paired structure. The result of these efforts was the "cloverleaf" structure shown in Figure 17-12. This structure is characterized by four base-paired regions and three loops and bends of unpaired regions. It is to be noted that the extraordinary nucleotides are all to be found in unpaired regions, either in loops

5 10 15 20 25 30 35

pGpGpGpCpGpUpGpUpG^{m}pGpCpGpCpGpUpApGpU^{h}pCpGpGpGpU^{h}pApGpCpGpCpG^{m}pCpUpCpCpCpUpUpIpGpC

40 45 50 55 60 65 70 75

pI^{m}pψpGpGpGpApGpApGpU^{h}pCpUpCpCpGpGpTpψpCpGpApUpUpCpCpGpGpApCpUpCpGpUpCpCpApCpCpA

Fragments produced by

pancreatic ribonuclease   takadiastase ribonuclease II

	pGpGpGpCp				
Cp (14x)	Up (6x)	ψp A	pGp	Gp(9x)	ApGp(2x)
GpCp(2x)	G^{m}pCp	GpUp(4x)	CpGp(4x)	CpG^{m}p	UpGp
I^{m}pψp	ApCp	ApGpU^{h}p	UpG^{m}p	UpApGp	U^{h}pCpGp
GpGpU^{h}p	G^{m}pGpC	ApGpCp	U^{h}pApGp	CpI^{m}pψpGp	
GpApUp	IpGpCp	GpGpTp	Tpψ CpGp	ApCpUpCpGp	
GpGpApCp	GpGpGpApGpApGpU^{h}p		ApUpUpCpCpGp	U^{h}pCpUpCpCpGp	
			CpUpCpCpCpUpUpIp	UpCpCpApCpCpA	

FIGURE 17-11.   Above: The exact nucleotide base sequence of the 77 nucleotides that make up the polynucleotide chain of alanine tRNA of yeast, as first determined by R. W. Holley. The chain carries a phosphate group esterified to its 5′-OH end. Below: The two sets of nucleotide fragments produced upon exhaustive digestion of the intact, purified alanine tRNA molecule with either one of two different enzymes. Pancreatic ribonuclease splits the internucleotide bond on the 3′-side of all pyrimidine nucleotides (C, U^h, Ψ, and T) between the phosphate group and the 5′-OH of the next nucleotide. Takadiastase ribonuclease T1 splits the internucleotide bond on the 3′-side of all guanine-like nucleotides (G, G^m, G^m, and I) between the phosphate group and the 5′-OH of the next nucleotide. Digestion with either nuclease yields only one fragment that carries a phosphate on its 5′-end, and hence represents the 5′-end of the chain, and only one fragment that carries no phosphate on its 3′-end, and hence represents the 3′-end of the chain. The task of reconstructing the entire sequence from the two sets of overlapping fragments was greatly simplified by the discovery that the whole tRNA molecule can be split chemically between nucleotides 38 and 39 into two nearly equally long fragments, which can be separated before nuclease digestion (the two fragments correspond to the upper and lower parts of the sequence as written here).

or in bends, of the cloverleaf. But where in this structure is the anticodon nucleotide triplet? Holley knew from the work to be recounted in the next chapter that the triplet codons pGpCpU, pGpCpC, or pGpCpA represent alanine in the genetic code. He therefore inferred that the sequence pIpGpC in the middle loop is the anticodon, since (as will be seen in the next chapter) it alone is capable of base pairing with all three of these codons. One further unpaired region of the tRNA molecule is made up of the last four nucleotides at the 3′-end of the polynucleotide chain at which alanine becomes attached.

  Soon after the first structural analysis of alanine-specific tRNA had been completed, the structure of tRNA species affined to other amino acids

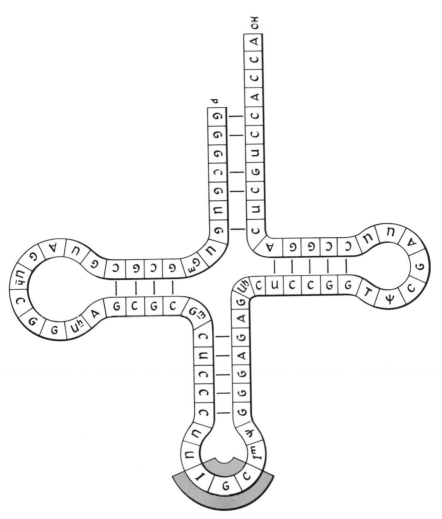

FIGURE 17-12.   The "cloverleaf" structure of yeast alanine tRNA, first inferred from considerations of optimum intramolecular A-U and G-C complementary base pairing of the linear nucleotide base sequence shown in Figure 17-11. The shaded area marking the pIpGpC nucleotide triplet is the anticodon. [After "The Nucleotide Sequence of a Nucleic Acid," by R. W. Holley, *Scientific American*, February 1966. Copyright © 1966 by Scientific American, Inc. All rights reserved.]

became known. Although each of these other tRNA polynucleotides proved to have its own distinctive nucleotide sequence, especially its characteristic anticodon, and some turned out to contain even more extraordinary nucleotides than those shown in Figure 17-10, all tRNAs nevertheless do appear to share in common some general structural features. First of all, they all carry the residue pG at their 5'-end. Second, they can all be described as a cloverleaf structure, in which the middle loop is the site of the anticodon. Finally, they all seem to carry the sequence pGpTp ΨpCpG in one of their lateral loops. The deeper meaning of these structural regularities still remains to be fathomed, however. It seems most probable, nevertheless, that the metabolic stability of tRNA derives from one of its structural features. In particular, one may envisage that the tRNA polynucleotide is not subject to breakdown because its pG 5' chain end, the apparent target of the breakdown enzyme that limits the lifetime of the wholly single-stranded mRNA, is hydrogen-bonded in the double-stranded stem part of the cloverleaf structure and hence protected against enzymatic attack.

AMINO ACID ASSEMBLY

We are now ready at last to consider the process of amino acid assembly, or the actual translation of mRNA into the corresponding polypeptide chain. Figure 17-13 presents a graphical summary of the facts to be set forth in the following paragraphs. We must note, first of all, that the mRNA molecule appears to be in contact with the 30S ribosomal subunit, since *in vitro* mixing experiments show that mRNA is bound by isolated 30S subunits but not by 50S subunits. The tRNA molecules, in contrast, appear to be in contact with the 50S ribosomal subunits, which by other *in vitro* mixing experiments can be demonstrated to possess two sites at which tRNA can be bound. The first of these sites is the *aminoacyl site*, at which a tRNA molecule carrying an amino acid can attach, provided that the anticodon borne by the tRNA molecule matches the codon displayed by that sector of the mRNA tape which happens to face the aminoacyl site. The second site is the *peptidyl site*, which accepts tRNA molecules only from the aminoacyl site. Growth of a peptide chain is initiated when the aminoacyl site of a ribosome displaying the messenger RNA codon corresponding to the first amino acid of the polypeptide chain is occupied by the corresponding tRNA and its attached amino acid. Transfer RNA and first amino acid now move to the peptidyl site, this movement being attended by advancement of the mRNA through the ribosome and by display of the next codon at the aminoacyl site. The corresponding tRNA bearing the second amino acid now occupies the aminoacyl site, bringing the α-amino group of the second amino acid into juxtaposition with the ester bond in which the α-carboxy group of first amino acid is linked to the 3' adenylic acid terminus of its own tRNA. The stage is now set for an enzymatically catalyzed exchange reaction, in which the α-amino group of the second amino acid forms a peptide bond to the α-carboxy group of the first amino acid, thereby splitting the ester bond that had linked the first amino acid to its tRNA. The two-amino acid peptide chain is thus attached to the second tRNA, which now moves from the aminoacyl site to the peptidyl site and displaces the free tRNA. This movement

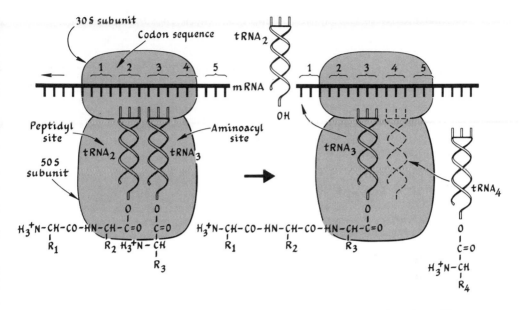

FIGURE 17-13.   Schematic summary of the amino acid assembly process in polypeptide synthesis. To simplify the illustration the tRNA molecules are shown as "hairpins" rather than "cloverleafs." The left-hand part of the figure represents a point in time at which the first two amino acids of a nascent polypeptide chain have been joined and are held to the peptidyl tRNA site of the ribosome by the tRNA molecule to which the second amino acid is still joined. The aminoacyl tRNA site is already occupied by the tRNA carrying the third amino acid called for by the third codon in the mRNA nucleotide sequence. The right-hand part of the drawing represents the same scene a few moments later. The third amino acid has been joined to the nascent polypeptide chain by means of an exchange reaction that concomitantly released the second amino acid from its tRNA. The tRNA of the third amino acid, which now carries the newly elongated polypeptide chain, has moved from aminoacyl to peptidyl tRNA site, from which the tRNA formerly holding the second amino acid has been expelled. The mRNA has advanced through the ribosome by one triplet codon, and the tRNA carrying the fourth amino acid called for by the fourth codon is ready to occupy the vacated aminoacyl tRNA site.

is again attended by advancement of the mRNA through the ribosome and by display of the third codon at the aminoacyl site. The corresponding tRNA bearing the third amino acid now occupies the aminoacyl site, bringing the α-amino group of the third amino acid into juxtaposition with the ester bond that links the two-amino-acid-peptide chain to the second RNA. Formation of the second peptide bond now proceeds, resulting in the attachment of the three-amino acid peptide chain to the third tRNA. Thus this process continues on and on, amino acid by amino acid, until the entire polypeptide chain encoded in the gene has been assembled. Finally, there must appear a special "chain-termination" nucleotide sequence at the aminoacyl site, to which no aminoacyl tRNA species responds. This sequence is recognized by some enzymatic component of the protein-assembly engine whose action splits the ester bond in which the now completed polypeptide chain is linked to the tRNA of the last amino acid that was added. In this way, the completed chain is released from its ribosomal site of assembly and is ready to assume the tertiary and quaternary structure of the protein as which it is to function. The ribosome, for its part, is now free to begin the assembly of another polypeptide chain.

PEPTIDE CHAIN INITIATION AND THE RIBOSOME CYCLE

The general schema of polypeptide assembly presented in Figure 17-13 had been worked out by about 1964. Soon thereafter additional details came to light concerning the mechanism by which growth of the peptide chain is actually initiated. First, it was discovered by F. Sanger and his colleagues that one species of tRNA molecules isolated either from yeast or *E. coli* carries a derivative of methionine at its 3'-chain terminus, namely N-formyl-methionine.

$$
\begin{array}{c}
\underset{\displaystyle}{\overset{\displaystyle O}{\underset{\|}{}}} \quad \underset{\displaystyle}{\overset{\displaystyle H}{\underset{|}{}}} \quad \underset{\displaystyle}{\overset{\displaystyle H}{\underset{|}{}}} \quad \underset{\displaystyle}{\overset{\displaystyle O}{\underset{\|}{}}} \\
H-C-N-C-C-O^- \\
\underset{|}{CH_2} \\
\underset{|}{CH_2} \\
\underset{|}{S} \\
CH_3
\end{array}
$$

N-formylmethionine

The presence of this methionine derivative on tRNA molecules at first seemed surprising, since the blockage of the methionine α-amino group by the formyl (CHO) group would prevent incorporation of the amino acid into growing polypeptide chains. Soon it was realized, however, that there is one site in a growing polypeptide chain into which formylmethionine *can* be introduced—namely, the amino-terminal end at which growth of the chain is initiated. Thus it was proposed that the assembly of polypeptide chains begins with a formylated amino acid. This made it possible to envisage that upon attachment of the first tRNA molecule bearing a formylated chain-initial amino acid to the ribosomal aminoacyl site, the ribosomal polypeptide assembly engine receives the same signal that it would otherwise receive at later stages of chain growth whenever the free α-amino group of the incoming amino acid has been joined in a peptide bond to the α-carboxy group of the last amino acid to have been added. For that signal evidently sets in motion the translocation of the tRNA and its attached chain from ribosomal amino-acyl site to peptidyl site and the attendant advance of the mRNA tape through the ribosome by one nucleotide triplet codon.

An exhaustive search was then made for *E. coli* tRNA species bearing formylated derivatives of amino acids other than methionine. Since this search proved unsuccessful, these notions of polypeptide chain initiation could be saved from abandonment only by supposing that growth of *all E. coli* polypeptide chains is initiated by formylmethionine. But this supposition seemed difficult to accept, inasmuch as it was known that most proteins of *E. coli* do not, in fact, carry formylmethionine at the amino terminus of their polypeptide chains. Studies of the synthesis of the coat-protein component of an *E. coli* phage carried out in the laboratories of J. D. Watson and N. D. Zinder soon resolved this dilemma. The primary structure of that coat protein was known to begin with the sequence alanine-serine at its amino terminus. But analysis of the same coat protein synthesized in a reaction mixture consisting of phage mRNA and the components required for polypeptide assembly showed that the *in vitro* product carried the sequence formylmethionine-alanine-serine at its amino terminus. It could be concluded, therefore, that even though in bacterial protein synthesis formylmethionine is laid down at the starting end of *all* polypeptide chains, the chain-initial formylmethione residue is later cut away from the nascent chain to expose another amino acid at the amino terminus. That is to say, the amino-terminal residue of a mature protein is not necessarily the same amino acid as that with which growth of the polypeptide chain had actually begun.

Further work by Sanger's colleagues elucidated the mechanism of formation of the chain-initiating formylmethionine-tRNA complex. The most simple idea—that an aminoacyl-tRNA synthetase attaches formylmethionine to its affined tRNA—turned out not to be true. Instead, it was found that there are present in *E. coli* two distinct species of tRNA, both of which accept methionine in the usual attachment reaction catalyzed by the methio-nyl-tRNA synthetase. The complex of methionine and only one of these two different methionine-affined tRNA species, however, can act as the substrate of another highly specific enzyme that catalyzes the formylation of the α-amino group of the attached methionine residue. This formylating enzyme works neither on free methionine nor on methionine attached to the other tRNA species. Thus peptide-chain initiation depends on the presence of a special species of methionine-affined tRNA whose structure allows enzymatic formylation of the α-amino group of its attached amino acid.

All of the preceding discussions concerning the function of ribosomes in polypeptide assembly have assumed tacitly, as was indeed long assumed by molecular biologists, that one 30S and one 50S ribosomal subunit are permanently wedded to form an intact 70S ribosome that spawns many a protein molecule in its lifetime. By 1967, however, it had dawned on the students of polyribosome dynamics that the union between any given pair of 30S and 50S subunits might last only as long as that 70S ribosome happens to work on one given mRNA tape. That is to say, the two subunits might dissociate once they have completed translation of, and been released from, the mRNA molecule, and might seek new partners before attaching to and starting to service another messenger. By means of heavy-light isotope-transfer experiments similar to those reported in Figure 16-4 with which Meselson, Brenner, and Jacob had first proven the validity of the mRNA concept in T4 phage-infected *E. coli*, Meselson was able to show the ephemeral nature of the union between 30S and 50S subunits. The results of this experiment are presented in Figure 17-14. Panel A of Figure 17-14 shows a

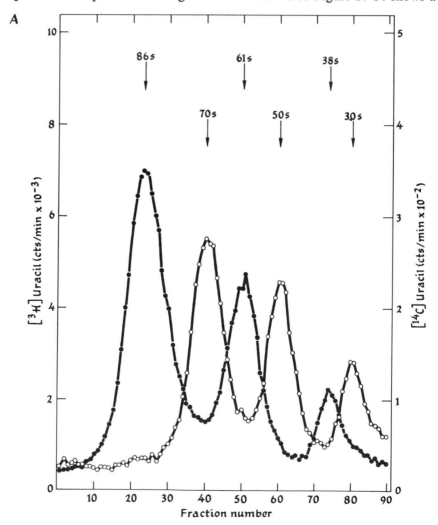

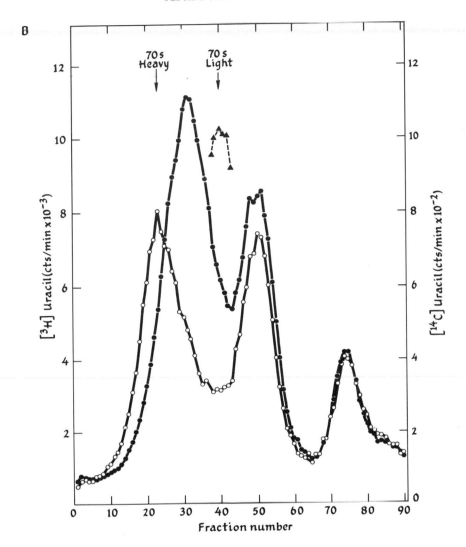

FIGURE 17-14.   Proof of cyclic dissociation of 70S ribosomes in protein synthesis. Sucrose density-gradient-sedimentation profiles of mixed pairs of ribosome extracts of *E. coli*. The Mg^{++} concentration of the sucrose solutions was 0.01 M, causing partial dissociation of 70S ribosomes into their 50S and 30S subunits. **A**. One extract was obtained from a culture grown in heavy (D_2O, ^{15}N and ^{13}C) ^3H-uracil-labeled medium and the other from a culture grown in light (H_2O, ^{14}N and ^{12}C) ^{14}C-uracil-labeled medium. **B**. One extract was obtained from a culture grown in heavy ^{14}C-uracil-labeled medium and the other from a culture first grown in heavy ^3H-uracil-labeled medium and then transferred to and allowed to grow for 3.5 generations in light nonlabeled medium. Filled circles = ^3H radioactivity; open circles = ^{14}C radioactivity; triangles (in **B**) = position of light reference 70S ribosomes. [After R. O. R. Kaempfer, M. Meselson, and H. J. Raskas, *J. Mol. Biol.* **31**, 281 (1968).]

sucrose density-gradient sedimentation profile of a mixture of ribosomes extracted from two *E. coli* cultures. One culture was grown in a heavy medium, composed of deuterium oxide, D_2O, instead of water; the heavy isotopes ^{15}N and ^{13}C as main nitrogen and carbon sources; and ^{3}H-uracil as a radioactive label. The other culture was grown in a normal light medium, composed of H_2O, ^{14}N, ^{12}C, and ^{14}C-uracil as a radioactive label. As can be seen, the ^{14}C-labeled ribosomes extracted from the light culture form three bands with typical sedimentation velocities of 70S, 50S, and 30S, corresponding to normal light 70S ribosomes and to some dissociated 50S and 30S subunits. The ^{3}H-labeled ribosomes extracted from the heavy culture can be seen to form three faster-moving bands with sedimentation velocities of 86S, 61S, and 38S, corresponding to heavy 70S ribosomes and to heavy dissociated 50S and 30S subunits. Panel B of Figure 17-14 shows a similar sucrose density-gradient profile of another mixture of ribosomes extracted from two *E. coli* cultures. One of these cultures was grown as before in ^{3}H-uracil-labeled heavy medium, then transferred to nonradioactive light medium and allowed to grow for 3.5 more generations in that medium before its ribosomes were extracted. The other culture was simply grown in ^{14}C-uracil labeled heavy medium. As can be seen, after 3.5 generations of growth in light medium the ^{3}H-labeled heavy 50S and 30S ribosomal subunits still sediment at the same high velocity as the ^{14}C-labeled heavy 50S and 30S subunits of the nontransferred heavy culture. But the ^{3}H-labeled, formerly

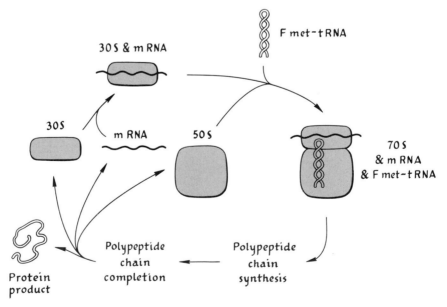

FIGURE 17-15.   Schematic summary of the cyclical employ of ribosomal subunits in protein synthesis.

heavy 70S ribosomes of the transferred culture now sediment at a velocity about half-way between that of fully heavy and fully light 70S ribosomes. These results demonstrate that whereas the heavy 50S and 30S ribosomal subunits remained intact during post-transfer bacterial growth in the light medium, most of them had meanwhile combined with a light, more recently synthesized partner to form a half-heavy, half-light hybrid 70S ribosome.

How free 30S and 50S subunits actually join to form their transient union as a 70S ribosome was elucidated by M. Nomura who was able to show that although the ribosomal aminoacyl- and peptidyl-tRNA attachment sites are provided by the 50S subunit, the 30S subunit is capable of accepting formylmethionyl-tRNA. Once the 30S subunit has accepted formylmethionyl-tRNA, it can bind to mRNA, engage a free 50S subunit, and initiate synthesis of a peptide chain. Thus it appeared that the idle ribosomes pass through a stage in which they exist in the cell as free 30S and 50S subunits before they reassociate more or less at random to form the 70S ribosome at the moment of peptide chain initiation. Figure 17-15 presents a schematic summary of this cyclical fate of ribosomal subunits in protein synthesis.

MISSENSE SUPPRESSORS

Our earlier consideration of suppressor mutations—mutations that restore a wild or pseudowild phenotype to a mutant individual—showed that there exist both *indirect-intergenic* as well as *direct-intragenic* suppressors. In 1960, in a single sentence of a 30-page review on "Gene Action," Yanofsky and Patricia St. Lawrence posited the existence of a third general mode of suppressor action, which is *direct* on the one hand, but *intergenic* on the other: "Also, possible as suppression mechanisms are minor changes in the amino acid activating and transfer systems (such that e.g., the activating enzyme that normally couples a specific [transfer]-RNA to glycine substitutes alanine 1 per cent of the time)." They were led to this proposal by an observation Yanofsky had made with Trp+ reverse mutants of Trp− *missense* auxotrophs, such as *trp*A23. As was seen in Figure 14-19, this mutant is unable to synthesize tryptophan because a mutation has replaced the normal glycine by arginine as the 210th amino acid residue of the tryptophan synthetase A protein. Among the Trp+ reverse mutants of *trp*A23 there were, as was set forth in Chapter 14, both true reverse mutants, to which the normal glycine had been restored, as well as pseudo-true reverse mutants, in which the noxious arginine residue had been replaced by an acceptable serine residue and which thus restored activity to the tryptophan synthetase A protein. But, as was *not* mentioned in Chapter 14, Yanofsky had also found a third class of Trp+ reverse mutants of *trp*A23 that could be shown to form *two* functionally and structurally distinguishable types of tryptophan synthetase A protein molecules: one, by far the most abundant type, was the unaltered, catalytically inert mutant protein; the other type, constituting only a few percent of the total A protein, was evidently normal, fully active, wild-type protein. Transductional tests with phage P1 soon showed that this third class of reverse mutants not only still carried its original trpA23 mutation but that the mutation that gave rise to the production of a small amount of normal A protein was situated well outside that region of the *E. coli* genome where the *trp* genes reside. (Later tests located that mutation halfway around the circular chromosome from trpA near minute 78 on the map of Figure 10-21.) The third class did not, therefore, owe its character to a true reverse mutation but rather to an *intergenic* suppressor mutation. The action of this suppressor, furthermore, is *direct*, in that it does repair the primary genetic lesion, albeit with less than complete efficiency. Its capacity for repair was found to be quite restricted, however; no other then-known mutation in the *trp*A gene responded to its cure—not even trpA46, in which, as was shown in Figure 15-15, the same normal glycine at position 210 has been replaced by glutamic acid. It was concluded, therefore, that this suppressor mutation allows the generation of wild-type A protein *by reducing the fidelity of translation of the mRNA*. That is to say, it allows the occasional insertion of glycine at position 210 when the corresponding trpA23 mutant codon actually calls for the insertion of arginine. The suppressor does not, however, allow the insertion of glycine in response to the allelic trpA46 mutant codon, which calls for the insertion of glutamic acid.

The proposal that this suppressor intervenes in mRNA translation was investigated in more detail by Yanofsky and his colleagues in the next few years. In particular they attempted to ascertain which components of the machinery of protein synthesis—ribosomes, aminoacyl-tRNA synthetases, or tRNA—are altered by the suppressor mutation. For this purpose, they extracted and purified these components from both normal and suppressor mutant cells and added them in all possible combinations to reaction mixtures capable of *in vitro* polypeptide synthesis. They then examined whether, in accord with the original proposal, one of the components of the suppressor mutant cell causes the occasional insertion of glycine into a polypeptide chain site at which the corresponding mRNA codon can be inferred to specify arginine. These experiments soon revealed that whereas both ribosomes and aminoacyl-tRNA synthetases of the suppressor mutant are normal, the mutant contains an *abnormal species of tRNA that accepts'glycine from the glycyl-tRNA synthetase but which inserts its glycine in response to an arginine codon*. Although definitive proof has not yet been supplied, it appears likely that the suppressor mutation results in the conversion of the anticodon structure of a minor fraction of the glycine tRNA molecules from glycine codon-to-arginine codon-complementarity.

It seems surprising, on first thought, that missense suppressor mutations of the kind described here exist, since they must cause translation of sense into missense at a large number of genetic sites other than the single mutant site they happen to repair. A highly efficient missense suppressor would, therefore, be expected to be lethal to the cell in which it occurred. Possibly, the suppressor studied by Yanofsky is not lethal because its efficiency of suppression is quite low. In the next chapter we will encounter a much more efficient class of suppressors, which closely resemble the missense suppressors in the nature of their action but achieve the suppression of only three special types of codons.

Bibliography

ORIGINAL RESEARCH PAPERS

Berg, P., and E. J. Ofengand. An enzymatic mechanism for linking amino acids to ribonucleic acid (RNA). *Proc. Natl. Acad. Sci. Wash.*, **44**, 78 (1958).

Hoagland, M. B., E. B. Keller, and P. C. Zamecnik. Enzymic carboxyl activation of amino acids. *J. Bio. Chem.*, **218**, 345 (1956).

Zamecnik, P. C., E. B. Keller, J. W. Littlefield, M. B. Hoagland, and R. B. Loftfield. Mechanism of incorporation of labeled amino acids into protein. *J. Cell. Comp. Physiol.*, **47**, Suppl. 1:81 (1956).

SPECIAL TEXTS, MONOGRAPHS, AND REVIEWS

Berg, P. Specificity in protein synthesis. *Ann. Rev. Biochem.*, **30**, 293 (1961).

Crick, F. H. C. On protein synthesis. In *The Biological Replication of Macromolecules*, Symposium of the Society for Experimental Biology XII, Cambridge Univ. Press, London, 1958, p. 138.

Ingram, V. M. *The Biosynthesis of Macromolecules*. Benjamin, New York, 1966.

18. Genetic Code

It was one thing to have proven, as did Crick and Brenner in 1961 by means of their formal genetic experiments with T4 phage frameshift mutants (recounted in Chapter 13), that the genetic code does represent each of the 20 standard amino acids by nucleotide triplet codons. It was quite another thing, however, to *break* that code and to discover which of the 64 possible nucleotide triplets listed in Table 18-1 designates which amino acid. By convention these codons are represented by three capitals letters; the one at the left stands for the 5'-end, and the one at the right stands for the 3'-end of the nucleotide triplet. For instance, the entry UAG in Table 18-1 represents the nucleotide triplet pUpApG.

AMINO ACID REPLACEMENTS

The easiest and least sporting way to break an unknown code is to compare a coded message, the cryptogram, with its corresponding cleartext. Thus the finding in 1799 of the Rosetta Stone, which bears both cryptic hieroglyphic as well as plain alphabetic inscriptions, allowed M. Champollion to decipher the ancient Egyptian writing by comparison with its Greek equivalent. And the most straightforward procedure for breaking the genetic code would

TABLE 18-1. The Sixty-four Codons of the Genetic Code

UUU	UCU	UAU	UGU
UUC	UCC	UAC	UGC
UUA	UCA	UAA	UGA
UUG	UCG	UAG	UGG
CUU	CCU	CAU	CGU
CUC	CCC	CAC	CGC
CUA	CCA	CAA	CGA
CUG	CCG	CAG	CGG
AUU	ACU	AAU	AGU
AUC	ACC	AAC	AGC
AUA	ACA	AAA	AGA
AUG	ACG	AAG	AGG
GUU	GCU	GAU	GGU
GUC	GCC	GAC	GGC
GUA	GCA	GAA	GGA
GUG	GCG	GAG	GGG

have been to compare a DNA or mRNA polynucleotide cryptogram with its corresponding polypeptide cleartext. In contrast to the situation in ordinary cryptanalysis, however, where the cryptogram is usually easy and the cleartext usually hard to obtain, the situation in molecular genetics was just the reverse. Here, a waxing number of protein amino acid sequences became known after Sanger's first structural analysis of insulin in 1954, but fifteen years were to pass before a nucleotide sequence encoding any of these polypeptide chains was finally delivered. The longest natural polynucleotide sequences that had been worked out meanwhile were those of the comparatively short and presumably nontranslated tRNA molecules; and these sequence analyses depended critically on the structural landmarks provided by the various extraordinary nucleotides peculiar to tRNA. Since these nucleotides are present neither in DNA nor in mRNA, the problem of how to determine the sequence of several hundred nucleotides in such macromolecules long awaited its technical solution. Finally, in 1969 Sanger and his collaborators did succeed in determining the sequence of an RNA molecule known to encode a polypeptide chain of known amino acid sequence, but, as will be set forth in this chapter, by then the genetic code had long been broken.

A different cryptanalytic approach to the genetic code, other than comparing cryptogram and cleartext, had become available in the late 1950's upon the finding that at some particular site of a mutant protein an amino acid β had replaced an amino acid α extant at the corresponding site in the wild-type protein, and upon the understanding of the molecular basis of spontaneous and induced mutation set forth in Chapter 13. For instance, if the mutation responsible for the replacement of α by β had been induced by a mutagen that causes base transitions, it would follow that the codons representing α and β have two nucleotides in common and that they both have as a third nucleotide base either a pyrimidine or a purine. If, furthermore, an amino acid γ had replaced α in another mutant protein induced by a transition mutagen, it would follow that the codons representing β and γ have one nucleotide in common and differ from each other in the other two nucleotides. Thus by surveying the known amino acid replacements in mutant proteins and by assessing the nucleotide interchanges likely to have given rise to the mutation in the corresponding genes, it is possible in principle to construct a network of amino-acid-codon relationships. And by analyzing such a network the genetic code might have been broken.

Intensive efforts to break the code in this way were begun in 1960 by H. G. Wittmann and by A. Tsugita and H. Fraenkel-Conrat. Their work focused upon the protein of mutants of the tobacco mosaic virus. Tobacco mosaic virus will figure prominently in the considerations of the next chapter, and for the purpose of the present discussion it suffices to note that the nucleic acid of that virus is encased by a protein coat made up of 2150 identical polypeptide chains, each chain being 158 amino acids in length. The primary structure of that polypeptide chain had been worked out by 1960 and is presented in Figure 18-1. Large numbers of both spontaneous and mutagen-induced mutants of tobacco mosaic virus were isolated in order to ascertain and catalogue the amino acid replacements that had occurred in their proteins. A summary of the diversity of replacements that were actually found in this way is presented also in Figure 18-1. It is to be noted that almost all observed replacements concerned only one amino acid residue at a time, offering the first direct proof of the long-held assumption that the genetic code is nonoverlapping—that is, each nucleotide forms part of but a single codon.

We may now consider a sample of these data on amino acid replacements observed after mutagenesis with nitrous acid. In accord with the chemical basis of the mutagenic action of nitrous acid set forth in Chapter 13, any amino acid replacement so produced can be expected to have resulted from the transition $A \rightarrow G$ or $C \rightarrow U$ in a codon of the mRNA coding for the protein of the mutant virus. The replacements shown on page 526 were found among the nitrous-acid-induced mutants in the protein.

<pre>
                                    10
AcSer-Tyr-Ser-Ile-Thr-Thr-Pro-Ser-Gln-Phe-Val-Phe-Leu-Ser-Ser-Ala-Trp-Ala-Asp-
                 Ile   His His Leu Met   Leu   Val
                                                                               Ala

20   30
Pro-Ile-Glu-Leu-Ile-Asn-Leu-Cys-Thr-Asn-Ala-Leu-Gly-Asn-Gln-Phe-Gln-Thr-Gln-
Thr Thr   Val Ser(4)   Ile   Ser
(3)Leu Met   Ala   Ala
                                                        Lys

          40   50
Gln-Ala-Arg-Thr-Val-Val-Gln-Arg-Gln-Phe-Ser-Gln-Val-Trp-Lys-Pro-Ser-Pro-Gln-
                              Lys   Leu
                              Gly(2)

          60   70
Val-Thr-Val-Arg-Phe-Pro-Asp-Ser-Asp-Phe-Lys-Val-Tyr-Arg-Tyr-Asn-Ala-Val-Leu-
Ala Ile(2)   Gly(2)   Ser(4)   Gly(2)   Ser(2)

          80   90
Asp-Pro-Leu-Val-Thr-Ala-Leu-Leu-Gly-Ala-Phe-Asp-Thr-Arg-Asn-Arg-Ile-Ile-Glu-
              Ala(4)   Asp

          100   110
Val-Glu-Asn-Gln-Ala-Asn-Pro-Thr-Thr-Ala-Glu-Thr-Leu-Asp-Ala-Thr-Arg-Arg-Val-
    Gly(2)   Arg   Met(3)

          120   130
Asp-Asp-Ala-Thr-Val-Ala-Ile-Arg-Ser-Ala-Ile-Asn-Asn-Leu-Ile-Val-Glu-Leu-Ile-
                          Gly   Val Ser   Thr
                                                      Val
                    140   150
Arg-Gly-Thr-Gly-Ser-Tyr-Asn-Arg-Ser-Ser-Phe-Glu-Ser-Ser-Ser-Gly-Leu-Val-Trp-
Gly(2)   Ile   Phe(3) Cys Lys   Phe Ser

Thr-Ser-Gly-Pro-Ala-Thr
Ile   Leu(3)
</pre>

FIGURE 18-1. Complete sequence of the 158-amino-acid polypeptide chain of the tobacco mosaic virus coat protein. The amino-terminal serine residue is acetylated—that is, has the chemical structure

$$CH_3CONHCHCO-$$
$$|$$
$$CH_2OH$$

and is represented by the symbol AcSer. The symbols written below certain amino acid residues of the chain designate replacements of the wild type amino acid at that site found in the coat protein of a large ensemble of mutant virus strains. Several different mutant strains were found to be examples of independent recurrences of a replacement at the same polypeptide site. In some of these strains the same mutant amino acid had replaced the wild-type amino acid residue (the number of such recurrences at any site is indicated in parentheses) and in other strains two or more different amino acids had replaced the wild-type amino acid.

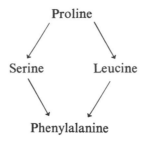

Proline

Serine   Leucine

Phenylalanine

This result meant that proline is designated by a codon of which at least two nucleotides are either A or C; the A → G or C → U transition of one of these nucleotides must generate a serine codon and the corresponding transition of the other nucleotide a leucine codon. In the phenylalanine codon, both of these two nucleotides must be either G or U. Similar deductions as to the generic relationship between the codons designating other amino acids could be made from their observed replacements in the mutant coat proteins. Meanwhile, data had been accumulating also for amino acid replacements observed in other mutant proteins, particularly in variants of human hemoglobin. And so it seemed likely that by collating the results of these extremely laborious analyses, the code would gradually be worked out. But in the summer of 1961, a startling discovery by M. Nirenberg (Figure 18-2) was to provide the means for a much more rapid decipherment of the code.

FIGURE 18-2.   Marshall Nirenberg (b. 1927). [Courtesy M. Nirenberg].

ARTIFICIAL MESSENGER RNA

In 1955, five years before the discovery of RNA polymerase now known to be responsible for transcription of RNA from DNA templates, another enzyme capable of effecting the *in vitro* synthesis of ribopolynucleotide chains had been found in bacteria by Marianne Grunberg-Manago and S. Ochoa. This enzyme, to which was given the name *polynucleotide phosphorylase*, catalyzes the polymerization of ribonucleoside *di*phosphates, according to the reaction

$$\left. \begin{array}{c} \end{array} \right\} P \left. \begin{array}{c} \end{array} \right\} P \left. \begin{array}{c} \end{array} \right\} P \left. \begin{array}{c} \end{array} \right\} + \overset{P}{\underset{P}{|}} \left. \begin{array}{c} \end{array} \right\} \rightleftarrows \left. \begin{array}{c} \end{array} \right\} P \left. \begin{array}{c} \end{array} \right\} P \left. \begin{array}{c} \end{array} \right\} P \left. \begin{array}{c} \end{array} \right\} P \left. \begin{array}{c} \end{array} \right\} + P \qquad (18\text{-}1)$$

In contrast to RNA polymerase, however, polynucleotide phosphorylase does not require the presence of any DNA template for its action and will randomly join into polynucleotide chains any mixture of ribonucleoside diphosphates presented to it. If it is presented with only one kind of nucleoside diphosphate—say, uridine diphosphate—a monotonous polyribonucleotide such as *polyuridylic acid* (or poly U), will be formed in the reaction mixture. If it is presented with a mixture of two kinds of nucleoside diphosphates—for example, uridine and cytidine diphosphate—a randomly assembled polyribonucleotide, namely poly UC, will be formed, in which the relative proportions of U and C in the polymer reflect their relative abundance in the nucleoside diphosphate substrate mixture. Finally, if all four nucleoside diphosphates are presented to this enzyme, poly UCAG will be formed, whose overall composition may resemble that of natural RNA but which will be devoid of any information content. The physiological role of polynucleotide phosphorylase still wants being accounted for, but it seems likely that it is concerned with the reverse of reaction (18-1)—that is, with breakdown rather than with the synthesis of RNA. In any case, in the hands of Nirenberg, these artificial, random ribopolynucleotides synthesized by polynucleotide phosphorylase were to become a surrogate Rosetta Stone of the genetic code.

Upon reading the first publications that announced the discovery of mRNA, Nirenberg decided to try to develop a cell-free reaction mixture in which it would be possible to direct the synthesis of a given polypeptide by introducing mRNA molecules encoding that amino acid sequence. Other workers had previously succeeded in achieving *in vitro* protein synthesis in reaction mixtures containing ribosomes, tRNA, and aminoacyl-tRNA synthetases, as well as ATP and ^{14}C- or ^{3}H-labeled amino acids. But these earlier systems had invariably contained their own endogenous mRNA and hence did not respond to the addition of exogenous mRNA. Nirenberg succeeded in destroying the endogenous mRNA present in these reaction mixtures and thus in making their polypeptide synthesis dependent on the addition of exogenous mRNA. This system, therefore, provided for an experimental readout of any putative mRNA. One day, Nirenberg added artificially synthesized polyuridylic acid to this reaction mixture instead of natural mRNA and obtained a most surprising result: this monotonous poly U polyribonucleotide, whose bases are all U, was found to stimulate the *in vitro* synthesis of an equally monotonous polypeptide—namely, polyphenylalanine, whose amino acid residues are all phenylalanine. This result could have but one meaning: a sequence of uridylic acids in the messenger polynucleotide—that is, a sequence of UUU codons—codes for phenylalanine.

Now, at one stroke, the breaking of the genetic code had become accessible to direct chemical methods, since it was evident that the nucleotide composition of the codons of other amino acids ought to be decipherable by examining the polypeptides assembled under the direction of other artificially synthesized RNA molecules. Thus it soon transpired that *polycytidylic*, or poly C, acid directs synthesis of *polyproline*, and hence that CCC designates proline, and that *polyadenylic*, or poly A, acid directs synthesis of *polylysine*, and hence that AAA designates lysine. (Polyguanylic acid could not be tested because of certain structural peculiarities of this polynucleotide, but later studies showed that GGG designates glycine.)

In this way four of the 64 codons were easily accounted for. In order to gain some insight into the meaning of the remaining 60 codons containing more than one kind of nucleotide, these experiments were continued by using artificially synthesized random ribopolynucleotides containing two, three, or four different nucleotide constituents. Table 18-2 presents a sample of Nirenberg's results obtained with poly UA, poly UC, and poly UG. It can be seen, first of all, that all three polymers do stimulate the incorporation of phenylalanine into polypeptides, which is in agreement with the expectation that the polynucleotide phosphorylase-catalyzed random synthesis of these polymers does generate a certain proportion of UUU codons. Second, it can be seen that poly UA also stimulates the incorporation of tyrosine and isoleucine, indicating that their codons contain both U and A. Comparison of the frequencies of incorporation of these two amino acids relative to phenylalanine with the expected frequencies of U_2A, UA_2, and A_3 triplets relative to the phenylalanine codon U_3 in the random poly UA chain indicates that the codons of tyrosine and isoleucine are likely to correspond to the empirical formula U_2A. Third, a similar analysis of the results of the amino acid incorporations stimulated by poly UC and poly UG led Nirenberg to conclude that the serine and proline codons are U_2C and UC_2, respectively (although the use of poly C showed that proline is also represented by C_3), that the valine and cysteine codons are U_2G, and that the tryptophan and glycine codons are UG_2. The incorporation of leucine is of particular

TABLE 18-2.   Polypeptide Assembly Stimulated by Three Artificial Polyribonucleotides

	Polynucleotide		
	UA	UC	UG
Relative base content	$U = 0.87$ $A = 0.13$	$U = 0.39$ $C = 0.61$	$U = 0.76$ $G = 0.24$
Probability of triplet relative to U_3	$U_3 \ - \ 100$ $U_2A = \ 13$ $UA_2 = \ 2.2$ $A_3 \ = \ 0.3$	$U_3 \ = \ 100$ $U_2C = \ 157$ $UC_2 = \ 244$ $C_3 \ = \ 382$	$U_3 \ = \ 100$ $U_2G = \ 32$ $UG_2 = \ 10.6$ $G_3 \ = \ 3.4$

	Polynucleotide			Inferred codon composition
Amino acid	UA	UC	UG	
Phenylalanine	100	100	100	U_3
Arginine	0	0	1.1	
Alanine	1.9	0	0	
Serine	0.4	160	3.2	U_2C
Proline	0	285	0	UC_2
Tyrosine	13	0	0	U_2A
Isoleucine	12	1.0	1.0	U_2A
Valine	0.6	0	37	U_2G
Leucine	4.9	79	36	U_2C, U_2G
Cysteine	4.9	0	35	U_2G
Tryptophan	1.1	0	14	UG_2
Glycine	4.7	0	12	UG_2
Methionine	0.6	0	0	
Glutamic acid	1.5	0	0	

The figures in the body of the table represent the percent incorporation into polypeptides of the amino acid listed relative to the incorporation of phenylanine in an analogous reaction mixture. The underlined figures were considered as being significantly above background by the authors.

From J. H. Matthaei, O. W. Jones, R. G. Martin, and M. Nirenberg, *Proc. Natl. Acad. Sci. Wash.* **48**, 666 (1962).

interest, since it is stimulated both by poly UC and poly UG. This means that leucine must be coded by at least two different codons, U_2C and U_2G. That is to say, the code contains more than one representation for the same amino acid. This conclusion is, of course, in accord with the inference drawn by Crick and Brenner from their study of the suppressibility of frame-shift mutations that only very few of the 64 codons can be nonsense. The data of Table 18-2 allow the conclusion, furthermore, that the codons of arginine, alanine, methionine, and glutamic acid are not composed of nucleo-tide triplets that can be generated by random combinations of U and A, U and C, or U and G.

These first codon assignments can now be compared with the prior in-ferences drawn from the study of amino acid replacements in the coat protein of nitrous-acid-induced mutants of tobacco mosaic virus. The codon assign-ment of UC_2 to proline, of U_2C to serine and to leucine, and of U_3 to phenylalanine is seen to be in excellent agreement with the earlier conclusion that the proline codon contains one more C (or A) and one less U (or G) than the serine and leucine codons, which in turn contain one more C (or A) and one less U (or G) than the phenylalanine codon.

After learning of Nirenberg's first success with poly U, Ochoa and his collaborators carried out another entirely analogous set of decoding experi-ments in which they used artificial random ribopolynucleotides. Through the parallel efforts of these two research groups the empirical formula of many amino acid codons became known within a year or so. But this work could not establish the actual *sequence* of the nucleotides within any codon, of course, except for the trivial cases of UUU, CCC, AAA, and GGG. Furthermore, it could not clearly identify cases where codons such as X_3 and X_2Y or X_2Y and XY_2 are synonyms.

It appeared, therefore, that in order to ascertain by *in vitro* polypeptide formation which structural isomer of a codon of given empirical formula designates which amino acid (for example, which of the three isomers UUG, UGU, or GUU of U_2G represents valine, which cysteine, and which leucine), artificial ribopolynucleotides of *defined* rather than random nucleotide sequence would have to be employed. Consequently, G. Khorana set himself the very difficult task of synthesizing such polyribonucleotides directly. He eventually succeeded in synthesizing alternating copolymers such as -UGUGUGUGUGUG-. Use of this particular alternating polynucleotide as messenger in the *in vitro* protein-synthesizing system gave rise to the forma-tion of the alternating polypeptide -valine-cysteine-valine-cysteine-, evidently directed by the codon sequence -UGU-GUG-UGU-GUG-. This result established that the U_2G leucine codon is *not* UGU, and that neither the UG_2 tryptophan nor the UG_2 glycine codons is GUG. It followed also either that U_2G valine is UGU and GUG is a cysteine synonym, or that U_2G cysteine is UGU and that GUG is a valine synonym. Furthermore, the formation of an alternating polypeptide from an alternating polynucleotide template proved definitively that the codon is composed of an *odd* number of nucleotides, thus eliminating the slight possibility, left by Crick and Brenner's proof of the triplet nature of the code, that the codon might actually consist of six rather than three nucleotides. By means of even greater organochemical virtuosity than that required for the synthesis of alternating ribopolynucleotides, Khorana managed to prepare ribopolynucleotides of repeating trinucleotide sequences, such as -UUGUUGUUGUUG-. Use of this particular artificial messenger in the protein-synthesizing system gave rise to three different monotonous polypeptides—namely, to polyleucine, polycysteine, and polyvaline. This result meant that, depending on the nucleotide at which translation happened to have started, this messenger was read as -UUG-UUG-UUG-UUG-, or as -UGU-UGU-UGU-UGU-, or as -GUU-GUU-GUU-GUU-, in direct confirmation of the earlier inference that the codons of leucine, cysteine, and valine are isomers of U_2G. It also provided another proof that the codon is a nucleotide triplet, or a multiple thereof.

In 1964, while Khorana was just completing the difficult synthesis of the first artificial ribopolynucleotides of defined sequence, Nirenberg made a second discovery that made possible a rapid culmination of the code-breaking efforts. He found that addition of simple *trinucleotides* to ribosomes would cause these ribosomes to bind that, and only that, aminoacyl-tRNA which carries the anticodon complementary to the trinucleotide added to the reaction mixture. For instance, ribosomes provided with the trinucleotide pUpUpU were found to bind selectively the phenylalanyl-tRNA complex from among a mixture of all other aminoacyl-tRNA species. Thus the short trinucleotide can take the place of mRNA on the ribosome and furnish to the ribosomal aminoacyl-tRNA binding site the specificity of recognition accorded to the incoming amino-acid-adaptor complex in normal polypeptide assembly. In this way the meaning of any codon triplet could be directly established by testing which aminoacyl-tRNA is bound in a set of homologous reaction mixtures containing ribosomes, the trinucleotide in question, and tRNA carrying one of the twenty [14]C- or ^{3}H-labeled amino acids. For instance, such tests soon resolved the identity of the three U_2G isomers, in that they showed that the UUG triplet is a leucine codon, that the UGU triplet is a cysteine codon, and that GUU is a valine codon. Since it was comparatively easy to secure the necessary quantities of all 64 trinucleotides shown in Table 18-1 (many of which were already in Khorana's possession anyway), and since these specific binding tests were much easier to carry out than any of the earlier decoding experiments, it was possible to work out the entire code within the next year. For about a dozen trinucleotides, however, the outcome of these binding tests was either negative or ambiguous (in that either no binding at all or binding of more than one kind of aminoacyl tRNA was observed), so that here the meaning of the codon had to be ascertained by one or more of the other code-breaking procedures.

THE CODE TABLE

The final result of the code-breaking efforts is presented in Table 18-3. The arrangement of this table, whose significance for biology has been compared to that of the Periodic Table of Elements for chemistry, was suggested by Crick. Each of the 20 protein amino acids is represented by a three-letter abbreviation, according to the schema of Figure 2-2. The nucleotide triplet codon corresponding to any given position in this table can be read off according to the following rules: The base of the first nucleotide of the codon is given by the large capital letter on the left, which defines a horizontal row containing four lines. The base of the second nucleotide is given by the medium-sized capital letter on the top, which defines a vertical column corresponding to 16 codons. The intersection of rows and columns defines a "box" of four codons, all of which carry the same bases in their first and second nucleotides. The base of the third nucleotide is given by the small capital letter on the right, which defines one line within any given horizontal four-line row.

Contemplation of this table reveals the following important features of the genetic code:

1. The code contains many synonyms, in that almost all amino acids are represented by more than one codon. Three amino acids—arginine, serine, and leucine—each have six synonymous codons; five—valine, proline, threonine, alanine, and glycine—each have four synonymous codons; one—isoleucine—has three synonymous codons; and nine—phenylalanine, tyrosine, histidine, glutamine, asparagine, lysine, aspartic acid, glutamic acid, and cysteine—each have two synonymous codons. Only two amino acids—methionine and tryptophan—are each represented by but a single codon.

2. The code has a definite structure, in that synonyms for the same amino acid are not randomly dispersed over the table. As is evident, synonymous codons are almost always in the same box, and hence differ from one another only in the last of their three nucleotides. (The only exceptions to this rule are arginine, serine, and leucine, which are represented by six synonymous codons, and hence are necessarily spread over more than one four-codon box.) The existence of a wide spectrum of [G] + [C] content in the DNA of bacterial species, ranging from a low of 26% for *Welchia perfringens* to a high of 74% for *Streptococcus griseus*, which was seen in the data of Table 8-1, may now be interpreted in the light of the code table. Some of these differences in [G] + [C] content can evidently be attributed to a differential use of synonyms by these organisms, since for almost every amino acid there is one codon that carries G or C and another that carries A or T as its third nucleotide. Thus by exclusive choice of one or the other kind of synonym,

TABLE 18-3.   The Genetic Code

	U	C	A	G	
U	Phe	Ser	Tyr	Cys	U
	Phe	Ser	Tyr	Cys	C
	Leu	Ser	Non2	Non3	A
	Leu	Ser	Non1	Trp	G
C	Leu	Pro	His	Arg	U
	Leu	Pro	His	Arg	C
	Leu	Pro	Gln	Arg	A
	Leu	Pro	Gln	Arg	G
A	Ile	Thr	Asn	Ser	U
	Ile	Thr	Asn	Ser	C
	Ile	Thr	Lys	Arg	A
	Met	Thr	Lys	Arg	G
G	Val	Ala	Asp	Gly	U
	Val	Ala	Asp	Gly	C
	Val	Ala	Glu	Gly	A
	Val	Ala	Glu	Gly	G

two bacterial species encoding exactly the same protein sequence information in their DNA could differ by as much as 33% in the $[G] + [C]$ content of their DNA's. Differences in $[G] + [C]$ content greater than this would have to be attributed to differences in primary structure of the encoded proteins. As can be seen from Table 18-3, a more frequent use of proline, arginine, alanine, and glycine in protein primary structure would contribute to a high $[G] + [C]$ content, whereas a more frequent use of phenylalanine, tyrosine, isoleucine, methionine, asparagine, and lysine would contribute to a low $[G] + [C]$ content. Analyses of the amino acid composition of the total protein of various bacterial species have shown that some of the variation in $[G] + [C]$ content of their DNA is indeed to be explained by such differences in primary protein structure.

3. The code contains three *nonsense* codons, UAG, UAA, and UGA, labeled respectively Non1, Non2, and Non3, which do not represent any amino acid at all. The existence of these nonsense codons, Non1 having been given the name *amber* and Non2 the name *ochre*, was first inferred from the growth behavior of some T4 phage mutants described in Chapter 12. The nonsense codons will form the subject of further discussions in later parts of this chapter. It will suffice to state here that UAG, UAA, and UGA nucleotide triplets do not promote the specific ribosomal attachment of any aminoacyl-tRNA in the binding test and that artificial ribopolynucleotides in whose nucleotide sequence there is an abundant occurrence of any of these three codons do not serve as messenger templates for the *in vitro* synthesis of long polypeptide chains.

Upon the availability of the genetic code in its final form, the tables could be turned on the theoretical analysis of observed amino acid replacements in mutant proteins. Rather than employing such replacement data for breaking the code, they could now be used for checking the fundamental postulate that mutations that lead to single amino interchanges derive mainly from single base substitutions in the hereditary polynucleotide. For instance, the amino acid replacements identified by Yanofsky in his proof of the colinearity of the *E. coli* tryptophan synthetase A gene and A-protein can be examined in this light. As can be seen by comparing the data presented in Figure 14-19 with Table 18-3, every observed amino acid replacement can be accounted for by a simple mutational substitution of one nucleotide base by another. For instance, in the mutation *trp*A23, normal glycine (codon GG·; the dot signifies that glycine can be coded by *any* of the four bases in the third position) at position 210 of the A-protein chain was replaced by the mutant arginine (codon AG_G^A). This mutation is evidently attributable to replacement of the normal guanine in the first codon position by adenine. In Chapter 14 we discussed the two types of Trp$^+$ reverse mutations of *trp*A23 at position 210; one restores the normal glycine and the other replaces the noxious arginine with the inocuous serine. These two Trp$^+$ reverse mutations must have arisen in the following manner: the reverse mutant that carries the normal glycine codon GG· arose by restoration of guanine to the first codon position of the mutant arginine AG$_G^A$ codon, whereas the reverse mutant that carries serine (codon AG$_C^U$) arose by replacement of adenine or guanine in the third position of AG$_G^A$ by uracil or cytosine.

Finally we may consider mutant *trp*A46, in which the same 210th glycine was replaced by glutamic acid (codon GA$_G^A$). Evidently this mutation arose by replacement of guanine in the *second* position of the glycine codon GG· by adenine. Now, the nonidentity of the two *genetic* sites of the *trp*A46 and *trp*A23 mutations had been inferred from the appearance of Trp$^+$ transductants in P1 phage transductions in which *trp*A46 is the donor strain and *trp*A23 the recipient Trp$^-$ strain. It follows that the appearance of these Trp$^+$ transductants *demonstrates genetic recombination at the finest conceivable level:* the recombinant *trp*A gene must have obtained G as first nucleotide of the recombinant glycine GG· codon from the glutamic acid GA$_G^A$ codon of the *trp*A46 parent and G as the second nucleotide of the recombinant codon from the arginine AG$_G^A$ codon of the *trp*A23 parent.

The general validity of the postulate that single amino acid replacements occur mainly by substitution of single nucleotides can be demonstrated in the following manner. Whereas there are $21 \times 20 = 420$ conceivable replacements of one amino acid (or of a nonsense codon) by another, the mutational interrelations implicit in the structure of the genetic code allow only 170 of these replacements to occur as a result of single-nucleotide base substitutions. As shown in Table 18-4, a total of 70 of these replacements had, by 1967, already been reported to occur in various mutant proteins. Moreover, these 70 replacements constituted so overwhelming a proportion of *all* the replacements observed that the few replacements that required the substitution of two or more bases could be readily explained as a double mutation.

DIRECTION OF THE CODE

If we consider two adjacent amino acid residues in a polypeptide chain, such as

Glycine (GG·)   Serine (UC·)

TABLE 18-4. Amino Acid Replacements as the Result of Single-base Changes in Codons from Table 18-3

Amino acid	Possible replacements*
ala	asp *glu* *gly* pro ser thr *val*
arg	cys gln *gly* his *ile* leu *lys* met non pro *ser* *thr* trp
asn	*asp* his ile *lys* *ser* thr tyr
asp	*ala* *asn* glu *gly* his tyr val
cys	arg *gly* non phe ser trp tyr
gln	*arg* *glu* his leu lys *non* pro
glu	*ala* *asp* gln *gly* *lys* non *val*
gly	ala *arg* *asp* *cys* *glu* non ser trp *val*
his	*arg* asn *asp* *gln* leu pro *tyr*
ile	arg asn leu lys *met* phe ser *thr* *val*
leu	*arg* gln his *ile* met non *phe* pro ser trp val
lys	arg *asn* gln *glu* ile met non thr
met	arg ile *leu* lys thr val
non	arg cys *gln* *glu* gly *leu* *lys* *ser* *trp* *tyr*
phe	cys ilu *leu* ser *tyr* val
pro	ala arg gln his *leu* *ser* *thr*
ser	ala *arg* asn cys *gly* ile *leu* non *phe* pro thr trp tyr
thr	*ala* arg asn *ile* *lys* *met* pro ser
trp	arg cys gly leu *non* ser
tyr	asn asp *cys* his non *phe* ser
val	*ala* *asp* *glu* *gly* ile leu *met* phe

*Italicized amino acids have been observed as replacements in mutations.
From A. Sadgopal, *Advances in Genetics*, **14**, 325 (1968).

then nothing that has been said so far here of the genetic code would indicate whether the corresponding polynucleotide coding for this segment has the structure

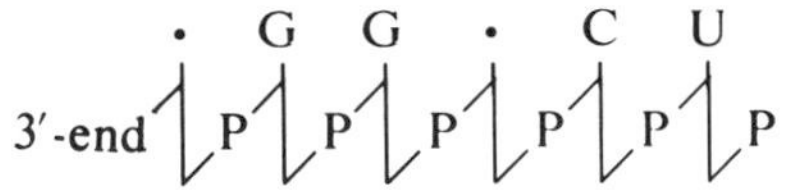

$$\cdot \quad G \quad G \quad \cdot \quad C \quad U$$

3'-end P P P P P P

or the structure

$$G \quad G \quad \cdot \quad U \quad C \quad \cdot$$

5'-end P P P P P P

But the general schema of the heterocatalytic function of DNA presented in Figure 17-6 actually demands that the second of these two alternatives be true. Inasmuch as under that schema translation of mRNA occurs concomitantly with its transcription, *the direction in which mRNA is translated (and hence moves across its client ribosomes) must be the same as that in which it is synthesized.* And since, as was set forth in Chapters 16 and 17, RNA chains grow from their 5′-end and protein chains grow from their amino-terminal end, it would follow that of two amino acid residues the one closer to the amino-terminus of the polypeptide is represented by a codon closer to the 5′-end of the corresponding mRNA.

Experiments designed to test this prediction were carried out by Ochoa and his colleagues in 1965, soon after the code table had become known in its final form. For this purpose, poly AC containing an average of about 20 residues of A per residue of C was synthesized *in vitro* from a 20:1 mixture of adenosine and cytidine diphosphates by action of polynucleotide phosphorylase. The resulting random polymer was then subjected to hydrolysis by pancreatic ribonuclease, an enzyme that splits the internucleotide bond between the 3′-phosphate group of a *pyrimidine* nucleotide and the 5′-hydroxyl of its neighbor (see also Figure 17-11). This procedure thus gave rise to short polynucleotide chains of average length 21 nucleotides. All of these short chains carried only A residues except for one C residue at their 3′ end, as shown in Figure 18-3. Or, put in another way, the 20 nucleotides of the average chain corresponded to a sequence of five AAA codons followed by one AAC codon. These chains were introduced as artificial mRNA in the *in vitro* protein-synthesizing system and the resulting polypeptide chains analyzed. As was to be expected from the code of Table 18-3, where AAA can be seen to stand for lysine and AAC for asparagine, the polypeptides formed contained both lysine and asparagine, the former being several times more abundant than the latter. Determination of the primary structure of

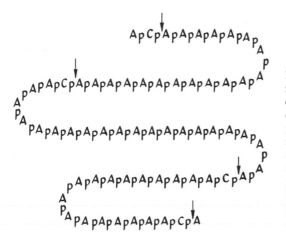

FIGURE 18-3. Symbolic representation of a poly-AC chain synthesized by random catalysis of polynucleotide phosphorylase from a 20:1 mixture of ppA and ppC. The arrows indicate the internucleotide bonds split by pancreatic ribonuclease and hence of the 3′-ends of the short polynucleotide chains generated by this procedure.

these polypeptides by methods similar to those described in Chapter 3 in connection with the amino acid sequence analysis of insulin revealed that they contained a run of lysine residues at their amino-terminal end and a single residue of asparagine at their carboxy-terminal end (Figure 18-4). This result demonstrated, therefore, that translation of the artificial messenger began at the 5'-end with its AAA codons and ended at the 3'-end with its single AAC codon.

An entirely different study made by G. Streisinger, A. Tsugita, and their colleagues also confirmed the prediction that the translation of mRNA must proceed from the 5'-end of the polynucleotide chain toward its 3'-end. This work focused upon the effect of *frameshift* mutations in the lysozyme gene of phage T4 on the primary structure of the lysozyme protein. As was discussed in Chapter 13, in connection with Crick and Brenner's proof of the general nature of the genetic code, such frameshift mutations were thought to be the result of the insertion or deletion of single nucleotides into or from the DNA and hence to cause the mistranslation of the entire genetic message beyond the mutant site. Crick and Brenner proposed that the noxious effect of such a frameshift mutation can be counteracted by a second mutation of opposite sign near the first mutation, since the correct reading frame would thus be restored to the genetic message. But the nucleotide sequence *between* the first and second mutations would still be translated in the wrong frame, and hence give rise to an amino acid sequence entirely different from that present in the wild-type protein in the corresponding sector. After Streisinger and Tsugita had constructed such a corrected frameshift double mutant of T4 phage, carrying two juxtaposed frameshift mutations of opposite sign in its lysozyme gene, they isolated the lysozyme protein formed by that mutant phage in *E. coli* host cells and compared its amino acid sequence with that of the lysozyme formed by the normal T4 wild-type phage. This analysis revealed that the lysozyme formed by a corrected frameshift double mutant embodies an abnormal polypeptide segment, whose sequence is shown below along with that of the corresponding segment of wild-type lysozyme. (It is to be borne in mind that in the writing of these amino acid sequences, the leftmost amino acid residue is that closest to the amino-terminus and the rightmost that closest to the carboxy terminus of the polypeptide chain.)

Double mutant: -thr-lys-val-his-his-leu-met-ala-ala-lys-

Wild type: -thr-lys-ser-pro-ser-leu-asn-ala-ala-lys-

The finding that the juxtaposition of two mutations has engendered the replacement of one block of five contiguous amino acids by another thus provided direct proof for the verity of Crick and Brenner's purely speculative proposal of the frameshift effect of the class of mutations on which their

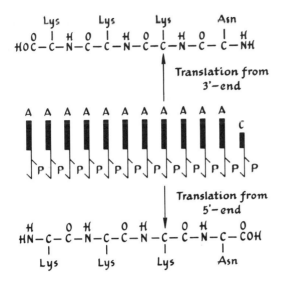

FIGURE 18-4.   The two possible alternatives for the direction of translation of the nucleotide sequence of RNA templates in the amino acid assembly process and the two alternative polypeptide chains to which translation of ApAp...pApCp chains would give rise.

analysis of the nature of the code had been based. In the example considered here, it could be concluded, furthermore, that the two frameshift mutations of opposite sign must have occurred in the codons specifying the leftmost serine and alanine residues of the wild type lysozyme sequence.

By using the genetic code table, it is possible to infer the sequence of 30 nucleotides in the relevant portion of the lysozyme gene that specifies the ten-amino-acid segment of the normal lysozyme protein, provided that due allowance is given to the possibility that any one of several synonymous codons might be present. It is to be borne in mind, furthermore, that the order in which that codon sequence is to be assembled depends on whether the codon of the amino acid closest to the amino-terminus of the lysozyme polypeptide is closest to the 5'-end or closest to the 3'-end of the polynucleotide chain. The inferential nucleotide sequence must, however, satisfy another condition—namely, that upon deletion of a nucleotide from (or insertion into) the leftmost serine codon and insertion of a nucleotide into (or deletion from) the leftmost alanine codon, a new nucleotide sequence is generated that codes for the five contiguous amino acid replacements observed in the mutant lysozyme polypeptide. This second condition turns out to be an extremely restrictive one, which not only determines completely the choice of synonymous codons at every site, but, more importantly, also allows for only one codon order—namely, that demanded by translation of the mRNA proceeding from the 5'-end towards the 3'-end. This condition demands, moreover, that the two frameshift mutations represent a deletion of A from the AGU serine codon and an insertion of G between the G and C of the GC· alanine codon. That uniquely possible polynucleotide sequence, and the consequence of the two frameshift mutations, is shown in Figure 18-5.

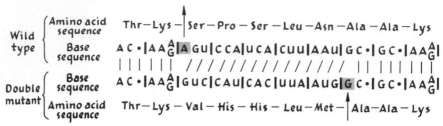

FIGURE 18-5.   Effect of two closely linked frameshift mutations of opposite sign in a sector of the T4 phage lysozyme gene. The wild-type nucleotide base sequence written here to represent the known amino acid sequence of the wild-type protein is the only one that can be chosen from among the many synonyms of the genetic code table which will generate the known mutant amino acid sequence by two frameshift mutations of opposite sign. These two frameshift mutations, furthermore, can only represent a *deletion* of A from the first position of a wild-type AGU serine codon and the *insertion* of a G between the wild-type AUG methionine and GC· alanine codons fifteen nucleotides down the gene. [After E. Terzaghi, Y. Okada, G. Streisinger, J. Emrich, M. Inouye, and A. Tsugita, *Proc. Natl. Acad. Sci. Wash.* **56**, 500 (1966).]

THE ANTICODON AND tRNA

As soon as the results of the first decoding experiments had indicated that the genetic code does contain synonymous codons (e.g., that, as was inferred from Table 18-2, leucine is coded by both $U_2 C$ and $U_2 G$), it became a matter of keen interest to ascertain how such synonymous codons are decoded in the polypeptide assembly process. Since fractionation studies of tRNA extracts, such as those presented in Figure 17-8, had only recently shown that more than one species of tRNA may be affined to the same amino acid, it appeared most plausible that every codon is in correspondence with its own tRNA species—the one that carries the complementary anticodon. The first test of this notion was made by Weisblum, Benzer, and Holley in 1962. They resolved the leucine-affined tRNA into two fractions, I and II, as shown in Figure 18-6, added one or the other of the two tRNA fractions to *in vitro* protein-synthesizing reaction mixtures containing either poly UC or poly UG as artificial messengers, and examined the incorporation of ^{14}C-leucine into polypeptides. The result of this experiment was that leucine tRNAI promotes the incorporation of ^{14}C-leucine only in the reaction mixture containing poly UC, whereas leucine tRNAII promotes ^{14}C-leucine incorporation only in the mixture containing poly UG. Hence it could be concluded that the two leucine tRNA species do carry different anticodons, each complementary to a different leucine synonym.

Once the code table had been established in final form and it had become clear that of the 64 possible codons 61 represent an amino acid, one might

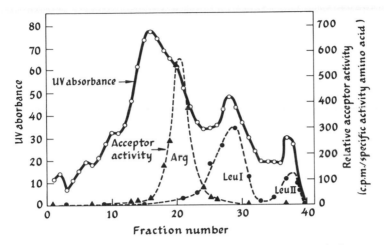

FIGURE 18-6.   Separation of two leucine tRNA species of *E. coli*. An extract of the total tRNA of *E. coli* was fractionated by means of the countercurrent distribution technique (see Figure 17-9). Each fraction was then assayed for its absorption of UV light at 260 mμ, or total nucleic acid content, and for its capacity to accept radioactively labeled arginine or leucine. The arginine acceptor capacity can be seen to be localized in a single Arg peak, whereas the leucine acceptor capacity is seen as distinct peaks, Leu I and Leu II. [After B. Weisblum, S. Benzer, and R. W. Holley, *Proc. Natl. Acad. Sci. Wash.* **48**, 1450 (1962).]

have expected that every cell contains 61 species of tRNA, each endowed with one of the 61 anticodons. And in apparent confirmation of that expectation, more and more tRNA species were being discovered upon further efforts to fractionate tRNA extracts from such biological sources as *E. coli* and yeast. But when the coding properties of these various tRNA species were investigated by means of Nirenberg's nucleotide-triplet-tRNA binding test, it turned out that many individual tRNA species recognize more than one of a set of synonymous codons. Furthermore, it was found that of the several tRNA species affined to the same amino acid, many show the same codon recognition pattern. For instance, the pure species of yeast alanine tRNA, of which Holley determined the primary structure shown in Figure 17-11, responds to GCU, GCC, and GCA, or to three of the four alanine codons GC·. Or, of the two species of yeast valine tRNA resolved by the fractionation procedure shown in Figure 17-9, one species responds to GUU, GUC, and GUA and the other species to GUA and GUG of the four GU· valine codons. Some tRNA species, however, do respond to only a single codon, such as the *E. coli* leucine tRNAII resolved by the fractionation procedure of Figure 18-6, which responds only to UUG of the six synonymous leucine codons UU$_G^A$, CU·.

In 1965 Crick made a proposal that not only explained this at-first-sight perplexing differential range in codon responses of various tRNA species but also provided an insight into the general structure of the code. This proposal, to which Crick gave the name "wobble hypothesis," envisaged that on the ribosomal scene only the *first two* nucleotide bases of the mRNA codon are constrained to form the standard Watson-Crick base pairs between A and U and between G and C, (as well as between I and C) with the corresponding nucleotide bases of the tRNA anticodon. There can, however, be some "wobble," or play, in the codon-anticodon interaction between the *third* nucleotide bases. That is to say, wobble makes possible the formation of more than one type of base pair at the third codon position. From structural considerations based on a study of molecular models, Crick inferred that third-position wobble would permit the formation of nonstandard base pairs between G and U, between I and U, and between I and A (Figure 18-7). Hence the following relation can be established between the base of the tRNA anticodon and the range of bases in the *third* position of the mRNA codon that it can recognize:

Third-position anticodon base	Third-position codon base
∩	A or G
Ɔ	G
Ɐ	U
⅁	U or C
I	U, C, or A

This schema explains how the yeast alanine tRNA, whose anticodon consists of the nucleotide triplet Ɔ⅁I, can respond to the three alanine codons GCU, GCC, and GCA. In order to be able to recognize the fourth alanine codon GCG, however, yeast can be expected to contain a second alanine tRNA species with anticodon Ɔ⅁∩ or ƆƆƆ. The coding response of the two species of yeast valine tRNA would indicate that the anticodon of the one responding to GUU, GUC, and GUA is ƆⱯI and the anticodon of the other responding to GUA and GUG is ƆⱯ∩. Finally, the restricted coding response of *E. coli* leucine tRNAII to only UUG suggests that its anticodon is ⱯⱯƆ.

The wobble hypothesis evidently explains why synonymous codons are in the same "box" of the code table. According to that hypothesis, no tRNA anticodon can give an exclusive response to either a C or an A in the third codon position. Hence a codon of type XYC *must* be synonymous with XYU. Thus even though the wobble hypothesis envisages the exclusive response

Anticodon

Codon

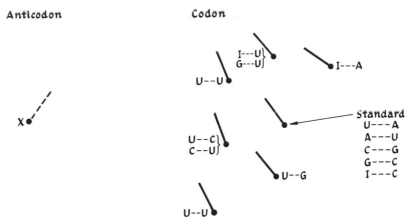

FIGURE 18-7. The structural basis of Crick's "wobble hypothesis." The point X represents the 1'-carbon atom of the ribose and the dashed line the bond linking that atom to the ring nitrogen atom of a nucleotide base of the anticodon. The other points and their solid lines indicate the position of the 1'-carbon atom and its bond to the ring nitrogen of the corresponding nucleotide base of the *codon* for the various base pairs listed. (Pairs with I in the codon have been omitted for simplicity.) The wobble hypothesis envisages that the three distant "nonstandard" positions above, below, and to the right of the standard Watson-Crick base pairs are acceptable for the anticodon-codon match in the third codon position. The three close "nonstandard" positions to the left of the standard base pairs are not acceptable. [After F. H. C. Crick, *J. Mol. Biol.* **19**, 548 (1966).]

of a third-position anticodon Ɣ to a third position codon U, and hence permits the existence of a tRNA species capable of responding only to the XYU codon, that codon cannot represent an amino acid different from that represented by the XYC codon. Similarly, a codon of type XYA must be synonymous with either XYG, or with XYU and XYC. Here, however, the exclusive response of third-position anticodon Ɔ to third-position codon G, and its consequent enablement of a tRNA species capable of responding only to the XYG codon, does offer the possibility that XYG is the sole representative of an amino acid. For this would mean that XYA would be synonymous with XYU and XYC, and would be recognized in the decoding process by a tRNA species whose anticodon carries I rather than ∩ in its third position. These implications of the wobble hypothesis are thus in agreement with the finding that AUG and UGG are the sole codons of methionine and tryptophan. But since this reasoning obviously does not pertain to *nonsense* codons, the wobble hypothesis is not vitiated by the Non3 codon UGA being synonymous with neither the cysteine codons UG_C^U nor the tryptophan codon UGG.

One important exception to the rule that the tRNA anticodon is highly selective with respect to the first two nucleotide bases of the codon but more permissive with respect to the third is the formylmethionine tRNA, which is responsible for polypeptide chain initiation in *E. coli*. As was set forth in Chapter 17, there exist two separable species of tRNA capable of accepting methionine from the methionyl-adenylate-tRNA synthetase complex. But the methionine of only one of these methionyl-tRNA species is converted into formylmethionine by action of the special formylation enzyme. The other, or ordinary, methionyl-tRNA species serves for the incorporation of methionine into the interior of growing polypeptide chains and can be shown to respond to the codon AUG in *in vitro* protein synthesis stimulated by artificial mRNA or in the nucleotide triplet binding test. The formylmethionyl-tRNA, however, will initiate polypeptide chain growth in response to either the methionine codon AUG or the valine codon GUG, and appears to have some weak affinity even for the leucine codons UUG and CUG. Hence the anticodon of that special tRNA species seems to be permissive with respect to the first nucleotide base of the codon and selective with respect to the second and third nucleotide bases.

The reader may, by now, be wondering how polyphenylalanine chain growth was initiated in Nirenberg's celebrated *in vitro* protein-synthesizing reaction mixture when poly U, the artificial mRNA which he used, does not embody any ·UG codon to which the chain-initiating formylmethionine tRNA can respond. The answer to this puzzling question turned out to be that Nirenberg's reaction mixture contained 0.02 M Mg^{++}. At this relatively high Mg^{++} concentration 30S and 50S ribosomal subunits can assemble on the mRNA template to form the working 70S ribosome and initiate polypeptide chain growth at any nucleotide triplet without the intervention of formylmethionyl-tRNA. At two- to fourfold lower Mg^{++} concentrations, however, (i.e., 0.005 to 0.01 M) the chain-initiating union of 30S and 50S ribosomal subunits proceeds only after the 30S subunit has found an AUG or GUG triplet on the mRNA template and has bound formylmethionyl-tRNA. That is to say, in reaction mixtures containing lower Mg^{++} concentrations artificial polymers such as poly U or poly UC are very poor stimulants of *in vitro* protein synthesis, whereas under these same conditions natural mRNA or artificial polymers such as AUGAn, which contain the AUG codon, are excellent stimulants. Furthermore, the rate of polypeptide synthesis directed at low Mg^{++} concentrations by polymers containing the AUG codon can be shown to be strongly dependent on the presence of formylmethionyl-tRNA in the reaction mixture, whereas poly-U-directed polyphenylalanine synthesis at high Mg^{++} concentrations does not show this dependence. Thus it can be concluded that in the translation of mRNA of *E. coli* and of phages growing on *E. coli*, the AUG triplet (and possibly also the GUG triplet) near the 5′-end of the polynucleotide chain plays the role of a *phasing codon* that determines the triplet reading frame in which the message is to be read, codon by codon. The extent to which this mechanism of polypeptide chain initiation is in general operation still remains to be established. At present it would appear that it does obtain in protein synthesis proceeding on the 70S ribosomes of prokaryotes and of self-reproducing cytoplasmic organelles of eukaryotes, such as mitochondria. It seems doubtful, however, that it is operative in protein synthesis proceeding on the cytoplasmic ribosomes of higher eukaryotes, since no convincing evidence about the involvement of formylmethionine-tRNA in their chain-initiation reactions has yet been found.

NONSENSE CODONS

As was set forth in Chapter 13, from 1953 onward Benzer had accumulated a large collection of rII mutants of phage T4. These mutants, it will be recalled, can grow on ordinary strains of E. coli, on which they form plaques of r mutant morphology, but they cannot grow on K strains (i.e., E. coli carrying the lambda prophage) on which the r^+ wild-type phage is able to grow. In 1960 Benzer suddenly discovered that, although no rII mutant ought to be able to grow on any strain K bacterium, his large collection of rII mutants included some ambivalent specimens that can grow on one but not on another variety of strain K. Further study of this unexpected finding revealed the existence of three distinct subsets of ambivalent mutants; the members of each subset can grow on one of three strain K varieties but not on the other two. Each subset included about one out of thirty of the hundreds of mutant sites that Benzer had by then identified in both rIIA and rIIB genes.

Benzer first interpreted the behavior of his ambivalent rII mutants in terms of Yanofsky and St. Lawrence's then recently proposed mechanism of genetic suppression, which was set forth in Chapter 17. He suggested that his ambivalent rII mutants have sustained missense mutations and that each of the three permissive K strains that defines one subset of ambivalent phage mutants carries a different missense suppressor mutation. The phage-mutant members of that subset would, therefore, be able to grow on that and only that permissive K strain whose missense suppressor happens to insert an acceptable amino acid in response to the mutant codon calling for a noxious amino acid. In this way, activity would be restored to the enzyme produced by the mutant rII phage gene.

Further studies carried out in collaboration with his student S. Champe soon caused Benzer to revise this explanation of ambivalence. Among Benzer's collection of rII mutants there was one in which a long DNA deletion had removed the boundary between the contiguous rIIA and rIIB genes. The consequence of this mutation is to weld the normally separate rIIA and rIIB polypeptide chains into a single polypeptide chain (Figure 18-8,A). This monster polypeptide is devoid of rIIA gene function (and hence the phage is of rII mutant phenotype), but it can supply the rIIB gene

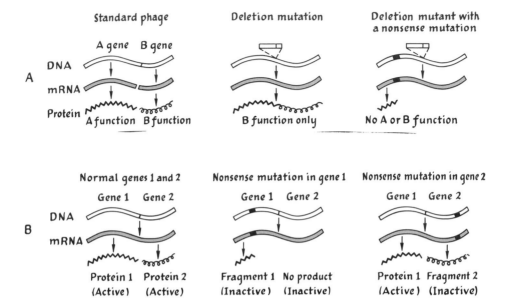

FIGURE 18-8.   **A.** Long-range effect of a nonsense mutation in a T4rII deletion mutant. In the standard wild-type phage, the rIIA and rIIB genes are shown to be transcribed onto separate mRNA molecules, which are translated into separate proteins. In the deletion mutant, a single mRNA molecule is formed that is translated into a single protein capable of performing the rIIB gene function but not the rIIA gene function. In the deletion mutant that also contains a nonsense mutation in the rIIA gene, a protein fragment is produced that can perform neither rIIA nor rIIB function. [After A. Garen, *Science*, **160**, 149 (1968). Copyright 1968 by American Association for the Advancement of Science.] **B.** The *polarity* effect of nonsense mutations in genes that are normally transcribed onto the same mRNA molecule. This drawing shows a hypothetical example of two such cotranscribed genes. The direction of DNA transcription, and hence also of mRNA translation, is assumed to be from left to right. [After A. Garen, *Science* **160**, 149 (1968). Copyright 1968 by American Association for the Advancement of Science.]

function in mixed-infection complementation tests with ordinary rIIB point mutants (see Chapter 13). Benzer and Champe examined the effect on the rIIB gene activity of this deletion mutant of various additional point mutations situated in the rIIA gene. They found that whereas most such additional rIIA mutations leave the rIIB gene activity of the monster polypeptide intact, ambivalent rIIA mutations abolish it. The following inferences were therefore made: The ambivalent mutations represent changes of a codon from sense to *nonsense*, rather than to missense. Since the mRNA transcribed from rIIA and rIIB genes is translated from left to right (as drawn in Figure 18-8,A), the presence of a nonsense codon in the rIIA portion of the deletion mutant will arrest polypeptide assembly at that point, and the rIIB segment cannot be synthesized at all. The permissive K strains, therefore, contain *nonsense* suppressors, rather than missense suppressors. These nonsense suppressors insert an amino acid in response to a mutant nonsense codon, to which there ordinarily corresponds no amino acid at all. The nonsense suppressors of the three permissive K strains differ with respect to the particular nonsense codon they mistake for a sense codon and/or with respect to the amino acid they mistakenly insert. The ambivalent rII mutants of the three subsets in their turn differ with respect to the particular nonsense codon they contain and/or the amino acid that is compatible with enzyme activity at the site of the polypeptide chain normally specified by that mutant codon. These experiments and ideas of Benzer and Champe were to open up the comparative study of sense and nonsense in the genetic code.

Upon accession to the idea that some mutations lead to the generation of a nontranslatable nonsense codon in the middle of a gene, a variety of hitherto unexplained bacterial mutant phenotypes could be accounted for. For instance, as was mentioned in Chapter 6, Yanofsky had found inactive missense tryptophan synthetase A protein in only some of his *E. coli* Trp⁻ auxotrophs bearing a point mutation in the *trp*A gene. In others, he could not detect any missense A protein at all. Since among this second mutant class were many whose true reversion to the Trp⁺ wild type could be induced by base-analog mutagens (and hence derived from base substitutions), their lack of any A-protein-like material could not be attributed to a frameshifting nucleotide insertion or deletion. Instead, it came to be recognized that this class carries nonsense mutations that prevent the synthesis of full-length A-protein polypeptide chains. Support for that explanation could be adduced by A. Garen, who showed that one of the nonsense suppressors of Benzer's ambivalent T4rII mutants also restores full length to mutant polypeptide chains of the *E. coli alkaline phosphatase* enzyme. Another previously perplexing mutant phenotype was represented by *polar* point mutations that had been observed to affect the expression of more than one gene (and which, had they been known at the time of the 1946 Cold Spring Harbor Symposium, would have dampened, albeit quite wrongly, the fervor of the champions of the one-gene–one-enzyme theory). For instance, as was mentioned in Chapter 5, Gal⁻ mutants of *E. coli* were long known in which a single mutation has abolished formation of the three enzymes galactokinase, uridyl transferase, and UDP-Gal epimerase, which are active in galactose metabolism. These enzymes were later found to be encoded in three contiguous genes, *gal*K, *gal*T, and *gal*E (see Figure 10-21), and the sites of the polar mutations affecting all three enzymes were mapped in the *gal*E gene. Another example of polar mutations was encountered in the enzymes of tryptophan biosynthesis, where there exist base-substitution mutations in the *trp*E gene of *E. coli*, which not only abolish the formation of the corresponding anthranilate synthetase protein but also reduce greatly the amount of tryptophan synthetase formed by the closely linked nonmutant *trp*A and *trp*B genes. Such polar mutations were now identified as nonsense mutations in a gene that is transcribed onto the same mRNA molecule as one or more other closely linked genes and normally translated before them (Figure 18-8,B). Thus the translation block posed by the nonsense codon in an upstream gene interferes with the further processing of that mRNA molecule and prevents ribosomes from initiating polypeptide assembly on the downstream genes.

As was related in Chapter 12, in 1960 a class of conditional lethal mutants of T4 phage, the *amber* or *am* mutants, were discovered. An *am* mutant, it will be recalled, has sustained its mutation in any one of the many genes of the phage and as a consequence is unable to synthesize the normal product of that gene (and hence fails to grow) upon infection of the ordinary nonpermissive *E. coli* host strain. Upon infection of the permissive K strain, however, the amber mutant *can* synthesize that gene product (and, in consequence, *can* grow). We are now at last ready to consider the explanation of this strange phage-mutant phenotype.

The first clue to the true nature of these *am* mutants was provided by the discovery that the permissive K strains used in the isolation of *am* mutants are precisely those that also contain nonsense suppressors for ambivalent T4*r*II mutants. The idea gained currency, therefore, that the *am* phenotype is attributable to the mutational generation of a nonsense codon in any gene of the T4 phage genome. The nonsense codon cannot be translated in the ordinary nonpermissive *E. coli* host strain but can be translated as sense in the permissive K strain, thanks to its possession of a nonsense suppressor. This notion also provided an explanation for the observation mentioned in Chapter 13 that whereas temperature-sensitive, *ts*, mutations located in phage genes coding for multiple subunit enzymes can show *intragenic complementation*, *am* mutations never do so. For unlike the homologous *missense* polypeptide chains of *ts* mutants, the *incomplete* polypeptide chains synthesized by *am* mutants in the restrictive host obviously cannot complement each other to form a catalytically active quaternary protein structure.

In 1964, S. Brenner (Figure 18-9) and his colleagues were able to adduce direct proof not only for this explanation of the *am* phenotype but also for the more general proposition that nonsense codons interrupt the polypeptide assembly process; in addition, they also provided, for good measure, a demonstration independent of that already described in Chapter 14 of the colinearity of nucleotide sequence in DNA and amino acid sequence in protein. In their experiments, they worked with an ensemble of ten T4 *am* mutants, all of whom carry their mutation in gene 23 (of the map of Figure 13-6), in which the primary structure of the T4 head protein is encoded. By

FIGURE 18-9.   Sydney Brenner (b. 1927). [Courtesy of May Brenner.]

means of genetic crosses like those described in Chapter 13, they were able to construct the fine-structure map of the relative order of these *am* mutations within the head-protein gene shown in Figure 18-10. Meanwhile, studies of the primary structure of the normal wild-type T4 phage-head protein had shown that it is a single polypeptide chain whose digestion with the enzymes trypsin and chymotrypsin gives rise to eight distinct fragments that are easily separable on the basis of their rate of migration in an electric field—that is, by electrophoresis. In order to compare the character of the head-protein gene product formed by these various phages in the absence of nonsense suppressors, cultures of the nonpermissive *E. coli* strain were infected with each of the ten head-protein gene *am* mutants and with the *am*$^+$ wild type. Radioactive ^{14}C-labeled amino acids were then added to the culture medium about ten minutes after infection, and the radioactively labeled protein that had been formed was extracted from the infected cells sometime later. The protein extract was subjected to digestion by trypsin and chymotrypsin, and finally, the resulting radioactive peptide fragments were separated by electrophoresis. It is possible to identify the eight characteristic head-protein peptide fragments in the enormously complex mixture generated by digestion of the total protein extract of the infected cells because the head protein is the destination of more than half of all the labeled amino acids assembled into polypeptide chains at later stages of intracellular T4 phage growth. The result

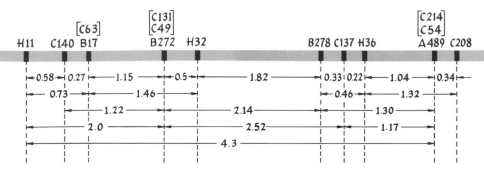

FIGURE 18-10.   Fine-structure genetic map of an ensemble of *am* mutations in gene 23 (head-protein gene) of T4 phage. The numbers within the arrows indicate the frequency of *am*⁺ recombinants produced in a cross between the *am* mutant pairs. [After A. S. Sarrabhai, A. O. W. Stretton, S. Brenner, and A. Bolle, *Nature* **201**, 13 (1964).]

of this experiment is presented schematically in Figure 18-11, where it can be seen that the eleven phages can be ordered with respect to the number of head-protein peptide fragments that can be recovered from their infected bacteria. For instance, whereas all eight fragments are found in bacteria infected with the *am*⁺ wild type, none of the normal eight fragments, and only one very short, abnormal fragment, are found in bacteria infected with the *am*-H11 mutant, and only two normal fragments are found in bacteria infected with the *am*-H32 mutant. Since it can be inferred that the fewer the number of normal peptide fragments found here, the shorter the head-protein polypeptide chain that is synthesized by the mutant gene, it follows that these ten amber mutants do carry nonsense codons at various positions of the head-protein gene at which polypeptide chain growth is precociously terminated. Furthermore, since comparison of Figures 18-10 and 18-11 shows that the genetic map order of these nonsense codons is the same as the order of lengths of the incomplete polypeptide chains to which they give rise, it can be concluded that codon order and amino acid order are colinear. That is to say, the nucleotide sequence of the head protein gene is evidently translated from left to right, as drawn in Figures 18-10 and 18-11.

The structure of the nonsense codons came to be known in 1965. By that time most of the 64 triplets of the code table had been deciphered as sense codons, so that it was clear that there could not be very many nonsense codons representing no amino acid. Among the few codons then still available for assignment were UAA, UAG, and UGA. In order to identify the nonsense codons, Garen and Brenner employed similar strategies. Both isolated a large number of *reverse* mutants—Garen from a nonsense mutant in the *E. coli* alkaline phosphatase gene (*phoA*) and Brenner from an *am* mutant in the T4 head-protein gene—and ascertained for each reverse mutant the amino acid present at the polypeptide site that corresponds to the reverted nonsense codon. The results of Garen's study are presented in

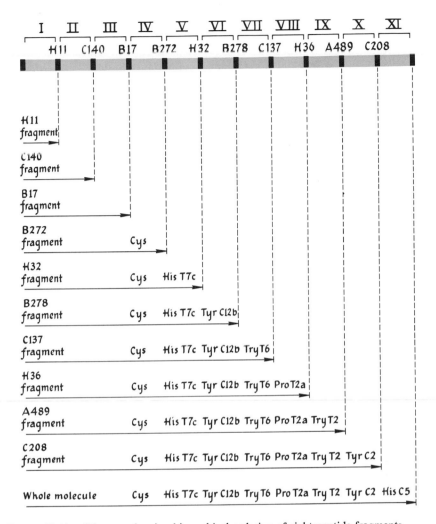

FIGURE 18-11. Diagram showing hierarchical ordering of eight peptide fragments ("Cys," "HisT7c," "Tyr C12b," "TryT6," "ProT2a," "TryT2," "TyrC2," "HisC5") generated by enzymatic digestion (trypsin and chymotrypsin) of the T4 phage head protein. The order was determined on the basis of their appearance in nonpermissive *E. coli* infected with each of ten of the ensemble of gene 23 *am* mutants charted on the map in Figure 18-10. The map order of the ten mutants is also shown at the top of this diagram, and the eleven map intervals between them have been assigned consecutive Roman numerals. Each arrow symbolizes the length of the head-protein polypeptide chain found to be synthesized by a given mutant. Three of the mutants, B17, C140, and H11, give rise to none of the normal eight peptide fragments and form only progressively shorter versions of the "Cys" fragment. [After A. Sarrabhai, A. O. W. Stretton, S. Brenner, and A. Bolle, *Nature* **201**, 13 (1964).]

Figure 18-12, where it can be seen that the nonsense codon had arisen at a codon that specifies tryptophan in the wild-type alkaline phosphatase polypeptide chain. Upon reverse mutation, that nonsense codon was found to

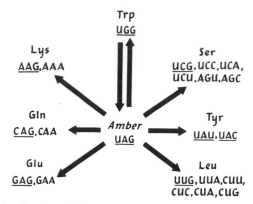

FIGURE 18-12. Identification of the *amber*, or Non 1, nonsense codon as UAG. A Pho⁻ mutant of *E. coli* (lacking the enzyme alkaline phosphatase) was isolated which carried an *am* mutation in its *pho*A gene. From this Pho⁻ mutant, a set of Pho⁺ reverse mutants was isolated whose recovery of alkaline phosphatase was not attributable to suppressor mutations. Amino acid sequence analysis of the alkaline phosphatase protein of these reverse mutants showed that they carried the variety of amino acid replacements shown in this diagram at the same polypeptide site at which the wild-type alkaline phosphatase carries a tryptophan residue. Each amino acid is represented here by its synonymous codons, and those codons that are connected to UAG by a single base change are underlined. It is evident that UAG is the only nucleotide base triplet that has this connection to at least one of the codons for each of the seven amino acid replacements found. [After A. Garen, *Science* **160**, 149 (1968). Copyright 1968 by American Association for the Advancement of Science.]

change into codons specifying the wild-type tryptophan, or serine, or tyrosine, or leucine, or glutamic acid, or glutamine or lysine. Given the codon assignments indicated in Table 18-3 for these amino acids, the only codon related to their codons by single base changes is UAG, or Non1 of that table. Since Brenner reached the analogous conclusion that the nonsense codon of T4 *am* mutants is UAG, Non1 is now generally referred to as the *amber codon*. On the basis of its particular response to another kind of nonsense suppressor and its slightly different pattern of reverse mutation, both Garen (using the *E. coli pho*A gene) and Brenner (using T4 phage and the *E. coli lac*Z gene) identified UAA as another nonsense codon, Non2, or the *ochre codon*. Finally, in 1967, Brenner and Crick identified UGA as the third nonsense codon, Non3, on the basis of the reverse mutation pattern of a class of T4rIIA ambivalent mutants that qualify as nonsense mutants by Benzer and Champe's deletion test.

NONSENSE SUPPRESSION

Study of the suppressive capacity of the suppressors found in various *E. coli* strains for nonsense mutations in both phage and bacterial genome showed that there exist at least five different suppressor, or *sup*, genes, whose distinguishing properties are summarized in Table 18-5. Three of these suppressors, *sup*D, *sup*E, and *sup*F suppress only mutants carrying the amber, or Non1, codon UAG, whereas two of them, *sup*C and *sup*G, suppress mutants carrying either UAG or the ochre, or Non2, codon UAA. None of them suppress mutants carrying the Non3 codon UGA, although suppressors active on Non3 are now known. By means of conjugational and transductional crosses between bacterial strains differing in the nature of their suppressors, it was possible to map the position on the *E. coli* chromosome of these five *sup* genes, with the results shown in Table 18-5. As can be

TABLE 18-5.   Properties of Some Nonsense Suppressor Genes in *E. coli*

Suppressor gene	Approximate map position (minute)	Nonsense codon	Amino acid inserted	Efficiency of suppression (percent)
*sup*D	38	UAG	serine	28
		UAA	none	0
*sup*E	16	UAG	glutamine	14
		UAA	none	0
*sup*F	25	UAG	tyrosine	55
		UAA	none	0
*sup*C	25	UAG	tyrosine	16
		UAA	tyrosine	12
*sup*G	16	UAG	basic amino acid	5
		UAA	basic amino acid	6

After A. Garen, *Science*, **160**, 149–159 (1968).

seen, these five genes are found at three widely spaced chromosomal regions. It is possible to determine the efficiency with which these suppressors work by measuring the extent to which they restore the synthesis of intact polypeptide chains encoded in a gene carrying a nonsense codon. For instance, Brenner compared the relative amounts of complete T4 phage head protein formed in host bacteria carrying the suppressor by phages carrying either a normal or *am* mutant gene 23. In this way it was found that the efficiency of the five suppressors ranges from 5 to about 60%. Finally, the primary structure of the polypeptide chain completed thanks to the action of the suppressor can be determined, in order to ascertain the nature of the amino acid residues that the suppressor inserts in response to the nonsense codon. As is shown in Table 18-5, each suppressor gene causes the insertion of its own characteristic amino acid. Thus a suppressor gene not only controls the capacity of a cell to translate a nonsense codon but also assigns a characteristic coding specificity of that codon. The mutant polypeptide chain whose completion any given suppressor rendered possible will thus be physiologically active only if the amino acid inserted in response to the nonsense codon is acceptable at that site.

Studies of the biochemical basis of the nonsense suppressors paralleled (in fact, preceded) those carried out on missense suppressors described in Chapter 17. By examining the suppressor capacity of various components of the machinery for protein synthesis extracted from *E. coli* carrying suppressor genes, it could be shown in the laboratories of Watson, of Garen, and of Zinder that the *sup*D suppressor engenders an abnormal species of serine tRNA whose anticodon allows it to insert serine in response to an UAG codon. This abnormal species is present in addition to the normal serine tRNA species, which display the normal coding pattern. Work carried out in Brenner's laboratory led to the analogous conclusion that an additional, abnormal species of *tyrosine* tRNA is the product of the *sup*F gene. It can be seen in Table 18-5 that *sup*F is located near minute 25 of the *E. coli* map, or in close linkage to the *trp* genes. As will be set forth in more detail in Chapter 20, it is possible to isolate a restricted transducing phage (formally analogous to λdg) that has incorporated the *trp* gene region of the *E. coli* genome, or the closely linked *sup*F, into its DNA. Brenner and his colleagues extracted the DNA of this transducing phage and demonstrated that this DNA forms a specific RNA-DNA hybrid with the abnormal species of tyrosine tRNA. Thus it could be concluded that the *sup*F gene encodes the nucleotide sequence of the abnormal tyrosine tRNA species.

It was possible to isolate and purify the abnormal tyrosine tRNA encoded into the *sup*F gene and subject its primary structure to nucleotide-sequence analysis by the methods described in Chapter 17. This analysis showed that the anticodon loop of the suppressor tRNA species does contain the ⱯՈƆ anticodon complementary to the UAG amber codon to which it responds. The anticodon of the normal *E. coli* tyrosine tRNA was found to consist of ⱯՈ⅁, which, according to the wobble hypothesis, should respond to the synonymous tyrosine codons UAU and UAC. Thus the mutation that created *sup*F arose by a simple base substitution at the anticodon site of a normal tyrosine tRNA gene. The identification of this base substitution, made as late as 1968, was the first *direct* chemical proof that mutations really do result from changes in nucleotide sequence.

The wobble hypothesis accounts for the fact manifest in Table 18-5 that suppressors such as *sup*C active in the suppression of the ochre or Non2 codon UAA also suppress the amber, or Non1, codon UAG. For the anticodon of their abnormal tRNA species would be expected to have the structure ⱯՈՈ, which wobble permits to respond to *both* UAA and UAG.

It appears likely at present that one or more of the three nonsense codons is employed at the end of a gene as the *natural* signal for polypeptide chain termination. (As was set forth in Chapter 17, there must exist some nucleotide sequence in the mRNA template that, upon its display at the ribosomal aminoacyl site, signals release of the completed polypeptide chain from the peptidyl site.) It is somewhat difficult to understand, however, how a cell could tolerate the acquisition of an efficient nonsense suppressor of the natural chain termination signal. For that reason the UAA ochre codon has become the favorite candidate for the natural chain termination signal, since all ochre suppressors encountered so far work with relatively low efficiency.

EVOLUTION OF THE CODE

Although most of the codons were first deciphered by use of the protein-synthesizing machinery of *E. coli*, later extensions of nucleotide-triplet-binding experiments to aminoacyl-tRNA species extracted from organisms of a wide taxonomic range—prokaryota and eukaryota (including humans) —showed that the code table is of universal validity. This result had already been foreshadowed by the discovery of the mutant amino acid replacements summarized in Table 18-4. All of these replacements could be accounted for by single base substitutions of the codon assignments in Table 18-3, even though the mutant proteins studied included not only *E. coli* enzymes but also the tobacco mosaic virus coat protein and human hemoglobin. This universality of the code among contemporary organisms shows that the code has remained immutable over long evolutionary periods. Two extreme alternative explanations for the high antiquity of the code immediately come to mind. One, the "stereochemical theory," states that, contrary to the chain of reasoning that had led Crick to postulate the amino acid adaptor in the first place (see Chapter 17), there does exist a steric relation between a codon (or anticodon) nucleotide triplet and the amino acid side chain that it designates. Hence the code table would derive both its structure and its evolutionary stability from the preordained stereochemistry of its elements. The other explanation, the "frozen accident theory," states that the structure of the code table evolved by chance but that once the complete codon assignments had been made in a common ancestor cell of all present-day living forms, further evolutionary diversification of the code became well-nigh impossible. After all, it stands to reason that any mutation engendering a change in codon assignments is bound to be lethal to the individual in which it occurs. For that individual, whose genes would still record their polypeptide sequence information in terms of the unchanged code, would suddenly synthesize *all* of its proteins in a missense, and hence generally inactive form.

As far as the stereochemical theory is concerned, it seems unlikely that the steric relations that it posits are still involved in the code-recognition processes of modern organisms. The experiment reported in Chapter 17, in which alanine attached to cysteine-tRNA was inserted as cysteine, shows that the amino acid side chain is not seen by the mRNA codon in the polypeptide assembly step; and the acceptance of tyrosine by the *sup*F tRNA, which bears the amber anticodon ∀∩Ɔ, shows that the side chain is not recognized by the anticodon in the amino acid activation step. It is entirely possible, however, that such stereochemical interactions were of importance at an early stage in the history of life, before highly specific aminoacyl-tRNA synthetases, or even authentic tRNA molecules, were on the scene. According to the notions put forward by Crick and by C. Woese, polypeptide synthesis in those faraway days might well have been a rather imprecise process in which the low degree of functional specificity demanded of the proteins of protoorganisms was satisfied by inserting into a given site of the polypeptide chain any one of a *group* of structurally similar amino acids. For instance, one and the same codon might have designated alanine *or* glycine and another threonine *or* serine. Even at that stage, however, when because of their ambiguity fewer than twenty different codons were required, the codon was likely to have been a nucleotide *triplet*. For, so Crick argues, it would have been extremely difficult to bring two successive amino acids into the juxtaposition required for their assembly if, as under a doublet code, the nucleic acid template were to advance only two nucleotides per amino acid residue incorporated. It is possible, however, that only the *first two* nucleotides of the ambiguous primitive triplet codons actually figured in the recognition process (in which case all triplets belonging to the same box of Table 18-3 necessarily would have been synonyms). The stage was thus set for an evolutionary differentiation of the hitherto ambiguous codons into unambiguous codons, thus allowing a concomitant refinement in the precision of polypeptide assembly, which, in turn, allowed for the formation of aminoacyl-tRNA synthetases of sufficient functional specificity to differentiate such look-alikes as alanine and glycine in the decoding process. To allow for the unambiguous coding of twenty amino acids (including the putative evolutionary latecomers, methionine and tryptophan, which are thought to have "invaded" the AU· and UG· boxes), the third nucleotide of the codon was included in the recognition process, albeit at the lower level of specificity accounted for by the wobble hypothesis. The whole code table would thus have come to be filled in and to evolve the final lexicographic character with which, eons later, it finally entered human consciousness.

Bibliography

HAYES

Chapter 14.

ORIGINAL RESEARCH PAPERS

Benzer, S., and S. P. Champe. Ambivalent *r*II mutants of phage T4. *Proc. Natl. Acad. Sci. Wash.*, **47**, 1025 (1961); An active cistron fragment. *J. Mol. Biol.*, **4**, 288 (1962).

Benzer, S., and S. P. Champe. A change from nonsense to sense in the genetic code. *Proc. Natl. Acad. Sci. Wash.*, **48**, 1114 (1962).

Crick, F. H. C. Codon-anticodon pairing: The wobble hypothesis. *J. Mol. Biol.*, **19**, 548 (1966).

Crick, F. H. C. The Origin of the Genetic Code. *J. Mol. Biol.*, **38**, 367 (1968).

Crick, F. H. C., L. Barnett, S. Brenner, and R. J. Watts-Tobin. The general nature of the genetic code. *Nature*, **192**, 1227 (1961).

Gamow, G. Possible relation between deoxyribonucleic acid and protein structure. *Nature*, **173**, 318 (1954).

Lengyel, P., J. F. Speyer, and S. Ochoa. Synthetic polynucleotides and the amino acid code. *Proc. Natl. Acad. Sci. Wash.*, **47**, 1936 (1961).

Nirenberg, M. W., and J. H. Matthaei. The dependence of cell-free protein synthesis in *E. coli* upon naturally occurring or synthetic polyribonucleotides. *Proc. Natl. Acad. Sci. Wash.*, **47**, 1588 (1961).

Sarabhai, A. S., O. W. Stretton, S. Brenner, and A. Bolle. Colinearity of gene with polypeptide chain. *Nature*, **201**, 13 (1964).

SPECIALIZED TEXTS, MONOGRAPHS, AND REVIEWS

Garen, Alan. Sense and nonsense in the genetic code. *Science*, **160**, 149–159 (1968).

Jukes, T. H. *Molecules and Evolution.* Columbia Univ. Press, New York, 1966.

Sadgopal, Anil. The genetic code after the excitement. *Advances in Genetics*, **14**, 325–404 (1968).

Woese, Carl R. *The Genetic Code.* Harper & Row, New York, 1967.

The Genetic Code. XXXI Cold Spring Harbor Symposium for Quantitative Biology, Cold Spring Harbor, New York, 1966.

19. Genetic RNA

The discussions of the preceding chapters have been subsumed under the dogma that DNA is *the* genetic material charged with carrying out both autocatalytic and heterocatalytic functions. The role of RNA was conceived of as playing second fiddle to DNA in the heterocatalytic function, since, as was set forth in Chapter 16, RNA polynucleotide chains arise merely through the transcription of the nucleotide sequences of DNA chains. But although this dogma seems to be a reasonable approximation to the facts of cellular life, it is not of universal validity. For as the matter to be set forth in this chapter shows, there exists at least one very widespread class of self-reproducing biological entities—the RNA viruses—in whose life RNA rather than DNA plays the role of the genetic material charged with both basic hereditary functions.

THE TOBACCO MOSAIC VIRUS

The story of the experimental study of RNA viruses begins in 1886, when A. Mayer showed that the mosaic blight of tobacco plants is an infectious disease that can be transferred from diseased to healthy plants (Figure 19-1). Mayer observed, furthermore, that the infectious agent present in the leaf juice of diseased plants is inactivated by brief boiling of the juice. Shortly thereafter, D. Iwanowsky found that the infectious agent must be very small, because it

FIGURE 19-1.   Three leaves from the tobacco plant *Nicotiana sylvestris*. The normal leaf shown on the top is taken from a healthy plant. The generally discolored leaf shown in the middle is taken from a moribund plant infected throughout with the normal, wild-type strain of tobacco mosaic virus, or TMV. The leaf on the bottom shows *local lesions*, which were caused by rubbing a suspension of mutant TMV particles onto the leaf. Unlike the TMV wild type, the mutant TMV is unable to spread throughout the plant, and its multiplication and toxic action is confined to a small leaf area surrounding the focus of infection. A local lesion of TMV on a tobacco leaf is, therefore, analogous to a phage plaque on a bacterial lawn (see Figure 11-1), and the number of local lesions appearing on a leaf is proportional to the TMV concentration in the solution rubbed on the leaf. Furthermore, the TMV virus particles present in a local lesion constitute a clone, which can be isolated by excising the lesion from the leaf. [Courtesy of H. Fraenkel-Conrat.]

can pass through the pores of bacteriological filters fine enough to retain all then-known types of bacteria. Though the title of "Father of Virology" has sometimes been claimed in behalf of Iwanowsky for this discovery, it was really M. W. Beijerinck who first realized that the agent of the tobacco mosaic disease, or *tobacco mosaic virus* (TMV), represents a special, hitherto unknown class of organisms. For in 1899 Beijerinck proposed that TMV, which can neither be seen in ordinary microscopes nor grown on artificial bacteriological media, is a self-reproducing *subcellular* form of life, a *contagium vivum fluidum*. In the wake of these first stirrings of virus research, an increasing number of self-reproducing pathogenic agents was discovered during the first two decades of this century—agents which shared with TMV the properties of filterability, invisibility, and refractoriness to *in vitro* culture and were therefore relegated to the order of viruses. Among these viruses were the causative agents of not only several other plant diseases but also of such animal diseases as measles, rabies, poliomyelitis, yellow fever, and smallpox.

For the first 30 years after their discovery, the interest in viruses was principally of medical or agricultural orientation, and a tremendous effort was expended in the practical direction of the prevention and cure of virus diseases. This effort was highly successful, in that the agents of many of the most dreaded diseases known to have plagued mankind since ancient times were finally identified and brought under control. But the study of viruses during that period contributed relatively little to the science of biology—that is, to our understanding of basic life processes. Meanwhile, biochemists were developing methods for purifying and isolating proteins, work which (as was set forth in Chapter 4) culminated in the demonstration that enzymes are proteins. Inspired by the success of that work, W. M. Stanley (Figure 19-2) decided in the early 1930's to attempt to purify and isolate TMV by these methods, since some indirect evidence had suggested that this most classical of all viruses may be a protein. By 1935, Stanley succeeded in obtaining from the juice of infected tobacco plants a paracrystalline precipitate that appeared to be a protein endowed with the infectious power of TMV. Shortly thereafter, F. C. Bawden and N. W. Pirie showed that a TMV particle is not really a "pure" protein but contains also about 5% RNA.

Stanley's discovery that TMV is a protein particle caused a tremendous sensation in its time, since it then appeared to many biologists that viruses must be "living molecules" rather than "organisms" if their structure is so simple that they can be crystallized like so much table salt. Indeed, this discovery, perhaps more than any other, marks the birth of molecular genetics, because the finding that an agent endowed with the power of self-reproduction —that supreme attribute of living forms—can be of such low complexity as TMV fired the imagination. It suddenly opened the prospect that the nature of biological self-reproduction might someday be understood in chemical terms.

FIGURE 19-2. Wendell M. Stanley (1904–1971). [Courtesy of W. M. Stanley.]

Just as the development of the electron microscope in the late 1930's had finally made possible the direct visualization of hitherto invisible phage particles, so did it permit the first direct visualization of TMV (Figure 19-3, A). These pictures showed that the TMV particle is a rod 18 mμ in diameter and 300 mμ in length. Meanwhile, X-ray diffraction studies of TMV crystals, initiated by J. D. Bernal and I. Fankuchen in 1941, showed that the TMV particle is assembled from a large number of structurally identical protein subunits. And in 1954, one year after his discovery of the double-helical structure of DNA, J. D. Watson proposed that the TMV rod is a helical assemblage of these protein subunits. Finally, the detailed structure of the TMV virus shown in Figure 19-3,B was worked out. According to this picture, each TMV particle consists of 2130 identical protein molecules, each molecule consisting of a polypeptide chain of 158 amino acids. Lateral aggregation of these oblong protein molecules by formation of ionic bonds between charged amino acid side chains produces a helix with a hollow central core of diameter 20 Å. Furthermore, the structure of the protein subunits is such that their assembly generates a helical groove winding around the long axis of the rod. This helical groove accommodates the one single-stranded molecule of RNA, about 6400 nucleotides in length, that is carried by each TMV particle.

These structural insights explained an earlier finding by G. Schramm that upon treatment with mild alkali, and hence elimination of the positive charges carried by the basic amino acids, the TMV particle falls apart into its

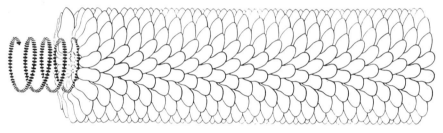

FIGURE 19-3.   Structure of the TMV particle. **A.** An electron micrograph of two rod-shaped TMV particles magnified about 800,000 times, one seen head-on and the other sideways. This picture shows that the TMV rod is a helical array of subunits that form a central, hollow core. [Micrograph by R. W. Horne. From "The Structure of Viruses," *Scientific American*, January 1963.] **B.** Diagrammatic representation of the quaternary TMV structure, based on X-ray crystallographic, electronmicroscopic and biochemical evidence. The coat protein subunits, here shown as radially arranged white structures, surround a single molecule of RNA, here shown as a black helix. [After "The Genetic Code of a Virus," by H. Fraenkel-Conrat, *Scientific American*, October 1964. Copyright © 1964 by Scientific American, Inc. All rights reserved.]

protein subunits and loses its infectivity. Further study of this reaction led H. Fraenkel-Conrat and R. C. Williams to the discovery that upon returning a solution containing a mixture of the disassembled protein subunits and purified TMV-RNA molecules to neutral pH, the virus particles not only reconstitute their original structure but also regain their erstwhile infectivity for tobacco plants (Figure 19-4). This experiment, like the crystallization of TMV 20 years earlier, made a big impact in its time, because it appeared that a "living molecule" had been put together *in vitro* from "nonliving" constituents. In a less dramatic vein, it also provided strong support for the basic tenets of molecular genetics that the secondary, tertiary, and quaternary structure of protein is a consequence of the primary structure of their polypeptide chains and that the formation of complicated macromolecular

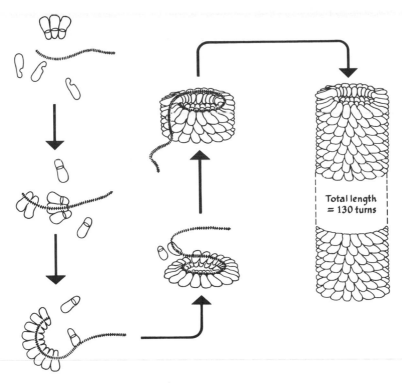

FIGURE 19-4.   Reconstitution of a TMV virus particle from its isolated coat-protein subunits and its RNA molecule. [After H. Fraenkel-Conrat, *Design and Function of the Threshold of Life*. Academic Press, New York, 1962.]

aggregates proceeds by spontaneous assembly of simpler subunits. It thus foreshadowed the discovery by Edgar and Wood of the much more complex spontaneous assembly of the T-even phage tail (described in Chapter 12).

INFECTIOUS RNA

The romantic notion that the reconstitution of intact TMV from its protein and RNA components was actually the *in vitro* creation of a "living molecule" was weakened, however, by a perplexing observation. Although the solution of free TMV protein subunits was entirely devoid of any infectivity, the same could not be said for the solution of purified TMV RNA. For the RNA extract invariably contained a low but significant amount of material capable of causing the mosaic disease in tobacco plants. This fact led Fraenkel-Conrat to suppose in 1956 that the free TMV RNA alone is capable of initiating virus growth. Thus the increase in infectivity observed in his TMV reconstitution experiment was to be explained by the more prosaic notion that the incorporation of the viral RNA molecule into a protective helical array of viral protein merely increases its chance of actually infecting the tobacco plant.

In that same year, A. Gierer and Schramm independently reached the same conclusion and demonstrated conclusively that the RNA *is* the infectious component of the TMV virus particle. For they were able to show that upon centrifugation of a solution of RNA molecules extracted from TMV virus particles, the infective agent sediments at the same velocity as the RNA molecules, or about ten times more slowly than the intact TMV particles. Furthermore, the infectivity of the solution is immediately abolished upon addition of traces of ribonuclease enzyme, a treatment that has no effect on the infectivity of intact TMV particles. Hence it could be concluded that the infectivity of the RNA solution does not derive from the presence of residual intact TMV particles that happen to have contaminated the RNA extract. Instead, it followed that RNA plays the role of the genetic material of TMV. Thus TMV, just like the T-even phage, can be made to pass through a stage in which its nucleic acid molecule is the only material link between successive generations. Hence RNA, no less than DNA, must be endowed with the capacity to engender its own self-replication and to give expression to the genetic information that it encodes.

That in the life of TMV RNA plays a hereditary role fully equivalent to that played by DNA in the life of the T-even phages could also be demonstrated by genetic studies. For direct chemical modification of TMV RNA by muta- gens such as nitrous acid was found to give rise to mutant virus strains. Since the complete sequences of the 158 amino acids of the normal TMV protein had been worked out meanwhile, it became possible to examine the protein of the mutant viruses for alterations of its primary structure. Such examinations revealed that, as has been described in Chapter 18, many of the chemically evoked TMV mutant strains do carry viral protein in which one amino acid had been replaced by another at some definite site of the polypeptide chain. Thus there could be no doubt that the primary structure of the TMV protein is encoded in the nucleotide sequence of the TMV RNA. But since the 6400 nucleotides of the TMV RNA molecule provide information of the polypep- tide sequence of a total of $6400/3 = 2130$ amino acids, it could be inferred that the 158-amino-acid-long protein carried by the virus particle is undoubt- edly not the only protein of which the TMV RNA has genetic knowledge.

With the discovery of the infectivity of TMV RNA in 1956, it had become clear that the secondary role relegated to RNA as a gene transcript re- quired some revision. In particular, it seemed urgent to fathom how the genetically competent viral RNA manages to go about its own replication. Unfortunately, TMV did not appear to be a very promising material for an experimental approach to this problem. First, the reproduction of TMV had to be studied in intact tobacco plants (or in excised tobacco leaves) of whose myriad cells only a tiny fraction are actually infected and support virus multiplication at any time. Hence it was extremely difficult to study the biochemical events attending the intracellular reproduction of TMV in the same way that the growth of T-even phages had been studied in massively infected cultures of *E. coli*. Second, only a tiny fraction of the TMV particles applied to a tobacco leaf actually manage to succeed in infecting the plant. Hence, in contrast with the study of T-even phages, in which nearly every virus particle can infect a sensitive *E. coli* cell, it was very difficult to follow the intracellular fate of that rare parental TMV particle that actually managed to give rise to progeny viruses.

TWO MALE-SPECIFIC PHAGES

In 1961, a new era began for the study of the mechanism of reproduction of RNA viruses, thanks to the discovery by Zinder and his student T. Loeb of a novel class of phages. It had occurred to Zinder and Loeb that it would be very useful for experiments on bacterial conjugation to have available a phage whose host range is restricted to male (i.e., F^+ or Hfr) strains of *E. coli* and which is unable to infect female (i.e., F^- strains). So they set out to isolate a *male-specific* phage from New York sewage and soon found two such phages that form plaques only on F^+ or Hfr strains (Figure 19-5). To these phages they assigned the names f1 and f2. Later studies showed that both f1 and f2 owe their predilection for male bacteria to a peculiarly fastidious feature of their adsorption process: rather than attaching directly to the surface of their host cell, the virus particles of these two phage strains fasten themselves to the *sex-pili* that male bacteria have on their outside (Chapter 11). Since female F^- bacteria lack the sex pilus, neither f1 nor f2 can infect them. Though the details of the mechanisms by which f1 and f2 penetrate their *E. coli* host still remain to be worked out, it now seems clear that the phage particles dissemble upon contact with the sex-pilus and appear to inject their nucleic acid moiety into the cell, either through the central hole of the sex-pilus or along its surface.

Study of the physical and chemical properties of f1 and f2 showed that they are two radically different phage types, neither of which had hitherto been encountered by molecular geneticists. But soon after the discovery of f1 and f2 in New York, numerous other male-specific phages came to be isolated in other parts of the world, most of which were found to bear a close resemblance to either f1 or f2. Thus such newly-found phage types as fd, M13, and F12 are close cousins of f1, whereas others, such as MS2, fr, and R27, are close cousins of f2. In the remainder of this chapter these two male-specific

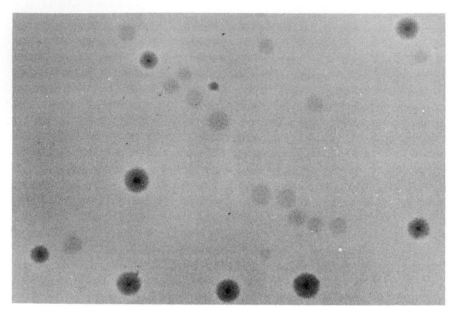

FIGURE 19-5.   Plaques of a mixture of f1 and f2 phages on a lawn of male *E. coli* carrying the F sex factor. The clear plaques are formed by the RNA phage R17 of the f2 family, which lyse their host cell after a latent period. The turbid plaques are formed by the DNA phage M13 of the f1 family, which do not lyse their host cell and are continuously secreted during bacterial growth. The intracellular multiplication of f1 phage merely reduces the rate of growth of the infected bacterium, and hence the turbid f1 plaques indicate areas of slower growth of the bacterial lawn rather than bacterial lysis. [Courtesy of David Pratt.]

phage families will be referred to collectively as f1 and f2, and it must be remembered that many of the following statements about f1 and f2 are actually based on experiments carried out on the cadet members of their clans. Phage Qβ, a historically important and slightly more distant cousin of f2, will find special mention, however.

The male-specific phage f1 turned out to be a very small phage carrying one single-stranded, circular DNA molecule only about 4100 nucleotides in length, or about 80% as long as that carried by the small phage ϕX174 (Chapter 12). In contrast to the compact spherical shape of ϕX174, however, the f1 phage was found to be a long slender rod (Figure 19-6) that contains about twice as much protein per particle as does ϕX174. The intracellular replication of the f1 DNA could be shown to proceed in a manner entirely analogous to that of the ϕX174 DNA, in that soon after its sex-pilus-mediated injection into the host cell, the single-stranded, or "plus"-strand viral DNA of f1 is converted into a double-stranded replicative form, or RF, by synthesis of a complementary "minus" polynucleotide strand. The RF then serves as template for both transcription of messenger RNA molecules and for

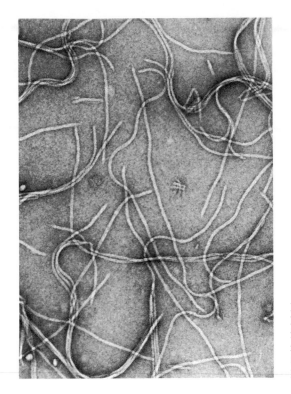

FIGURE 19-6.   The male-specific DNA phage M13 of the f1 family. The length of the rod-shaped particle is about 850 mμ. [Courtesy of David Pratt.]

symmetric or asymmetric replication to generate additional RF molecules or daughter DNA plus strands. The messenger RNA molecules are translated into the f1 phage proteins, whereas the daughter DNA plus strands are encapsulated into phage protein to serve as the genetic material of an f1 progeny phage particle. What sets f1 most strikingly apart from ϕX174, however, and indeed from all other phage types mentioned in the preceding chapters, is the manner in which its progeny particles are liberated from the infected bacteria. For f1-infected *E. coli* do not lyse at the end of the latent period but instead *secrete* mature f1 progeny particles through their cell wall. Indeed, the infected bacteria continue to grow and divide while secreting vast numbers of f1 phage particles, and the only overt manifestation of ongoing f1 multiplication in an infected *E. coli* culture is a reduction in the net rate of bacterial growth. It is one of the ironies of the history of phage research that violent controversies raged in the 1930's over whether bacteria liberage phages by lysis or by secretion and that this argument was finally settled—for good, it was then thought—in favor of the adherents of lysis by the one-step growth experiment. But as the work on f1 finally showed 30 years later, the adherents of secretion had not been all that far off the mark.

Loeb and Zinder's first examination of the other male-specific phage, f2, brought an even bigger surprise, however. This phage turned out to be a small spherical particle of about the same diameter as ϕX174, namely 25 mμ (Figure 19-7), and, like ϕX174, to contain about 70% protein by weight. But unlike ϕX174, or any other then known phage, *f2 was found to contain RNA instead of DNA as its nucleic acid.* Thus the discovery of f2 seemed like the answer to a molecular geneticist's prayer, for the methodology previously developed for the study of the reproduction of DNA phages could be applied

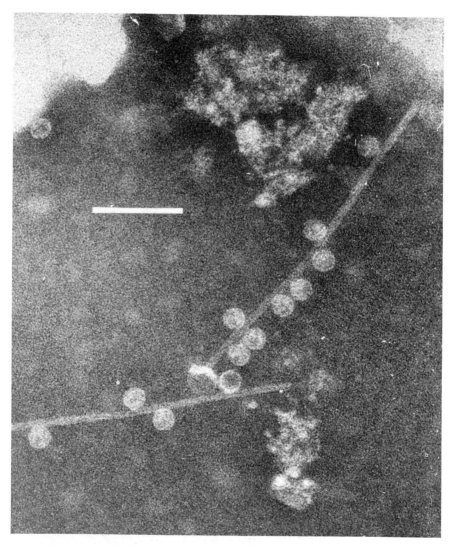

FIGURE 19-7.   Spherical particles of the male-specific phage f2, adsorbed to the sex-pili of an Hfr strain of *E. coli*. The length of the white bar corresponds to 100 mμ, or to the length of a T-even phage head. [Electron micrograph taken by R. C. Valentine and Germaine Cohen-Bazire.]

to the vexing problem of how RNA manages to act as genetic material. This opportunity was immediately grasped by Zinder, who did not even bother to use the male-specific phages for the purpose he had originally set out to isolate them, and instead devoted all his efforts to the chemistry and biology of f2 phage. Dozens of other workers soon followed his lead, and within the next few years the f2 phage became one of the most thoroughly understood of all viruses.

THE f2 PHAGE

Physical and chemical studies of the f2 phage particle showed that it contains one single-stranded (noncircular) RNA molecule about 3300 nucleotides in length. (Thus the f2 RNA carries only about half as much genetic information as the TMV RNA.) The nucleotide base composition of f2 phage RNA is $[A] = 0.23$; $[G] = 0.26$; $[U] = 0.26$; and $[C] = 0.25$. The f2 phage protein represents a spherical assemblage of 180 identical protein molecules, each molecule representing a polypeptide chain of 129 amino acids. Analysis of the total amino acid sequence of the protein of f2 and its cousins MS2 (isolated in California) and fr (isolated in Germany) revealed that the MS2 protein differs from that of f2 only at polypeptide site 88, where the protein f2 contains leucine and the MS2 protein methionine, whereas the fr protein differs from that of f2 at 21 of the 129 polypeptide sites.

In addition to these 180 identical protein subunits, each f2 particle contains one molecule of a qualitatively different polypeptide chain composed of about 320 amino acids. This minor protein molecule appears to play an essential role in the reaction by means of which f2 invades its host cell, for otherwise intact phage particles lacking it do not adsorb to the sex-pilus. The 129 amino acid, major phage protein, and the 320-amino acid minor phage protein together require about 1400 of the 3300 nucleotides of the f2 RNA for specification of their primary structure, thus allowing for the coding of at most one or two additional polypeptide chains by the genetic material of the phage.

Following adsorption to the sex-pilus of its *E. coli* host cell and injection of its RNA molecule, the parental f2 phage begins its intracellular growth. After a latent period of about 50 minutes, the first infected bacteria lyse and liberate a crop of 1000 to 10,000 infective f2 progeny particles per cell, depending on the conditions of growth. Study of the biochemical events that take place in f2-infected *E. coli* showed that, in contrast with the T-even phages, intracellular phage multiplication does not interfere with the synthesis of host DNA, host RNA, and host protein until nearly the end of the latent period. This finding made it important to ascertain the extent to which such host-directed synthesis might play any role in f2 phage reproduction. In particular, it was important to learn whether, by any chance, the genetic function of the viral RNA depends upon a "reverse-transcription" of its genetic information to an intracellular DNA species, which then presides over the synthesis of f2 progeny RNA and protein, in accord with the notions worked out for the heterocatalytic function of DNA. These possibilities, however, were soon eliminated from consideration when it was shown that f2 phage multiplication proceeds more or less normally in bacteria treated with inhibitors of either DNA replication or DNA transcription into RNA. It was thus shown that the viral RNA is a fully autonomous genetic material that does not require the assistance of DNA template chains at any stage of its twin hereditary functions.

Just as it had been seen earlier that *E. coli* whose cell envelope has been rendered more permeable by partial lysozyme digestion can be infected with DNA molecules extracted from ϕX174 phages, so it was found that similarly treated *E. coli* can also be infected with the RNA molecule extracted from f2 phages. Upon uptake of the pure phage RNA from the culture fluid by the permeable bacteria, a normal crop of mature progeny virus particles is produced. Once introduced in this direct manner into its host cell, f2 RNA also can initiate phage growth in normally f2-resistant F$^-$ female bacteria, showing that the "male-specificity" of f2 devolves entirely from the essential role of the sex-pilus in the invasion process. That is, the presence of the F sex factor is not required for the intracellular phage growth.

RNA REPLICATION

Knowledge of the mechanism of replication of the single-stranded DNA molecule of ϕX174 phage inspired students of RNA phages to search f2-infected cells for the presence of a double-stranded replicative form of the phage RNA, analogous to the double-stranded RF DNA. For this purpose, bacteria were infected with parental f2 phages whose RNA carried a radioactive label. Total RNA was then extracted from the infected bacteria at various stages of intracellular phage growth and the properties of the radioactively labeled parental phage RNA present in such extracts were examined. These studies showed that the parental phage RNA in extracts made at the very earliest stages of infection resembled the coiled, single-stranded RNA recoverable from the f2 phage particle, in that it is very sensitive to digestion by ribonuclease and that upon centrifugation it sediments with a velocity of 27S. In extracts made a few minutes later, however, the labeled parental RNA can be seen to have been converted into a double-stranded form that is resistant to ribonuclease digestion and that, because of its stiff helical conformation, sediments at the slower velocity of 14S. In extracts made later than 10 minutes after the onset of intracellular phage growth, the labeled parental RNA is found in a partially double-stranded form that is partially sensitive to ribonuclease digestion and that before digestion sediments at

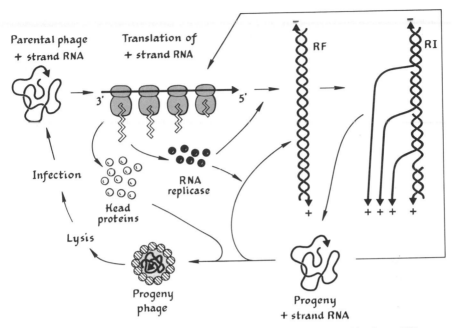

FIGURE 19-8.   General schema of intracellular reproduction of an RNA phage. RF =
double-stranded replicative form consisting of a plus (+) strand RNA and a minus (−)
strand RNA assembled by action of the phage RNA replicase enzyme on the plus strand
as template. RI = replicative intermediate consisting of a minus strand RNA and
several nascent plus strands being assembled in tandem on the minus strand as template.

velocities up to 30S and after digestion sediments at the slower velocity of
14S. On the basis of these results, the following image of the replication of
the f2 RNA was developed (Figure 19-8), which resembles in its essential
features the mechanism of DNA replication of ϕX174. At the outset of its
replication, the single parental RNA, or "plus" (+), strand serves as template for the synthesis of a complementary "minus" (−) RNA strand to
produce a double-stranded RNA RF. Next, the newly formed minus strand
of the RF serves as template for the synthesis of a complementary daughter
RNA plus strand. But long before growth of the first daughter plus strand
is even complete, synthesis of a second and then of third and fourth daughter
plus strands commences upon the RF, converting it into a new form called
"replicative intermediate," or RI. In this RI, only the most recently formed
part of each nascent plus strand is hydrogen-bonded to its complementary
part of the minus template strand, the remainder having been expelled from
the double helix by growth of the next plus strand in its wake. Thus the RI
consists of a ribonuclease resistant "backbone" of double-stranded RNA and
of several single-stranded, lateral RNA "tails." This structure, being much
heavier, sediments more rapidly than RF, but after digestion of its

ribonuclease-sensitive single-stranded RNA "tails," the RI is reduced to a form closely resembling the RF. Once growth of a nascent daughter plus strand and expulsion from the RI is complete, the plus strand has the option of either serving as template for synthesis of a complementary minus strand and generation of a new RF or of being encapsulated into phage coat protein and thus gaining the status of genome of a progeny phage particle. The former of these options predominates during the first 15 minutes after infection, during which time there proceeds a gradual accumulation of RF's and RI's within the host cell, and hence a gradual increase in the number of structures competent to synthesize progeny RNA molecules. But later stages of the latent period, when the intracellular concentration of phage coat protein subunits has attained a high level, the latter of these options comes to predominate. The number of RF's and RI's (and hence of minus strands) per cell has by this time reached its final level, and the number of intracellular mature progeny phages steadily increases until lysis brings the reproductive processes to term.

In order to ascertain whether the RNA replication process outlined in Figure 19-8 is catalyzed by a special enzyme system, extracts of f2 phage-infected *E. coli* were tested for their capacity to carry out the synthesis of phage RNA. These experiments soon showed that within a few minutes after the outset of intracellular f2 phage growth there does appear a phage-specific RNA *replicase* within the infected cells which upon its extraction can effect some *in vitro* synthesis of both plus and minus strands of phage RNA in reaction mixtures containing phage RNA template molecules and the four ribonucleoside triphosphates ATP, GTP, CTP, and UTP. Thus it appeared that part or all of the 1700 nucleotides of the f2 RNA that are not used for specifying the amino acid sequence of the two phage proteins encode the structure of the RNA replicase, whose formation proceeds at the earliest stages of phage development. Eventually, Sol Spiegelman and his collaborators found that an RNA replicase enzyme complex can also be isolated from extracts of *E. coli* cells infected with the RNA phage Qβ. Though the general structure and life cycle of Qβ is rather similar to those of f2, the properties of both protein and RNA of Qβ show substantial departures from those of the other members of the f2 family. And, as Spiegelman was able to show in 1965, the purified Qβ replicase is not only capable of synthesizing a few phage RNA plus and minus strands in the *in vitro* reaction mixture but polymerizes the ATP, GTP, CTP, and UTP substrate molecules added to the reaction mixture into *infectious* RNA molecules which, upon gaining entrance to an *E. coli* host cell give rise to a brood of intact, infectious Qβ phages. The result of such an *in vitro* RNA replication experiment is shown in Figure 19-9, where it can be seen that during a 4-hour incubation period of a reaction mixture initially containing Qβ RNA plus strands there has been a nearly 100-fold increase in the total amount of RNA in the system, while at the same time the number of RNA molecules

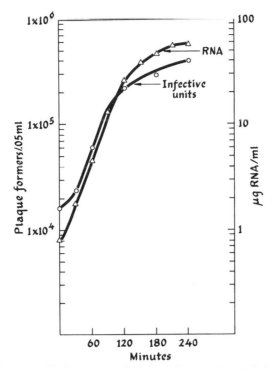

FIGURE 19-9.   Kinetics of *in vitro* phage RNA replication. The right-hand ordinate shows the total concentration of RNA present at various times in a reaction mixture after 0.8 μg/ml of plus strand RNA extracted from Qβ phage particles was added to 160 μg/ml of an RNA-free, replicase-enriched protein extract of *E. coli* infected with Qβ phage. The reaction mixture also contained 26 μg/ml each of the four replicase substrates ATP, GTP, CTP, and UTP. The left-hand ordinate shows the concentration of phage RNA molecules that are capable of infecting lysozyme-treated *E. coli* host cells and giving rise to a phage plaque upon plating the infected host cells on agar seeded with sensitive indicator bacteria. Since both ordinates represent logarithmic scales, the straight-line portions of both curves reflect an exponential growth process. This implies that replica Qβ RNA molecules synthesized in the reaction mixture can serve as templates in their own right for the assembly of further replica RNA molecules in accord with the general kinetics of exponential growth described in equations 2-3 and 2-4. The replication process can be seen to deviate from exponential kinetics only after a considerable fraction of the added nucleotide substrates have been converted into replica RNA chains and the added replicase enzyme approaches saturation by these chains. [After S. Spiegelman, I. Haruna, I. B. Holland, G. Beaudreau, and D. Mills. *Proc. Natl. Acad. Sci. Wash.* **54**, 919 (1965).]

capable of initiating phage growth upon presentation to lyzozyme-treated *E. coli* has risen 20-fold. Subsequent chemical analyses of the RNA products present in this reaction mixture at various stages have shown that here too the replication proceeds through the formation of double-stranded, complementary RF and RI structures. Thus with this development it really began to appear that a "living molecule" had finally been put together *in vitro* from nonliving constituents, albeit by use of "living" RNA plus-strand templates and replicases.

PHAGE PROTEIN SYNTHESIS

The demonstration of the formation of RF and RI described in the preceding section, in which *E. coli* cells were infected with f2 phages carrying a radioactive label in their RNA, had required the preparation of *protein-free* RNA extracts from the infected bacteria. But sucrose density-gradient sedimentation of *crude* extracts of such infected bacteria still containing intact ribosomes revealed that the single-stranded parental phage RNA recovered at the earliest stages of intracellular phage growth is, in fact, not "free" but is present in association with *polyribosomes*. That is to say, the very first act of the parental phage RNA is to engage the host cell ribosomes and to serve as a messenger RNA, so as to allow immediate translation of its encoded polypeptide chains. This stands to reason, of course, for since the phage-induced RNA replicase is required for the formation of RF and RI, the heterocatalytic function of the parental RNA must necessarily precede its autocatalytic function in the infected cell. Thus the parental plus strand can serve as template for the synthesis of the first minus strand, and hence enter its first RF, only after some ribosomes have already translated its replicase gene. At later stages of intracellular growth, the nascent daughter plus strands, once expelled from the RI on which they arose, have an additional option, besides the two mentioned in the preceding section: they can also engage host cell ribosomes and serve as messengers for the translation of phage proteins, according to the schema presented in Figure 19-8. Thus a simplified version of the heterocatalytic DNA function obtains in the life of RNA viruses like f2, in which the transcription step of that function does not exist; here, the genetic material serves *directly* as the messenger in the translation of its polypeptide sequence information.

A direct proof of the capacity of the f2 RNA plus strand to act as a messenger was adduced when Zinder and his collaborators added the RNA extracted from mature virus particles to a reaction mixture capable of *in vitro* protein synthesis, of the type developed by Nirenberg in connection with his code-breaking experiments (Chapter 18). They found, first of all, that the presence of the phage RNA stimulates the polymerization of the radioactively labeled amino acids of the reaction mixture into polypeptide chains. More importantly, however, they showed that a large fraction of these polypeptide chains have the same primary structure as the f2 phage coat protein. This experiment, done within a year of Nirenberg's discovery of the coding properties of synthetic polyribonucleotides, was the first demonstration of the *in vitro* synthesis of a natural protein by a natural messenger RNA.

Later experiments showed also that the tertiary structure of the polypeptides synthesized in this manner must resemble that of the phage coat protein, in that these artificial coat proteins have the capacity to associate with the phage RNA present in the reaction mixture.

Detailed study of the artificial f2 coat protein synthesized *in vitro* under the direction of the f2 RNA plus strand revealed, however, a slight, albeit important difference between it and the natural coat protein: whereas the coat protein isolated from mature f2 particles carries alanine as its amino-terminal amino acid, the artificial coat protein carries formylmethionine (followed by alanine) at its amino terminus. This discrepancy led to the discovery (described in Chapter 17) that in *E. coli* growth of most (or all) peptide chains is initiated by laying down formylmethionine at the amino-terminus in response to the chain-initiation codon AUG. Evidently Nirenberg's original *in vitro* protein synthesizing reaction mixture contains the necessary ingredients for carrying out the natural process of peptide chain initiation but lacks the enzymatic apparatus present in the f2-infected *E. coli* cell that splits off the chain-initial formylmethionine residue from the nascent polypeptide and exposes the alanine residue next in line as the amino-terminus of the mature coat protein.

RNA-PHAGE GENETICS

The biochemical work on the life cycle of the f2 phage family has been effectively complemented by the isolation and study of phage mutants. These mutants were mainly of the same two conditional lethal types by means of which the circular map of the T-even phage was charted: (a) temperature-sensitive (or "missense") mutants, which do not grow at a high, or restrictive, temperature at which the wild-type phage is able to grow, and which do grow at a low, or permissive, temperature; and (b) amber (or "nonsense") mutants, which can grow only on host strains which themselves carry a suppressor mutation that engenders insertion of an acceptable amino acid into the growing peptide chain in response to the mutant nonsense UAG (and UAA and UGA) codons. These phage mutants arise spontaneously or can be induced by base analogs during intracellular f2 phage growth, thanks to copy errors either upon synthesis of the RNA minus strand in the formation of RF or upon synthesis of the RNA plus strand by RI. Mutants can also be induced by treating the mature f2 virus particles, or the infective RNA extracted from them, with chemical mutagens —for instance with nitrous acid.

On the basis of the phenotype they manifest under the restrictive growth conditions, these mutants can be placed in three distinct groups. Group I fails to synthesize any minus strands and hence gives rise to neither RF nor RI. These mutants can be inferred to carry their mutation in the gene that codes for the primary structure of the phage RNA replicase. Group II, in contrast, not only manifests normal RNA replication under restrictive conditions but even gives rise to structurally intact, albeit noninfective progeny particles. These mutants have been shown to carry their mutation in the gene that codes for the primary structure of that 320-amino-acid polypeptide of which each f2 phage particle carries one molecule and whose function appears to be required for invasion of the host cell. Group III, finally, fails to form normal coat protein, since its mutants carry their mutation in the coat-protein gene. Furthermore, under restrictive conditions Group III mutants synthesize abnormally high amounts of RF and RI, since phage RNA plus strands produced in the infected cell are not encapsulated by viral phage protein and are ever free to offer themselves as templates for the synthesis of new minus strands. Complementation experiments in which bacteria were infected jointly with two different f2 mutants under restrictive conditions have shown that the three phenotypic groups also constitute three unambiguous complementation groups: mixed infection of bacteria with any two phage mutants belonging to different phenotypic groups or to the same phenotypic group leads to normal or to abnormal phage growth respectively. These results lead to the belief that the nucleotide sequence of the f2 phage RNA comprises no more than three genes. It should be noted that none of the mixed infection experiments revealed the appearance of any genetic recombinants of f2 phages. The deeper meaning of this negative result is, however, unclear, since most stocks of f2 mutants contain about 0.1 % wild type revertants. Thus the high mutation rate of the RNA genome renders difficult the search for rare genetic recombinants. It is also possible, of course, that genetic recombination of RNA genomes does not occur at all and that this process is reserved to poly*deoxy*ribonucleotides.

Bibliography

ORIGINAL RESEARCH PAPERS

Fraenkel-Conrat, H. The role of the nucleic acid in the reconstitution of active tobacco mosaic virus. *J. Am. Chem. Soc.*, **78**, 882 (1956).
Gierer, A., and G. Schramm. Infectivity of ribonuclcic acid from tobacco mosaic virus. *Nature*, **177**, 702 (1956).
Loeb, T., and N. D. Zinder. A bacteriophage containing RNA. *Proc. Natl. Acad. Sci. Wash.*, **47**, 282 (1961).
Spiegelman, S., I. Haruna, I. B. Holland, C. Beaudreau, and D. R. Mills. The synthesis of a self-propagating and infectious nucleic acid with a purified enzyme. *Proc. Natl. Acad. Sci. Wash.*, **54**, 919 (1965).
Stanley, W. M. Isolation of a crystalline protein possessing the properties of tobacco mosaic virus. *Science*, **81**, 644 (1935).

20. Regulation of Gene Function

An *E. coli* bacterium contains a total of about 10^7 protein molecules, and, as was estimated in Chapter 8, the *E. coli* DNA encodes the primary structure of about 3000 different polypeptide chains. Now if all of the corresponding 3000 genes were given equal service in the heterocatalytic function of DNA, and hence if all possible protein species were synthesised in equal amounts, then each *E. coli* bacterium would contain about $10^7/3000 = 3000$ copies of every protein molecule of which it has genetic knowledge. But detailed analyses of the relative intracellular abundance of various *E. coli* proteins have shown that some of them weigh in with as few as 10 molecules per cell, whereas others weigh in with as many as 500,000 molecules per cell. The ribosomal proteins are somewhere in the (logarithmic) upper part of this nearly 100,000-fold range, in that there are present about 10,000 molecules per cell of each species of ribosomal protein. Thus it follows that *E. coli* certainly does *not* synthesize all the polypeptide chains encoded in its DNA in equal amounts. This conclusion stands to reason, since in the life of the cell some proteins are obviously needed in much greater quantities than others. Indeed, for harmonious and economic function of the cellular ensemble of enzymes it is absolutely essential that each gene produces neither more nor less than the "optimal" amount of its product. This chapter will set forth the insights gained by molecular geneticists as to how this vital regulation of gene function is achieved.

ENZYME INDUCTION

It was known by the turn of this century that the enzymatic properties of microbes depend on the medium on which they have grown, in that microbes can be "trained" for, or *adapted* to, growth in a variety of different environments. Thus in 1900, F. Dienert found that yeast contains the enzymes of galactose metabolism when it depends on lactose or galactose for its carbon and energy source, and that these enzymes are lost when the yeast is transferred to a glucose substrate, where they are not required for growth. In the 1930's, H. Karström studied the formation of several enzymes of carbohydrate metabolism in bacteria and divided them into two classes: *adaptive* enzymes, which are formed only in the presence of their substrates in the medium, and *constitutive* enzymes, which are formed regardless of the nature of the medium. In 1938, J. Yudkin published a review of enzyme adaptation among microbes in which he proposed a simple, mass-action model of adaptation. He suggested, in line with some then-current notions of the mechanism of enzyme synthesis, that enzymes are formed from enzyme precursors, with which they exist in equilibrium in the cell. Yudkin supposed, furthermore, that for constitutive enzymes, the equilibrium favors the enzyme, E, and that for adaptive enzymes, the equilibrium favors the precursor, P. Combination of the adaptive enzyme with its substrate, S, to form the enzyme-substrate complex, E-S, however, displaces the equilibrium to lead to enzyme accumulation. Thus

$$P \underset{k_2}{\overset{k_1}{\rightleftharpoons}} E \xrightarrow[S]{} E\text{--}S$$

Adaptive enzymes: $k_2 \gg k_1$

Constitutive enzymes: $k_2 \ll k_1$

$$(20\text{-}1)$$

Yudkin's review opened the modern era of the study of adaptive enzymes, even though the theory he proposed later turned out to be wrong.

One of the enzymes of *E. coli* that had been recognized as belonging to the adaptive class is β-galactosidase. This enzyme, as was mentioned in Chapter 3, catalyzes the hydrolysis of its "natural" substrate lactose (Figure 20-1) as well as of other β-galactoside compounds. Thus *E. coli* was known to possess β-galactosidase activity only when growing in a medium containing lactose as the carbon and energy source, and to lack this enzyme when growing in media in which a natural sugar other than lactose is provided. In 1946, Jacques Monod (Figure 20-2) began a study of the adaptive formation of the *E. coli* β-galactosidase, which, in the course of the next 15 years,

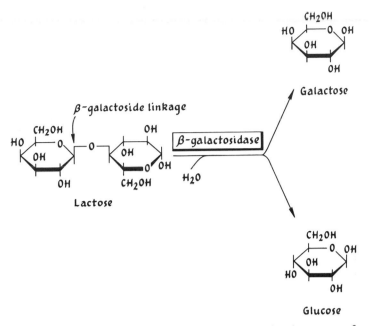

FIGURE 20-1. The hydrolysis of lactose, catalyzed by the enzyme β-galactosidase.

FIGURE 20-2. Jacques Monod (b. 1910).

was to provide the answer to the problem of the regulation of bacterial gene function. In the first phase of this study, the work of Monod and his collaborators brought into focus and defined the true nature of enzyme adaptation. As a result of this sharpened focus, a 1953 manifesto signed by its then leading students rechristened enzyme adaptation "enzyme induction," and defined as *inducers* those compounds (e.g., lactose) to whose presence a cell responds with the formation of an enzyme.

Study of the nature of inducers to whose presence *E. coli* responds with the formation of β-galactosidase showed that some β-galactosides are inducers without being substrates of the enzyme, whereas others are substrates without being inducers. This finding led to the conclusion that action of the enzyme on the inducer is neither necessary nor sufficient to induce enzyme synthesis. Indeed, this first suggested (although it did not rigorously prove) that, contrary to Yudkin's proposal, the enzyme is not the target of its inducer in the induction process. The most important practical consequence of these studies, however, was the discovery that nonmetabolizable sulfur analogs of ordinary β-galactosides, such as methyl- and isopropylthiogalactoside, are inducers of high potency (Figure 20-3). These sulfur analogs made possible a meaningful study of the true kinetics of the induction process under *gratuitous* conditions in media that contain a carbon and energy source other than lactose, e.g., glycerol, in which the rate of cell growth does not depend on the intracellular concentration of β-galactosidase and in which the inducer concentration remains invariant. By means of such studies it was found that under gratuitious conditions the cells form β-galactosidase at a constant and maximal rate, p, within a few minutes after a saturating amount of inducer has been added to the growth medium. That is, $\Delta Z/\Delta B = p$, where Z is the amount of enzyme and B the bacterial protein in the culture, and ΔZ and ΔB the increments in these amounts formed during any time interval. As soon as the inducer is removed from the medium, p falls to a value less than a thousandth of that reached when a maximal inducer concentration is present. Thus it became apparent that during bacterial growth, a constant

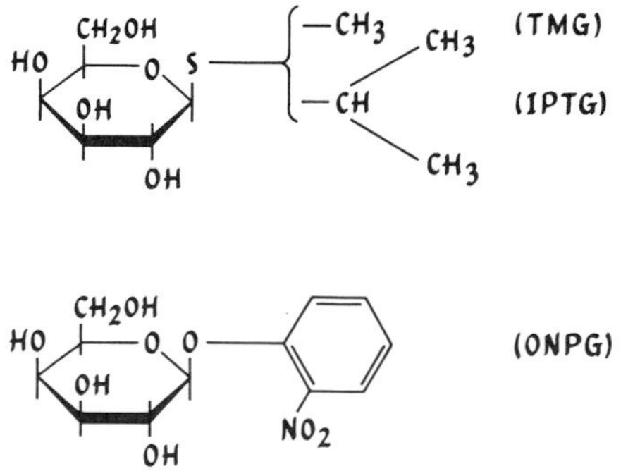

FIGURE 20-3.   Structure of two nonmetabolizable, or gratuitous, inducers of *E. coli* β-galactosidase: thiomethyl galactoside (TMG) and isopropyl thiogalactoside (IPTG), and of the chromogenic enzyme substrate o-nitrophenyl galactoside (ONPG) used in the enzyme assay.

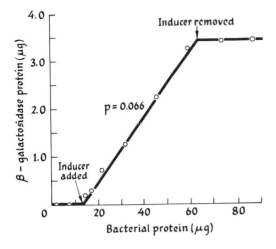

FIGURE 20-4. Induction kinetics of β-galactosidase in *E. coli* under gratuitous conditions. The parameter *p* expresses the ratio of the increments in the mass of enzyme and of total protein in the culture. [After J. Monod, A.M. Pappenheimer, Jr., and G. Cohen-Bazire, *Biochim. Biophys. Acta* **9**, 648 (1952).]

proportion *p* of new protoplasm is β-galactosidase protein, the value *p* depending on the inducer concentration in the growth medium (Figure 20-4). Once purification, isolation and physicochemical study of β-galactosidase had shown it to be a rather large protein, consisting of four identical polypeptide chains each 1173 amino acids in length, and once its specific enzymatic activity per milligram of protein was established, *Z* and *B* could be expressed in the same units, and *p* became the fraction of total cellular protein synthesis devoted to the manufacture of β-galactosidase. As seen in Figure 20-4, *p* has the value of 0.066 in the presence of maximal inducer concentration; that is, in maximally induced cells, 6.6 per cent of all protein manufactured consists of β-galactosidase.

The induction kinetics seen in Figure 20-4 strongly suggested that the induced appearance of β-galactosidase is the result of *de novo* enzyme synthesis, rather than of substrate-induced conversion of any pre-existing intracellular enzyme precursor. To prove this point, an earlier isotope tracer experiment—one that had finally ended precursor-conversion notions in the domain of phage multiplication—was adapted to the problem of enzyme induction. For this purpose, *E. coli* cells were first grown for several generations in a medium supplemented with radiosulfur ^{35}S and devoid of any galactoside inducer. The radioactive bacteria were then transferred to a nonradioactive medium lacking ^{35}S, and β-galactosidase formation was initiated by adding an inducer to the culture. Isolation and purification of the β-galactosidase from these bacteria revealed that the enzyme formed under this labeling program contained no [35]S. Hence the enzyme cannot have been derived from a protein that existed in the cell before addition of the inducer, because the sulfurylated amino acids—cysteine and methionine—in the putative precursor protein would have contained [35]S. Thus, instead of causing the conversion of a ready-made precursor, the inducer appears to instruct the cell to start assembling the β-galactosidase poly-peptide directly from amino acid building blocks. It now seemed clear that the explanation of this instruction mechanism should provide important clues for the understanding of protein synthesis.

Not long after it was found that induced formation of β-galactosidase is in fact a *de novo* protein synthesis, it was discovered that the activity of this enzyme, although necessary, is not sufficient for utilization of lactose by intact *E. coli* cells. A second enzymatic activity, quite distinct from hydrolysis of the galactoside bond, was found to exist. This activity allows the lactose to enter the cell. Formation of the enzyme responsible for this activity, which was called *galactoside-permease*, is also inducible, because the presence of galactoside inducers is required for its synthesis. Thus *E. coli* grown in the absence of lactose are not only unable to ferment it, but until synthesis of an adequate permease level has been induced, they do not even take up lactose from the medium. Some years after the discovery of the permease, a third distinct enzymatic activity pertaining to lactose metabolism in *E. coli* was identified. The responsible enzyme, *galactoside transacetylase*, catalyzes the transfer of an acetyl group from an acetyl donor molecule to the galac-tose moiety of β-galactosides. As in the synthesis of galactosidase and per-mease, synthesis of the transacetylase enzyme proceeds only in the presence of β-galactoside inducers. What physiological role, if any, is actually played by the transacetylase is not known, except that its presence is *not* required for apparently normal lactose fermentation.

GENETIC CONTROL OF ENZYME FORMATION

Not long after Monod began his studies on the physiology of lactose fer-mentation in *E. coli*, J. Lederberg started to work out its genetic basis. By 1948 Lederberg had isolated a collection of lactose-negative (Lac⁻) mutants of *E. coli*, which are unable to ferment lactose. Further work on these mutants showed that some of them owe their Lac⁻ character to an inability to produce β-galactosidase. [In order to test for the presence of the enzyme in these mutants, Lederberg had invented the assay method that makes use of the colorless lactose analog o-nitrophenyl-β-galactoside, or ONPG (Figure 20-3), as substrate for β-galactosidase. As had been previously explained in connection with the experiment of Figure 4-1, hydrolysis of ONPG yields the intensely yellow o-nitrophenol, whose color attests to catalytic activity of the enzyme.] In succeeding years more and more Lac⁻ mutants were isolated, characterized, and their mutant sites located on the bacterial genetic map by conjugation and transduction experiments. The physiological basis of some of these Lac⁻ mutants appeared to be compli-cated, whereas that of others was rather simple. The simplest were of two types: LacZ⁻,Y⁺,A⁺ mutants that lack galactosidase (Z) but possess permease (Y) and transacetylase (A); and LacZ⁺,Y⁻,A⁺ mutants that lack permease but possess galactosidase and transacetylase. (A LacZ⁺,Y⁻ mutant lacking permease is called *cryptic*, because its capacity to hydrolyze lactose is not shown by the intact cell and becomes manifest only in a cell extract.) It was also possible to find LacA⁻ mutants, even though mutants lacking trans-acetylase are not of the Lac⁻ type and hence cannot be detected on EMB-lactose indicator agar. Study of all these mutant types made it possible to identify three closely linked and probably contiguous *lac* genes on the *E. coli* chromosome (Figure 20-5; see also Figure 10-27): the galactosidase gene *lacZ*, the permease gene *lacY*, and the acetylase gene *lacA*, each of which specifies the primary structure of an enzyme protein.

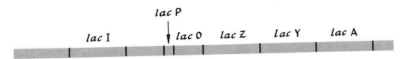

FIGURE 20-5.   Relative order of the *E. coli* genes coding for enzymes of lactose fermentation and of the genetic sites that participate in the control of synthesis of these enzymes.

Among Lederberg's mutants there also was one in which formation of galactosidase did not depend on the presence of inducer: this *constitutive mutant* synthesized the enzyme in the absence of lactose or other galactosides. Such constitutive mutants are now designated as LacI⁻, in contradistinction to the inducible LacI⁺ wild type. Later studies showed that LacI⁻ mutants are constitutive for synthesis of their permease and transacetylase as well as for their galactosidase—that is, the mutation from LacI⁺ to LacI⁻ has altered the functional regulation of all three *lacZ*, *lacY* and *lacA* genes. Genetic mapping of various LacI⁻ mutations showed that they pertain to a separate gene, the *lacI* regulator gene, which is closely linked to, but distinct from, the other three *lac* genes (Figures 20-5 and 10-27).

THE REPRESSOR

How is it that constitutive LacI⁻ mutants synthesize the enzymes of their *lac* genes in the absence of any exogenous inducer? One obvious possibility seemed to be that the mutation from LacI⁺ to LacI⁻ leads to the formation of an *internal inducer* that allows the cells to synthesize their *lac* gene enzymes without externally added galactosides. Search for such internal inducers in constitutive mutants was, however, without success. When, in the late 1950's, the work of Wollman, Jacob, and Hayes (recounted in Chapter 10) revealed that in bacterial conjugation the DNA of a male Hfr donor bacterium migrates into the female recipient cell to form a transient merozygote endowed with the female cytoplasm, an experiment became possible whose result led to the abandonment of the internal inducer idea. In this experiment, A. B. Pardee, Jacob, and Monod mated normal LacI$^+$,Z$^+$ Hfr donor bacteria grown in the absence of any inducer to LacI$^-$,Z$^-$ F$^-$ recipients in an inducer-free medium, and followed the appearance of galactosidase activity in the culture. The result of this experiment is shown in Figure 20-6. Before mating, there is no galactosidase in the culture, since the inducible LacI$^+$,Z$^+$ donor bacteria have not been exposed to inducer, and the constitutive LacI$^-$,Z$^-$ recipient bacteria cannot form the enzyme at all. After about one hour of conjugation, however, by which time the DNA bearing the normal *lac*I,Z genes of the Hfr donor bacteria has entered the F$^-$ cytoplasm, galactosidase synthesis begins and proceeds for about an hour, after which time it stops. This experiment showed that the merozygote formed upon entrance of the *lac*I,Z donor genes into the LacI$^-$ recipient cytoplasm acts at first as a constitutive cell, since the donor *lac*Z gene forms galactosidase in the absence of inducer. But as soon as the donor *lac*I gene has had an opportunity to express itself, constitutive function of the *lac*Z gene ceases, and the zygote is converted to the inducible state, which requires the presence of inducer for further galactosidase synthesis. Hence the *lac*I gene of the inducible LacI$^+$ wild type appears to give rise to a cytoplasmic product, the *repressor*, which exerts a negative effect on *lac* gene enzyme synthesis, or, in the words of Pardee, Jacob, and Monod "inhibits information transfer from structural gene (or genes) to protein." Galactoside inducers then exert their effect by neutralization, or inactivation, of this repressor, and LacI$^-$ constitutive mutants can dispense with external inducers, not because they possess an internal inducer, but because they lack an active repressor.

Relatively little progress was made in identifying the *lac*I gene product, or repressor, for many years following the first postulation of its existence, although some further information about its nature did emerge from the ensemble of known *lac*I gene-mutant phenotypes. Both temperature-sensitive and nonsense-codon mutations in the *lac*I gene were found—that is, mutants whose character is constitutive (LacI$^-$) at temperatures higher than 37°C or in the absence of a nonsense-codon suppressor gene in their genetic background and whose phenotype is inducible (LacI$^+$) at low temperatures or in the presence of a nonsense-codon suppressor gene. It seemed most plausible, therefore, that the *lac*I gene product is a protein. Moreover, because other *lac*I gene mutants, called LacI^s, were found that need much higher inducer concentrations for enzyme induction than does the LacI$^+$ wild type (or cannot be induced at all, and hence appear as Lac$^-$ phenotypes), and because the heat stability of the repressor in yet other *lac*I gene mutants is raised in the presence of inducer, it could reasonably be inferred that the *lac*I gene product, as originally postulated, does interact with the inducer.

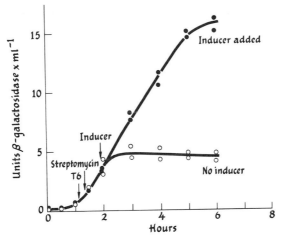

FIGURE 20-6.   Appearance of β-galactosidase in transient merozygotes formed by the conjugation of LacI$^+$, Z$^+$, Str^s, Tsx^s Hfr donor *E. coli* with LacI$^-$, Z$^-$, Str^r, Tsx^r F$^-$ recipient *E. coli*, both grown in the *absence* of inducer before and during conjugation. After enough time had passed for transfer of the *lac* genes from donor to recipient, the Hfr donor cells were killed selectively by addition of T6 phage and streptomycin to the mating mixture (see arrows). In the main experiment no inducer was ever added to the mating mixture (open circles). To a control sample of the mating mixture (filled circles), inducer was added after the Hfr cells had been killed. The continued synthesis of β-galactosidase in this control sample after the third hour shows that the eventual arrest of β-galactosidase synthesis in the main experiment is not attributable to an *intrinsic* incapacity of the merozygotes to continue to form β-galactosidase. This result thus proves that by the third hour, the merozygotes have changed from the constitutive to the inducible state. [After A. B. Pardee, F. Jacob, J. Monod, *J. Mol. Biol.* **1**, 165 (1959).]

Finally, in 1967 W. Gilbert and B. Müller-Hill succeeded in identifying the repressor product of the *lac*I gene. For this purpose, they examined concentrated extracts of *E. coli* for a protein capable of binding radioactively labeled isopropylthiogalactoside, or IPTG (see Figure 20-3). This examination revealed the presence of a tiny amount of such a protein, corresponding to a binding capacity of only about 20 to 40 molecules of IPTG per cell. By monitoring the IPTG binding power of various fractions of the extract, this protein could be purified about 100-fold over its initial concentration, and by measuring its sedimentation rate in a centrifugal field this protein was found to correspond to a molecule containing about 1000 amino acids. Further study showed the protein to consist of four subunits, each of which is a polypeptide chain containing about 250 amino acids and capable of binding one molecule of IPTG. Hence there are present about 5 to 10 molecules of this protein per *E. coli* cell. That this IPTG-binding protein is, in fact, the repressor product of the *lac*I gene, and not some other protein that happens to bind β-galactosides, was indicated by the observation that this protein cannot be detected in LacI⁻ constitutive mutants.

OPERATOR AND OPERON

Granted that the *lac*I gene repressor inhibits information transfer unless neutralized by a galactoside inducer, what, then, is the target in the cell toward which the repressor directs its negative role? In reply to this question, Jacob and Monod made a most important proposal: they declared that the three closely linked genes *lac*Z, Y, and A, which are subject to coordinate control by the *lac*I gene repressor, form an *operon*, in that these genes share a second gene of regulation, their operator, or *lac*O. The operator, whose genetic site is at one extremity of the operon linkage group (Figure 20-5), is the target of the repressor. The operator can exist in either of two states, open or closed. It is open as long as it is free of repressor, and it is closed as soon as it has combined with the repressor (the repressor, for its part, can combine with the operator only if it has not interacted with inducer). Closing of the operator prevents transcription of the messenger RNA of all genes of the operon, and, *a fortiori*, synthesis of any of the corresponding enzyme proteins. Figure 20-7 presents a schematic summary of these notions.

Taking the protein nature of the repressor for granted, long before it had been proven, Jacob and Monod then showed how the interaction with the inducer might affect its interaction with the operator. They proposed that the repressor is an allosteric protein that possesses two specific sites. One of these sites is affined to all, or part, of the exact nucleotide sequence of the operator gene cognate to the repressor, and the other site is affined to the

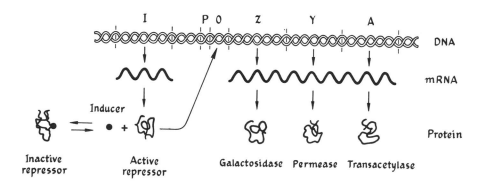

FIGURE 20-7. The original operon model of Jacob and Monod, as proposed for the regulation of the *lac* genes of *E. coli* in 1961.

cognate inducer molecule. Combination of repressor and inducer at the second allosteric site reduces the operator affinity of the first site and thus frees the operator from the repressor.

Jacob and Monod were able to adduce convincing genetic proof of their topological notions of the operon by isolating constitutive mutants that evidently carry mutations in the operator locus. For this purpose, they had constructed a variety of partial diploid bacteria that carried two sets of *lac* genes in their genome, one set on the bacterial chromosome and another set on an F' sex factor. Table 20-1 presents a comparison of induced and non-

TABLE 20-1.   Relative Concentrations of Galactosidase and Galactoside Transacetylase in *E. coli* with Various Haploid and Diploid Genomes

Genome	Galactosidase (*lac*Z)		Galactoside-transacetylase (*lac*A)	
	Noninduced	Induced	Noninduced	Induced
LacI$^+$,Z$^+$,A$^+$	0.1	100	1	100
LacI$^-$,Z$^+$,A$^+$	100	100	90	90
LacI$^-$,Z$^+$,A$^+$/ F LacI$^+$,Z$^+$,A$^+$	1	240	1	270
LacI^s,Z$^+$,A$^+$	0.1	1	1	1
LacI^s,Z$^+$,A$^+$/ F LacI$^+$Z$^+$,A$^+$	0.1	2	1	3
LacO^c,Z$^+$,A$^+$	25	95	15	100
LacO$^+$,Z$^-$,A$^+$/ F LacO^c,Z$^+$,A$^-$	180	440	1	220
LacI^s,O$^+$,Z$^+$,A$^+$/ F LacI$^+$,O^c,Z$^+$,A$^+$	190	219	150	200

From F. Jacob and J. Monod, *J. Mol. Biol.* **3**, 318 (1961).

induced concentrations of galactosidase and transacetylase in a variety of diploid and haploid bacteria carrying an assortment of *lac* gene mutations Rows 1 and 2 show induced and noninduced concentrations of the two enzymes in haploid bacteria of LacI$^+$,Z$^+$,A$^+$ inducible and LacI$^-$,Z$^+$,A$^+$ constitutive phenotypes. It can be seen that in the inducible strain the presence of inducer raises the concentration of galactosidase by more than 1000-fold and that of transacetylase by more than 100-fold, whereas in the constitutive LacI$^-$ strain both enzymes achieve these concentrations in the absence of any inducer. Row 3 confirms the inference made on the basis of the experiment with transient zygotes (Figure 20-6) that the *lac*I gene produces an active repressor, since the $LacI^+/F\ LacI^-$ heterozygous diploid is of inducible, and not constitutive, phenotype. (The higher induced concentrations of both enzymes in the diploid bacterium reflect a gene dosage effect, because, compared to the haploid, the partial diploid carries twice as many *lac*Z and *lac*A genes relative to the rest of its genome.) Rows 4 and 5 show that the $LacI^s$ mutant carries a mutant repressor of greatly reduced affinity for the inducer galactoside: the $LacI^s/F\ LacI^+$ diploid attains only very low concentrations of both enzymes upon induction. Rows 6 and 7 show the enzyme concentrations in a new kind of mutant, $LacO^c$, or *operator-constitutive*. In a haploid, the $LacO^c$ phenotype is only partly constitutive, in that only part of the maximum enzyme concentrations are achieved in the absence of inducer. In a heterozygous $LacO^+/F\ LacO^c$ diploid, however, the effect of the $LacO^c$ mutation is evidently confined to the expression of the genes coresident with it on the F' sex factor: synthesis of galactosidase by the *lac*Z gene carried by the $LacO^c$ sex factor is partially constitutive, whereas synthesis of transacetylase by the *lac*A gene carried by the $LacO^+$ bacterial chromosome remains inducible. (The *lac*A mutant gene of the $LacO^c$ sex factor bears a mutation in the transacetylase structure, and hence cannot contribute to synthesis of that enzyme under any conditions.) Finally, Row 8 shows that the $LacO^c$ mutation appears to correspond to a change in the structure of the operator gene that has reduced its sensitivity to closure by the repressor, since synthesis of the enzymes in the $LacI^s$, $O^+/F\ LacI^+, O^c$ diploid is not impeded by the presence of the mutant $LacI^s$ repressor. The behavior of the ensemble of genotypes shown in Table 20-1 is thus in full accord with that to be expected from the model of Figure 20-7.

Further studies on the nature of $LacO^c$ mutants showed that they correspond to deletions of genetic material situated between the *lac*I and *lac*Z genes. This finding agrees with the idea that the repressor carries an affinity for the nucleotide sequence of the *lac*O gene, since genetic deletion of that sequence would evidently remove the repressor target from the operon and render function of its genes constitutive. Another dramatic demonstration of the verity of Jacob and Monod's general topological considerations of operon structure was provided by a later finding that very long deletions of the $LacO^c$ type, which remove all genetic material between the *lac*Z gene and the *pur* genes that govern purine synthesis (Figure 10-21), not only free the *lac* operon genes from control by galactoside inducers but also place their expression under the control of metabolites in the pathway of purine biosynthesis.

Once Gilbert and Müller-Hill had managed to isolate the *lac* operon repressor, they were able to put to a critical test Jacob and Monod's proposal that the *lac*O gene is its target. For this purpose they extracted the *lac*I gene repressor from bacteria whose protein had been labeled with radiosulfur ^{35}S and presented the radioactive repressor protein to various kinds of DNA molecules in the presence or absence of IPTG inducer in *in vitro* reaction mixtures. These mixtures were then subjected to sucrose density-gradient sedimentation in order to ascertain whether any of the radioactive repressor protein (which, in its free state sediments five times more slowly than the large DNA molecules used in this experiment) had become attached to, and hence would sediment at the same high velocity as, the DNA molecules. The results of this experiment were the following: (1) The repressor does not attach to DNA molecules lacking all genes of the *lac*-operon. (2) The repressor does attach to DNA molecules carrying the *lac*-operon genes, but only if IPTG inducer has been omitted from the *in vitro* reaction mixture. The presence of IPTG prevents the binding of the repressor to DNA. (3) The repressor does *not* attach to DNA molecules carrying the *lac*-operon genes if the operon is of the operator-constitutive LacO^c mutant phenotype. Thus these results directly confirmed the original proposal that the functional regulation of the *lac* genes *is* achieved by an inducer-sensitive interaction between the repressor protein and the operator gene.

Study of the adaptive formation of *E. coli* enzymes that govern the utilization of carbohydrates other than lactose has shown that the expression of genes coding for the primary structure of these enzymes can also be fathomed in terms of the operon concept. Thus the induction by galactose of three closely linked genes coding for the three enzymes that effect the conversion of galactose to glucose appears to occur by the neutralization of an operator-affined repressor by galactose. Similarly, there exist inducible operons of genes that govern the utilization of the sugars arabinose, maltose, and rhamnose. The regulation of these three operons, however, appears to be more complicated than that of the *lac* and *gal* operons, in that here there has become manifest an involvement of regulatory proteins that exert a *positive* effect on gene expression in the presence of the inducer. (This is to be contrasted with the *lac* and *gal* gene repressors, which merely fail to exert their negative effect in the presence of inducer.) But it seems unlikely that the eventual clarification of these positive regulatory effects will engender a fundamental revision of the schema of Figure 20-7.

ENZYME REPRESSION

Positive enzymatic adaptation, or induction of synthesis of an enzyme by the presence of its substrate (or of a structural analog of the substrate) had been known for many years, when *negative* enzymatic adaptation, or *enzyme repression*, was discovered. Here the synthesis of an enzyme, instead of being induced by its substrate, is repressed by the presence of the *product* of the reaction that it catalyzes. The existence of enzyme repression was first uncovered in Monod's laboratory in 1953, when it was found that in *E. coli* the synthesis of tryptophan synthetase, the protein product of the *trp*A and *trp*B genes, is repressed by the presence of tryptophan in the growth medium. The thrifty aspect of this phenomenon is as apparent as was that of the induction of β-galactosidase by lactose, for it would be patently uneconomic for *E. coli* to synthesize enzymes catalyzing the terminal step in tryptophan biosynthesis as long as that amino acid were present, ready-made, in the environment. Within the next few years, many other examples of enzyme repression were encountered, most of them pertaining to bacterial enzymes that participate in the biosynthesis of amino acids and in the hydrolysis of phosphorylated organic compounds.

Further study of the repression of enzymes in the pathway of tryptophan biosynthesis in *E. coli* showed that the presence of tryptophan exerts a coordinate control over the expression of all five contiguous genes, *trp*A,B,C, D,E (see Chapter 14). That is, the intracellular levels of tryptophan synthetase (*trp*A, *trp*B), IGP synthetase (*trp*C), phosphoribosyl anthranilate transferase (*trp*D), and anthranilate synthetase (*trp*E), all wax and wane directly with the rise and fall of the concentration of tryptophan available to the cell. Since the coordinate repression of these enzymes appears to be formally analogous to the coordinate induction of the lactose enzymes, Jacob and Monod proposed that the *trp* genes also constitute an operon. The only necessary modification of the model that they devised (Figure 20-7) to explain enzyme repression was to propose that in the repressible *trp* operon the repressor is in the inactive state (in which it cannot combine with the operator) in the absence of tryptophan and changes into the active state (in which it can combine with the operator) in the presence of tryptophan. This proposal drew support from the discovery of a variety of constitutive *E. coli* mutants that produce maximal levels of the *trp* gene enzymes even in the presence of a high tryptophan concentration. One class of such mutants, the $TrpR^-$ type is evidently the regulatory analog of constitutive $LacI^-$ mutants, in that $TrpR^-$/F $TrpR^+$ partial diploids show normal, repressible behavior. These mutants carry their mutation in the *trp*R gene that can be inferred to encode the structure of a repressor that exerts a negative effect on *trp* gene enzyme synthesis. $TrpR^-$ mutants are constitutive because they either lack the repressor altogether or because their mutant repressor cannot be activated by tryptophan. Furthermore, unlike the *lac*I gene, which was seen to be closely linked to the *lac* operon, the *trp*R gene is located near minute 0 of the *E. coli* genetic map (Figure 10-27) and is thus separated by a quarter of the bacterial chromosome from the *trp* genes whose function it controls. A second class of mutants that are constitutive in their synthesis of *trp* gene enzymes is evidently the regulatory analog of constitutive $LacO^c$ mutants. These mutants carry their mutation at one extremity of the block of *trp* genes—namely, either just within or just without the *trp*E gene. Hence the target of the *trp*R gene repressor is an operator region at the *trp*E gene end of the *trp* operon, whose closure by the tryptophan-activated repressor prevents transcription of the set of *trp*E, D,C,B,A genes.

It was thus concluded that enzyme induction and enzyme repression are but opposite sides of the same coin, with the main difference between the two phenomena apparently residing in the nature of the repressor. More recent experiments on the mechanism of the repression of enzymes of amino acid biosynthesis, however, have led to the unexpected discovery that for amino acids to repress formation of their biosynthetic enzymes, they must first be activated by their cognate aminoacyl-tRNA synthetase and possibly become attached to their corresponding tRNA. This fact came to light when it was found that bacteria synthesize their normally repressible enzymes of amino acid biosynthesis even in the presence of high levels of the relevant amino acid if activation of that amino acid proceeds at a subnormal rate. Thus for the *his* operon of repressible genes encoding the structure of enzymes active in the biosynthesis of histidine, constitutive mutants have been isolated that carry a defective histidinyl-tRNA synthetase. And for the *trp* operon, a third class of regulatory mutants is known, which carries its mutation in the *trp*S gene that encodes the structure of tryptophanyl-tRNA synthetase (located at minute 64 of the *E. coli* genetic map). The requirement for amino acid activation in enzyme repression might mean merely that the relevant repressor is activated, not by binding the free amino acid, but by binding a derivative of the activated amino acid, such as aminoacyl-tRNA. It is also possible, however, that the amino acid intervenes in the activation of the repressor in an even more indirect manner; for instance, the product of some reaction involving aminoacyl-tRNA might determine whether the repressor is in the active or inactive state. Which of these various alternatives actually obtains will surely become known as soon as the repressor protein of an operon of amino acid biosynthesis has been isolated and its behavior characterized.

RNA-DNA HYBRIDIZATION

In addition to explaining an ensemble of facts known at the time of its formulation and being further supported by data such as those presented in Table 20-1, the operon model of regulation of gene function shown in Figure 20-7 led to an important prediction: the presence of galactoside inducers should raise not only the concentration of *lac* gene enzymes in the cell but also the concentration of specific mRNA molecules homologous to the DNA of the *lac* operon. Experimental tests of this prediction became possible once the techniques of forming specific molecular hybrids of mRNA and its homologous DNA had been developed, as described in Chapter 16. Such hybrids provide a means by which the amount of mRNA corresponding to a particular gene can be specifically assayed in the presence of the myriad of other mRNA species that exist in the same cell.

The following experiment became possible with the development of the RNA-DNA hybridization technique. Two cultures of *E. coli* were grown, one in the presence and the other in the absence of IPTG inducer. Both cultures were then exposed briefly to radioactively labeled uracil, so as to introduce the radioactive isotope into the uracil bases of all messenger RNA molecules. The total labeled messenger RNA was then extracted from the bacteria of both cultures, and each extract was added to a hot solution of melted DNA of the F-*lac* sex-factor episome that had incorporated the *E. coli lac* genes, as described in Chapter 10. The two solutions were then cooled slowly in order to allow formation of hybrid helices between any DNA and RNA strands of complementary base sequence. After cooling, both solutions were filtered and the amount of radioactivity retained on the two filters compared. The result of this experiment was that much more radioactivity was registered by the filter containing the RNA-DNA hybrids derived from the induced culture than was registered by the filter containing the RNA-DNA hybrids derived from noninduced culture. That is to say, in complete agreement with the prediction of Jacob and Monod's model, the presence of a galactoside inducer was found to induce the bacteria to synthesize greater quantities of both the *lac* gene enzymes and the messenger RNA molecules transcribed from the *lac* gene DNA.

Similar hybridization experiments were eventually carried out by Yanofsky and his colleagues with RNA extracted from *E. coli* either starved for tryptophan or growing in the presence of tryptophan, and hence either derepressed or repressed for their *trp* operon genes. Both derepressed and repressed cultures were labeled with ^{3}H-uracil for 3 minutes. The radioactive RNA extract was then hybridized with DNA extracted from phage particles of phage $\phi80$ (a phage strain related to *lambda*) and of three of its transducing derivatives, ptA–B, ptC–E, and ptA–E, each of which carries a different portion of the *trp* operon. (The $\phi80$ prophage attachment site is near minute 25 of *E. coli* genetic map, and thus just beyond the *trp*A gene. By a crossover mechanism similar to that outlined in Figure 14-9 for the origin of λdg, $\phi80$pt phages arise that are capable of restricted transduction of the *trp* genes.) Thus the normal $\phi80$ DNA carries none of the *trp* genes; ptA–B DNA carries only the operator-distal *trp*B,A genes; ptC–E DNA carries the operator-proximal *trp*E,D,C genes; and ptA–E DNA carries the entire *trp* operon. The result of this experiment is presented in Table 20-2. The data of Table 20-2 show that 0.05 to 0.07 % of the total labeled RNA extracted from repressed or derepressed bacteria (which, in this experiment, do not carry the $\phi80$ prophage) hybridizes to the normal $\phi80$ DNA. This small amount of hybridization is a measure of the amount of *E. coli* mRNA transcribed from a few bacterial

TABLE 20-2.   Detection of Tryptophan Operon Messenger RNA

RNA source	DNA source	Radioactive RNA hybridized (percent)	Difference between pt and $\phi80$ values (percent)
	$\phi80$	0.07	—
trp operon	ptA–B	0.08	0.01
repressed	ptC–E	0.08	0.01
	ptA–E	0.09	0.02
	$\phi80$	0.05	—
trp operon	ptA–B	0.14	0.09
derepressed	ptC–E	0.49	0.44
	ptA–E	0.58	0.53

From F. Imamoto, J. Ito, and C. Yanofsky, *Cold Spring Harbor Symp. Quant. Biol.* **31**, 235–249 (1966).

genes sufficiently homologous in nucleotide sequence to some genes of the normal $\phi80$ phage for formation of RNA-DNA hybrids. The fraction of labeled RNA extracted from *repressed* bacteria that is hybridizable to the ptA–E DNA is only slightly higher than this small background value, indicating that here no more than 0.02% of the total RNA synthesized during the 3-minute labeling period constitutes *trp* operon mRNA. In contrast, 0.58% of the labeled RNA extracted from *derepressed* bacteria (or 0.53% above background) can be seen to be hybridizable to the ptA–E DNA, demonstrating that, in accord with the operon model, derepression causes a more than 25-fold increase in the rate of *trp* mRNA synthesis. Finally, the finding that only 0.09 and 0.44% above background of the labeled RNA extracted from derepressed bacteria is hybridizable to ptA–B and ptC–E DNA, respectively, and hence that the *trp*A,B genes make up $0.09/0.53 = 0.17$ and the *trp*C,D,E genes make up $0.44/0.53 = 0.83$ of the total nucleotide sequence of the *trp* operon, is in excellent accord with independent gene size estimates derived from fine-structure mapping data and from measurements of polypeptide chain lengths.

The availability of ptA–E DNA as a specific *trp* mRNA detector also made possible a direct test of the view set forth in Chapter 16 that mRNA molecules are polygenic transcripts carrying the nucleotide sequence of several contiguous genes. For this purpose, the ^{3}H-labeled RNA extracted from derepressed bacteria was subjected to sucrose density-gradient sedimentation. The RNA present in samples collected from various parts of the sucrose density gradient was then hybridized to ptA–E DNA in order to ascertain the position of radioactive *trp* mRNA in the gradient, and hence the sedimentation velocity of *trp* mRNA. The result of such an experiment is presented in Figure 20-8, where it can be seen that *trp* mRNA molecules manifest a rather broad distribution over the sucrose gradient. The peak of that distribution can be seen to coincide with the sedimentation band formed by the large amount of 23S ribosomal RNA contributed to the extract by the bacterial ribosomes (and detected in the samples by its absorption of UV light). Significant amounts of *trp* mRNA are found in advance of the 23S rRNA band, at a position corresponding to a sedimentation velocity of about 33S. This sedimentation velocity corresponds to RNA molecules approximately 6700 nucleotides in length, which is just the length estimated for the entire *trp* operon on the basis of the polypeptide chain lengths of the *trp* enzymes. Thus it can be concluded that the *trp* genes are transcribed onto a single *trp* mRNA

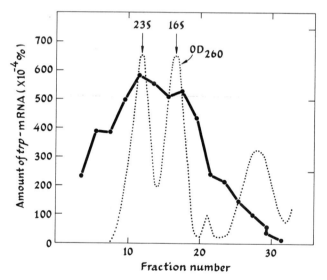

FIGURE 20-8.   Demonstration of the polygenic character of the *trp* operon mRNA. In this experiment, a culture of *E. coli* was allowed to assimilate ^{3}H-uracil for 14 minutes after derepression of the *trp* operon genes was initiated. The total bacterial RNA was then extracted from the culture and subjected to sucrose density-gradient sedimentation. Each fraction collected from the centrifuged gradient was analyzed for its UV-absorption (curve labeled OD_{260}), and hence total RNA content, and for the fraction of the total ^{3}H-labeled RNA capable of forming specific molecular hybrids with ptA–E DNA extracted from ϕ80 phages capable of transducing the entire *trp* operon. (solid points). The two largest peaks of the OD_{260} curve correspond to the sedimentation bands of 23S and 16S rRNA in the sucrose gradient. The broad peak of hybridizable ^{3}H-labeled RNA corresponds to the sedimentation profile of *trp* mRNA. The *trp* mRNA molecules that sediment in advance of the 23S rRNA band constitute nearly complete polygenic transcripts of all five genes of the *trp* operon. [After F. Imamoto, *Proc. Natl. Acad. Sci. Wash.* **60**, 305 (1968).]

molecule. The wide range in sedimentation velocities (and hence in molecular sizes) manifest in Figure 20-8 is attributable to the fact that at the moment of extraction of the cell, most nascent *trp* mRNA molecules had not yet reached their ultimate length and that most completed *trp* mRNA molecules had already reached the end of their functional lifetime and were already in the process of degradation.

THE PROMOTOR

The foregoing considerations have shown that genes encoding the structure of enzymes of related function are assembled as operons on the bacterial chromosome, so that their expression can be controlled coordinately according to the needs of the cell through the interaction of a repressor with a common operator gene. Although the repressor-operator dialectic is undoubtedly of the widest regulatory significance, it cannot be the *only* mechanism governing the quantitative aspects of the heterocatalytic function of the genetic material. An example of a gene whose expression is *not* controlled by the level of active repressor in the cell is the *lac*I gene, which itself encodes the primary structure of a repressor protein. For the very low rate of synthesis of the *lac* gene repressor protein, which maintains only about ten repressor molecules per cell is invariant and unaffected by the presence or absence of inducers of the *lac* operon. Since as few as ten active repressor molecules per cell suffice to close the *lac*O operator gene, there is, of course, never any *need* for the cell to synthesize repressor at a higher rate. Hence a low and invariant rate of expression of the *lac*I gene appears to be programmed into *E. coli* DNA by another mechanism. A similar conclusion can be reached for the rate of expression of other genes whose products are needed in invariant amounts.

Recent studies have shown that one mechanism of built-in, fixed control of gene expression devolves from the presence of *promotor* genes. These promotor genes appear to be genetic sites at which molecules of RNA polymerase enzymes first attach to their DNA template in order to start transcription of the polygenic genetic message into messenger RNA. These sites were referred to as transcription "starting points" in Chapter 16. One such promotor gene, *lac*P, has been identified at the operator end of the *lac* operon (Figure 20-5). Mutations in the *lac*P gene greatly reduce the maximal rate of synthesis of all three *lac* operon enzymes by inducible LacI$^+$ strains in the presence of high concentrations of IPTG inducer or by constitutive LacI$^-$ mutants. Thus, despite maximal opening of the *lac*O operator gene, enzyme synthesis proceeds at a reduced rate because the mutation in the *lac*P gene has reduced the frequency with which RNA polymerase enzyme molecules can start transcription of the *lac* operon genes. The *lac*I repressor gene, which, according to the schema of Figure 20-7 does not itself form part of the *lac* operon, depends for its transcription on another promotor gene. The structure of this promotor gene is such that only about 1 RNA polymerase molecule per bacterial generation period attaches itself there and starts transcription of the *lac*I gene.

The molecular basis of the interaction of promotor and RNA polymerase has not yet been elucidated, although, as was set forth in Chapter 16, it appears that for its attachment reaction the enzyme must recognize certain specific structural features of the double-stranded DNA polynucleotide chain, and that of the three different polypeptide subunits of the enzyme, the σ subunit appears to carry out this recognition process. Thus our present notion of the quantitative control of the heterocatalytic function is that every gene depends for its transcription on a promotor gene. The purine and pyrimidine base sequence of that promotor gene determines the frequency with which RNA polymerase molecules attach to it, and hence sets the *maximum* rate at which the relevant gene can be transcribed into messenger RNA. For some genes, such as *lac*I, that maximum rate is always the actual rate. For other genes, however, such as *lac*Z,Y,A, the maximum rate is achieved only as long as their operator gene is in the repressor-free open state. Closure of the operator prevents either attachment to the promotor or further progress along the DNA template of RNA polymerase molecules, which would otherwise have effected transcription, and hence caused synthesis of the protein product, of the relevant genes.

PROPHAGE IMMUNITY AND INDUCTION

The operon model of genetic regulation was to provide also an explanation of the nature and mode of action of the immunity repressor of temperate phages. As was set forth in Chapter 14, the *c*I gene of the temperate phage *lambda* gives rise to an immunity repressor whose presence in the lysogenic bacterium not only holds in check the endogenous *lambda* prophage but endows the cell with immunity to superinfection by exogenous *lambda* phages. The postulation of this *lambda* immunity repressor actually preceded by one or two years the postulation of the *lac*I gene repressor, and efforts to identify the nature and mode of action of these two very different repressors proceeded more or less concurrently. First, the discovery of λcI temperature-sensitive mutants (which can establish the lysogenic response only at low temperatures and whose immunity repressor is inactivated at temperatures above 37°C) and of λcI *amber* mutants (which can establish a lysogenic response only in mutant host bacteria carrying an *amber* suppressor gene) suggested that the immunity repressor is a protein. Second, just when Gilbert and Müller-Hill's attempts to isolate the *lac*I gene repressor finally succeeded, M. Ptashne also managed to isolate the λcI gene immunity repressor. For this purpose, Ptashne treated bacteria carrying the *lambda* prophage in a manner that greatly reduced their endogenous rate of protein synthesis and then superinfected these immune bacteria with exogenous *lambda* phages in the presence of radioactively labeled amino acids. His hope was that under these conditions an appreciable proportion of the residual protein synthesis would be attributable to formation of the *c*I gene product by the superinfecting phages, since host-protein synthesis had been inhibited by pre-infection treatment and synthesis of most vegetative phage proteins by the superinfecting phages would be held in abeyance by the endogenous immunity repressor. This hope was realized, in that extraction and chromatographic fractionation of the radioactively labeled protein from these infected bacteria did reveal that one fraction could be identified as the *c*I gene product. For this fraction was present only upon infection of bacteria with λcI$^+$ phages carrying the normal repressor gene and was absent upon infection with λcI *amber* mutants. Study of the sedimentation velocity of this protein fraction in sucrose density gradients revealed that its molecular weight corresponds to that of a polypeptide chain about 200 amino acids in length, or comparable to that of one of the four polypeptide subunits of the *lac*I gene repressor.

What is the target of the λcI gene immunity repressor? Early studies by Kaiser and Jacob had already shown that the repressor target must be one or more genes in what they referred to as the "immunity" region of the *lambda* genetic map situated between what were later called genes N and O and including the *c*I gene itself (Figure 20-9). This conclusion followed from studies of the temperate phage 434. Phages 434 and *lambda* are mutually *heteroimmune*, in that each can grow normally on lysogenic bacteria carrying the other's prophage. That is to say, neither phage is sensitive to the action of the other's immunity repressor. Kaiser and Jacob found that *hybrid* phages produced by genetic recombination between *lambda* and 434 are not sensitive to the λcI repressor as long as they do not carry the N-*c*I-O region of the *lambda* parent, even though all the rest of their genome is of *lambda* provenance. Once Ptashne had managed to isolate the *c*I gene immunity repressor he directly confirmed the conclusion that its target lies in the immunity region of the *lambda* genome. In experiments essentially analogous to (and actually preceding by a few months) those that demonstrated the binding of the *lac*I gene repressor to *lac* operon DNA, he was able to show, first of all, that upon mixing DNA extracted from normal *lambda* virus particles with radioactively labeled repressor protein, the repressor is bound to the *lambda* DNA. Second, Ptashne found that the λcI gene repressor is *not* bound to DNA extracted from λ-434 hybrid phage particles lacking the *lambda* immunity region.

Subsequent genetic studies showed that there are *two* operators, V_1 and V_2, within the immunity region of lambda, each being the formal equivalent of the *lac*O gene. Closure of one of these operators by the immunity repressor

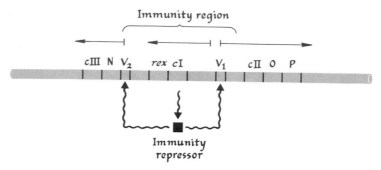

FIGURE 20-9.   The region of the *lambda* phage chromosome that governs the control of vegetative phage multiplication and its repression in prophage immunity. The horizontal arrows indicate the origins and directions of mRNA transcription. The symbols V_1 and V_2 denote operator sites; their closure by the immunity repressor prevents transcription of the operons whose expression they control.

prevents expression of gene N, and closure of the other operator prevents expression of the two contiguous genes O and P (Figure 20-9). Identification of these two operators became possible upon the isolation of *virulent* mutants of *lambda*, which not only fail to give the lysogenic response upon infection of a nonlysogenic bacterium but which are even capable of growth on lysogenic bacteria carrying the *lambda* prophage. These *lambda* mutants owe their virulence to an insensitivity to the *cI* gene repressor, thanks to mutational alterations of both V_1 and V_2 operator genes that have reduced their capacity to bind the normal repressor. Virulent mutants of *lambda* are therefore formally analogous to LacO^c mutants.

The overall regulation of *lambda* phage development can now be described in the following terms. In the presence of the *cI* gene repressor, transcription of mRNA from genes N, O, and P is held in abeyance. And since, as was seen in Chapter 14, the presence of the protein products of these three genes is required for the expression of all the other *lambda* genes, direct action of the *cI* gene repressor on the operators of genes N, O, and P can thus repress indirectly almost the entire *lambda* genome. This conclusion finds support from molecular hybridization experiments with *lambda* phage DNA of radioactively labeled RNA extracted from lysogenic bacteria carrying the *lambda* prophage and superinfected with exogenous *lambda* phages. It can be seen in such experiments that the only *lambda*-specific mRNA synthesized under these conditions is that homologous to the immunity region and hence responsible for the production of the *cI* gene-repressor protein itself and of the protein coded by the adjacent *rex* gene responsible for the restriction of T4*r*II mutant phage growth on K12(λ) bacteria. Upon inactivation of the *cI* gene repressor attending the induction of a lysogenic cell carrying the *lambda* prophage, or upon infection of a nonlysogenic cell with *lambda* phage particles, mRNA is free to be transcribed from N, O, and P genes. As soon as the N, O, and P proteins appear after translation of that mRNA, transcription of other *lambda* genes commences, and vegetative phage development is under way.

CONTROL OF RIBOSOME SYNTHESIS

As was seen in Figure 2-14, the generation time T of *E. coli* depends on the composition of the medium in which the bacteria are growing. Thus in nutrient broth T was seen to have the value 20 minutes in contrast to the value $T = 50$ minutes in the simple synthetic medium. The exact reason why bacteria grow more rapidly in the nutrient broth has not yet been fully ascertained, but it undoubtedly devolves from the fact that in broth the bacteria find ready-made in their environment many substances such as amino acids and purines and pyrimidines, which they must home-make in the nutritionally sparse synthetic medium. Thus in broth the bacteria need to devote chemical energy to the synthesis of their basic building blocks at a much lower rate and can, instead, spend energy at a much higher rate than in the synthetic medium on the polymerization of these building blocks into nucleic acids and proteins. In the mid-1950's Ole Maaløe (Figure 20-10)

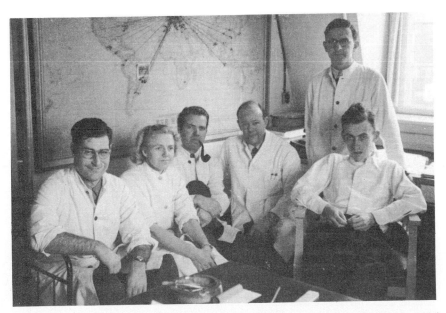

FIGURE 20-10.   At the State Serum Institute of Copenhagen, 1951: Ole Maaløe (b. 1914), 3rd from left; Niels Jerne (b. 1911), 5th from left; James Watson, 6th from left.

carried out a systematic study of the relative content of protein, DNA, and RNA of bacteria growing in different media at different rates. This study was to lead to some deep insights into the mechanisms by which the cells actually achieve their different rates of growth. The results of one such series of analyses are presented in Table 20-3 as the dry weight, the DNA content (or average number of genomes) per cell, the amount of protein per genome, and the number of ribosomes per genome for bacteria growing in four different media with generation times T ranging from 25 to 300 minutes.

The first conclusion that can be drawn from the data of Table 20-3 is that the shorter the generation time, the greater the total dry weight as well as the number of genomes per cell. The most plausible explanation of this fact would appear to be that as the bacteria are transferred from nutritionally poor to nutritionally rich media, some step of the cell-division process—for instance, formation of the septum separating the two daughter cells—cannot accelerate in step with the overall growth rate. Thus, let the time required for formation of a septum be S minutes (where $S > T$), and let the number of growing septa per μg bacterial dry weight be the same in all four media. Then the average dry weight per cell would be proportional to S/T. Thus the ratio in dry weights per cell W_1/W_2 for growth in two different media would be given by

$$W_1/W_2 = S_1 T_2/S_2 T_1, \quad \text{or} \quad S_1/S_2 = W_1 T_1/W_2 T_2 \quad (20\text{-}2)$$

Hence we may reckon from the data of Table 20-3 that the ratio of S_1 for growth in broth to S_2 for growth in proline synthetic medium would be

$$S_1/S_2 = 77 \times 10^5 \times 25/16 \times 10^5 \times 300 = 0.4$$

That is to say, the 12-fold decrease in the generation times T engendered by growth in broth would be accompanied by only a 2.5-fold decrease in the time S required for septum formation.

The second conclusion implicit in the data of Table 20-3 is that the bacteria contain essentially the same amount of protein per DNA genome in all four media. That is to say, the genomes of both rapidly and slowly growing bacteria cause the same number P of amino acids to be polymerized into protein per generation time T. Hence the rate of protein synthesis per bacterial genome, P/T, increases as the generation time decreases. Let us now consider how this increasing rate of protein synthesis could be achieved. If the bacteria contain an average of R ribosomes per genome, and if we suppose that each of these R ribosomes is engaged in the assembly of one polypeptide chain growing at the rate of a amino acids per second (or $60a$ amino acids per minute), then it follows that

$$P/T = 60a \times 1.5R \quad (20\text{-}3)$$

TABLE 20-3. Content of DNA, Protein and Ribosomes, and Peptide Chain Growth Rate in Bacteria at Different Rates of Growth (37°C)

Growth medium	T (min)	Dry weight per cell ($\mu g \times 10^5$)	DNA genomes per cell	Protein amino acid residues per genome	Ribosomes per DNA genome	Peptide chain growth rate (amino acids per sec)**
Nutrient Broth	25	77	4.5	540,000,000	15,500	16
Glucose synthetic*	50	32	2.4	490,000,000	6800	17
Glycerol synthetic†	100	21	1.7	470,000,000	4200	14
Proline synthetic†	300	16	1.4	450,000,000	1450	14

Data from O. Maaløe and N. O. Kjeldgaard, *Control of Macromolecular Synthesis*, Benjamin, New York, 1966. The bacteria are *Salmonella typhimurium*, a close relative of *E. coli*.
* A medium similar to that formulated in Table 2-1.
† The glucose of the recipe of Table 2-1 is replaced by either glycerol or proline.
** Calculated from equation 20-3.

(The factor 1.5 takes into account that during the generation time T, the number of ribosomes per genome increases from R to $2R$.) Thus the bacteria could increase their rate of protein synthesis P/T by increasing a or R or both. The sixth column of Table 20-3 shows that the number of ribosomes per genome does increase in inverse proportion to the generation time. One may now ascertain the magnitude of the polypeptide chain growth rate a in the four growth media by solving (20-3) for a:

$$a = P/(90TR) \tag{20-4}$$

and substituting into (20-3) the experimental values of T, P, and R. The result of these calculations is shown in the last column of Table 20-3, where it can be seen that despite the 12-fold range in generation times, a has a very nearly constant value of about 15 amino acids per second. That is, the rate of polypeptide chain growth—the rate at which amino acids are added to a nascent polypeptide chain—is the same in the four growth media. Hence the third and most-important conclusion implicit in these data is that bacteria vary their rate of overall protein synthesis by varying their content of ribosomes, and thus the number of nascent peptide chains under construction at any moment, rather than varying the peptide chain growth rate.

Maaløe's indirect estimate of the peptide chain growth rate of 15 amino acids per second at 37°C has meanwhile been confirmed by direct measurements of the time required to complete the synthesis of a polypeptide chain of known length. For instance, it has been found that 80 seconds elapse between the initiation and completion of synthesis of the 1173-amino-acid-long polypeptide chain of E. coli β-galactosidase, which leads to an estimate of $a = 1173/80 = 15$ amino acids per second. The agreement between these two independent estimates of a suggests that the assumption made in the derivation of equation (20-3)—that every bacterial ribosome is actually engaged in protein synthesis—is not far from the truth, though in view of the cyclic association and dissociation of 30S and 50S ribosomal subunits set forth in Chapter 17, this cannot be entirely true.

The finding that peptide chains grow at the constant rate of 15 amino acids per second can now be compared with the observation mentioned in Chapter 16 that nascent RNA chains grow at the rate of about 43 nucleotides per second. Since the triplet nature of the genetic code specifies that three messenger RNA nucleotides must pass through a ribosome for every amino acid added to the nascent polypeptide chain, it follows that transcription and translation of messenger RNA are dynamically compatible processes. That is, as demanded by the schema shown in Figure 17-6, RNA chain growth proceeds at just such a rate that three nucleotides are added to the growing end of the messenger as one amino acid is added to each of the nascent polypeptide chains being simultaneously translated from the nascent mRNA.

Since, as we have just seen, the parameters P (protein per genome) and a (peptide chain growth rate) of equation (20-3) are independent of the generation time T, it follows from (20-3) that the rate of ribosome synthesis per genome, or R/T, depends on T according to

$$R/T = \text{constant}/T^2 \qquad (20\text{-}5)$$

This equation states that the rate at which both ribosomal RNA and ribosomal proteins are transcribed and translated from the bacterial DNA increases with the inverse square of the generation time. That is to say, the ribosomal genes are expressed at a rate $12^2 = 144$ times faster upon growth in broth than upon growth in proline synthetic medium. The details of this most important bacterial regulatory process have yet to be elucidated. It is entirely possible that the ribosomal genes comprise one or more operons whose expression is controlled by one or more repressors whose activity is in turn controlled by the variable intracellular level of one or more metabolites. It seems equally possible, however, that some other, as yet unknown mechanism is responsible for the matching of the number of ribosomes per genome with the requirements for overall protein synthesis dictated by the bacterial growth rate.

Bibliography

HAYES

Chapter 23.

ORIGINAL RESEARCH PAPERS

Gilbert, W., and B. Müller-Hill. Isolation of the *lac* repressor. *Proc. Natl. Acad. Sci. Wash.*, **58**, 2415 (1965).

Hogness, D. S., M. Cohn, and J. Monod. Studies on the induced synthesis of β-galactosidase in *E. coli*: The kinetics and mechanism of sulfur incorporation. *Biochim. Biophys. Acta*, **16**, 99 (1955).

Lederberg, J. Gene control of β-galactosidase in *E. coli*. *Genetics*, **33**, 716 (1948).

Monod, J., A. M. Pappenheimer, Jr., and G. Cohen-Bazire. La cinétique de la biosynthèse de la β-galactosidase (lactase) chez *E. coli*. La spécificité de l'induction. *Biochim. Biophys. Acta*, **9**, 648 (1952).

Monod, J. and G. Cohen-Bazire. L'effet d'inhibition spécifique dans la biosynthèse de tryptophan desmase chez *Aerobacter aerogenes*. *Compt. rend.* **236**, 530 (1953).

Pardee, A. B., F. Jacob, and J. Monod. The genetic control and cytoplasmic expression of inducibility in the synthesis of β-galactosidase by *E. coli*. *J. Mol. Biol.*, **1**, 165 (1959).

Ptashne, M. Isolation of the λ repressor. *Proc. Natl. Acad. Sci. Wash.*, **57**, 306 (1967); The λ phage repressor binds specifically to λ DNA. *Nature*, **214**, 232 (1967).

SPECIALIZED TEXTS, MONOGRAPHS, AND REVIEWS

Jacob, F. and J. Monod. Genetic regulatory mechanisms in the synthesis of proteins. *J. Mol. Biol.*, **3**, 318 (1961).

Maaløe, O., and N. O. Kjeldgaard. *Control of Macromolecular Synthesis.* Benjamin, New York, 1966.

Stent, G. S. The Operon: on its third anniversary. *Science*, **144**, 816 (1964).

Zipser, D., and J. Beckwith. *The lac Operon.* Cold Spring Harbor Lab. Quant. Biol., New York, 1970.

21. Ramifications

The foregoing chapters have shown that molecular genetics has lifted the "real core of genetic theory" from "the deep unknown" in which, according to H. J. Muller, classical genetics had left it. The main thrust for this surfacing operation came from the study of microscopic bacteria and viruses, whose genetic material is now known to a degree of structural and functional detail that will satisfy most persons interested in the basic features of biological self-reproduction. Thus if organic evolution had not passed the threshold of eukaryotic cell organization, our narrative would now have reached its Happy End. (Although if evolution *had* failed to advance beyond the humble prokaryotes, this story would, in all probability, not have been told in the first place.) But since the plant and animal kingdoms are populated by complex macroscopic clonal ensembles made up of a myriad of highly differentiated eukaryotic cells, it would be unreasonable to conclude this story before relating how and to what extent the lessons of the molecular genetics of prokaryotes pertain also to their more highly evolved and much more complicated fellow creatures. Admittedly, the presentations of the preceding chapters have made occasional reference to eukaryotes; indeed, some of the crucial evidence pertaining to this or that aspect of the auto- and heterocatalytic functions of DNA had been adduced from observations made with organisms as high on the evolutionary scale as mammals (including man). But the reader may have looked in vain for a commentary on the pertinence of the fundamental insights gained from the study of *E. coli* and its phages to the life of probably more familiar, and certainly more conspicuous, life forms. This last chapter will therefore provide a cursory survey of some of the ramifications of molecular genetics so far as they extend to, or give an account of, the nature of eukaryotic life. As will be seen, it is the existence of complex multicellular organisms rather than of prokaryotes that poses the most of the remaining unsolved deep problems of biology. These problems are therefore unlikely to be solved by future studies of *E. coli*. But it can be confidently expected that whatever the nature of the eventual solutions of these problems, they will rest on the intellectual bedrock of molecular genetics.

CHROMOSOME STRUCTURE

As was set forth briefly in Chapter 1, the discovery of chromosomes in the nuclei of eukaryotic cells and of the manner of their distribution over the nuclei of daughter cells in mitosis and meiosis eventually provided a concrete rationale for Mendel's laws of heredity. And the later work of T. H. Morgan and his school established that the eukaryotic chromosome constitutes a linear array of genes, of which, as was set forth in Chapter 2, DNA is a major constituent. Although so far there has been no explicit statement to that effect here, it has been taken for granted implicitly ever since DNA was demonstrated to be the genetic material of bacteria that DNA is also the genetic material of eukaryotes. That is to say, the genes of eukaryotes are also embodied in a stretch of double-helical polynucleotide chains into which the amino acid sequence of a polypeptide chain is inscribed via the universal genetic code. But as was already adumbrated in Chapter 2, the state of organization of the DNA in the eukaryotic chromosome is rather different from the more or less "free" form in which it exists in the bacterial cell. Miescher's first isolation of "nuclein" from pus cells and salmon sperm had shown a century ago that in eukaryotic nuclei the negatively charged, or acidic, DNA exists in association with an approximately equal weight of positively charged, or basic, proteins. At about the turn of the century, A. Kossel clarified not only the nature of the fundamental chemical constituents of DNA but also that of the basic proteins with which DNA is combined. The most prominent class of such basic proteins is formed by the *histones*, which are polypeptide chains about 50 to 200 amino acids in length. The histones owe their positive charge to their content of the three basic amino acids arginine, lysine, and histidine, which carry a second amino group in their side chains (see Figure 2-2) and make up nearly 25 per cent of the total amino acid residues. This high relative content of basic amino acids characteristic of histones can be compared with the compositional data of Table 4-1, where it may be seen that in β-galactosidase and tryptophan synthetase A protein of

E. coli, and also in bovine insulin, the basic amino acids range in amount from only 8 to 12% of the total amino acid residues. The association of chromosomal DNA and chromosomal histone appears to proceed by formation of an ionic bond between the phosphate groups of the polynucleotide chain and the side-chain amino groups of the polypeptide chain. Together DNA and histone make up about two-thirds of the weight of most chromosomes, the remaining weight usually being attributable to nonhistone proteins and RNA.

Despite their intensive study for more than eighty years, it cannot be said that the structural details of eukaryotic chromosomes have as yet been firmly established. Among the difficulties (but also fascination) presented by chromosome investigations at the molecular level is the variety of aspects that a given chromosome of a given organism may present both at different stages of the mitotic and meiotic cell cycles as well as in different tissues. Furthermore, the chromosomes of one organism may possess some features that are dramatically different from those seen in another organism. The description of a "typical" eukaryotic chromosome must, therefore, necessarily be as idealized an oversimplification as the picture of the "typical" eukaryotic cell. Hence the following streamlined account of what some students of chromosome structure presently believe to be the organizational plan of the genetic material of higher forms is more in the nature of a tentative, partial image than of universal gospel truth.

Eukaryotic chromosomes undergo a morphological life cycle that has some formal resemblance to the life cycle of phages. In *metaphase* of the cell division cycle (see Figure 1-7) the chromosome can be seen in the visible light microscope as a compact structure (Figure 1-6). The DNA is metabolically quiescent at this stage, serving as template neither for auto- nor heterocatalytic functions; it is densely compressed to form a tight parcel that can be moved to the site of the future nucleus of one of the two daughter cells. Thus the metaphase chromosome is analogous to the quiescent DNA molecule of the infective, resting phage particle, packaged for transit to a future host cell. In *interphase* of the cell division cycle the chromosome generally vanishes from view in the ordinary light microscope. Indeed, for many years there was some doubt whether the linear continuity of the chromosome is preserved during interphase. More recently, electron microscopy of interphase nuclei has shown that the continuity *is* preserved, and that the DNA merely distends from its tight metaphase package and fills the nuclear volume as ultrathin fibrils 200 to 300 Å in diameter, the appearance of which suggests that they are individual histone-coated DNA double helices. In interphase, the DNA is metabolically active, serving as template for its own replication in preparation for the next cell division and also for the transcription of messenger RNA. Thus the interphase chromosome is analogous to the vegetative phage DNA molecule as it is directing the events of intracellular phage growth. The molecular processes involved in the periodic condensation and distension of the chromosomal DNA still remain to be worked out, but it seems likely that histone-DNA and histone-histone interactions lie at the root of the interphase–metaphase transformation in chromosome morphology.

A eukaryotic chromosome, such as one of the human chromosomes, contains about 10^8 DNA base pairs, or some 30 times more than the 3×10^6 base-pair *E. coli* genome. The length of the double-helical DNA contained by the human chromosome is about $40,000\mu$, and since in its condensed metaphase aspect that chromosome is only about 4μ in length, it follows that the DNA must be very tightly packed. The topological relation between the chromosome and its DNA complement has not yet been firmly established. But the simplest view compatible with the available evidence is that the chromosomal DNA complement is simply one single, continuous, $40,000\mu$-long double helix that extends the length of the chromosome and that is coiled and supercoiled in the tight packing process. That is not to say that such an enormously long double helix is necessarily a covalently linked continuum; its two complementary polynucleotide strands might carry occasional staggered single breaks, or it might be a linear articulation of several different DNA molecules joined end to end by (purely hypothetical) protein "linkers." In any case, as was shown by an experiment carried out by H. J. Taylor in 1957, the eukaryotic chromosome behaves as if it *were* composed of a single, semiconservatively replicated DNA double helix (see Figure 9-2).

CHROMOSOME REPLICATION AND RECOMBINATION

Taylor's experiment consisted of exposing rapidly dividing root cells of the bean plant *Vicia faba* to ³H-thymidine and allowing incorporation of the labeled nucleotide precursor into the DNA replicas for a time equivalent to about one cell-division period. The labeled thymidine was then replaced by non-labeled thymidine, and the cells were allowed to grow for one more cell-division period. Samples of the root cells, taken both at the end of the original labeling period and at the end of the second, post-labeling division period, were mounted on a microscope slide, overlaid with a photographic film, and stored to allow production of an autoradiograph like the one discussed in connection with Cairns' study of the replication of the *E. coli* DNA complement (see Figure 9-12). The results of this experiment are presented in Figure 21-1, where two microscope fields show nuclei with metaphase chromosomes. In these metaphase nuclei the two sister chromosomes generated by the DNA replication step of the preceding interphase are situated next to each other, in preparation for their impending separation and transport to the opposite poles of the dividing cell in the coming *anaphase* (see Figure 1-7). The black grains visible above the sister chromosomes were formed by exposure of the photographic film by electrons emanating from decaying radioactive ³H

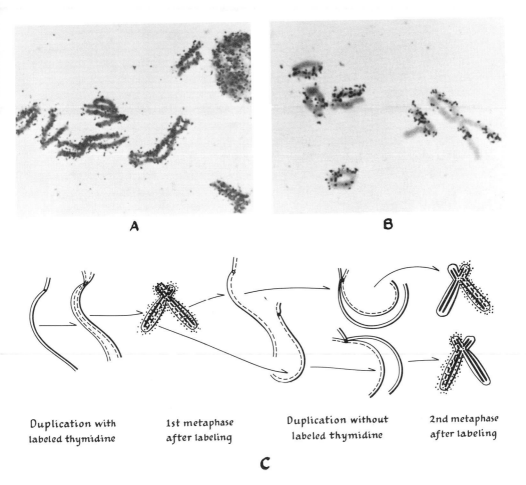

Duplication with
labeled thymidine

1st metaphase
after labeling

Duplication without
labeled thymidine

2nd metaphase
after labeling

C

FIGURE 21-1.   Semiconservative replication of eukaryotic chromosomes.
A. Autoradiograph of metaphase chromosomes of a nucleus in a root cell of the
bean *Vicia faber*, taken after one DNA replication cycle during which
^{3}H-thymidine was present in the solution perfusing the root cells.
B. Autoradiograph of a similar group of chromosomes after one DNA
replication cycle in the presence of ^{3}H-thymidine and one additional replication
cycle in the absence of radioactive DNA precursors. (Magnification 3200 times.)
C. Diagrammatic representation of the distribution of DNA strands during
chromosome duplication. Although individual DNA strands are not resolved
by microscopic examination of the autoradiographs, they are shown here
schematically as lines in order to indicate the manner of their distribution that
accounts for the observed labeling patterns of parts A and B. Broken lines
represent labeled strands; unbroken lines represent unlabeled strands. The dots
represent grains in the autoradiographs. [From J. H. Taylor, *Molecular Genetics.*
Academic Press, New York, 1963, Part I, pp. 74–75.]

atoms incorporated into the chromosomal DNA. Hence the grain density above any chromosome is proportional to the amount of ^{3}H-thymidine it contains. As is evident from the autoradiographs of Figure 21-1, the two sister chromosomes examined at the end of the original labeling period contain equally labeled DNA molecules, whereas of the two sister chromosomes examined after the second, post-labeling division period, one is labeled and the other is not. The interpretative cartoon of Figure 21-1 shows that this is precisely the labeling pattern to be expected from semiconservative replication of a chromosome composed of a single double-helical DNA molecule.

An autoradiographic experiment similar in principle to that presented in Figure 21-1 was later carried out with rapidly dividing cells of the Chinese hamster, except that in that experiment the labeling of both the chromosomal DNA and of the chromosomal protein was followed in successive division periods. Although the DNA label was found to follow the same semiconservative distribution that Taylor had previously found in the bean root, the chromosomal protein label was found to behave rather differently. In the first metaphase after exposure to ^{3}H-labeled amino acids, the protein of both sister chromosomes showed an equal level of labeling; but after three more division cycles in the absence of ^{3}H-amino acids the protein label had practically vanished from the metaphase chromosomes. Thus it could be concluded that the proteins, in particular the histones, are only transient constituents of the chromosome, whose material as well as informational continuity is provided only by the DNA molecule.

It should be noted that the autoradiograph in Figure 21-1,B, taken after the second post-labeling division period, shows that there have occurred *reciprocal exchanges* of DNA between the two sister chromosomes. As can be seen, in one sector the first of the two sister chromosomes does and the second does not contain the label, whereas in another sector it is the second that does and the first that does not contain the label. These sister chromosome exchanges can be readily attributed to the same mechanism of daughter DNA strand exchange that we have previously considered in Chapter 15 in connection with the post-replication repair of UV lesions of the *E. coli* DNA. That is to say, according to a proposal made by H. L. K. Whitehouse in 1963, after replication of the DNA molecule of the parental chromosome there may occur, as is shown in Figure 21-2, staggered breaks of the two nascent, complementary daughter DNA strands, followed by a limited replicative extension along the parental template strands of the broken ends. The two overlapping sectors of the daughter strands may then pair and thus effect a crosswise reunion of daughter DNA strands of the two future sister chromosomes. Breakage of the parental DNA strands would then allow resolution of the topological impasse created by the crosswise reunion of the daughter strands into a pair of recombinant "joint molecules." Finally, excision of redundant sectors of the daughter strands, followed by repair replication and ligase

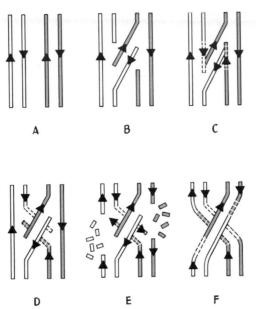

FIGURE 21-2.   The Whitehouse model of eukaryotic chromosome crossover in DNA molecules. The polynucleotide strands of the two parental molecules are distinguished by white and shaded lines. The arrowheads indicate the $5' - 3'$ polarity of the strands as well as the direction of synthesis of the new DNA. The interrupted lines represent newly synthesised DNA of complementary base sequence to the paired strand of opposite polarity. The fragments in **E** indicate DNA breakdown of the outer, unpaired parental strands. For simplicity, the base pairs that unite the strands of the double helices by hydrogen bonding are not shown. **A**. Homologous regions of the parental DNA molecules. **B**. Breakage at different sites, of two parental strands of opposite polarity, and their crosswise pairing to form a hybrid duplex region. **C**-Synthesis of new DNA strands by copying unpaired regions of parental strands. **D**. Separation from their parental template strands and crosswise pairing of newly synthesised DNA strands. **E**. Excision of redundant, unpaired regions of parental strands. **F**. Repair DNA synthesis and covalent bonding of strands of the same polarity. [After Whitehouse, *Nature*, **199**, 1034 (1963).]

action, would convert the two "joint molecules" into a pair of covalently linked recombinant double helices.

It does not seem far-fetched to imagine that the process depicted in Figure 21-2 is responsible also for the genetic recombination by crossover of homologous (rather than sister) chromosomes which occurs in the first of the two cell divisions of meiosis (Figure 1-13). That is to say, it seems probable that the actual molecular event of breakage and reunion of the DNA molecules of the two homologous chromosomes occurs during the interphase preceding the first meiotic division, while their homologous DNA molecules are being replicated and are in their extended form. The point-by-point alignment or *synapsis* of the homologous chromosomes which becomes manifest upon the prophase of the first meiotic division would, therefore, be the aftermath rather than the prelude to the genetic exchanges of chromosomal DNA molecules. That is, these genetic exchanges have resulted in an entanglement of the homologous chromosomes, so that condensation of the two pairs of sister chromosomes that are to assume the compact prophase form must occur side-by-side. The *chiasmata* (or chromosomal crossover points) (see Figure 1-13) that are visible under the microscope at this stage of meiosis are, from this point of view, merely the souvenir rather than the cause of chromosome recombination, whose molecular processes occurred long before in the preceding interphase.

CHROMOSOMAL REDUNDANCY

The picture of the "typical" eukaryotic chromosome as being composed of a single enormously long DNA molecule needs elaboration from at least two points of view. One important departure from this picture is provided by the *giant chromosomes* present in some tissues of the larvae of certain species of insects, such as the fruit fly Drosophila and the midge Chironomus (Figure 21-3). A giant chromosome is more than ten times longer, is almost a hundred

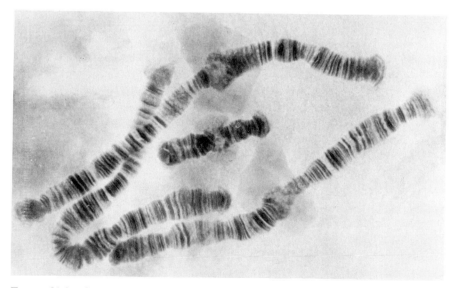

FIGURE 21-3.   Set of four giant chromosomes from a cell in the salivary gland of the midge *Chironomus tentans*, at a magnification of about 700 diameters. The enlarged regions on two of the long chromosomes are nucleoli. Chromosome IV is the shortest of the four; its enlarged region, or puff, reflects a massive transcription of mRNA. The banding pattern on each of the four chromosomes is visible in corresponding giant chromosomes from entirely different tissues. [Photograph courtesy Wolfgang Beermann. From "Chromosome Puffs," *Scientific American*, April 1964.]

times thicker and contains a thousand times more DNA than the corresponding normal chromosome found in other tissues of the same animal. The giant chromosome appears to represent a lateral association of a thousand identical sister DNA double helices, all of which have descended from the single parental double helix of a normal precursor chromosome by ten successive replication cycles ($2^{10} = 1024$). Thus a cell containing such giant chromosomes carries its genetic information in a thousand-fold redundancy.

A second example of chromosomal redundancy came to light only in 1966. Whereas in the giant chromosomes the *entire* DNA complement is multiplied a thousand-fold over its normal chromosomal content, even the single DNA molecules thought to be carried by some *normal* chromosomes of eukaryotic organisms turn out to be highly redundant. This redundancy is only partial, however, in that some gene-length DNA nucleotide sequences are present in but a single copy, whereas others are present in hundred-, thousand-, or even millionfold multiplicity. This discovery was made by R. J. Britten, who had undertaken a refined study of the kinetics of renaturation of denatured DNA, previously mentioned in Chapter 8. As was seen in Figure 8-11, the rate at which the single complementary DNA strands reassociate to reform an intact double helix decreases with the genetic complexity of the organism whose DNA has been denatured. For the greater the number of different nucleotide base sequences present in the denatured DNA sample, the lower the probability per unit of time that any given single polynucleotide chain might collide with a chain of complementary sequence. Figure 21-4 presents Britten's graphical summary of the experimental proof of this contention. This graph presents the fraction of the dissociated complementary DNA strands that have reassociated to form a stable double helix within a "normalized time" $C_0 t$. (Since the rate of reassociation of the complementary strands is obviously proportional to the total DNA concentration C_0, the product of C_0 times the actual elapsed time t provides a concentration-independent, normalized time parameter which allows direct comparison of experiments carried out at widely different DNA concentrations.) As can be seen, the normalized time $C_0 t$ required for reassociation to reach *half* completion covers a 10^9-fold range for the five polynucleotide samples studied. The most rapid reassociation rate is observed for the complementary, monotonous synthetic polynucleotides polyuridylic and polyadenylic acids, which can be thought of as a "genome" represented by a single base pair. The time required for half-reassociation for T4 phage DNA is seen to be 10^5 times, for *E. coli* DNA 5×10^6 times, and for calf-thymus DNA 10^9 times greater than that required for polyuridylic and polyadenylic acids. Since the genome of T4 comprises 2×10^5, that of *E. coli* 3.2×10^6 and that of the calf 3.2×10^9 nucleotide base pairs, this finding demonstrates that the rate of strand reassociation of DNA molecules is a quantitive measure of their sequence complexity.

It must be noted, however, that the calf-thymus DNA used in the renaturation experiment of Figure 21-4 is actually a fractionated sample from which

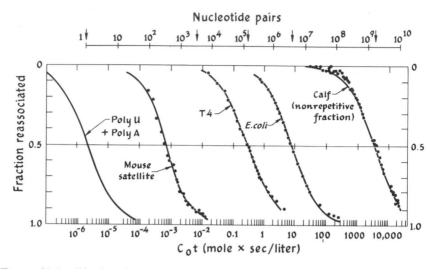

FIGURE 21-4.   Kinetics of reassociation of dissociated double-stranded DNA of diverse provenance. The genome size in terms of nucleotide base pairs of each DNA source is indicated by the arrows on the logarithmic scale at the top. The denatured DNA samples were fragmented by mechanical shear to chain lengths of about 400 nucleotides and incubated at a temperature near 60°C, previously ascertained as being optimal for their reassociation into double helices. The UV-absorption at 260 mμ of the solutions was measured periodically, in order to follow the strand-reassociation process by means of the 40% decrease in UV-absorbance attending reformation of the double helix (see Chapter 8). The "fraction reassociated," registered on the ordinate, is estimated from the UV-absorbance A as $(A_d - A)/(A_d - A_i)$, where A_d and A_i are the UV-absorbances of the corresponding completely denatured and completely intact DNA samples, respectively. The abscissa registers the "normalized time" C_0t, which is the product of the initial concentration of the denatured DNA sample (in moles of nucleotides per liter) in the reaction mixture multiplied by the number of seconds elapsed since the solution was heated to the reassociation temperature, plotted on a logarithmic scale. As is evident, the C_0t time required for half-reassociation is proportional to the number of nucleotide base pairs in the genome of the DNA source. [After R. Britten, *Science* **161**, 529 (1968). Copyright 1968 by American Association for the Advancement of Science.]

a minor but substantial component has been removed. Figure 21-5 presents the renaturation kinetics of *unfractionated* calf-thymus DNA, from which it can be seen that calf-thymus evidently has two components. The strands of one of these components, which amounts to about 40% of the total, reassociate very rapidly; half of them are back together by the time C_0t has the value of 0.03 molar seconds. The strands of the other component, which amounts to about 60% of the total, reassociate at the very slow rate seen in Figure 21-4; half of them are back together only after C_0t has reached the value of 3,000 molar seconds. Thus, according to the previously inferred inverse relation between strand association rate and sequence complexity, it can be concluded that the rapidly reassociating 40% fraction is composed

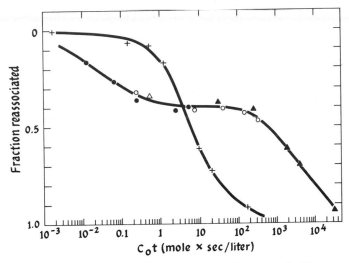

FIGURE 21-5.   Kinetics of reassociation of unfractionated calf-
thymus DNA in four experiments with four different DNA
concentrations: open triangles, 2 μg/ml; closed circles, 10 μg/ml;
open circles, 600 μg/ml; closed triangles 8600 μg/ml. The crosses
show the renaturation kinetics of *E. coli* DNA at 43 μg/ml under the
same conditions. [After R. Britten, *Science* **161**, 529 (1968).
Copyright 1968 by American Association for the Advancement of
Science.]

of DNA molecules each of which is present in $3000/0.03 = 10^5$-fold redundancy
per genome. The reassociation kinetics of denatured mouse DNA are qualitatively similar to those shown for calf DNA in Figure 21-5, except that here
there is manifest a 10% component ("satellite") that renatures even more
rapidly than the redundant fraction of calf DNA. As can be seen in Figure
21-4, half of the mouse "satellite" DNA fraction has reassociated before
C_0t has the value of 0.001 molar seconds, which indicates that the molecules
of this fraction must be present in more than 10^6-fold redundancy per genome.
A subsequent survey of the DNA of a wide range of eukaryotic organisms,
drawn from both plant and animal kingdoms, showed that the presence of
redundant DNA sequences is a well-nigh universal phenomenon, although
the proportion of redundant DNA, as well as the multiplicity of the redundant sequences, is subject to wide, species-specific variations. No evidence for
a significant fraction of redundant DNA sequences has been found in any
prokaryotic organism or virus, however.

The development of the renaturation analysis held out promise of an
explanation for the long-known and surprising fact that the DNA genome of
eukaryotic organisms covers a thousandfold span, ranging from 10^8 base
pairs for sponges and jellyfish through 10^9 base pairs for annelid worms
and crabs, through 3×10^9 base pairs for fish and mammals to 10^{11} base

pairs for some amphibia, such as the salamander *Amphiuma*. Though it would seem reasonable that the simpler invertebrate animals would have fewer genes than the more highly evolved vertebrates, the quantitative relations within the vertebrate orders, particularly the very high DNA content of the amphibian genome, are rather difficult to fathom from the straightforward viewpoint of organismal complexity. (It is also surprising that the genome of as highly evolved an organism as the chordate *Asidea astra* contains no more base pairs than the genomes of sponges and jellyfish.) Hence a more plausible explanation was to attribute the wide span in amount of genomic DNA to the differential preponderance of redundant polynucleotide sequences in the chromosomes of different species. To test whether salamanders do owe their prodigous DNA content to an overwhelming redundancy of nucleotide sequences, Britten studied the renaturation kinetics of *Amphiuma* DNA molecules. He found that 80% of that DNA is indeed composed of molecules capable of rapid reassociation and hence highly redundant. The remaining 20%, however, renatured so slowly that the $C_0 t$ value for half-complete reassociation was 20,000 molar seconds, or seven times greater than the corresponding time for calf DNA. Thus it appears that the high DNA content of *Amphiuma* cannot be attributed simply to an overwhelming redundancy of its nucleotide sequences: instead the salamander chromosomes do seem to contain a greater number of unique, nonrepeated sequences than do those of the mighty mammals.

The significance of the observed nucleotide-sequence redundancy of eukaryotic organisms still remains to be fully fathomed. Two particular situations in which the manifold duplication of some particular genes does, or might, play some significant role will be considered later in this chapter. But to neither of these situations can we attribute the generality and extent of repeated DNA sequences among eukaryotes. Britten has proposed that these repeated sequences are the reflection of evolutionary processes. He envisages that there occurs a sudden, or saltatory, manifold replication of some chromosomal nucleotide sequence in a germ-line cell of a eukaryotic organism. The many copies of the replicated sequence are then transmitted to the descendants of that organism, and in this transmission process they are free to accumulate mutations, most of which would have been lethal if the organism had carried but a single nonredundant copy of the reduplicated sequence. In this way an ensemble of novel nucleotide sequences can arise gradually and nondetrimentally, and as soon as one of these sequences happens to generate a novel functional polypeptide chain of adaptive value, it will spread through the interbreeding population by natural selection. Although it seems plausible that novel genes should arise by such a process of nucleotide sequence replication followed by mutational drift of redundant genes, it is not evident why a millionfold or even a thousandfold, rather than a mere two- or threefold, replication of an individual sequence should attend the first step in this process.

NUCLEAR-CYTOPLASMIC RELATIONS

To a first approximation, the two stages of the heterocatalytic function of the eukaryotic DNA can be said to take place in topologically complementary cellular compartments: transcription of DNA polynucleotide sequences into mRNA (as well as into rRNA and tRNA) molecules occurs within the space bounded by the nuclear membrane (i.e., in the nucleus), whereas translation of mRNA into polypeptide chains occurs without (i.e., in the cytoplasm). Thus RNA molecules arise in eukaryotes, just as they do in bacteria, through template service of the chromosomal DNA for the RNA polymerase-catalyzed assembly of ribonucleoside triphosphates. And hence the RNA that is found to be associated with eukaryotic chromosomes consists of *nascent* RNA molecules that were in the process of being transcribed at the moment the chromosome was examined. In contrast to the situation in bacteria, however, where (as was seen in Chapter 17) translation of the head of the nascent mRNA molecule not only starts but may even end before the tail has been transcribed, in eukaryotes the nascent mRNA must be transported from its chromosomal site of origin to its cytoplasmic destination where it is to service ribosomes in protein synthesis. Hence in eukaryotes transcription and translation are unlikely to be closely coupled processes. The nature of mRNA transport remains to be discovered, although it is known that in the nucleus the nascent mRNA molecules are associated with protein and reach the cytoplasm in the form of nucleoprotein particles. Once in the cytoplasm, the lifetime of a eukaryotic mRNA molecule can be much longer than the fleeting existence of bacterial mRNA molecules. For instance, the mRNA molecules directing hemoglobin synthesis in rabbit-blood reticulocytes have lifetimes of days rather than minutes, although other mRNA species in other rabbit tissues have much shorter lifetimes.

The ribosomes of the eukaryotic cytoplasm resemble those of prokaryotes as far as their general structure and function are concerned, but the eukaryotic ribosomes are somewhat larger than prokaryotic ribosomes (they sediment with a specific velocity of 80S and contain about 10% more RNA per particle than do the 70S ribosomes of *E. coli*). They also show some behavioral differences, such as being insensitive to the action of such drugs as chloramphenicol, which are potent inhibitors of the function of bacterial ribosomes. Furthermore, it appears that on eukaryotic ribosomes polypeptide chain growth is not initiated by formylmethionyl tRNA, as was described for *E. coli* in Chapter 17, although the nature of the alternative initiation mechanism that does obtain here has not yet been worked out.

To a second approximation, however, the topological compartmentalization of transcription and translation in eukaryotes is not all that clear. First, it has been reported that *some* protein synthesis does proceed also in the nucleus. Examples of the observations upon which these reports are based are that isolated nuclei were found to be capable of incorporating labeled amino acid into proteins and that autoradiographic examination of cells exposed to short pulses of ^{3}H-labeled amino acid revealed the intracellular presence of labeled proteins at a time thought to be too early for these proteins to have been formed in the cytoplasm and transported into the nucleus. Second, transient RNA-labeling of a variety of animal cells has shown that a considerable fraction of the nascent RNA molecules never reach the cytoplasm and are broken down in the nucleus after a brief lifetime. The meaning of this phenomenon remains to be explained, but it might reflect the existence of mRNA molecules consecrated to the synthesis of (putative) nuclear protein on (putative) nuclear ribosomes.

Although the *nucleolus* (see Figure 1-5) was one of the first nuclear land-marks seen by early nineteenth-century explorers of cell structure, its function long remained mysterious. It was shown in the 1930's that formation of the nucleolus during interphase depends on a specific chromosomal locus, the *nucleolus organizer*, and that the number of nucleoli visible per nucleus is equal to the number of nucleolus organizer loci present in the cell genome. By 1940, Brachet had found that the nucleolus is a domain of high concentration of RNA, and after the role of ribosomes in protein synthesis had been elucidated in the 1950's, the notion became established that the nucleolus is the site of their formation. In support of this view it could be shown that a deletion mutation of the nucleolus organizer locus results in an incapacity for ribosome synthesis of an amphibian embryo carrying that mutation in the homozygous state. Furthermore, RNA-DNA hybridization tests carried out by Spiegelman demonstrated that the nucleolus organizer is that sector of the chromosome in which the nucleotide sequence of rRNA is inscribed. In these experiments, chromosomal DNA was extracted from three different Drosophila mutant strains that carried the abnormal numbers of 1, 3, and 4 nucleolus organizer loci per diploid genome, as well as from normal Drosophila males and females that carried 2 nucleolus organizers. The various DNA extracts were then tested for their capacity to hybridize with rRNA extracted from Drosophila ribosomes, in a manner entirely analogous to that described in the legend of Figure 16-6. The result of this experiment is presented in Figure 21-6, where it can be seen that the hybrid acceptor capacity for rRNA of any DNA extract is exactly proportional to the number of nucleolus organizer loci the corresponding Drosophila genome is known to contain. The final plateau value, or maximum rRNA acceptor capacity, of the normal Drosophila DNA can be seen to be 0.27%. Since the haploid Drosophila genome consists of about 5×10^8 nucleotide base pairs, and since the rRNA has a total chain length of about 5000 nucleotides, it can be inferred that each nucleolus organizer locus embodies a repetition $0.27 \times 10^{-2} \times 5 \times 10^8 / 5000 = 300$ times that of the rRNA nucleotide sequence. The nucleolus organizer thus provides one accountable instance of chromosomal DNA

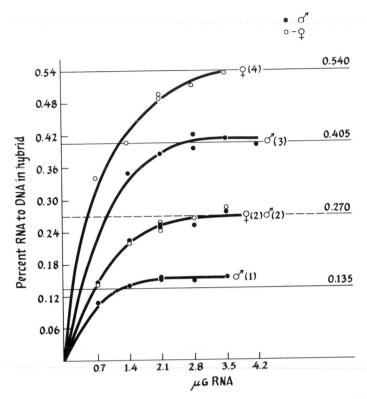

FIGURE 21-6.   Hybridization of rRNA extracted from Drosophila ribosomes to DNA extracted from Drosophila flies carrying the dosages of the nucleolus organizer region indicated on each curve in parentheses. The broken horizontal line at the ordinate value 0.270 % is assumed to be the correct estimate for the rRNA hybridization capacity of the DNA of normal male and female flies carrying two nucleolus organizer regions. The solid horizontal lines indicate the hybridization plateau values expected for the abnormal dosages 1, 3, and 4 nucleolus organizer regions on the basis of the normal plateau value of the dotted horizontal line. The abscissa registers the amount of rRNA added to the hybridization mixture containing a fixed amount of DNA. [After F. M. Ritossa and S. Spiegelman, *Proc. Natl. Acad. Sci. Wash.* **53**, 737 (1965).]

sequence redundancy, but it is unlikely that this comparatively modest amount of sequence repetition can make much of a contribution to the overall DNA strand reassociation kinetics presented in Figure 21-5.

But in amphibian oocytes [the cells that undergo meiosis to become eggs (see Figure 1-8)] the nucleolus organizer DNA (itself containing hundreds of copies of the rRNA sequence) is multiplied about a thousandfold to produce a thousand *extrachromosomal* nucleoli. Thus here the DNA sequences encoding the rRNA are present in the 10^5 to 10^6-fold redundancy noticed by Britten in his experiments. This selective replication of DNA templates bearing the rRNA sequence information allows for rapid synthesis of the massive numbers of ribosomes with which the amphibian egg is to be endowed. The extrachromosomal nucleoli were isolated intact from the oocyotes and subjected to direct electron-microscopic examination by O. L. Miller. Miller's pictures show that the nucleolus organizer replica is present as a circular, double-helical DNA molecule coated with protein. At periodic intervals along the circular DNA molecule matrix elements can be seen, a few of which are shown in the frontispiece. These matrices are nothing less than a direct visualization of the genetic transcription process. Each matrix consists of about 100 thin fibrils connected by one end to the DNA core and increasing in length from one end of the element to the other. The sensitivity of these fibrils to RNAse and proteolytic enzymes, as well as their radioactive labeling pattern as followed by autoradiography, indicates that they are nascent RNA molecules coated with protein. Since the length of DNA core supporting one matrix element corresponds to that of a 5000- to 6000-nucleotide base-pair double helix, it seems reasonable to infer that each such core segment is a gene into which the rRNA nucleotide sequence is encoded. Thus it can be concluded that just as the nascent mRNA molecules are associated with protein in the nucleus, so are the nascent rRNA molecules part of nuclear nucleoprotein particles. These ribosomal precursors are then transported from nucleolus to cytoplasm, where they undergo a maturation process and achieve the status of functional ribosomes.

MITOCHONDRIA

A further complication for the first-approximation view that the nucleus is the exclusive province of the genetic transcription process is presented by cytoplasmic organelles such as mitochondria (Figures 2-8 and 21-7). Mitochondria are the seat of the main energy-yielding reactions of the eukaryotic cell, such as citric acid cycle and oxidative phosphorylation (see Chapter 3). Because discussion of the details of mitochondrion structure and function would carry us beyond the intended scope of this text, suffice it to say that mitochondria had long been suspected of being self-reproducing entities endowed with their own genetic continuity. Thus mitochondria were observed to increase in number by elongation and division, in a manner entirely analogous to the reproduction of bacteria, which they resemble in external size and shape (though not in internal structure). Furthermore, hereditary variations of mitochondrial structure and function were found whose transmission to the offspring organism does not obey the Mendelian segregation pattern of nuclear genes. Instead, the genetic elements responsible for these variations can be demonstrated to reside in the mitochondria themselves. Isolated mitochondria were found to be capable of incorporating radioactive amino acids into proteins and subsequently shown to contain such elements of the protein synthesizing machinery as ribosomes, tRNA, and aminoacyl

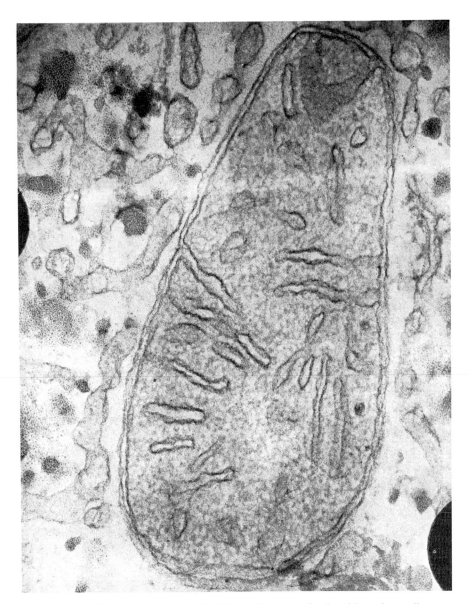

FIGURE 21-7.   Electron micrograph of a thin section of a mitochondrion of a rat-liver cell enlarged some 100,000 times. The enclosed spaces visible inside the mitochondrion are *cristae*, formed by flattened infoldings of the outer membrane. The cristae are thus filled with parts of the surrounding cytoplasmic fluid. One crista lying almost in the plane of the section forms the wide **V** at the top. Other cristae, seen near the center, cut at right angles, project like fingers from the outer membrane. The connections to the outer membrane do not show in those cristae that happen to have been cut at oblique angles. [Micrograph courtesy Michael Watson. From "Energy Transformation in the Cell," by A. L. Lehninger, *Scientific American*, May 1960.]

tRNA synthetases. Finally, it was discovered in 1963 that mitochondria contain their own DNA complement, in the form of a double-helical, circular molecule some 2×10^4 nucleotide base pairs in length. In many organisms, the mitochondrial DNA has a $[G] + [C]$ content different from that of the nuclear chromosomal DNA, and hence the former can be separated from the latter on the basis of a differential buoyant density in CsCl gradient sedimentation. (Upon separating nuclear from mitochondrial DNA it can be shown that mitochondrial DNA is *not* mainly responsible for the redundant nucleotide sequences discovered by Britten.) More detailed study of the ribosomes of mitochondria later revealed that they have the characteristics of *prokaryotic* ribosomes, rather than of the eukaryotic ribosomes present in their cytoplasmic surround: mitochondrial ribosomes sediment at the velocity of 70S, are sensitive to the inhibitory action of chloramphenicol, and initiate polypeptide chain growth by means of formylmethionyl tRNA. Upon these discoveries the idea gained currency in the late 1960's that mitochondria are degenerate prokaryotic microorganisms that live in symbiosis with their eukaryotic host, in whose cytoplasm they perform the yeoman service of degradative metabolism. It is apparent, however, that since the mitochondrial DNA complement of only 2×10^4 nucleotide base pairs amounts to less than 1 % of that of the *E. coli* genome, it can hardly encode the polypeptide chain sequences of all of the many different protein species that make up the mitochondrion. Hence the mitochondrion can be built up only in part from self-made components and must consist mainly of exogenous members synthesized on cytoplasmic ribosomes under the direction of the nuclear chromosomal DNA.

At least two other kinds of organelles found in the cytoplasm of eukaryotes appear, like mitochondria, to be degenerate prokaryotic symbionts. One of these is the *chloroplast*, the component of plant cells in which there proceed the main events of photosynthesis—the phenomenon by which the radiant energy of sunlight is converted into the chemical energy of ATP. Another such organelle is the *centriole*, which directs the migration of sister chromosomes to opposite poles of the cell during mitosis (see Figure 1-7). Thus a eukaryotic cell may be thought of as an empire directed by a republic of sovereign chromosomes in the nucleus. The chromosomes preside over the outlying cytoplasm in which formerly independent but now subject and degenerate prokaryotes carry out a variety of specialized service functions.

CELL DIFFERENTIATION

Chapter 2 began with a homily which preached that before asking such questions as how the genes of the parental Drosophila germ cells manage to give rise to an entire offspring fly, one must inquire into the processes by means of which the genes preside over the formation of *cellular* structures and components in the successive cycles of cell growth and division. Having now laboriously completed this inquiry we can at last ask how the genes of a fertilized Drosophila egg *do* manage to give rise to an entire offspring fly made up of billions of highly differentiated cells. This question is an epitome of the goal of the venerable discipline of *embryology*, founded by Aristotle. Since its beginnings in the fourth century B.C. embryology has accumulated a vast amount of descriptive detail about the developmental processes that lead from fertilized egg to mature multicellular organism. But until now, relatively few unifying principles have come to light that can explain how different members of the cell clone descended from the egg assemble and differentiate in an orderly manner into the bewildering spectrum of specialized cell types. Now, in retrospect, it seems evident that these processes hardly *could* have been explained before the autocatalytic and heterocatalytic functions of DNA were understood, though it is not certain, of course, that the presently known principles of molecular genetics, albeit necessary, are also sufficient for solving the enigma of the embryo.

The first coherent embryological theory was developed by A. Weismann in the 1890's at the time when, as was set forth in Chapter 1, he elaborated the chromosomal theory of inheritance that led to the later rediscovery of Mendel's laws. Weissmann proposed that cell differentiation attending embryonal development is the consequence of an *unequal partition* of the hereditary determinants in successive cell divisions. That is to say, he envisaged that cell differentiation is the consequence of *nuclear differentiation*, which is, in turn, the consequence of a progressive *loss* of parts of what would now be called the parental genome. That genome would be preserved intact only during cell divisions in the *germ line*, so that the germ cells eventually produced by meiosis in the sexually mature adult can pass on to the offspring the complete genetic patrimony. With the rise of classical genetics in the first part of this century and the demonstration that every body cell has exactly the same chromosome complement as the fertilized egg that spawned it, Weissmann's theory came into general disfavor. But, as a matter of fact, no really *critical* evidence had been adduced during all that time which proved that a covert genetic differentiation of the nucleus is not, after all, responsible for the overt phenotypic differentiation of the cell. Finally, in the 1950's, tests of Weissmann's theory in its most general form became possible when techniques were developed by means of which the nucleus of the differentiated body cell of an amphibian could be transplanted to an egg of the same species whose own nucleus had been previously inactivated by UV-irradiation. The first experiments of this type did seem to lend support to the theory of nuclear differentiation, in that it was found that the only UV-irradiated eggs capable of giving rise to a normal embryo, and eventually to an adult animal, were those to which the nuclei of fertilized eggs or of cells at the very earliest stages of embryonic development had been transplanted. Transplantation of nuclei from differentiated body cells of more mature embryos, however,

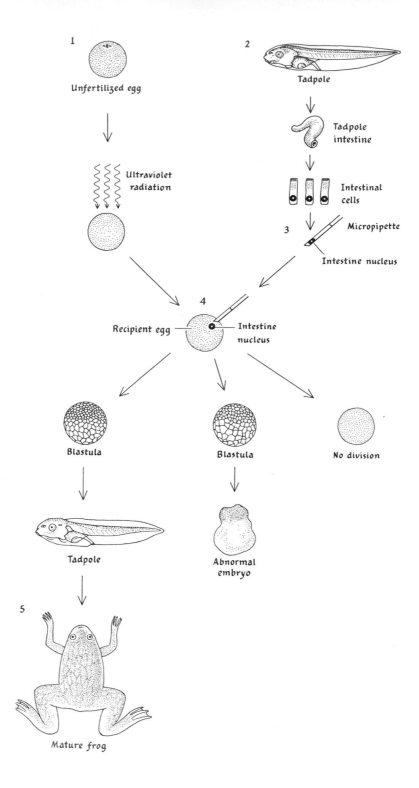

1 Unfertilized egg

2 Tadpole

Tadpole intestine

Intestinal cells

3 Micropipette

Intestine nucleus

Ultraviolet radiation

4 Recipient egg — Intestine nucleus

Blastula

Blastula

No division

Tadpole

Abnormal embryo

5 Mature frog

invariably seemed to lead to abortive embryonal development. But in 1964 J. B. Gurdon was able to show that it *is* possible to obtain normal adult frogs from frog eggs into which the nuclei of differentiated cells of the tadpole intestine had been transplanted (Figure 21-8). Thus Gurdon's experiment finally disposed of the nuclear differentiation theory, in that it showed that the nucleus of the differentiated intestinal cell still carries all the genes necessary for instructing the frog egg how to produce a whole frog.

Thus it can be concluded that cell differentiation is not the consequence of a permanent change in character of the cell genome but must instead derive from a *differential expression* of the myriad of genes embodied by that genome. That is, the explanation of embryonic development must be sought in terms of mechanisms of the *regulation* of gene function, such as those discussed for prokaryotes in Chapter 20. An example of just such differential gene expression in development was already mentioned earlier in this chapter in relation to the thousandfold replication of the nucleolus organizer DNA in amphibian oocytes. It is to be noted that this mechanism of regulation of gene function by facultative replication of individual genes to increase a specific RNA transcription template capacity does not appear to be in use among prokaryotes (and hence was *not* discussed in Chapter 20). After doing further nuclear transplantation experiments Gurdon was also able to shed some light on how the genes of the embryonal cells are instructed in the process of differential expression. These experiments were based on previous observations that had shown that during the first ten cell divisions in the development of the frog embryo (Figure 21-9) there occurs practically no synthesis of RNA in the nuclei, while the actively dividing cells are rapidly replicating their DNA and make do for their protein synthesis with the enormous, RNA-rich cytoplasm of the egg parceled out among them. In other words, during that earliest embryonic stage all protein synthesis proceeds according to plans laid down in the egg by the maternal genome before fertilization, without any informational contribution by the paternal genome.

FIGURE 21-8.   Transplantation of intestinal cell nuclei into frog eggs. The procedure starts with the preparation of an unfertilized frog egg (1) for receipt of the nucleus of another cell, by destroying the egg cell nucleus through exposure to UV light. Next, the intestine is taken from a tadpole that has progressed far enough in its development to feed (2) and part of the intestinal tissue is dispersed to yield a suspension of single cells. A single intestinal cell is then drawn into a micropipette. This operation breaks the cell wall and liberates the cell nucleus (3). The nucleus is then injected into the UV irradiated egg (4), which is allowed to develop. Only about 1% of such transplants are successful, in that the impregnated egg develops into a mature frog (5), all of whose cell nuclei are descendants of the donor intestine nucleus. In the remaining, or unsuccessful, transplants the impregnated egg either gives rise to an abnormal embryo or does not divide at all. [From "Transplanted Nuclei and Cell Differentiation," by J. G. Gurdon, *Scientific American*, December 1968. Copyright © 1968 by Scientific American, Inc. All rights reserved.]

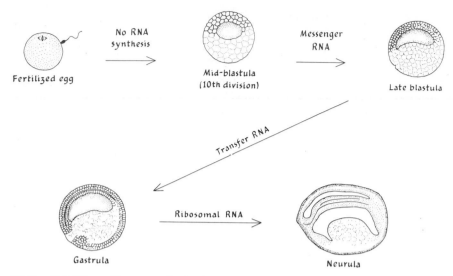

FIGURE 21-9. The earliest stages in the development of the frog egg. As the normal embryo develops, its cells are engaged in rapid DNA synthesis throughout the developmental stages shown here. Before the mid-blastula stage, however, virtually no RNA synthesis occurs. Synthesis of mRNA begins after mid-blastula; of tRNA, after late blastula; and of rRNA, only after the gastrula stage. [From "Transplanted Nuclei and Cell Differentiation," by J. G. Gurdon, *Scientific American*, December 1968. Copyright © 1968 by Scientific American, Inc. All rights reserved.]

After the tenth division, or at the mid-blastula stage, there begins synthesis of new mRNA species in the nuclei of the approximately 1000-cell embryo, followed by the onset of synthesis of new tRNA a few divisions later at the late blastula stage. Synthesis of rRNA (and formation of new ribosomes) begins only at the gastrula stage, by which time the embryo differentiates into the three primary cell types *ectoderm*, *endoderm*, and *mesodern*. (Since up to this stage of development the cells make use exclusively of the dowry of maternal ribosomes provided in the egg, embryos whose genome is homozygous for a deletion mutation of the nucleolus organizer develop normally at first. It is only after this point that the genetic incapacity to synthesize new ribosomes exerts its lethal effect.)

Gurdon then followed the synthesis of DNA and RNA in embryos descended from frog eggs into which nuclei from later developmental stages had been transplanted. In one such experiment, brain cells of the adult frog were used as nuclear transplant donors. Adult brain cells no longer divide, and hence their nuclei no longer replicate the chromosomal DNA; there is, however, a very active RNA synthesis in such brain cells. Within a short time of their transplantation to the egg, the brain-cell nuclei stopped their active RNA synthesis and started to replicate their DNA—that is, they began to behave as egg nuclei rather than brain-cell nuclei. (This finding helps to explain why the first attempts at transplantation of nuclei from differentiated cells failed to give rise to normal embryos. Apparently, the post-transplantation induction of rapid DNA synthesis and division in replicatively quiescent donor nuclei leads to lethal chromosomal abnormalities unrelated to the processes of normal embryogenesis.) In another such experiment it was shown that after RNA synthesis has been shut off following transplant into an egg of a late-stage nucleus previously active in genetic transcription, synthesis of the three classes of RNA resumes in the resulting embryo according to exactly the same temporal pattern as in a normal embryo. It can be concluded, therefore, that the differential gene expression in development results from a mutual interaction between nucleus and cytoplasm: the fertilized egg starts its life with a cytoplasm programmed to hold transcription of the chromosomal DNA in check. As development proceeds, the character of the cytoplasm changes so as to allow expression of some part of the genome. This expression, in turn, causes further changes in the cytoplasm, which, in turn, cause further changes in the nature of gene expression.

The existence of giant chromosomes in tissues of insect larva has made possible a direct visualization by W. Beermann and U. Clever of the pattern of differential gene expression in development. Inspection of the four giant chromosomes of the Chironomous salivary gland shown in Figure 21-3 reveals that three of them have large protuberances, or *puffs*. The puffs of the two long chromosomes are nucleoli in which active transcription of rRNA from the nucleolus organizer DNA is proceeding. The puff of the short chromosome, labeled chromosome IV, is an area within which there proceeds massive transcription of mRNA. Plate IV shows a more detailed micrograph of chromosome IV stained with a dye that colors DNA blue and RNA reddish-violet. It is evident from this picture that the puff contains an abundance of RNA. Examination of the fine-structure of such puffs under the electron microscope reveals that they owe their appearance to a local unraveling of the thousand parallel, tightly coiled, protein-coated DNA double strands from which the giant chromosome is built. Each of these strands can be seen as the core of fibrillar matrix elements that resemble the nucleolar transcription matrices of oocytes, shown in the frontispiece. Thus a puff is a directly visible manifestation of the active transcription of a particular set of genes situated in the puffed sector of the chromosome. With this insight it became feasible to "see" the dynamics of differential gene expression in Chironomous by examining the puffing pattern of giant chromosome IV in the salivary glands at various times during the metamorphosis of the larva into the adult insect. The results of this study are presented schematically in Figure 21-10, which shows the presence or absence of puffs at four particular sites of chromosome IV at successive stages of metamorphosis. As can be seen, the puffs at sites A and B wax and wane twice in alternating phase during the

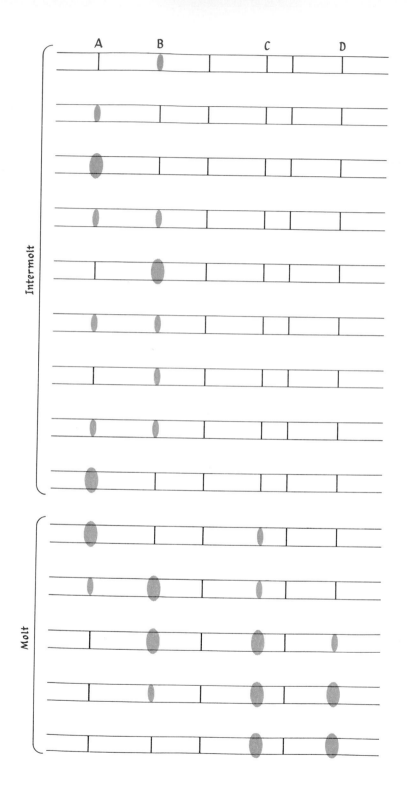

period under observation, whereas the puffs at sites C and D appear only at the end of the period.

How *does* the mutual interaction of cytoplasm and nucleus manage to control the differential expression of the eukaryotic cell genome in embryonal development? No conclusive answer to this question can as yet be supplied; indeed, the search for its answer will surely be one of the main preoccupations of molecular geneticists in the coming years. It is likely that the operon system of gene regulation of prokaryotes will form at least a *part* of that answer. For the operon is capable, in principle at least, of providing for the most complicated regulatory circuits by articulation of relatively few inducer-repressor-operator elements. Indeed, even the few control elements of the *lac* operon shown in Figure 20-7 allow the bacterial cell to exist in alternative, permanently differentiated states. For instance, if a very low concentration of a *lac* operon inducer, such as IPTG, is added to the growth medium of an *E. coli* culture, none of the *lac* enzymes is induced; in the absence of the *lac*Y gene permease the dilute external inducer does not reach a sufficiently high intracellular concentration to neutralize the *lac*I gene repressor. But if a high concentration of inducer is first added to initiate induction of the *lac* enzymes, including the permease, and if only thereafter is the inducer concentration reduced to a low level, then the bacteria do continue their synthesis of the *lac* enzymes. For once the permease is present, it allows enough of the dilute external inducer to accumulate intracellularly to keep on neutralizing the repressor. Hence under one given condition of growth (i.e., low inducer concentration), the *E. coli* cell can have its *lac* operon either permanently turned on or permanently turned off, depending only on its previous history. It is evident that the existence of such bacterial differentiation plays havoc with the fundamental distinction made in Chapter 6 between *adaptation* and *mutation* as antithetical explanations for the change in property of bacterial cultures. By the criterion of the permanency of change posited in Chapter 6, the continued synthesis of *lac* enzymes in low inducer concentration after transient exposure to high inducer concentration would have to be classified as a mutation, even though it is now clear that the permanent change observed here does not result from any change in *genetic* information.

Although eukaryotes would fain remember the operon—that excellent control device invented by their humble prokaryotic antecedents—it is likely that the embryology of the future will uncover other regulatory processes

FIGURE 21-10. Sequential pattern of puff formation at four bands (A,B,C,D) of the giant chromosome IV in the salivary gland of *Chironomous tentans* (see Plate IV). Starting from the top, horizontal bars depict schematically the aspect of chromosome IV at successive stages before and during the molt in which the insect larva is transformed into the pupa. Two bands are shown also that do not puff at all during these developmental stages. [From "Chromosome Puffs," by W. Beermann and V. Clever, *Scientific American*, April 1964. Copyright © 1964 by Scientific American, Inc. All rights reserved.]

that have no counterpart in prokaryotes. One such process appears to govern the wholesale blockage of the transcription of thousands or tens of thousands of linked chromosomal genes, or even of entire chromosomes. The covering of chromosomal DNA with histones would recommend itself as a reasonable explanation of this phenomenon, for which explanation there is even some direct experimental support. Another regulatory process not hitherto found in prokaryotes, but one that definitely plays a role in embryonic development, is the control of protein synthesis at the level of mRNA translation rather than, as under the operon control system, at the level of DNA transcription. For though no mRNA synthesis takes place during the earliest stages of embryonic development, the character of protein synthesis is not constant during that period. This demonstrates that some mRNA species present in the egg temporarily remain quiescent while others serve as templates for polypeptide chain assembly.

IMMUNE RESPONSE

One special, and long-mysterious, instance of cell differentiation—the immune response of vertebrate animals—now seems close to being solved, thanks in part to the application of principles of molecular genetics. Study of the immune response began with Edward Jenner's discovery of vaccination in the closing years of the eighteenth century, though by then it had long been known that persons once having recovered from the primary attack by an infectious disease are immune against further, secondary attacks by that same disease. At the end of the nineteenth century the search for an explanation of this remarkable phenomenon led Emil Behring to the discovery of a special class of protein molecules in the blood serum—the *antibodies*. These molecules were recognized to be capable of specifically combining with, and hence of neutralizing, the type or virus or bacterium that had been responsible for the disease in the infected animal. Immunity is thus attributable to the presence of antibodies whose appearance has been elicited by the primary infection. Soon after the discovery of their connection with resistance to infectious disease, it was found that specific antibodies are also formed in response to injection into the bloodstream of nonliving materials, such as dead bacteria, bacterial toxins, and snake venoms. And within another few years it was shown that introduction into the bloodstream of *any* foreign protein, or *antigen*, whether noxious or innocuous, results in a few days in the appearance of antibodies specifically directed against that, and only that antigen. But protein molecules obtained from an individual's own tissues (or from those of another individual carrying exactly the same genes, such as his identical twin) do not act as antigen in that individual. Antibody formation is thus not just a mere defense reaction against infectious disease, but is a phenomenon of rather wide biological significance: a mechanism for the recognition of nonself. The capacity for antibody formation among all vertebrates at least as high on the evolutionary ladder as the bony fishes indicates the high antiquity of this process. One plausible explanation that has been advanced for the fundamental biological "purpose" of the immune response among the highest eukaryotic organisms is that it furnishes a mechanism for the removal of abnormal proteins of *endogenous* origin. For instance, the immune reaction would recognize as foreign, and would attempt to eliminate from the body, any abnormal, and hence potentially noxious, variant cell in which a mutation in the chromosal DNA has resulted in the synthesis of a mutant type of protein molecule. If it were not for this house-keeping function among the billions of mutable cells of a vertebrate animal, every fish, frog, bird, or mammal might well have died before it reached maturity.

By the 1930's the work of Karl Landsteiner and others had shown that a staggering diversity of specific antibody molecules—at least a million different varieties—can be synthesized by any one individual. But how does the antigen manage to give rise to the synthesis of precisely that one of the myriad of possible antibody proteins that happens to have the specific capacity to combine with it and it alone? In answer to this question Pauling and others proposed in the early 1940's that the antigen is taken up by the specialized blood cells responsible for antibody production. In these blood cells, the freshly formed, as-yet nonfunctional antibody polypeptide chains were thought to *fold* around the antigen and thus be molded into a specific tertiary structure complementary to that of the antigen. This theory thus readily explained how there could exist as many different kinds of antibody molecules as there are different kinds of antigens. But this theory could *not* account for the recognition of antigens as either self or nonself, and eventually had to be abandoned when it was shown that the blood cells that actually produce the antibody do not, in fact, contain any antigen. In any case, the notion of molding a given polypeptide chain into any one of a myriad of possible configurations ran afoul of the fundamental principle of molecular genetics that states that the form and functional specificity of any protein molecule is determined wholly by the amino acid sequence of its polypeptide chains. And by the mid-1960's it had been shown that antibody proteins are no exception to this principle: antibodies directed against different antigens do differ in their amino acid sequence, and the specific, antigen-complementary form of any antibody protein is but a direct consequence of its amino acid sequence. More specifically, the antibody protein molecule was found to be a quaternary structure consisting of two pairs of identical polypeptide chains, or of a pair of "light" chains about two-hundred amino acids in length and a pair of "heavy" chains about four-hundred amino acids in length (Figure 21-11). The antibody molecule has two antigen-complementary sites, the stereospecific conformation of each of which derives from apposed sectors of

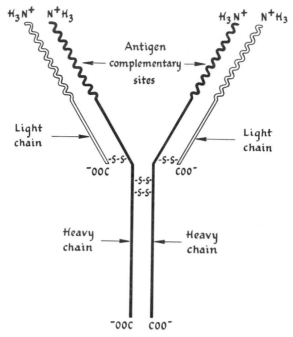

FIGURE 21-11.   Schematic diagram of the quaternary protein structure of an antibody molecule. Each of a pair of identical "light" polypeptide chains of about 200 amino acid residues is held to one of a pair of identical "heavy" polypeptide chains of about 400 amino acid residues through a disulfide (—S—S—) bond formed between cysteine residues of the chains. The two "heavy" chains are, in turn held to each other by a pair of another disulfide bonds. The segment of each chain drawn as a wavy line represents the *variable* region of about 100 amino acid residues in which differences in primary structure occur among antibodies carrying sites complementary to different antigens. The straight-line segment of each chain represents the *constant* region in which antibody molecules of different antigen specificity circulating in an animal may have the same primary structure.

one light and one heavy chain. Amino acid sequence analysis of the polypeptide chains of different antibody molecules has shown that the first hundred amino acids from the amino terminus of both light and heavy chains constitute a *variable* region in which differences in primary structure occur. The remaining hundred light-chain, or three hundred heavy-chain, amino acids constitute a *constant* region in which different antibody molecules have the same primary structure. Thus the antigen-complementary site is evidently formed by the tertiary structures of the variable regions of light and heavy chains. The problem of the immune response can thus be rephrased in terms of the question of how introduction of an antigen into an animal gives rise to the production of light and heavy antibody polypeptide chains whose variable regions have just that primary structure which will generate a quaternary structure stereospecifically complementary to that antigen.

In 1954, thought about the immune response received a new orientation when N. K. Jerne (Figure 20-10) proposed that the role of the antigen is not to specify the structure of its affined antibody at all, but rather to cause the selective stimulation in the animal of a system that was already synthesizing that antibody protein before the antigen ever appeared on the scene. The mature animal does not, however, contain any system capable of synthesizing antibody molecules which have any affinity for the self-proteins of that same animal. Thus the antigen is recognized either as being "foreign" by encountering spontaneous antibody molecules with which it can combine or as "self" by failing to encounter any spontaneous antibody molecules with which it can combine. Within a short time after this proposal, F. M. Burnet improved Jerne's selective theory of antibody formation by envisaging that what the antigen selectively stimulates is the proliferation of a type of blood cell whose genetic constitution allows that cell to synthesize one and only one type of antibody molecule—namely, the one to which the antigen happens to be affined. There must, therefore, exist in any individual animal as a great diversity of genetically different blood cells as there are antigens that can elicit an immune response in that animal. Burnet thought that this diversity is generated by mutations in the line of descent of blood cells during embryonic development and growth of the animal. Once the quaternary structure and the nature of the variability of the antibody molecule had become known, Burnet's theory could be rephrased to state: the mutations that the antigen selects occur in those sectors of the genes specifying the primary structure of light and heavy chains in which the amino acid sequence of the variable region is encoded. Figure 21-12 presents the result of amino acid sequence analyses of the variable region of the light chain of a set of different human

```
 1    2    3    4    5    6    7    8    9   10   11   12   13   14   15   16   17  18
Asp–Ile–Gln–Met–Thr–Gln–Ser–Pro–Ser–Ser–Leu–Ser–Ala–Ser–Val–Gly–Asp–Arg–
      Val  Leu  Leu                      Thr  Thr       Val       Leu Arg
                                              Phe
```

```
19   20   21   22   23   24   25   26   27   28   29   30   31   32   33   34  35......
Val–Thr–Ile–Thr–Cys–Gln–Ala–Ser–Gln–Asp–Ile–Ser–Ile–Phe–Leu–Asn–Trp–
Ile            Ala       Arg                 Ser Asx Asn Thr Trp       Ala
                                             Lys Lys Tyr
                                             Asn
```

```
91   92   93   94   95   96....104  105  106  107  108  109  110
Tyr–Asp–Asn–Leu–Pro–Arg–  Leu– Glu–Ile–Lys–Arg–Thr–Val
Phe Glu  Thr             Leu  Val Asp Phe       Gly
                         Tyr          Val
                         Pro          Leu
```

FIGURE 21-12.   Amino acid replacements found at various sites of the variable region of the light chain of different species of human antibody molecules. The symbol "Asx" stands for "Asn or Asp." Amino acid residue number 1 is the amino-terminus of the polypeptide chain. [Adapted from M. Cohn, in O. J. Plescia and W. Braun, eds., *Nucleic Acids and Immunology*, Springer, New York, 1968.]

antibody molecules. As can be seen, the data of Figure 21-12 bear a strong resemblance to the amino acid replacements in the coat protein of mutants of the tobacco mosaic virus presented in Figure 18-1; the light chains differ from each other at various polypeptide sites, and, as can be ascertained by reference to the genetic code of Table 18-3, the amino acid replacements extant at these sites can generally be attributed to simple nucleotide interchanges in a single position of a triplet codon. Thus the nature of the variability of the primary structure of the antibody proteins does appear to be in accord with Burnet's mutation hypothesis.

An alternative hypothesis of the origin of the genetic diversity of antibody-producing cells was later proposed. This alternative hypothesis envisages that the germ-line genome of vertebrate animals contains a few homologous genes in which several different versions of the variable regions of light and heavy chains are encoded. During embryonic development and growth of the animal there is then supposed to occur a frequent process of crossing-over between these homologous genes, which results in a "scrambling" of their nucleotide sequences. The kind of antibody molecule a cell produces would then depend on which of the myriad of possible kinds of scrambled genes it happens to have inherited from its parental cell. Actually, it is not a necessary feature of the selective theory of antibody formation that the blood cells whose proliferation the antigen stimulates selectively be *genetically* different. Thus it is also formally admissible that the chromosomes of every cell contain a million different genes encoding the primary structure of the variable regions of light and heavy chains of every possible antibody protein molecule. Owing to the process of selective control of gene expression in cell differentiation, only one pair of these genes might be expressed, and hence only one type of antibody molecule be produced by any one particular cell line. The adherents of this multiple-gene theory found succor in the discovery of the million-fold repeated DNA sequences, such as the mouse "satellite" DNA described in this chapter. For since the variable regions of the million different antibody protein molecules are more similar than they are different in their primary structure, it would be expected that the putative million DNA nucleotide sequences encoding these similar polypeptide chains reveal themselves as redundant sequences in the DNA-strand reassociation test.

At the present time no definitive proof or disproof of any of these theories of the origin of the structural diversity of spontaneous antibody molecules is available, and it is not even excluded that the true explanation will embody a potpourri of the mutation, scrambling, and multiple-gene hypotheses. Moreover, many details are still lacking in the understanding of how the antigen manages to induce the selective proliferation of the special blood cells to whose spontaneously synthesized antibody protein it is affined. Nevertheless, it seems unlikely that the present general view of the immune response could be very far off the mark. And whatever be processes that might generate antibody diversity and cause the selective proliferation of cells synthesizing antibodies that happen to fit the antigen, they bid fair to lie at the root of other cellular-differentiation phenomena.

THE BRAIN

But even after the enigma of the embryo has been solved, and the control of the differential expression of the eukaryotic cell genome in development has been fully understood, there would still remain one major frontier of biological inquiry for which reasonable molecular mechanisms still cannot even be imagined—the brain. Its fantastic attributes continue to pose as hopelessly difficult and intractably complex a problem as did the hereditary mechanism a generation ago. And the brain does, of course, present the most ancient and best-known of paradoxes in the history of human thought—the relation of mind to matter. It seems likely that in the coming years, students of the nervous system, rather than geneticists, will form the vanguard of biological research. One of the as-yet deeply mysterious aspects of the brain—namely, how in its development there arise the vast number of specific and genetically controlled interconnections between its constituent nerve cells—appears to be a problem toward which molecular genetics could be directed with some profit. But since in the establishment of that network specific recognition of one nerve-cell surface by another undoubtedly plays a key role, it is not unlikely that the solution of this problem will require a fundamentally new insight into the structure and function of cell surfaces. At a yet higher level of complexity, there remains to be understood the logic of the nerve-cell network—that is to say, the manner in which its circuits acquire, process, store, and emit information. In the study of these problems there occurred an important breakthrough in the early 1960's, when it was discovered that relatively small ensembles of nerve cells, such as those receiving the visual stimuli from small parts of the retina of the vertebrate eye, make a yes-no analysis of the signals they receive in terms of preprogrammed questions and thus send to the brain pre-evaluated and abstracted information rather than raw data. In the brain, in turn, this information is abstracted further and further in successive stages by other small nerve-cell ensembles. Heuristically, this discovery could mean for the study of the brain what the one-gene–one-enzyme theory meant for the study of the gene: if a limited number of interconnected nerve cells are capable of doing in a small way what the brain does in a big way, then there is hope of ultimately finding out how the brain does what it does. Or, as David H. Hubel, one of those mainly responsible for this development, put it in 1963: "The areas [of visual perception] we study can be understood in terms of comparatively simple concepts, such as the nerve impulse, convergence of many nerves on a single cell, excitation and inhibition. Moreover, if the connections suggested by these studies are remotely close to reality, one can conclude that at least some parts of the brain can be followed relatively easily, without necessarily requiring higher mathematics, computers or a knowledge of network theories."

But we may ask whether scientific study of the nervous system can *ever* resolve the mind-matter paradox. Is it, in fact, likely that consciousness, the unique attribute of the brain that appears to endow its ensemble of atoms with self-awareness, will ever be explained? As Bohr said in 1932, in his "Light and Life" lecture (quoted also in Chapter 1), the principle of complementarity would be of help in fathoming the nature of this problem in physical terms: "The recognition of the limitation of mechanical ideas in atomic physics would much rather seem suited to conciliate the apparently contrasting points of view which mark physiology and psychology. Indeed, the necessity of considering the interaction between the measuring instruments and the object under investigation in atomic mechanics corresponds closely to the peculiar difficulties, met with in psychological analyses, which arise from the fact that the mental content is invariably altered when the attention is concentrated on any single feature of it . . . Indeed, from our point of view, the feeling of the freedom of the will must be considered as a trait peculiar to conscious life, the material parallel of which must be sought in organic functions, which permit neither a causal mechanical description nor a physical investigation sufficiently thoroughgoing for a well-defined application of the statistical law of atomic mechanics." V. Weisskopf has summarized Bohr's attitude in the following terms: "The awareness of personal freedom in making decisions seems a straightforward factual experience. But when we analyze the process, and follow each step in its causal connection the experience of free decision tends to disappear. . . . Bohr, an enthusiastic skier, sometimes used the following simile, which can be understood perhaps only by fellow skiers. When you try to analyze a Christiania turn in all its detailed movements, it will evanesce and become an ordinary stem turn, just as the quantum state turns into classical motion when analyzed by sharp observation." This attitude would mean nothing less than that searching for a "molecular" explanation of consciousness is a waste of time, since the physiological processes responsible for this wholly private experience will be seen to degenerate into seemingly quite ordinary, workaday reactions, no more and no less fascinating than those that occur in, say, the liver, long before the molecular level has been reached. Thus, as far as consciousness is concerned, it is possible that the quest for its physical nature is bringing us to the limits of human understanding, in that the brain may not be capable of providing an explanation of itself. Indeed, Bohr ended his "Light and Life" lecture with the thought that "without entering into metaphysical speculations, I may perhaps add that any analysis of the very concept of an explanation would, naturally, begin and end with renunciation as to explaining our own conscious activity." Perhaps *this*, then, is the "other law" that molecular genetics failed to turn up: There exist biological processes which, though they clearly obey the laws of physics, can *never* be explained.

Bibliography

PATOOMB

T T Puck. The mammalian cell.

R. Dulbecco. The plaque technique and the development of quantitative animal virology.

H. Rubin. Quantitative tumor virology.

Niels K. Jerne. The natural selection theory of antibody formation; ten years later.

Werner E. Reichardt. Cybernetics of the insect optomotor response.

ORIGINAL RESEARCH PAPERS

Britten, R. J., and D. E. Kohne. Repeated sequences in DNA. *Science*, **161**, 529 (1968).

Burnet, F. M. A modification of Jerne's theory of antibody production using the concept of clonal selection. *Austral. J. Sci.*, **20**, 67 (1957).

Jerne, N. K. The natural selection theory of antibody formation. *Proc. Natl. Acad. Sci. Wash.*, **41**, 849 (1955).

Whitehouse, H. L. K. A theory of crossing-over by means of hybrid deoxyribonucleic acid. *Nature*, **199**, 1034 (1963).

SPECIALIZED TEXTS, MONOGRAPHS, AND REVIEWS

Ebert, J. D. *Interacting Systems in Development.* Holt, Rinehart and Winston, New York, 1965.

Markert, C., and H. Ursprung. *Developmental Genetics*, Prentice-Hall, Englewood Cliffs, N.J., 1970.

Woodridge, D. E. *The Machinery of the Brain.* McGraw-Hill, New York, 1963.

Index